Computational Financial Mathematics using MATHEMATICA®

Optimal Trading in Stocks and Options

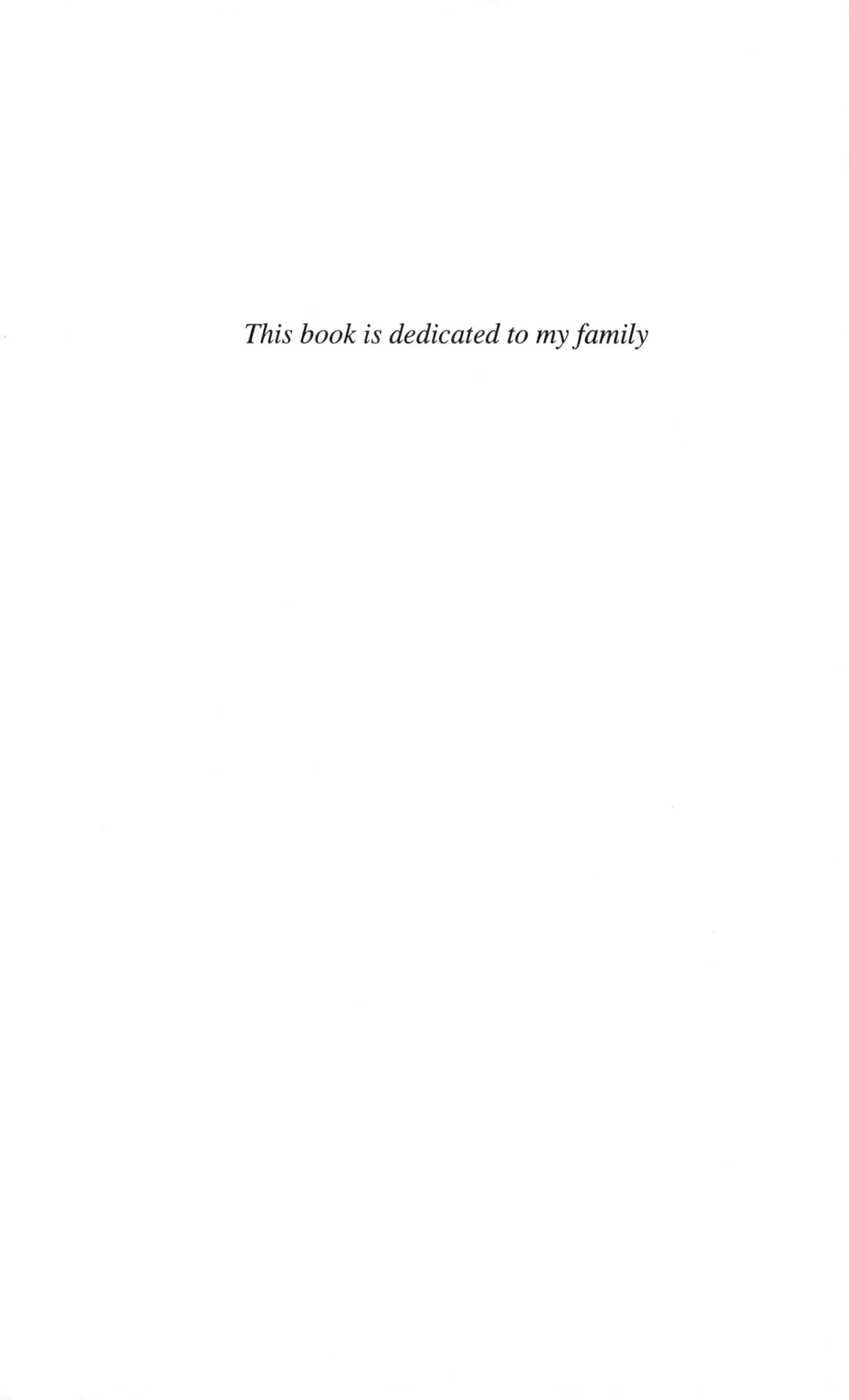

This book is dedicated to my family

Srdjan Stojanovic

Computational Financial Mathematics using MATHEMATICA®

Optimal Trading in Stocks and Options

Srdjan Stojanovic
University of Cincinnati
Department of Mathematical Sciences
Cincinnati, OH 45221-0025
USA

Library of Congress Cataloging-in-Publication Data

Stojanovic, Srdjan.
Computational financial mathematics using Mathematica : optimal trading in stocks and options / Srdjan Stojanovic.
p. cm.
Includes bibliographical references and index.
ISBN 0-8176-4197-1 (alk. paper) — ISBN 3-7643-4197-1 (alk. paper)
1. Finance–Mathematical models. 2. Mathematica (Computer file) 3. Securities. I. Title.

HG 106.S76 2002
332.63'2042–dc21 2002074746
CIP

AMS Subject Classifications: 35H10, 35J60, 35J85, 35K55, 35K85, 35Q80, 35R35, 49J20, 49L20, 60J60, 60J65, 60H30, 60H35, 62C10, 62M05, 65C05, 65C30, 65M06, 65N06, 68N15, 68W30, 92E11, 93E20

Printed on acid-free paper.

Birkhäuser 

ISBN 0-8176-4197-1 SPIN 10768773
ISBN 3-7643-4197-1

Typeset by the author in MATHEMATICA®
Printed in the United States of America.

9 8 7 6 5 4 3 2 1

Birkhäuser Boston • Basel • Berlin
A member of BertelsmannSpringer Science+Business Media GmbH

Contents

0 Introduction

"Although all mathematicians have denied it, the applications serve as the measure of worth of mathematics."

(**David Hilbert** at the meeting of the Society of German Scientists and Physicians in Königsberg in the fall of 1930, on the occasion of a presentation to Hilbert upon his retirement of an honorary citizenship of the town)

0.1 Audience, Highlights, Agenda

This book is addressed to students and professors of academic programs in financial mathematics, i.e., computational finance and financial engineering. It is based on the premise that such programs need to be computationally oriented, or at least, computationally motivated, while searching to define and deliver a kind of mathematical knowledge that can prove the value, or even indispensability, of mathematics to financial industry employers. Enabled by the *Mathematica*® computer platform, we offer an applied approach, which integrates and gives a fresh perspective on such mathematical disciplines as probability, ordinary, stochastic and partial differential equations, mathematical statistics, calculus of variations, optimal control of stochastic, ordinary and partial differential equations. We present some of the things that each of these disciplines can do for market analysis, investing and trading, as well as motivating further study of these subjects, and their integration into practice.

This work is also intended for mathematically inclined individual small investors and day traders, by addressing and completely solving problems around those financial instruments that are most often available to them: cash, stocks, and (American) stock options. Indeed, no exotic options are discussed at all. On the other hand, simple options are discussed in great and sometimes quite complicated detail. We derive and implement completely results such as the time-dependent Black–Scholes formula, fast numerical solvers for the price-dependent Black–Scholes and Dupire partial differential equations and obstacle problems (for American options) along with their optimal control as a method of finding implied volatilities. Also, we derive and implement completely optimal portfolio selection under all kinds of constraints with or without appreciation rate uncertainty, and so on.

Finally, mathematically oriented institutional professional investors and traders will also find many things that are new, different, and useful. In particular, some advanced mathematics is applied and implemented to address several sophisticated issues in what is in trading circles sometimes referred to as "statistical arbitrage." For example, among new results, two that might prove to be quite important are:

1. Dimension reduction for Monge–Ampère PDEs in advanced optimal portfolio hedging (optimal portfolio hedging under non-Log-Normal assets price dynamics, e.g., optimal options–portfolios), and numerical solutions (Section 8.2); and

2. Momentum trading model—in general, the concept of degeneracy, and in particular, that of hypoellipticity in computational financial mathematics, and numerical solutions (Section 8.3).

Mathematically advanced readers interested in Monge–Ampère PDEs or the concept of hypoellipticity in PDEs will further be served.

What is meant by Computational Financial Mathematics? In various branches of mathematics *solving* problems has evolved to mean so many different things. Here, on the contrary, the only problems that are considered solved are those for which it is known how to *compute* solutions. Computing solutions is done either *symbolically* (explicit solutions, which are the most preferable), *numerically* (most often by means of numerical solutions of associated differential equations and related problems–the second most preferable), or by means of Monte–Carlo *simulations* (the least preferable). Once a particular problem is solved in one way or another, the search for more efficient and/or precise and/or stable ways of computing continues. As a matter of fact, possibly the most striking results are obtained by a combination of those different kinds of methods: Section 8.2 is based on a powerful combination of symbolic and numerical methods. So the subject is defined by its goal, *computing* solutions of problems in finance, and not quite as much by its methodology.

In turn, the goal has a decisive influence on the agenda. Indeed, in such a framework the most important thing is to have a *correct* solution of the given problem. The second most important thing is to have the most computationally *efficient* solution. On the other hand, while having a rigorous mathematical proof is still the most preferable way of convincing ourselves of the truthfulness of a given solution or a mathematical claim, it is not of the highest priority. Put a bit differently, we seek and present mathematical truth using all means available. As a matter of fact, sometimes a sketchy proof combined with computational evidence is even more convincing of the correctness of the underlying claim than an involved mathematically rigorous attempt (see the section on Black–Scholes Hedging Implemented (3.2.2)). The other justification for such an agenda is that the search for a rigorous mathematical framework for specific problems quite often, even when achieving its own ends, does not lead to or even suggest how to find computational answers for the same.

The required background varies according to the objectives of the reader. Indeed, most of this book, especially in its electronic form, can be applied in actual trading without understanding all of the details: most of the advanced results are implemented completely ready to be executed by the reader (on the computer using

Mathematica®). On the other hand, those with ambitions to study mathematical intricacies of some of the mathematical subjects listed above will find a strong motivation to do so. A minimal mathematical background would be Calculus, some Differential Equations and Probability.

It might be helpful to try here to classify the parts of the book according to the mathematical sophistication level. If so, one possible, although necessarily incomplete classification could be the basics (Chapters 1–4), intermediate level (Chapters 5 and 7), and advanced level (Chapters 6 and 8), where even when advanced mathematics is on the table, the aim is always to be user-friendly. Also, for those who might be in a hurry to get to the results that interest them the most, the following diagram could be useful in pointing out both substantial dependencies and/or order of chapters.

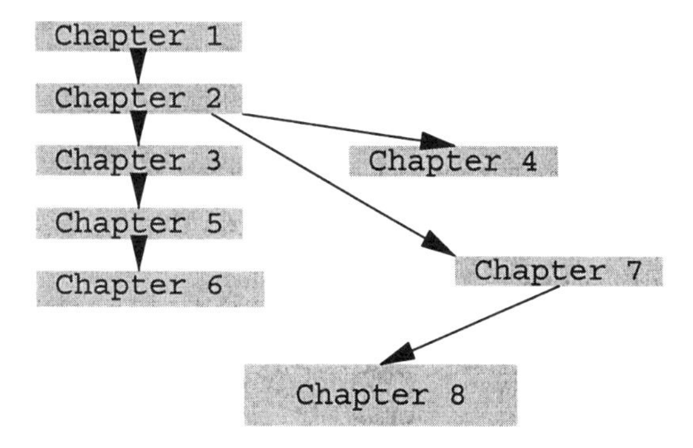

Although a great deal of effort has been made to avoid and correct mistakes, no claim is made that what follows is error-free. If the reader chooses to use the material presented in this book, whether it is right or wrong for actual investing and trading, he or she should not hold responsible either the author or the publisher for the risks undertaken.

This book is an expanded version of a course on financial mathematics that I have developed and taught regularly at the University of Cincinnati since the spring of 1998, and during the academic year 2001/02 at Purdue University. The course was always taught 100% of the time in the computer laboratory, and the same is strongly advised for any instructor who considers this text for adoption. The content of the course has grown over the years, so that the course given at Purdue was a full year (3-hours) sequence, and the same is planned now at the University of Cincinnati. This suggests that if all the mathematical and *Mathematica*® details are presented carefully, i.e., if mathematics is derived and if *Mathematica*® is programmed, there should be more than enough material in the book, even in the first 7 chapters, for a 2-semester or a 3-quarter sequence. Alternatively, a one quarter/semester course could possibly be based on the material presented starting from Chapter 1 until (and including) Section 5.3.

One common feature of my teaching experience so far has been the high level of student motivation for learning both computational financial mathematics as well as

Mathematica® programming. As a matter of fact, some of the students' ideas have been integrated into the work, as more elegant ways of programming are occasionally easier found by many than by one. Their enthusiasm also contributes to my confidence to make a prediction that the computational platform way of presenting, teaching, learning, experiencing, and ultimately discovering mathematics will become quite popular in the very near future.

0.2 Software Installation

The book is accompanied by a CD-ROM. The print has obvious advantages such as reading convenience, while the electronic version has its own advantage, namely, executability of the code, and color display for pictures. Also, to avoid the repetition of *Mathematica*® definitions, once defined, functions that will be used over and over again are usually stored in several packages-files in a subdirectory called CFMLab (which stands for Computational Financial Mathematics Laboratory). This directory is available on the accompanying CD-ROM, and it needs to be copied and pasted into the directory

```
In[1]:= ToFileName[
          {$TopDirectory, "AddOns", "Applications"}]

Out[1]= C:\Program Files\Wolfram Research\Mathematica\4.1\AddOns\
          Applications\
```

This is in the standard setup under Windows 98 if Mathematica® 4.1 is used. Under other operating systems the above will identify the right location. This yields:

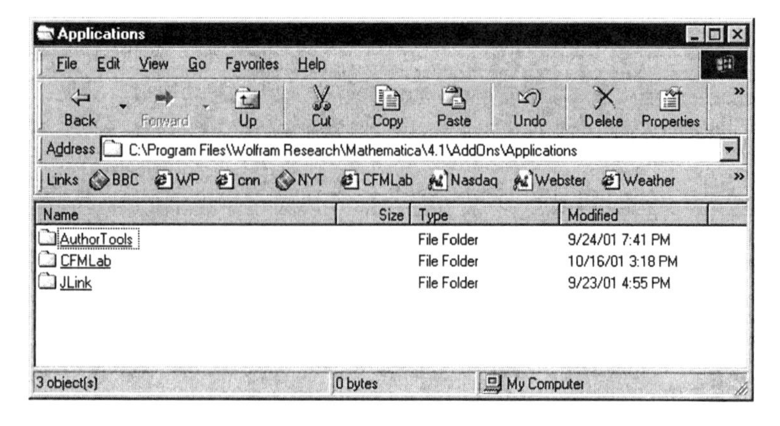

Once this is done, while being inside a *Mathematica*® session, pulling down the "Help" menu, one should select "Rebuild Help Index". Thereafter the book will be a part of the *Mathematica*® Help Browser, under Add-Ons | CFM Laboratory.

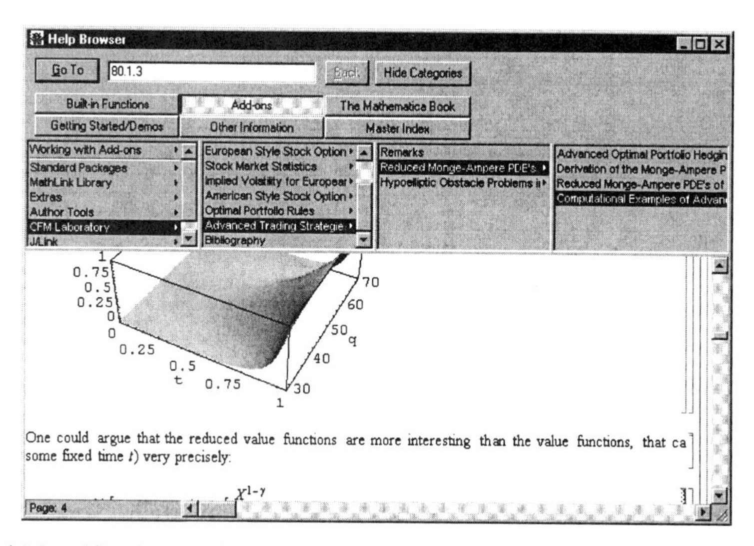

Format | Magnification can be used to adjust the font size. When executing various parts of the electronic book, the reader should find the place where the definitions start, usually identified by the

```
In[2]:= Clear["Global`*"]
```

command, and start the execution there. Otherwise all the necessary definitions might not be taken into account. It is further recommended that the reader start each chapter with a new *Mathematica*® Kernel, and execute the chapter consecutively. This is to avoid, for example, the shadowing of definitions. Quitting Kernel and starting over from the previous Clear["Global`*"] statement is the first thing to try whenever something does not execute properly.

Obviously, to uninstall the book one just has to cutout the CFMLab directory, and rebuild the help index again.

Finally, this is a book and not a software product, and no warranty is issued with regard to any software supplied with this book. In particular this implies that if any kind of software bug is discovered, it will not be the responsibility of the author or of the publisher.

Information about further developments might be available at the web site CFMLab.com. The reader may direct comments to the same address.

Srdjan Stojanovic
Cincinnati, Ohio
2002

0.3 Acknowledgments

Market data and the permission to use it in this publication was kindly supplied by OptionMetrics, LLC (http://www.OptionMetrics.com). I express my sincere gratitude to OptionMetrics.

I would like to thank Jin Ma for his early and continuing enthusiasm for this work, for arranging my visiting professorship at Purdue University during the academic year 2001/02, which has turned out to be so inspirational for the final shape this book has taken, and for many discussions we had during that time. I would like to thank Ann Kostant, Executive Editor of Birkhäuser, for her early and continuing enthusiasm for this work—it contributed so much to my own.

1 Cash Account Evolution

Solving Ordinary Differential Equations with Mathematica®

1.1 Symbolic Solutions of ODEs

Let's start fresh:

```
In[1]:= Clear["Global`*"]
```

where

```
In[2]:= ?Clear

    Clear[symbol1, symbol2, ... ] clears values and definitions
      for the symboli. Clear["form1", "form2", ... ] clears
      values and definitions for all symbols whose names
      match any of the string patterns formi. More...
```

Cash account holdings, which for all practical purposes are *risk free*, give to an investor a feeling of security which is greatly appreciated especially during periods of market stagnation or decline. Cash accounts pay small interest, as a matter of fact, very small compared to what can be gained in the market during periods of market advance. Nevertheless, to start introducing the language of this book, let us compute some things about cash accounts. Much more about this "language" can be found in [27,44,53,59].

Suppose an investor makes a deposit of P dollars into a cash account that pays interest rate r 100% per year, compounded continuously. How does the account balance evolve as a function of time t (measured in years)? As discussed already in Business Calculus, the account balance satisfies (the simplest possible) Ordinary Differential Equation (ODE; as compared to SDE and PDE standing for stochastic and partial differential equations)

$$y'(t) = r\, y(t) \tag{1.1.1}$$

(which says only that the rate of change of the balance is proportional to the balance itself, or alternatively and a bit more precisely, that the amount of change in the account balance is equal to the (interest rate) × (previous balance) × (elapsed time)) with an initial condition

$$y(0) = P. \tag{1.1.2}$$

This problem is very simple to solve, either by hand, or using *Mathematica*®. It may indeed be too simple. We note here that deciding what is the right level and type of simplicity/sophistication of the model is the major problem in any kind of applied mathematics, and in particular in financial mathematics. Simple models ignore many features of real problems which complex models attempt to address, while complex models are often not manageable, and it is difficult to estimate their parameters. So, how does *Mathematica*® handle the simplest possible ODE?

```
In[3]:= DSolve[{y'[t] == r y[t], y[0] == P}, y[t], t]
```

Out[3]= $\{\{y(t) \to e^{r\,t} P\}\}$

where we have used the *Mathematica*® built-in function DSolve. What is DSolve? *Mathematica*® answers, a bit too ambitiously (indeed, try solving your favorite PDE),

```
In[4]:= ?DSolve

DSolve[eqn, y, x] solves a differential equation for the
   function y, with independent variable x. DSolve[{eqn1,
   eqn2, ... }, {y1, y2, ... }, x] solves a list of
   differential equations. DSolve[eqn, y, {x1, x2, ... }]
   solves a partial differential equation. More...
```

In the above simple line of program there are several syntax details to observe. The *standard format type* for enclosing an argument of a function, such as $y[t]$ above, is the bracket parentheses [ ]. We shall also use the *traditional format type* in which an argument of a function can be enclosed by usual parentheses ( ) as well. Also, looking carefully, we see double equality signs ==, as opposed to what one might have expected. Indeed,

```
In[5]:= ? ==

lhs == rhs returns True if lhs and rhs are identical.
 More...
```

The novice to *Mathematica*® may be inclined to dislike the output of DSolve above, expecting to see only the solution

$$e^{r\,t} P.$$

Instead, the output is a *list* of substitution *rules*. Lists are one of the most important objects in *Mathematica*®:

```
In[6]:= ?List

{e1, e2, ... } is a list of elements. More...
```

Also, by a further analysis of the syntax of DSolve arguments above, we can see that the first argument, i.e., the equation, is actually not just an equation, but rather a list of two equations: ODE and the initial condition (initial condition is also an equation, although not a differential equation).

Elements of lists can be extracted using

```
In[7]:= ? Part

    expr[[i]] or Part[expr, i] gives the ith part of expr.
      expr[[-i]] counts from the end. expr[[0]] gives the
      head of expr. expr[[i, j, ... ]] or Part[expr, i,
      j, ... ] is equivalent to expr[[i]] [[j]] ... . expr[[
      {i1, i2, ... } ]] gives a list of the parts i1,
      i2, ...  of expr. More...
```

For example,

```
In[8]:= DSolve[{y'[t] == r y[t], y[0] == P}, y[t], t][[1]]
```

Out[8]= $\{y(t) \to e^{r\,t}\,P\}$

or, which is the same, but somewhat awkward looking

```
In[9]:= Part[DSolve[{y'[t] == r y[t], y[0] == P}, y[t], t], 1]
```

Out[9]= $\{y(t) \to e^{r\,t}\,P\}$

Notice also the syntax for multiplication

```
In[10]:= ? Times

    x*y*z or x y z represents a product of terms. More...
```

Going deeper into the syntax of the above DSolve output we see operator ->

```
In[11]:= ? ->

    lhs -> rhs represents a rule that transforms lhs to rhs.
     More...
```

Rule, or more descriptively, substitution rule can be *applied* using

```
In[12]:= ? /.

    expr /. rules applies a rule or list of rules in an
      attempt to transform each subpart of an expression
      expr. More...
```

For example,

```
In[13]:= a /. a → b
```

Out[13]= b

or, finding the unique solution of the above ODE (since the solution is unique, selecting the first substitution rule, means selecting the unique solution; if there were more than one solution, one would have to be more careful):

```
In[14]:= y[t] /.
           DSolve[{y'[t] == r y[t], y[0] == P}, y[t], t][[1]]
```

Out[14]= $e^{r t} P$

How to define a function, say $z(t)$, which is equal to the above solution? One way, a quite natural one, would be

```
In[15]:= z[t_] = y[t] /.
           DSolve[{y'[t] == r y[t], y[0] == P}, y[t], t][[1]]
```

Out[15]= $e^{r t} P$

One can notice two new objects. The first new object is the equality sign = (compare with == above), the meaning of which is

```
In[16]:= ? =

       lhs = rhs evaluates rhs and assigns the result to be the
          value of lhs. From then on, lhs is replaced by rhs
          whenever it appears. {l1, l2, ... } = {r1, r2, ... }
          evaluates the ri, and assigns the results to be the
          values of the corresponding li. More...
```

One should also compare the equality sign = to := to be used below soon:

```
In[17]:= ? :=

       lhs := rhs assigns rhs to be the delayed value of lhs.
          rhs is maintained in an unevaluated form. When lhs
          appears, it is replaced by rhs, evaluated afresh each
          time. More...
```

So, the reason we chose = as opposed to := is that we first need DSolve to find a solution, and only then is that solution named $z[t]$. The second new object above is the underscore sign _

```
In[18]:= ? _

       _ or Blank[ ] is a pattern object that can stand for any
          Mathematica expression. _h or Blank[h] can stand for
          any expression with head h. More...
```

An example could help here:

```
In[19]:= Clear[g]; g[x] = x^2; {g[x], g[y]}
```

Out[19]= $\{x^2, g(y)\}$

In[20]:= `Clear[g]; g[x_] = x^2; {g[x], g[y]}`

Out[20]= $\{x^2, y^2\}$

In[21]:= `Clear[g]; g[x_Integer] = x^2; {g[y], g[2.], g[2]}`

Out[21]= $\{g(y), g(2.), 4\}$

In[22]:=
```
Clear[g]; g[x : Blank[Integer]] = x^2;
  {g[y], g[2.], g[2]}
```

Out[22]= $\{g(y), g(2.), 4\}$

It is quite seductive in *Mathematica*® to keep going deeper and deeper—for example, trying to understand now what is the meaning of Head in the above definition of Blank—but sometimes one has to resist such a desire and try to stay on the main track, so we want go there this time.

We go back to the above defined solution $z[t]$. To draw pictures, we need numbers. For example, let

In[23]:= `r = .05; P = 1000; T = 7;`

then, the evolution of the account balance looks like this:

In[24]:=
```
p1 = Plot[z[t], {t, 0, T}, PlotRange → {0, z[T]},
    AxesLabel → {"t", "$"}, ImageSize → 160];
```

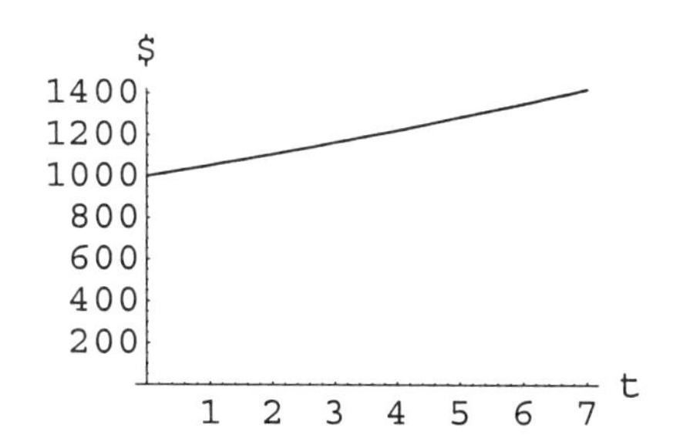

where

In[25]:= `? Plot`

```
Plot[f, {x, xmin, xmax}] generates a plot of f as a
   function of x from xmin to xmax. Plot[{f1, f2, ... },
   {x, xmin, xmax}] plots several functions fi. More...
```

In[26]:= `? PlotRange`

```
PlotRange is an option for graphics functions that
   specifies what points to include in a plot. More...
```

and

In[27]:=
```
? AxesLabel
```

```
AxesLabel is an option for graphics functions that
   specifies labels for axes. More...
```

If the above was too simple, now consider a more difficult problem, one when the interest rate is changing, as it happens in practice, at several moments over a considered time interval. The success or failure of the symbolic calculation will depend on the way the time dependency of the interest rate is introduced. *Mathematica*® has a function

In[28]:=
```
? UnitStep
```

```
UnitStep[x] represents the unit step function, equal to
   0 for x < 0 and 1 for x ≥ 0. UnitStep[x1, x2, ... ]
   represents the multidimensional unit step function
   which is 1 only if none of the xi are negative. More...
```

which can be used successfully to that end. For example, let the interest rate be

In[29]:=
```
R[t_] :=
   .06 - .005 UnitStep[t - 1] - .005 UnitStep[t - 1.25] -
    .005 UnitStep[t - 1.5] + .005 UnitStep[t - 3] +
    .005 UnitStep[t - 3.25] .05 - .005 UnitStep[t - 6]
```

which looks like

In[30]:=
```
Plot[R[t], {t, 0, T}, PlotRange → {0, Automatic},
    AxesLabel → {"t", "R[t]"}, ImageSize → 160];
```

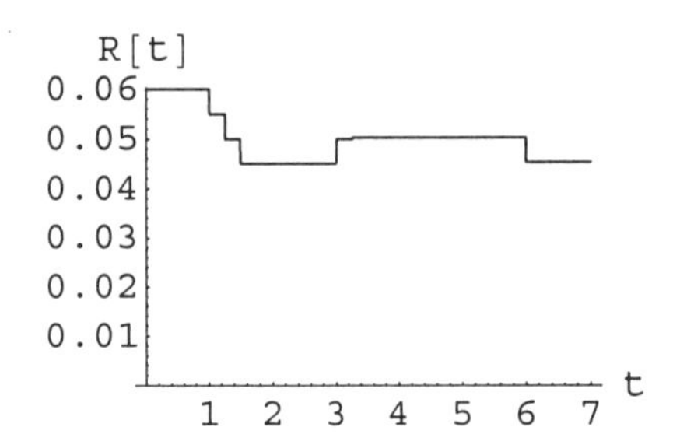

and we can see that the symbolic calculations still prevail here:

In[31]:=
```
s[t_] = y[t] /. DSolve[
      {y'[t] == R[t] y[t], y[0] == P}, y[t], t][[1]]
```

Out[31]= $1000\, e^{0.06\, t-0.005\,(1.\, t-6.)\,\theta(t-6.)+0.00025\,(1.\, t-3.25)\,\theta(t-3.25)+0.005\,(1.\, t-3.)\,\theta(t-3.)-0.005\,(1.\, t-1.5)\,\theta(t-1.5)-0.005\,(1.\, t-1.25)\,\theta(t-1.25)-0.005\,(1.\, t-1.)\,\theta(t-1.)}$

which looks like

```
In[32]:= p2 =
   Plot[s[t], {t, 0, T}, PlotRange → {0, Automatic},
    PlotStyle → RGBColor[1, 0, 0],
    AxesLabel → {"t", "$"}, ImageSize → 160];
```

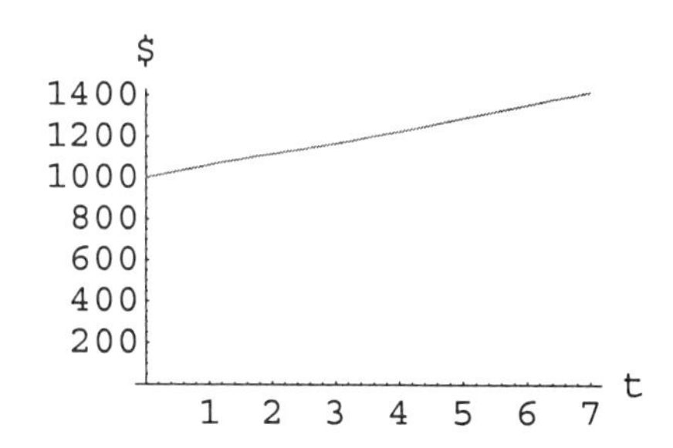

1.2 Numerical Solutions of ODEs

Let us develop a numerical scheme for solving the same problem. This of course is not needed—explicit solutions, like the one already computed above, are better than numerical ones; moreover, *Mathematica*® has a built-in numerical ODE solver. Our motivation to spend time doing this is that the same scheme will be modified later for solving stochastic differential equations, which is the basic mathematical construct used for the stock price evolution modeling.

The above differential equation

$$y'(t) = r\,y(t) \tag{1.2.1}$$

can be rewritten in the differential form as

$$dy = f(y)\,dt \tag{1.2.2}$$

and then approximated by $y_{n+1} - y_n = f(y_n)\,dt$, or

$$y_{n+1} = y_n + f(y_n)\,dt = g(y_n). \tag{1.2.3}$$

Consequently, taking into the account the initial condition $y(0) = P$, we get

$$y_{n+1} = g(g(g(g(\ldots g(P)\ldots)))).$$

How to implement this method in *Mathematica*® programming language? For example, let the right hand side be given by

```
In[33]:= Clear[f]; f[y_] := r y;
```

and also let

```
In[34]:= T = 20; K = 2000; dt = T / K;
```

Then, construct the above g as

```
In[35]:= Clear[g]; g[y_] := y + f[y] dt;
```

```
In[36]:= ? g
```

Global`g

```
g[y_] := y + f[y] dt
```

When programming, and in particular, because of its efficiency, when programming in *Mathematica*®, it is always advisable to check new constructs every step of the way, from the simple building blocks to the more complex ones as soon as they are introduced. So, in that spirit, check g:

```
In[37]:= g[y]
```

Out[37]= $1.0005\, y$

It works just as it is supposed to. Now, *Mathematica*® has a very useful set of iterative functions, one of which is called NestList:

```
In[38]:= ? NestList
```

```
NestList[f, expr, n] gives a list of the results of
   applying f to expr 0 through n times. More...
```

For example, using the above defined g,

```
In[39]:= NestList[g, 1000, 5]
```

Out[39]= {1000, 1000.5, 1001., 1001.5, 1002., 1002.5}

produces list of iterates $g(g(\ldots g(1000)\ldots))$. Once, NestList was checked, and it works properly, we can compute the full list with $K + 1$ elements, where K is now a large number (never do the first calculation with large K—it is always unpleasant to crash, or freeze a computer) suppressing the output since it is too large,

```
In[40]:= SolList = NestList[g, 1000, K];
```

Visualizing

```
In[41]:= ListPlot[Take[SolList, {1, -1, 10}],
           PlotRange → {0, SolList[[-1]]}, ImageSize → 150];
```

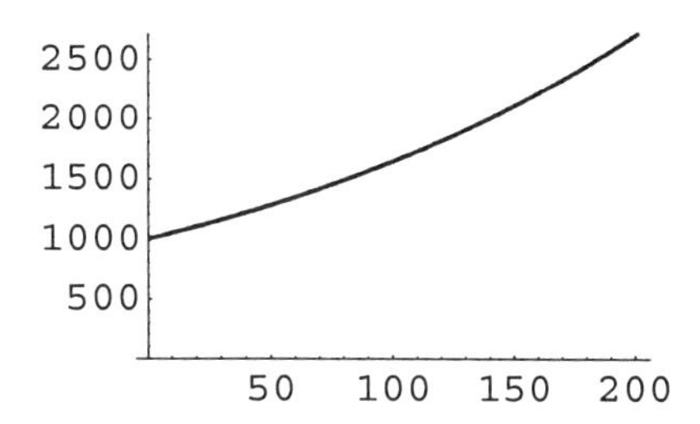

where

```
In[42]:= ? ListPlot

    ListPlot[{y1, y2, ... }] plots a list of values. The x
       coordinates for each point are taken to be 1, 2, ... .
       ListPlot[{{x1, y1}, {x2, y2}, ... }] plots a list of
       values with specified x and y coordinates. More...
```

Comparing this plot with the previous one, we can notice the difference in the argument: while before the argument was time (in years), now the argument is the number of iterations. Moreover, in the present calculation, tracking of time was not even attempted. In order to track the time and also be able to solve somewhat more general ODEs, namely the initial value problem of the form

$$\frac{dy}{dt} = f(t, y) \qquad y(0) = y_0 \tag{1.2.4}$$

we proceed as follows. As before, the above ODE is approximated with finite difference equation $y_{n+1} - y_n = f(t_n, y_n)\,dt$, or

$$y_{n+1} = y_n + f(t_n, y_n)\,dt = g(t_n, y_n) \tag{1.2.5}$$

So, beginning programming, let

```
In[43]:= Clear[f, g];
         f[t_, y_] := r y;
         g[t_, y_] := y + f[t, y] dt;
         G[x_] := {x[[1]] + dt, g[x[[1]], x[[2]]]};
```

It is interesting to carefully examine function G. Indeed, G is a function of a *single* variable x, while that variable x has two components: x[[1]] is time t, while x[[2]] is the value y. By the way, in *Mathematica*®, for example, a function of a single two-dimensional variable, and a function of two single-dimensional variables are *different* objects: *Mathematica*®, being a programming language, needs to be more precise than even Mathematics. We check G as soon as it is defined:

```
In[47]:= G[{t, y}]
```

Out[47]= $\left\{t + \frac{1}{100}, 1.0005\, y\right\}$

The equivalent, maybe somewhat more readable, alternative definition of G would be

```
In[48]:= G[{t_, y_}] := {t + dt, g[t, y]}
```

We check G again:

```
In[49]:=  G[{t, y}]
```

Out[49]= $\left\{t + \frac{1}{100}, 1.0005\, y\right\}$

Either way, we can start the iteration. So, this time the short iteration looks like

```
In[50]:= NestList[G, {0, 1000}, 5] // StandardForm
```

Out[50]//StandardForm=

$$\left\{\{0, 1000\}, \left\{\frac{1}{100}, 1000.5\right\}, \left\{\frac{1}{50}, 1001.\right\}, \left\{\frac{3}{100}, 1001.5\right\}, \left\{\frac{1}{25}, 1002.\right\}, \left\{\frac{1}{20}, 1002.5\right\}\right\}$$

while the full size iteration

```
In[51]:= SolList2 = NestList[G, {0, 1000}, K];
```

looks like this:

```
In[52]:= ListPlot[Take[SolList2, {1, -1, 10}],
           PlotRange → {0, SolList2[[-1, 2]]},
           AxesLabel → {"t", "$"}, ImageSize → 150];
```

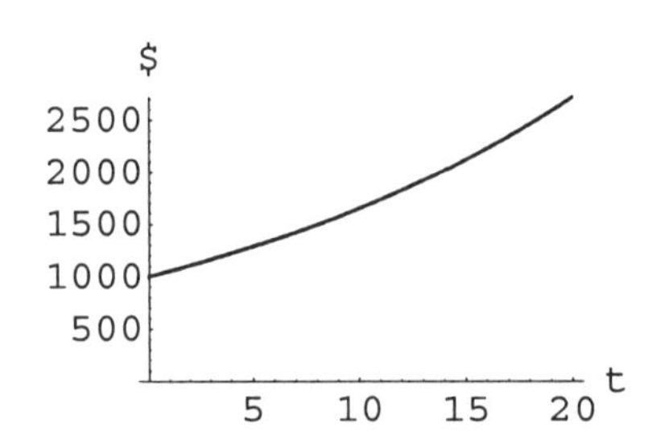

Notice an improvement: the argument is time t. Additionally, instead of having a solution only on discrete points, we can use interpolation to define the solution for all the values of the argument in the domain of definition. This is done by the built-in *Mathematica*® function

```
In[53]:= ? Interpolation
```

```
Interpolation[data] constructs an InterpolatingFunction
   object which represents an approximate function that
   interpolates the data. The data can have the forms
   {{x1, f1}, {x2, f2}, ... } or {f1, f2, ... }, where
   in the second case, the xi are taken to have values
   1, 2, ... . More...
```

So, compute

```
In[54]:= Sol = Interpolation[SolList2]

Out[54]= InterpolatingFunction[( 0.  20. ), <>]
```

It is interesting to take parts of Sol. For example,

```
In[55]:= Sol[[1, 1]]

Out[55]= {0., 20.}
```

are the beginning and end points of the interval of definition of the InterpolatingFunction Sol. Sol is a well-defined function: we can compute the value for any argument inside the interval of definition. For example,

```
In[56]:= Sol[15]

Out[56]= 2116.6
```

(The reader may try to compute the value for an argument outside of the interval of the definition to see what happens.) Furthermore, we can differentiate

```
In[57]:= Sol'

Out[57]= InterpolatingFunction[( 0.  20. ), <>]
```

or similarly

```
In[58]:= SolPrime[t_] = D[Sol[t], t]

Out[58]= InterpolatingFunction[( 0.  20. ), <>][t]
```

where

```
In[59]:= ? D

D[f, x] gives the partial derivative of f with respect
   to x. D[f, {x, n}] gives the nth partial derivative
   of f with respect to x. D[f, x1, x2, ... ] gives a
   mixed derivative. More...
```

Indeed, those two definition amount to the same thing; for example,

```
In[60]:= Sol'[10] == SolPrime[10]

Out[60]= True
```

We can also check whether the ODE we were solving actually holds (approximately, at some fixed time), i.e., whether the following numbers are close enough:

```
In[61]:= {SolPrime[t], Sol'[t], r Sol[t]} /. t → 10

Out[61]= {82.4052, 82.4052, 82.4258}
```

and visualize both functions on the same graph (they would coincide, i.e., the ODE holds), or to make a picture more interesting let's see how much they differ, and how they compare to the right-hand side of the ODE with the exact solution $e^{rt} P$ in it, if the discretization is *not* very fine, say

```
In[62]:= K = 5;

In[63]:= T = 20; dt = T / K; Clear[f]; f[t_, y_] := r y;
         NestList[G, {0, 1000}, K];
         Interpolation[%];
         Plot[
           Evaluate[{D[%[t], t], r %[t], r 1000 Exp[r t]}],
           {t, 0, T}, PlotRange → {0, r Sol[T]},
           PlotStyle → {Dashing[{}], Dashing[{.03, .03}],
             Dashing[{.09, .03}]}, ImageSize → 170];
```

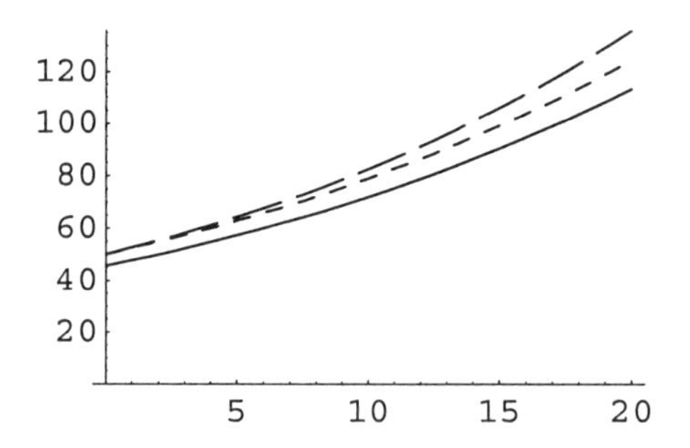

Several more new *Mathematica*® objects are introduced here. We leave plotting styles to the reader to investigate, and discuss only the symbol %, and function Evaluate.

```
In[66]:= ? %

%n or Out[n] is a global object that is assigned to be
  the value produced on the nth output line. % gives
  the last result generated. %% gives the result before
  last. %% ... % (k times) gives the kth previous result.
More...
```

Notice above that even though they all belonged to a single cell, each line, unless wrapped around, was counted as separate input line (there were 3 input lines, and %

can be used to refer to every previous one). It is relevant here to emphasize the precise definition of ";"

In[67]:= ? ;

```
expr1; expr2; ...  evaluates the expri in turn, giving
   the last one as the result. More...
```

Turning to the other important function

In[68]:= ? Evaluate

```
Evaluate[expr] causes expr to be evaluated even if it
   appears as the argument of a function whose attributes
   specify that it should be held unevaluated. More...
```

we see that in the above situation, what happens is that function Plot has attributes

In[69]:= `Attributes[Plot]`

Out[69]= {HoldAll, Protected}

where the first attribute means

In[70]:= ? HoldAll

```
HoldAll is an attribute which specifies that all arguments
   to a function are to be maintained in an unevaluated
   form. More...
```

So, the reason we have to Evaluate under Plot is that otherwise the derivative D[%[t],t] is not evaluated, and therefore Plot is not possible. The other alternative (not so much an alternative, but may help understanding what is going on) is to Clear the HoldAll Attribute of the Plot function; then Evaluate is not needed under Plot. For example,

In[71]:=
```
ClearAttributes[Plot, HoldAll];
Plot[D[x^2, x], {x, -2, 2}, ImageSize → 100];
SetAttributes[Plot, HoldAll];
```

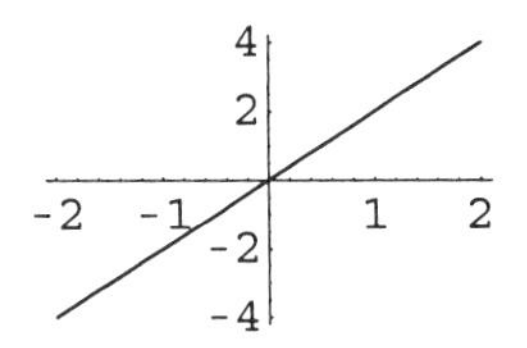

Once things work properly, it is advisable to pack several, or as many as it takes, lines of program into a single function (provided it is useful). So, in the case of the problem:

$$\frac{dy}{dt} = f(t, y)$$
$$y(t_0) = y_0 \qquad (1.2.6)$$

we define

```
In[74]:= ODESolver[f_, y0_, t0_, t1_, K_] :=
          Module[{dt, G, SolList},
            dt = (t1 - t0) / K;
          G[{t_, y_}] := {t + dt, y + f[t, y] dt};
          SolList = NestList[G, {t0, y0}, K];
          Interpolation[SolList]]
```

We have used a new function, called Module, which means

```
In[75]:= ? Module

       Module[{x, y, ... }, expr] specifies that occurrences
          of the symbols x, y, ...  in expr should be treated
          as local. Module[{x = x0, ... }, expr] defines initial
          values for x, ... . More...
```

One could think of Module, simply, as just parenthesis *enclosing* several lines, with an additional feature which is the *localization* of variables, which means that definitions made and declared inside the module are not going to interfere with the rest of the work.

Let us check whether newly defined function works properly, using the same data as before, i.e.,

$$f(t, y) = .05\, y$$

It works:

```
In[76]:= ODESolver[0.05 #2 &, 1000, 0, 20, 1000];
          Plot[%[s], {s, 0, 20},
            PlotRange → {0, %[20]}, ImageSize → 140];
```

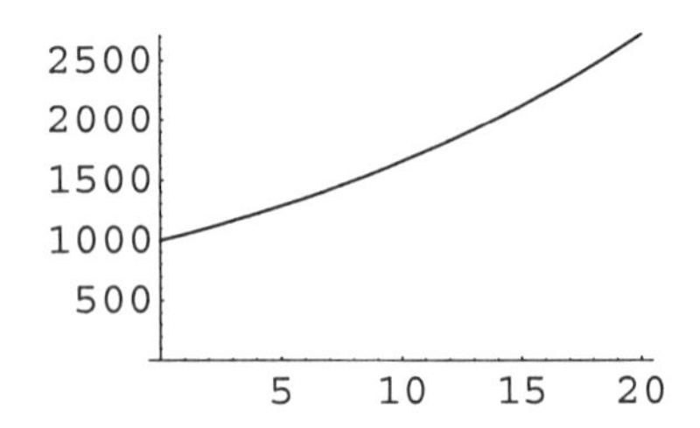

We emphasize that the first argument of the ODESolver has to be supplied as a *pure function*, i.e.,

```
In[78]:= ? Function

        Function[body] or body& is a pure function. The formal
           parameters are # (or #1), #2, etc. Function[x, body]
           is a pure function with a single formal parameter x.
           Function[{x1, x2, ... }, body] is a pure function
           with a list of formal parameters. More...
```

One can think of a pure function as a *function without arguments*. For example,

```
Sin
```

is a pure function. The same can be written also as

```
Sin[#] &
```

or even as

```
Function[x, Sin[x]]
```

This means that expressions

```
Function[{t, y}, 0.05 y]
```

and, the one found in the ODESolver above

```
0.05 #2 &
```

are equivalent. Notice

```
In[79]:= 0.05 #2 &[y, z]

Out[79]= 0.05 z
```

or even

```
In[80]:= Function[{t, y}, 0.05 y][y, z, 1, 2, 3, 9, q, w]

Out[80]= 0.05 z
```

while if we tried to submit only a single variable, it would not have worked. Pure functions, as we shall see later in this book, are very useful in *Mathematica*® programming.

To visualize the time dependency of the data, let the right-hand side in the ODE, be $f(t, y) = r(\sin(t) + 1)\,y$, or, written as a pure function

```
In[81]:= f := r (Sin[#1] + 1) #2 &
```

Then

```
In[82]:= Module[{T, r, Sol},
           T = 20; r = 0.05;
           Sol = ODESolver[f, 1000, 0, T, 1000];
           Plot[Sol[t], {t, 0, T},
            PlotRange → {0, Sol[T]}, AxesLabel → {"t", "$"},
            ImageSize → 150]]; // Timing
```

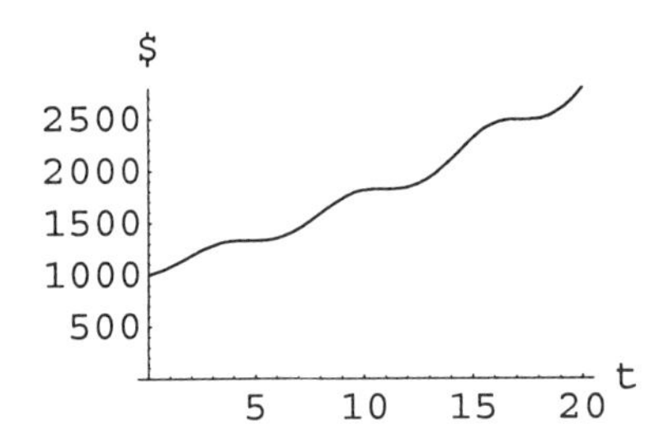

Out[82]= {0.11 Second, Null}

We used Timing, which is defined as

```
In[83]:= ? Timing
```

```
Timing[expr] evaluates expr, and returns a list of time
   used, together with the result obtained. More...
```

Of course, *Mathematica*® has already a built-in, better numerical ODE Solver, called NDSolve,

```
In[84]:= ? NDSolve
```

```
NDSolve[eqns, y, {x, xmin, xmax}] finds a numerical
   solution to the ordinary differential equations eqns
   for the function y with the independent variable x
   in the range xmin to xmax. NDSolve[eqns, y, {x, xmin,
   xmax}, {t, tmin, tmax}] finds a numerical solution
   to the partial differential equations eqns. NDSolve[
   eqns, {y1, y2, ... }, {x, xmin, xmax}] finds numerical
   solutions for the functions yi. More...
```

and indeed there was no need for the above function ODESolver:

```
In[85]:= Timing[Module[{T, r, s},
            T = 20; r = 0.05;
            s[t_] = y[t] /.
              NDSolve[{y'[t] == r (Sin[#1] + 1) #2 &[t, y[t]],
                  y[0] == 1000}, y[t], {t, 0, T}][[1]];
            Plot[s[t], {t, 0, T}, PlotRange → {0, Sol[T]},
             AxesLabel → {"t", "$"}, ImageSize → 150]];]
```

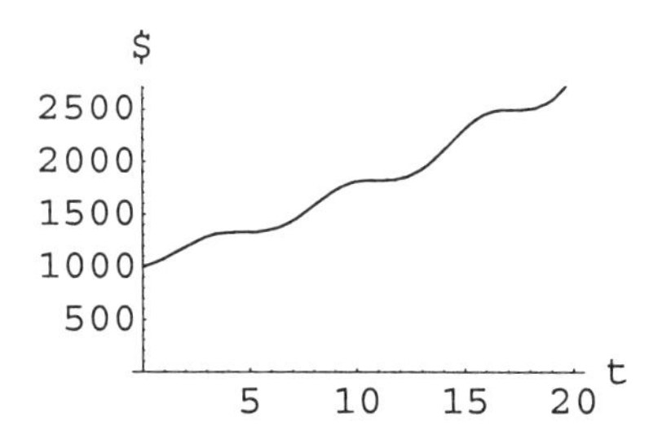

Out[85]= {0.05 Second, Null}

The point is, as said earlier, to develop a method that is going to be supplemented in the next chapter for solving and understanding stochastic differential equations, which are the main mathematical constructs used for modeling evolution of stock prices. Along the way we introduce various *Mathematica*® functions as well.

2 Stock Price Evolution

Modeling Stock Price Dynamics with Stochastic Differential Equations

2.1 What are Stocks?

There exists a fundamental difference in investing in risk-free instruments, such as cash accounts discussed in the previous chapter, and risky assets, such as stocks of publicly traded companies. The risk that an investor assumes when investing in stocks has to be understood and taken into consideration. Estimating how much risk is actually taken when investing, deciding in advance how much risk to take when investing, and then according to such a decision how to invest, are some of the central issues in computational financial mathematics.

Stocks, or shares, are certificates of partial ownership of a company. Ownership, even a partial one, is supposed to have its privileges, such as sharing of profits, if any, in the form of dividends. Stocks, i.e., the tiny pieces of publicly traded companies, have a price on the open market, and that price changes almost continuously, according to demand and supply, greed and fear, general ever-changing perception about the conditions and the direction of the whole economy, political stability, and many other reasons. Those fluctuations of the price can be attributed also to the constant search for fair value, i.e., fair price. In addition to those (random) fluctuations, there is more or less (locally in time) constancy, or at least stability, in appreciation, if a particular company is on the rise, or depreciation, if a company is on the decline. Notice also that if the price of the stock is accepted as its current value, then the stock market can be viewed as an instrument for valuation of whole companies and by extension, even of whole economies. For example, Microsoft Corporation stock, ticker symbol MSFT was traded, at 12:09 on August 5th, 1999, at USD 85 7/8, while there were total of 5,103,859,000 shares outstanding. This means that, at that moment, the whole company was (perceived as) worth, i.e., the Market Cap(italization) was

```
In[1]:= MSFTvalue =
          AccountingForm[(85 + 7/8) 5103859000, DigitBlock -> 3]
Out[1]//AccountingForm=
        438,293,891,625
```

or

In[2]:= N[%]

Out[2]= 4.38294×10^{11}

dollars, where

In[3]:= ?AccountingForm

```
AccountingForm[expr] prints with all numbers in expr
   given in standard accounting notation. AccountingForm[
   expr, n] prints with numbers given to n-digit precision.
 More...
```

and

In[4]:= ?DigitBlock

```
DigitBlock is an option for NumberForm and related
   functions which specifies the maximum length of blocks
   of digits between breaks. More...
```

2.2 Stock Price Modeling: Stochastic Differential Equations

2.2.1 Idea of Stochastic Differential Equations

In the previous chapter we solved numerically the general first-order ordinary differential equation

$$\frac{dy(t)}{dt} = a(t, y(t))$$
$$y(t_0) = y_0.$$

As already pointed out, the same equation can be written and understood in the differential form

$$dy(t) = a(t, y(t))\,dt$$
$$y(t_0) = y_0.$$

Can one use such a model, i.e., first-order ODEs, to compute the evolution, and thereby predict future stock prices?

Of course, not. First, the drift a in reality cannot depend only on time t and on the present value of the stock $y(t)$. On the other hand, even if one allows a to depend on many other things (interest rates, inflation, unemployment, exchange rates, etc.), the nature of that dependence is just too complicated, unpredictable, and ultimately unquantifiable. Even the most ambitious and successful deterministic models have to give up at some depth of understanding the particular phenomenon it attempts to describe. The rest, the unknown, the unquantifiable, is then modeled as random. Once it is modeled as random, then that randomness can be quantified. As a matter of fact, it is quite interesting, as we are going to see later, how much easier it is to quantify randomness than it is to quantify determinism in the presence of the unknown, i.e., in the presence of randomness.

Keeping in mind the way the above ODE was solved numerically, one can ask what happens if at each time step, in addition to the above deterministic drift $a(t, y)$, there is also some random effect. So the above equation is extended to

$$\begin{aligned} dy(t) &= a(t, y(t))\, dt + \sigma(t, y(t))\, dB(t) \\ y(t_0) &= y_0 \end{aligned} \tag{2.2.1}$$

where $dB(t)$ is the "differential" of the Brownian motion $B(t)$, and $\sigma(t, y)$ is a given function ($\frac{\sigma(t,y(t))}{y(t)}$ is called volatility). What is meant formally is that $dB(t)$'s are (independent) normally distributed random variables, with zero mean and standard deviation $\sqrt{dt}$ (variance dt), i.e.,

$$dB(t) \sim N\left(0, \sqrt{dt}\right). \tag{2.2.2}$$

The above equation is called a stochastic differential equation (SDE), or more appropriately but less efficiently a stochastic ordinary differential equation. To build the concept of SDEs precisely, we shall first recall the normal distribution from the perspective provided by *Mathematica*®.

2.2.2 Normal Distribution

```
In[5]:= Clear["Global`*"]; Off[General::"spell1"]
```

The mathematical knowledge that is the content of the *Mathematica*® computer platform is located generally in two locations. The basic knowledge, deemed over the time as indispensable, is part of the *Mathematica*® "Kernel", and it is available each time *Mathematica*® is in session. On the other hand, there is a large quantity of important content that is located in the so-called packages. The paradigm is that the new content is placed on the outside, i.e., in the packages, and called upon only if and when it is needed. There are packages that are supplied by *Mathematica*®, and among them the most important ones are the ones referred to as *standard Mathematica® packages*. Packages can also be written by users, and some will be supplied with and used in this book. So, for now, we shall need to import a standard *Mathematica*® package:

```
In[6]:= << "Statistics`NormalDistribution`"
```

where

```
In[7]:= ? <<

    <<name reads in a file, evaluating each expression in
      it, and returning the last one. More...
```

The same can be achieved by

```
In[8]:= Needs["Statistics`NormalDistribution`"]
```

where

```
In[9]:= ?Needs
```

```
Needs["context`"] loads an appropriate file if the
   specified context is not already in $Packages.
   Needs["context`", "file"] loads file if the specified
   context is not already in $Packages. More...
```

The *Mathematica*® object we are going to use is

```
In[10]:= ?NormalDistribution
```

```
NormalDistribution[mu, sigma] represents the normal (
   Gaussian) distribution with mean mu and standard
   deviation sigma. More...
```

Let us define the standard normal distribution shortly as

```
In[11]:= ND := NormalDistribution[0, 1]
```

The standard normal distribution has the density

```
In[12]:= PDF[ND, x]
```

Out[12]= $\frac{e^{-\frac{x^2}{2}}}{\sqrt{2\pi}}$

where

```
In[13]:= ?PDF
```

```
PDF[distribution, x] gives the probability density
  function of the specified statistical distribution
  evaluated at x.
```

and whose graph looks like

```
In[14]:= Plot[PDF[ND, x], {x, -5, 5}];
```

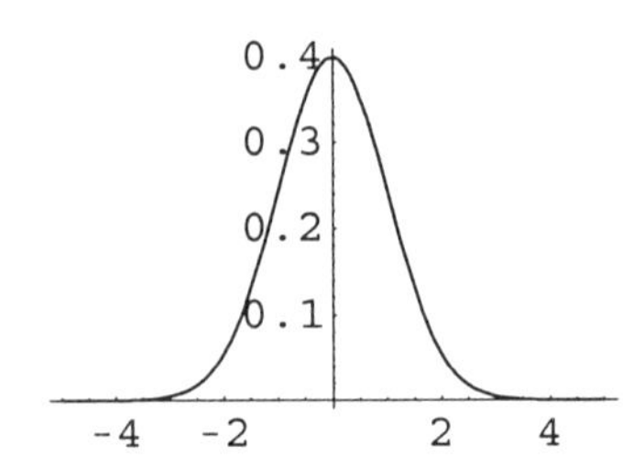

Also the cumulative distribution function

In[15]:= ? CDF

```
CDF[distribution, x] gives the cumulative distribution
  function of the specified statistical distribution
  evaluated at x.  For continuous distributions, this
  is defined as the integral of the probability density
  function from the lowest value in the domain to x.
  For discrete distributions, this is defined as the sum
  of the probability density function from the lowest
  value in the domain to x.
```

in the case of the standard normal distribution is equal to

In[16]:= CDF[ND, x]

Out[16]= $\frac{1}{2}\left(\operatorname{erf}\left(\frac{x}{\sqrt{2}}\right)+1\right)$

whose graph (compared with the constant function 1, and with the density) looks like

```
In[17]:= Plot[{1, CDF[ND, x], PDF[ND, x]},
           {x, -5, 5}, PlotStyle → {RGBColor[.5, .5, .5],
             RGBColor[0, 0, 0], RGBColor[.5, .5, .5]}];
```

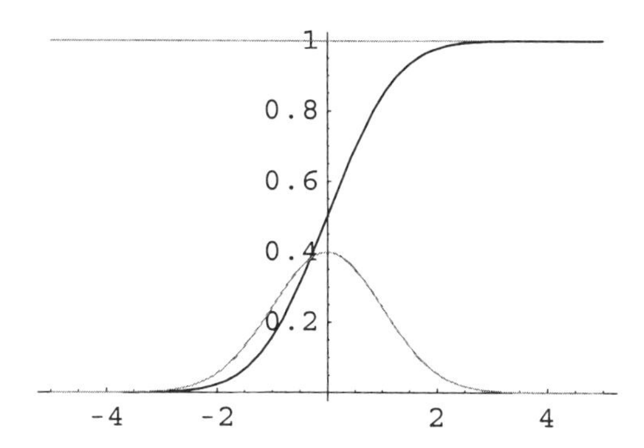

where

In[18]:= ? PlotStyle

```
PlotStyle is an option for Plot and ListPlot that specifies
   the style of lines or points to be plotted. More...
```

and

In[19]:= ? RGBColor

```
RGBColor[red, green, blue] is a graphics directive which
   specifies that graphical objects which follow are to
   be displayed, if possible, in the color given. More...
```

It is quite instructive for the development of proper intuition to experiment with standard normal random variables. For example, to generate a (pseudo) random number with such a distribution one can use

```
In[20]:= Random[ND]

Out[20]= 0.223211
```

where

```
In[21]:= ? Random

Random[ ] gives a uniformly distributed pseudorandom
   Real in the range 0 to 1. Random[type, range] gives
   a pseudorandom number of the specified type, lying
   in the specified range. Possible types are: Integer,
   Real and Complex. The default range is 0 to 1. You
   can give the range {min, max} explicitly; a range
   specification of max is equivalent to {0, max}. Random[
   distribution] gives a random number with the specified
   statistical distribution. More...
```

or to generate a list of such random numbers

```
In[22]:= Table[Random[ND], {20}]

Out[22]= {-0.213466, 0.432191, 0.392801, 1.26251, 0.0952652, -0.258423,
   -2.45324, -1.21499, 0.707917, -0.211041, 0.0379052,
   0.348389, 0.545783, 1.27341, 0.540497, 0.452756, 0.830841,
   0.44365, 0.0174794, -1.57809}
```

where

```
In[23]:= ? Table

Table[expr, {imax}] generates a list of imax copies of
   expr. Table[expr, {i, imax}] generates a list of the
   values of expr when i runs from 1 to imax. Table[expr,
   {i, imin, imax}] starts with i = imin. Table[expr,
   {i, imin, imax, di}] uses steps di. Table[expr, {i,
   imin, imax}, {j, jmin, jmax}, ... ] gives a nested
   list. The list associated with i is outermost. More...
```

Larger samples are even more instructive, but they need to be visualized to be appreciated:

```
In[24]:= Module[{SampleSize, x, y}, SampleSize = 1000;
   RandomSample = Table[Random[ND], {SampleSize}];
```

```
x = ListPlot[RandomSample, DisplayFunction →
    Identity, PlotStyle → RGBColor[0, 1, 0]];
y = Plot[{-3, -2, -1, 0, 1, 2, 3}, {x, 0,
    SampleSize}, DisplayFunction → Identity];
Show[x, y, DisplayFunction → $DisplayFunction,
 AxesLabel → {"", "X"},
 Ticks → {None, {-3, -2, -1, 0, 1, 2, 3}},
 PlotRange → All]];
```

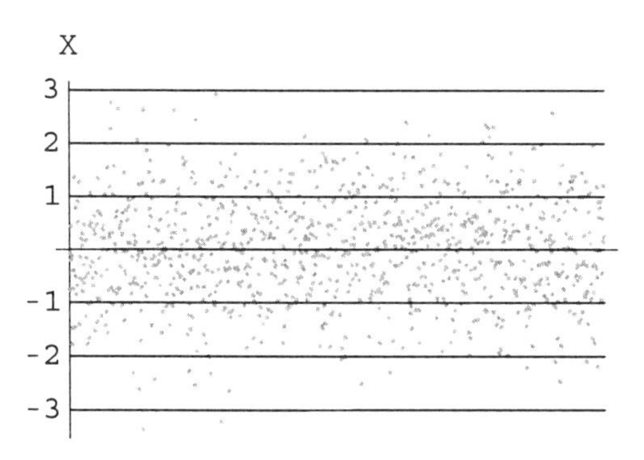

where

In[25]:= `? Show`

```
Show[graphics, options] displays two- and three-dimensional
   graphics, using the options specified. Show[g1,
   g2, ... ] shows several plots combined. More...
```

and

In[26]:= `? DisplayFunction`

```
DisplayFunction is an option for graphics and sound
   functions that specifies the function to apply to
   graphics and sound objects in order to display them.
  More...
```

Let us select in the above random sample the outcomes X, such that $|X| \leq a$, for various values $a > 0$. For example, if

In[27]:= `a = .005`

Out[27]= 0.005

(we take a small a in order to produce short output), then

In[28]:= `Select[RandomSample, Abs[#1] ≤ a &]`

Out[28]= {−0.00236407, 0.00475498, 0.00154544, −0.00435424}

where

```
In[29]:= ? Select
```

```
Select[list, crit] picks out all elements ei of list for
   which crit[ei] is True. Select[list, crit, n] picks
   out the first n elements for which crit[ei] is True.
  More...
```

Further, let's count the number of those realizations

```
In[30]:= Length[Select[RandomSample, Abs[#1] ≤ a &]]
```

Out[30]= 4

where

```
In[31]:= ? Length
```

```
Length[expr] gives the number of elements in expr. More...
```

and finally the relative frequency of those realizations:

```
In[32]:= N[ Length[Select[RandomSample, Abs[#1] ≤ a &]] / Length[RandomSample] ]
```

Out[32]= 0.004

Putting it all together, define a function

```
In[33]:= Quant[a_] :=
          N[ Length[Select[RandomSample, Abs[#1] ≤ a &]] / Length[RandomSample] ]
```

and compute it for $a = 1, 2, 3$:

```
In[34]:= Table[Quant[a], {a, 1, 3}]
```

Out[34]= {0.696, 0.963, 0.998}

Finally, compare realized frequencies with the theoretical values, i.e., with the probabilities $P(|X| \leq a)$

```
In[35]:= N[Table[CDF[ND, a] - CDF[ND, -a], {a, 1, 3}]]
```

Out[35]= {0.682689, 0.9545, 0.9973}

which is quite a good match.

2.2.3 Markov Random Processes

(Continuous time) random process (or stochastic process) is a collection of (possibly vector-valued) random variables $\{X(t)\}_{t_0 \le t < t_1}$, where index t is time. A random process is called a Markov process if, loosely speaking, it has no memory. A bit more precisely, for such a process the future evolution depends only on the current state and not on the past evolution. As a matter of fact, in applications, especially if one would like to draw on the wealth of mathematical knowledge in differential equations, and consequently be able to reach far-reaching conclusions, one is pretty much limited to consider Markov processes only. This is not a forbidding limitation if one allows for a very general state space—the dependencies on the past would be then imbedded into an extended state space (similarly to ODEs, when for example a one-dimensional 2nd order ODE, or even SDE like in Section 8.3, can be considered as a 1st order ODE, or SDE, with a two-dimensional state space). The mathematical description of a specific (Markov) random processes can be done in different ways, but it seems quite natural to emphasize two kinds of specifications. The first one is the "differential structure", i.e., the information about how the process evolves from one state to the other for all possible states. For example, in the case of 1st order ODEs, i.e., a deterministic (Markov) processes, this can be formulated as

$$dX(t) = a(t, X(t))\, dt$$

for some function $a(t, X)$, called drift. The other kind of specification would be the information about the initial condition, i.e., where exactly the process was initially (or at some other time). Where exactly the process was may not always be known, when this condition is relaxed to a condition about the probability distribution for the initial position. If initial position x_0 is known, this condition can be written as

$$X(t_0) = x_0.$$

The two specifications (differential and initial) now determine completely the probability structure of the process. That structure can be referred to as a probability (measure) P. The other, just as important a mathematical construct, the expectation, can be thought of as an average outcome, and can be denoted by E. For example one can attempt to compute expectations of various functionals of such a process $\{X(t)\}_{t \ge t_0}$, such as

$$E\, \phi(t, X(t))$$

which can be thought of as an average value of quantities $\phi(t, X(t))$ over many random realizations of $X(t)$, and in principle can be computed as an integral of $\phi(t, X(t))$ with respect to the probability P.

During the realization of a random process, additional information becomes available, such as the information that at some time, say $t = s$, the process was at the state z, i.e., $X(s) = z$ (for example, today's price of MSFT is such and such). This new information needs to be taken into consideration. One way to do this would be to consider a new process $\{X^{s,z}(t)\}_{t \ge s}$, with the same differential structure as $\{X(t)\}_{t \ge 0}$, for $t \ge s$, but with the initial condition $X^{s,z}(s) = z$, in which case one works with the same probability P and with the same expectation E. The other way to do this is to consider the conditional probability $P_{s,z}$ and the corresponding conditional expectation $E_{s,z}$, and to work with the original process $\{X(t)\}_{t \geq t_0}$. In conclusion

$$E\,\phi(X^{s,z}(t)) = E_{s,z}\,\phi(X(t))$$

for any function ϕ. The shorter notation for $\{X^{0,x_0}(t)\}_{t\geq 0}$ is $\{X^{x_0}(t)\}_{t\geq 0}$. Also, notice that $\{X^{0,x_0}(t)\}_{t\geq 0} = \{X(t)\}_{t\geq 0}$.

2.2.4 Brownian Motion

Brownian motion $\{B(t)\}_{t\geq 0}$ is a (Markov) random process with increments $dB(t)$ that are independent and normally $N(0, \sqrt{dt})$ distributed, and with the initial condition $B(0) = 0$. It was observed experimentally by Brown in 1827, and described mathematically by Wiener in 1923. It can be proved mathematically that such an intellectual construct does exist. We shall see many (approximations of) trajectories of Brownian motions. Brownian motion is probably the most important stochastic process of all—it is for random processes what (standard) normal distribution is for random variables.

Brownian motion can be generated (and understood) in the following way. Let us first introduce time scale. For example, let us consider the time interval (0, T), and assume that K time-subintervals suffice for precision.

```
In[36]:= Clear["Global`*"]; T = 1; K = 10; dt = N[T/K];
```

```
In[37]:= << "Statistics`NormalDistribution`"
```

We need to generate a list of (many) independent, normally $N(0, 1)$ distributed random variables. Let us start with only a few to see what is going on:

```
In[38]:= Normal01 =
          Table[Random[NormalDistribution[0, 1]], {K}]

Out[38]= {0.901186, 0.349125, -0.386838, 1.70817, 0.165606, -0.85725,
          -0.406916, 1.01588, 1.59473, 0.641195}
```

To make a list of $N(0, \sqrt{dt})$ distributed random variables one has only to multiply the above list with $\sqrt{dt}$

```
In[39]:= dB = √dt Normal01

Out[39]= {0.28498, 0.110403, -0.122329, 0.540171, 0.0523692, -0.271086,
          -0.128678, 0.321249, 0.504297, 0.202764}
```

A somewhat more direct way would be to instead define

```
In[40]:= dB =
          Table[Random[NormalDistribution[0, √dt]], {K}]
```

```
Out[40]= {-0.00114024, 0.626956, -0.232329, 0.0134333, -0.2332,
          -0.314167, 0.313856, 0.334457, -0.14027, 0.246727}
```

Either way, dB can be taken to be the list of increments of the Brownian motion. To construct a Brownian motion one has to add them up, keeping track of partial sums. This can be done using the *Mathematica*® function FoldList. Indeed,

```
In[41]:= ? FoldList

    FoldList[f, x, {a, b, ... }] gives {x, f[x, a], f[f[x,
      a], b], ... }. More...
```

So,

```
In[42]:= BrownianValues = FoldList[Plus, 0, dB]

Out[42]= {0, -0.00114024, 0.625815, 0.393486, 0.406919, 0.173719,
          -0.140448, 0.173408, 0.507865, 0.367595, 0.614322}
```

is an approximation of the Brownian motion trajectory, except that we did not keep track of time. One way of doing this would be to compute

```
In[43]:= TimeList = Range[0, T, dt]

Out[43]= {0, 0.1, 0.2, 0.3, 0.4, 0.5, 0.6, 0.7, 0.8, 0.9, 1.}
```

and then

```
In[44]:= StandardForm[BrownianMotionList =
           Transpose[{TimeList, BrownianValues}]]

Out[44]//StandardForm=
    {{0, 0}, {0.1, -0.00114024}, {0.2, 0.625815},
     {0.3, 0.393486}, {0.4, 0.406919}, {0.5,
      0.173719}, {0.6, -0.140448}, {0.7, 0.173408},
     {0.8, 0.507865}, {0.9, 0.367595}, {1., 0.614322}}
```

where

```
In[45]:= ? Transpose

    Transpose[list] transposes the first two levels in list.
      Transpose[list, {n1, n2, ... }] transposes list so
      that the levels 1, 2, ... in list correspond to levels
      n1, n2, ... in the result. More...
```

Once an approximate trajectory of the Brownian motion is computed, it can be visualized as

```
In[46]:= ListPlot[BrownianMotionList,
           PlotJoined → True, AxesLabel → {"Time", ""}];
```

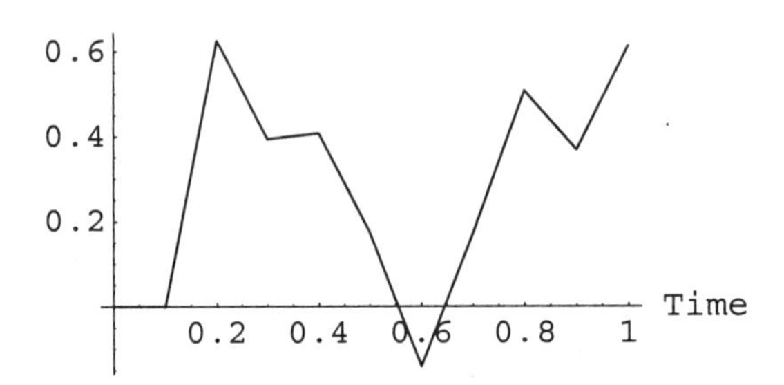

where

```
In[47]:= ? PlotJoined

    PlotJoined is an option for ListPlot that specifies
      whether the points plotted should be joined by a line.
     More...
```

The other way, possibly a more direct one, would be to fold both time and space at the same time. For example,

```
In[48]:= StandardForm[BrownianMotionList =
            FoldList[{dt + #1[[1]], #1[[2]] + #2} &, {0, 0}, dB]]

Out[48]//StandardForm=
    {{0, 0}, {0.1, -0.00114024}, {0.2, 0.625815},
     {0.3, 0.393486}, {0.4, 0.406919}, {0.5,
      0.173719}, {0.6, -0.140448}, {0.7, 0.173408},
     {0.8, 0.507865}, {0.9, 0.367595}, {1., 0.614322}}
```

So, putting together all that is essential in the above, let us define a Brownian Motion generator

```
In[49]:= BrownianMotion[T_, K_] :=
            Module[{dt, dB, BrownianMotionList}, dt = N[T/K];
             dB = Table[Random[NormalDistribution[0, √dt]],
                {K}]; BrownianMotionList =
              FoldList[{dt + #1[[1]], #1[[2]] + #2} &, {0, 0}, dB];
             Interpolation[BrownianMotionList,
              InterpolationOrder → 1]];
```

where

```
In[50]:= ? InterpolationOrder
```

```
InterpolationOrder is an option to Interpolation and
   ListInterpolation. InterpolationOrder-> n specifies
   interpolating polynomials of order n. InterpolationOrder ->
   {n1,n2,...} specifies interpolating polynomials of
   order n1, n2, ... for dimensions 1,2,..., respectively.
 More...
```

Then

```
In[51]:= With[{T = 1}, BM = BrownianMotion[T, 1000];
           Plot[BM[s], {s, 0, T}, AxesLabel → {"Time", ""}]];
```

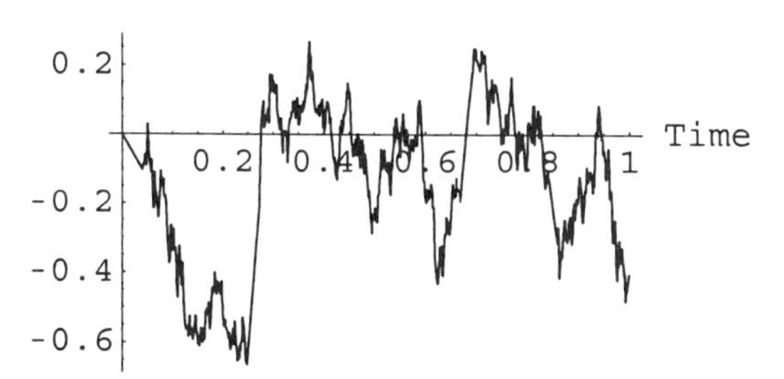

where

```
In[52]:= ? With
```

```
With[{x = x0, y = y0, ... }, expr] specifies that in
   expr occurrences of the symbols x, y, ... should be
   replaced by x0, y0, ... . More...
```

To grasp the dynamics of the Brownian motion, it is also quite instructive to see many trajectory realizations, and to compare them with square root functions; notice that $B(t) \sim N(0, \sqrt{t})$, $t > 0$, and therefore, as it was calculated in the previous section:

$$P(|B(t)| \le \sqrt{t}) = 0.682689,$$
$$P(|B(t)| \le 2\sqrt{t}) = 0.9545,$$
$$P(|B(t)| \le 3\sqrt{t}) = 0.9973$$

for any $t > 0$. Indeed

```
In[53]:= Module[{T = 1, SampleSize = 100, Sol, x, y},
           Sol = Table[BrownianMotion[T, 1000][s],
             {SampleSize}]; x = Plot[Evaluate[Sol],
             {s, 0, T}, DisplayFunction → Identity,
             PlotStyle → RGBColor[0, 1, 0]];
```

```
y = Plot[Evaluate[Table[k √z, {k, -3, 3}]],
   {z, 0, T}, DisplayFunction → Identity];
Show[x, y, DisplayFunction → $DisplayFunction,
  AxesLabel → {"", "B(t)"}, PlotRange → All]];
```

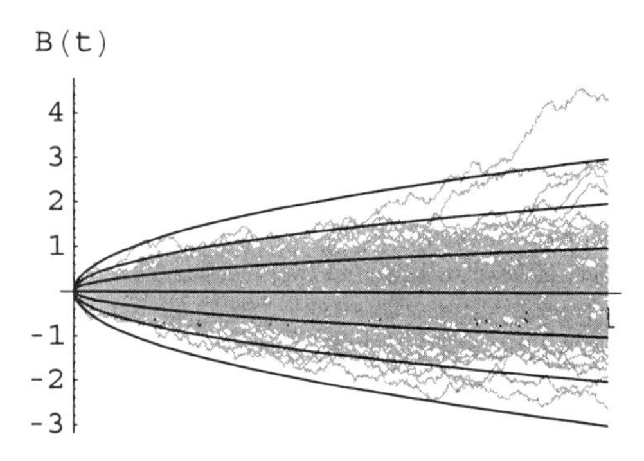

Finally, one should remember for what follows the fundamental properties of the Brownian motion:

$$\begin{aligned} &E_{s,B(s)}\, B(t) = B(s) \\ &E_{s,B(s)}(B(t) - B(s))^2 = t - s \end{aligned} \tag{2.2.3}$$

for $t \geq s$. Alternatively, as explained above, these properties can be written as

$$\begin{aligned} &E\, B^{s,z}(t) = z \\ &E\,(B^{s,z}(t) - z)^2 = t - s \end{aligned} \tag{2.2.4}$$

for $t \geq s$.

2.2.5 Stochastic Integral

A stochastic integral (with respect to Brownian motion) is a more general mathematical construct than Brownian motion, yet closely related. The stochastic (or, more precisely, Itô stochastic) integral of a random non-anticipative function (process) $f(t)$ with respect to the Brownian motion $B(t)$ is a stochastic process $\{I^{s,0}(t)\}_{t \geq s}$ denoted by

$$I^{s,0}(t) = \int_s^t f(w)\, dB(w)$$

or, written in the differential form

$$dI(t) = f(t)\, dB(t)$$

whose value $I^{s,0}(t)$ at time t is defined as the limit (in a proper sense) quantity of finite sums such as

$$\sum_{i=1}^{n} f(w_i)\,(B(w_{i+1}) - B(w_i)) \tag{2.2.5}$$

where $s = w_1 < w_2 < \ldots < w_{n+1} = t$, $n \to \infty$, $\max_{1 \le i \le n}(w_{i+1} - w_i) \to 0$. The non-anticipativness (with respect to the running Brownian motion) of the integrand $f(t)$, and this is very important, means that for any t, the random variable $f(t)$ may depend only on the realizations of the Brownian motion up until the time t, and possibly on some other sources of randomness, but does not depend on the future increments of $B(w)$, $w > t$. This property, philosophically very plausible, yields far-reaching mathematical and practical consequences, and actually the whole shape of the Itô stochastic calculus is its consequence. One can taste how the non-anticipativeness of the integrand works by examining the finite sum approximation above. Indeed, since f is non-anticipative, $f(w_i)$ and $(B(w_{i+1}) - B(w_i))$ are independent, and consequently

$$E\sum_{i=1}^{n} f(w_i)\,(B(w_{i+1}) - B(w_i)) = \sum_{i=1}^{n} E\,f(w_i)\,(B(w_{i+1}) - B(w_i))$$

$$= \sum_{i=1}^{n} E\,f(w_i)\,E\,(B(w_{i+1}) - B(w_i)) = \sum_{i=1}^{n} E\,f(w_i)\,0 = 0;$$

moreover

$$E\left(\sum_{i=1}^{n} f(w_i)\,(B(w_{i+1}) - B(w_i))\right)^2 = E\sum_{i=1}^{n} (f(w_i)\,(B(w_{i+1}) - B(w_i)))^2$$

$$+E\sum_{i \neq j}^{n} (f(w_i)\,(B(w_{i+1}) - B(w_i)))\;(f(w_j)\,(B(w_{j+1}) - B(w_j)))$$

$$= E\sum_{i=1}^{n} (f(w_i)\,(B(w_{i+1}) - B(w_i)))^2$$

$$+2\,E\sum_{i>j}^{n} f(w_i)\,(B(w_{i+1}) - B(w_i))\;f(w_j)\,(B(w_{j+1}) - B(w_j))$$

$$= E\sum_{i=1}^{n} (f(w_i)\,(B(w_{i+1}) - B(w_i)))^2$$

$$+2\sum_{i>j}^{n} E\,(B(w_{i+1}) - B(w_i))\;E(f(w_i)\,f(w_j)\,(B(w_{j+1}) - B(w_j)))$$

$$= E\sum_{i=1}^{n} (f(w_i)\,(B(w_{i+1}) - B(w_i)))^2$$

$$= \sum_{i=1}^{n} E\, f(w_i)^2\, E\,(B(w_{i+1}) - B(w_i))^2 = \sum_{i=1}^{n} E\, f(w_i)^2\,(w_{i+1} - w_i).$$

Those properties are preserved in passing the limit when the stochastic integral is defined, and they explicitly, and a bit more generally, hold as (compare with the Brownian motion in the last section):

$$\begin{aligned}
&E_{s,I(s)}\, I(t) = E_{s,I(s)}\Big(\int_0^t f(w)\, dB(w)\Big) \\
&= E_{s,I(s)}\Big(\int_0^s f(w)\, dB(w) + \int_s^t f(w)\, dB(w)\Big) \\
&= E_{s,I(s)}\Big(\int_0^s f(w)\, dB(w)\Big) + E_{s,I(s)}\Big(\int_s^t f(w)\, dB(w)\Big) \\
&= \int_0^s f(w)\, dB(w) = I(s)
\end{aligned} \tag{2.2.6}$$

$$\begin{aligned}
&E_{s,I(s)}(I(t) - I(s))^2 = E_{s,I(s)}\Big(\int_s^t f(w)\, dB(w)\Big)^2 \\
&= \int_0^s E_{s,I(s)}\, f(w)^2\, dw.
\end{aligned} \tag{2.2.7}$$

Intuitively, the stochastic integral is very similar to Brownian motion. It wanders up and down without partiality. The difference is only in the intensity of the wandering, which can be any (random, but non-anticipative) function of time. Consider some examples. Let

```
In[54]:= Clear["Global`*"];
         << "Statistics`NormalDistribution`";
```

```
In[56]:= s = 0; t1 = 10; K = 1000; dt = N[(t1 - s)/K];
```

As an example compute the stochastic integral of the Brownian motion itself:

$$I(t) = \int_0^t B(w)\, dB(w)$$

for $0 \le t \le t_1$. Compute a Brownian motion trajectory first:

```
In[57]:= dB = Table[
             Random[NormalDistribution[0, Sqrt[dt]]], {K}];
         B = FoldList[{dt + #1[[1]], #1[[2]] + #2} &, {0, 0}, dB];
         ListPlot[B, AxesLabel -> {"Time", "B(t)"},
           PlotJoined -> True];
```

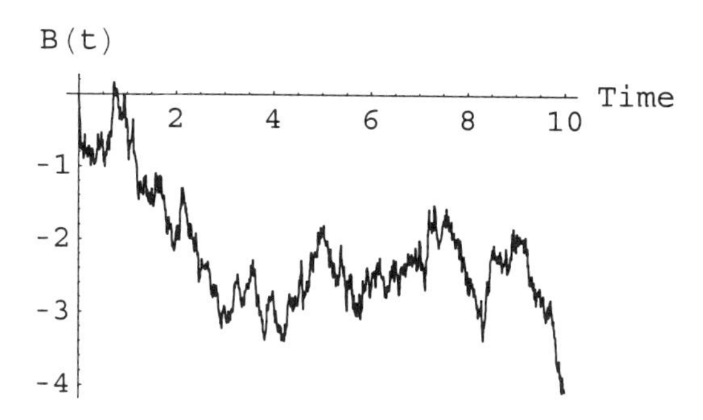

Next, compute the stochastic integral $\int_0^t B(w)\, dB(w)$

```
In[60]:= BdB = Drop[Transpose[B][[2]], -1] dB;
         ItoStochasticIntegral =
           FoldList[{dt + #1[[1]], #1[[2]] + #2} &, {0, 0}, BdB];
```

whose trajectory, corresponding to the same trajectory of the Brownian motion as above, looks like

```
In[62]:= ItoIntegralPlot = ListPlot[ItoStochasticIntegral,
            PlotJoined -> True, AxesLabel -> {"Time", "I(t)"},
            PlotStyle -> RGBColor[1, 0, 0], PlotRange -> All];
```

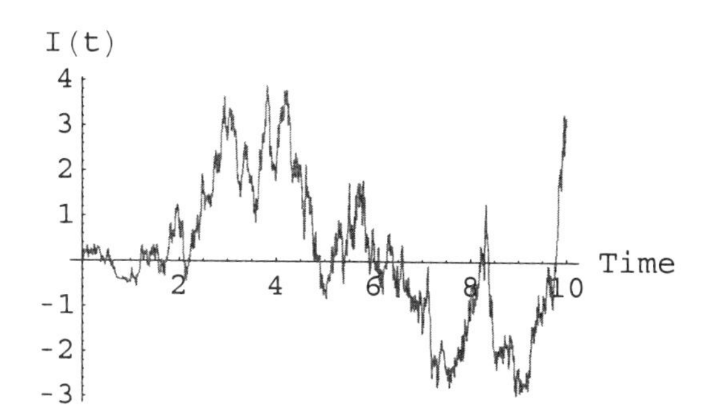

Notice how the stochastic integral flattens at times when the integrand passes through zero. Now we want to furthermore emphasize the above discussed crucial detail in the Itô definition of the stochastic integral. Namely, what if instead of

$$\sum_{i=1}^{n} B(w_i)\,(B(w_{i+1}) - B(w_i)) \tag{2.2.8}$$

we computed, for example,

$$\sum_{i=1}^{n} B(w_{i+1})\,(B(w_{i+1}) - B(w_i)). \tag{2.2.9}$$

In classical calculus, this would not cause a difference, i.e., the difference would be very small, and as the discretization becomes finer and finer, the difference would disappear. This does not happen in the case of the stochastic integral. Indeed, let

```
In[63]:= NonItoBdB = Drop[Transpose[B][[2]], 1] dB;
         NonItoStochasticIntegral = FoldList[
            {dt + #1[[1]], #1[[2]] + #2} &, {0, 0}, NonItoBdB];
```

Comparing the two trajectories

```
In[65]:= NonItoIntegralPlot =
           ListPlot[NonItoStochasticIntegral,
            PlotJoined → True, AxesLabel → {"Time", "I(t)"},
            PlotStyle → RGBColor[0, 0, 1],
            PlotRange → All, DisplayFunction → Identity];
         Show[ItoIntegralPlot, NonItoIntegralPlot,
           DisplayFunction → $DisplayFunction];
```

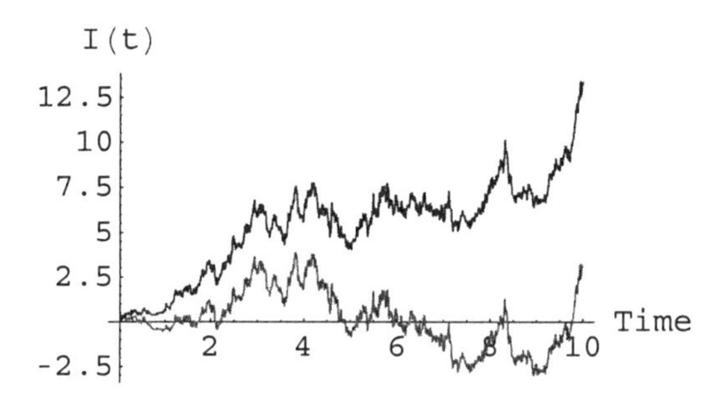

we see that the difference is quite significant, and moreover, that it will not diminish if discretization becomes finer. Actually it is interesting to see the difference between the two precisely :

```
In[67]:= ListPlot[
           Transpose[{Transpose[ItoStochasticIntegral][[1]],
             Transpose[NonItoStochasticIntegral][[2]] -
              Transpose[ItoStochasticIntegral][[2]]}],
           PlotJoined → True, AxesLabel → {"Time", ""},
           PlotRange → All];
```

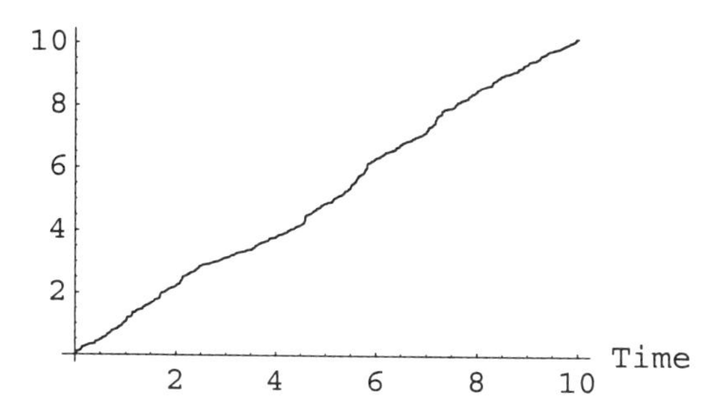

which is approximately equal to the deterministic function $g(t) = t$ (in the limit it becomes equal). The reason for such a somewhat surprising turn of events is the following simple argument:

$$\sum_{i=1}^{n} B(w_{i+1})\,(B(w_{i+1}) - B(w_i)) - \sum_{i=1}^{n} B(w_i)\,(B(w_{i+1}) - B(w_i))$$

$$= \sum_{i=1}^{n} (B(w_{i+1}) - B(w_i))\,(B(w_{i+1}) - B(w_i)) \tag{2.2.10}$$

$$= \sum_{i=1}^{n} (B(w_{i+1}) - B(w_i))^2 \approx \sum_{i=1}^{n} (w_{i+1} - w_i) = w_{n+1} - w_1 = t$$

as $n \to \infty$, $0 = s = w_1 < w_2 < \ldots < w_{n+1} = t$, $\max_{1 \le i \le n}(w_{i+1} - w_i) \to 0$.

2.2.6 Stochastic Differential Equations

As stated before we would like to introduce the meaning, and also to be able to solve an equation, more precisely, a (first-order) stochastic differential equation (SDE) of the form

$$\begin{aligned} d\,y(t) &= f(t, y(t))\,dt + \sigma(t, y(t))\,dB(t) \\ y(t_0) &= y_0. \end{aligned} \tag{2.2.11}$$

Solving an SDE may have different meanings at different times. One possible meaning would be, and occasionally but not often this is possible, to write $y(t)$ as an expression involving various random variables, such as $B(t)$, but not involving $y(t)$ itself. The other meaning, and this is the meaning that will be used in the present section, is to construct an algorithm that can generate (approximations of) particular trajectories of the solution process $y(t)$. We can refer to such a solution as a Monte–Carlo solution of an SDE.

One can notice on the right-hand side of the above equation two parts. The first one is the same as in an ordinary differential equation

$$f(t, y(t))\, dt.$$

The contribution of such a deterministic part can be thought of as a tendency of the movement up or down. The second term on the right-hand side of the SDE needs to be understood as a stochastic integral, written in the differential form,

$$\sigma(t, y(t))\, dB(t)$$

and its contribution can be understood as wandering up and down with no partiality. What makes the above expression an *equation* that needs to be solved is the presence of the sought-after process $\{y(t)\}$ on the right-hand side as well. So, in a way, SDEs combine ODEs and stochastic integrals. The solution of the SDE will be a random (Markov) process $\{Y^{t_0, y_0}(t)\}_{t \geq t_0}$ such that its "differential structure" is equal to

$$dY(t) = a(t, Y(t))\, dt + \sigma(t, Y(t))\, dB(t).$$

Instead of proving mathematically the existence and uniqueness of the solution of the SDE (both hold under certain assumptions in appropriate frameworks; see, e.g., [22,32,42]), we shall construct a solution numerically. The hope is that the development of, and subsequent experiments with, such a Monte–Carlo solver for SDEs will help in the understanding of the mathematical notion of the solution.

Similar to solving ordinary differential equation in the previous chapter, we discretize the above equation to

$$y_{i+1} = y_i + f(t_i, y_i)\, dt + \sigma(t_i, y_i)\, dB_i \tag{2.2.12}$$

which we solve similarly to constructing the stochastic integral in the previous section. Without going into too many details:

```
In[68]:= Clear["Global`*"];
         Needs["Statistics`NormalDistribution`"];
```

```
In[69]:= SDESolver[f_, σ_, y0_, t0_, t1_, K_] :=
           Module[{dt, dB, G, SolList}, dt = N[(t1 - t0)/K];
            dB = Table[Random[NormalDistribution[0, √dt]],
              {K}]; G[{t_, y_}, db_] :=
             {t + dt, y + f[t, y] dt + σ[t, y] db};
            FoldList[G, {t0, y0}, dB]];
```

As an example, let us solve

$$\begin{aligned} dy(t) &= y(t)\,((1 + \sin(4\pi t))\, dt + .1\, dB(t)) \\ y(0) &= 1 \end{aligned} \tag{2.2.13}$$

We produce a collection of random trajectories:

```
In[70]:= Sol = Table[Interpolation[SDESolver[
             #2 (Sin[4 π #1] + 1) &, .1 #2 &, 1, 0, 1, 1000],
             InterpolationOrder → 1][s], {40}];
         Plot[Evaluate[Sol], {s, 0, 1}, PlotRange →
            {0, Automatic}, PlotStyle → RGBColor[0, 1, 0]];
```

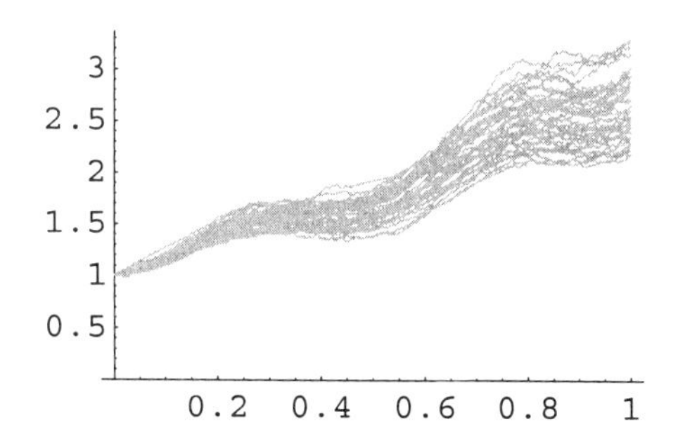

On the other hand, of course, a Brownian motion can be reproduced using this general SDE solver if we take $f = 0$, $\sigma = 1$, $t_0 = y_0 = 0$:

```
In[72]:= Module[{T = 1, SampleSize = 100, Sol, x, y}, Sol =
            Table[Interpolation[SDESolver[0 &, 1. &, 0, 0, 1,
                1000], InterpolationOrder → 1][s], {40}];
          x = Plot[Evaluate[Sol], {s, 0, 1},
            DisplayFunction → Identity,
            PlotStyle → RGBColor[0, 1, 0]];
          y = Plot[Evaluate[Table[k √z, {k, -3, 3}]],
            {z, 0, T}, DisplayFunction → Identity];
          Show[x, y, DisplayFunction → $DisplayFunction,
            PlotRange → All]];
```

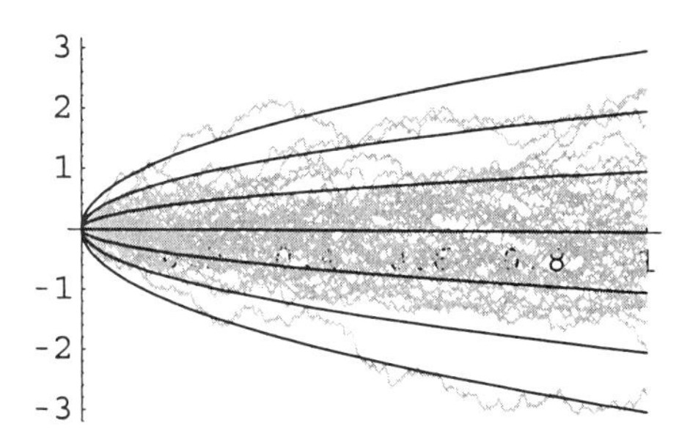

The reader is invited to change the data in the above SDESolver and experiment. For further reference, we shall store the above defined SDESolver (without the Interpolation at the end) in the package CFMLab`ItoSDEs`.

2.3 Itô Calculus

2.3.1 Itô Product Rule

Usually, in classical calculus, the derivative is introduced before integral, and certainly before differential equations. In stochastic, or Itô calculus, the order is reversed since the stochastic integral has a meaningful definition (and construction) as presented above, while the stochastic derivative, or more precisely, the stochastic differential is defined as a reverse operation, i.e., as an anti-stochastic-integral. So, the expression

$$d X(t) = a\, dt + b\, d B(t) \tag{2.3.1}$$

means that

$$X(t) = X(s) + \int_s^t a\, d\tau + \int_s^t b\, d B(\tau). \tag{2.3.2}$$

Having defined stochastic derivative, and knowing what can be done with the derivatives in classical calculus, it is natural to ask what the properties are of stochastic differentiation. Obviously, the stochastic derivative is linear:

$$\begin{aligned} d(X(t) + Y(t)) &= d X(t) + d Y(t) \\ d(c\, X(t)) &= c\, d X(t). \end{aligned} \tag{2.3.3}$$

Looking back at classical calculus, one naturally comes to the issue of the product rule. Does the usual product rule

```
In[73]:= ∂t (f[t] g[t])
```

Out[73]= $g(t)\, f'(t) + f(t)\, g'(t)$

extend to the Itô calculus? The answer is yes, but with a correction! The Itô product rule, or Itô stochastic product rule, reads as

$$d(X(t)\, Y(t)) = X(t)\, d Y(t) + Y(t)\, d X(t) + d X(t)\, d Y(t). \tag{2.3.4}$$

So, in stochastic calculus, as opposed to classical calculus, the term $d X(t)\, d Y(t)$ is not negligible. Notice that since $X(t + dt) \approx X(t) + d X(t)$ the same rule can be written also as

$$d(X(t)\, Y(t)) = X(t + dt)\, d Y(t) + Y(t)\, d X(t). \tag{2.3.5}$$

or, similarly,

$$d(X(t)\,Y(t)) = X(t)\,dY(t) + Y(t+dt)\,dX(t) \tag{2.3.6}$$

We shall confine ourselves to verify these claims with a representative example, and postpone the more complete justification for a discussion after our study of the vector case of the Itô chain rule. Let $X(t) = Y(t) = B(t)$ be a Brownian motion. We shall verify the last identity, i.e., we shall check whether the identity

$$d(B(t)^2) = B(t+dt)\,dB(t) + B(t)\,dB(t) \tag{2.3.7}$$

holds (approximately; actually in this example, the identity holds exactly even for fixed $dt \neq 0$). Indeed,

```
In[74]:= Clear["Global`*"]
         << "Statistics`NormalDistribution`"

In[76]:= s = 0; t1 = 10; K = 10; dt = N[(t1 - s)/K];

In[77]:= dB = Table[
            Random[NormalDistribution[0, √dt]], {K}];
         B = FoldList[Plus, 0, dB];

In[79]:= FoldList[Plus, 0, Drop[B, 1] dB];

In[80]:= FoldList[Plus, 0, Drop[B, -1] dB];

In[81]:= Chop[B^2 - (% + %%)]

Out[81]= {0, 0, 0, 0, 0, 0, 0, 0, 0, 0, 0}
```

Also notice, $dX(t)\,dY(t)$ in this example is equal to $dB\,dB = (dB)^2 = dt$ and this is not negligible, as announced above. We shall come back to the Itô product rule after the multidimensional Itô chain rule is discussed later in this chapter. This suggests that the stochastic product rule is not quite as elementary as implied above.

The product rule having been introduced and verified, the next question is whether the chain rule, which in classical calculus reads as

```
In[82]:= ∂t f[g[t]]

Out[82]= f'(g(t)) g'(t)
```

holds, or whether it can be modified to hold in stochastic calculus as well. The answer is affirmative again, and it is discussed in the next section.

2.3.2 Itô Chain Rule

2.3.2.1 Itô Theorem

```
In[83]:= Clear["Global`*"]
```

In classical calculus one of the most useful results is the so-called chain rule. The chain rule can be formulated in the following way. Suppose

$$\frac{dy}{dt} = a(t, y) \tag{2.3.8}$$

i.e.,

$$dy = a(t, y)\, dt \tag{2.3.9}$$

and let $g(t, y)$ be a sufficiently smooth function. Then the classical chain rule yields the following formula:

$$d(g(t, y)) = \frac{\partial g(t, y)}{\partial t}\, dt + \frac{\partial g(t, y)}{\partial y}\, dy$$

or more explicitly,

$$d(g(t, y)) = \left(\frac{\partial g(t, y)}{\partial t} + \frac{\partial g(t, y)}{\partial y}\, a(t, y) \right) dt. \tag{2.3.10}$$

Indeed, using *Mathematica*®, we get equivalently

```
In[84]:= ∂t g[t, y[t]] /. y'[t] → a[t, y]
```

Out[84]= $a(t, y)\, g^{(0,1)}(t, y(t)) + g^{(1,0)}(t, y(t))$

The question now is what happens if, instead of $dy = a(t, y)\, dt$, we have

$$dy = a(t, y)\, dt + \sigma(t, y)\, dB$$

The answer is given by the celebrated Itô Theorem.

Theorem:

Suppose

$$dy = f(t, y)\, dt + \sigma(t, y)\, dB \tag{2.3.11}$$

and let $g(t, y)$ be a sufficiently smooth function. Then

$$\begin{aligned} d(g\,(t, y)) &= \frac{\partial g(t, y)}{\partial t}\, dt + \frac{\partial g(t, y)}{\partial y}\, dy \\ &+ \frac{1}{2}\, \frac{\partial^2 g(t, y)}{\partial y^2}\, (dy)^2 = \frac{\partial g(t, y)}{\partial t}\, dt + \frac{\partial g(t, y)}{\partial y}\, dy \\ &+ \frac{1}{2}\, \frac{\partial^2 g(t, y)}{\partial y^2}\, \sigma(t, y)^2\, dt \end{aligned} \tag{2.3.12}$$

or more explicitly

$$dg(t, y) = \left(\frac{\partial g(t, y)}{\partial t} + \frac{\partial g(t, y)}{\partial y} f(t, y) + \frac{1}{2} \frac{\partial^2 g(t, y)}{\partial y^2} \sigma(t, y)^2 \right) dt + \frac{\partial g(t, y)}{\partial y} \sigma(t, y)\, dB. \quad (2.3.13)$$

Comment:

Notice the unexpected (according to the classical chain rule) term

$$\frac{1}{2} \frac{\partial^2 g(t, y)}{\partial y^2} \sigma(t, y)^2\, dt.$$

The presence of that term is the reason for the difference between classical and Itô stochastic calculus.

2.3.2.2 Monte–Carlo Simulation "Proof" of the Itô Theorem

First, we show that

$$dt\, dB = dt\, dB(t) = 0 \quad (2.3.14)$$

and that $(dB(t))^2$ is not zero, nor is it random, but rather

$$(dB)^2 = (dB(t))^2 = dt. \quad (2.3.15)$$

The more precise meaning of those two statements is

$$\sum_{i=1}^{n} (t_{i+1} - t_i)\,(B(t_{i+1}) - B(t_i)) \longrightarrow 0 \quad (2.3.16)$$

and (amazingly)

$$\sum_{i=1}^{n} (B(t_{i+1}) - B(t_i))^2 \longrightarrow (t - s) \quad (2.3.17)$$

as $n \to \infty$ and $s = t_1 < t_2 < \ldots < t_{n+1} = t$, $\max_{1 \le i \le n} (t_{i+1} - t_i) \to 0$. These two claims are usually understood and proved using mathematical arguments. Instead, we offer a convincing Monte–Carlo simulation verification. For the first identity, consider

```
In[85]:= << "Statistics`NormalDistribution`";
         t = 1; s = 0; K = 200; dt = N[(t - s)/K];

In[86]:= dB = Table[
            Random[NormalDistribution[0, √dt]], {K}];
         dtdBList = FoldList[{dt + #1[[1]], #1[[2]] + dt #2} &,
            {0, 0}, dB];
```

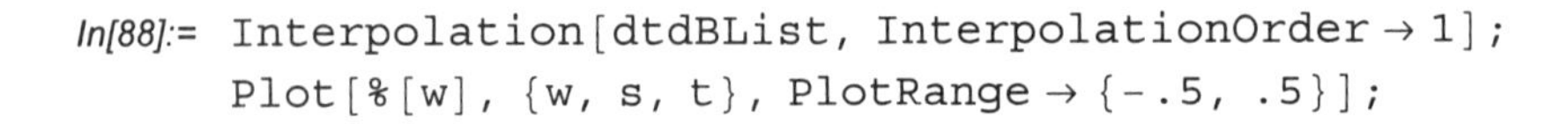

```
In[88]:= Interpolation[dtdBList, InterpolationOrder → 1];
         Plot[%[w], {w, s, t}, PlotRange → {-.5, .5}];
```

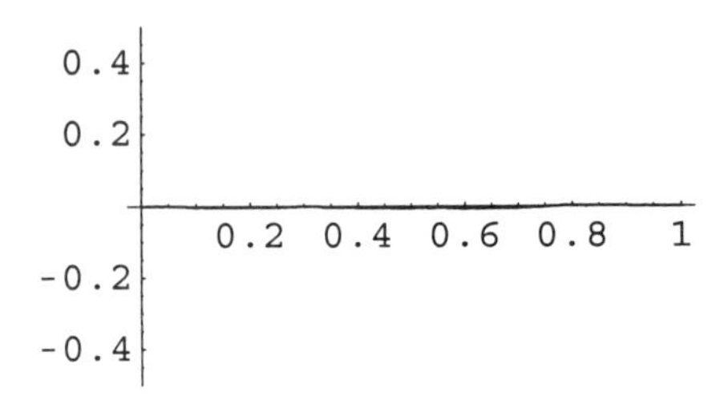

For the second, an even more interesting claim, the calculation (we compare the computed function with the function $g(t) = t - s = t$; we do not use same randomization as above, since we need more precise discretization to get a reasonable match) is

```
In[90]:= t = 1; s = 0; K = 5000; dt := N[(t - s)/K];
         dB = Table[
            Random[NormalDistribution[0, √dt]], {K}];
         dBSquareList := FoldList[
            {dt + #1[[1]], #2^2 + #1[[2]]} &, {0, 0}, dB];
         solving = Interpolation[dBSquareList,
            InterpolationOrder → 1];
         Plot[{w, solving[w]}, {w, s, t},
           PlotRange → {0, 1}, Frame → True, PlotStyle →
            {RGBColor[1, 0, 0], RGBColor[0, 0, 0]}];
```

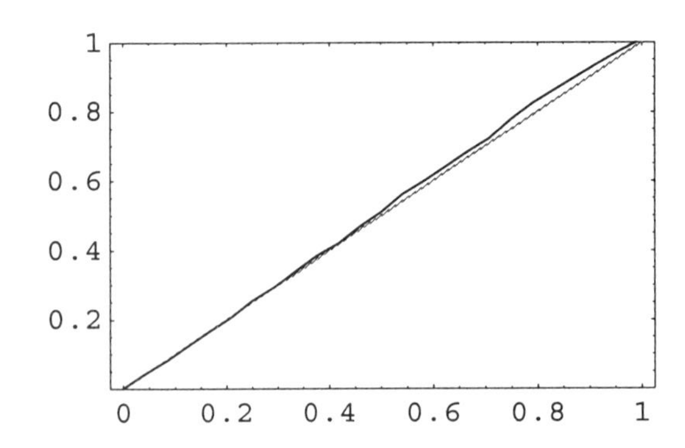

The reader can increase K to get a better match. The rest of the "proof" of the Itô theorem is the following formal calculation:

$$(dy)^2 = (f(t, y)\,dt + \sigma(t, y)\,dB)^2$$
$$= f(t, y)^2\,(dt)^2 + 2\,f(t, y)\,dt\,\sigma(t, y)\,dB + \sigma(t, y)^2\,(dB)^2$$
$$= \sigma(t, y)^2\,(dB)^2 = \sigma(t, y)^2\,dt$$

and consequently

$$dg(t, y) = \frac{\partial g(t, y)}{\partial t}\,dt + \frac{\partial g(t, y)}{\partial y}\,dy + \frac{1}{2}\,\frac{\partial^2 g(t, y)}{\partial y^2}\,(dy)^2$$
$$= \frac{\partial g(t, y)}{\partial t}\,dt + \frac{\partial g(t, y)}{\partial y}\,(f(t, y)\,dt + \sigma(t, y)\,dB)$$
$$+ \frac{1}{2}\,\frac{\partial^2 g(t, y)}{\partial y^2}\,\sigma(t, y)^2\,dt$$
$$= \left(\frac{\partial g(t, y)}{\partial t} + \frac{\partial g(t, y)}{\partial y}\,f(t, y) + \frac{1}{2}\,\frac{\partial^2 g(t, y)}{\partial y^2}\,\sigma(t, y)^2\right)dt$$
$$+ \frac{\partial g(t, y)}{\partial y}\,\sigma(t, y)\,dB$$

which concludes the argument.

Remark:

The conclusion of the previous theorem is quite interesting: *even though dB is random, $(dB)^2$ is not, and moreover even though it is a quadratic term it is not equal to 0 but to dt (when $dt \to 0$).* This is the reason for the difference between classical calculus and the stochastic, i.e., Itô calculus. The chain rule, as well as the product rule, had to be corrected by an additional term due to the fact that $(dB)^2 \neq 0$.

2.3.3 An Application: Solving the Simplest Stock Price SDE Model

```
In[95]:= Remove["Global`*"]
```

We shall need SDESolver. To that end import our package

```
In[96]:= << "CFMLab`ItoSDEs`"
```

The simplest SDE modeling the stock price evolution is

$$\begin{aligned} dS(t) &= a\,S(t)\,dt + \sigma\,S(t)\,dB(t) \\ S(0) &= p \end{aligned} \tag{2.3.18}$$

Experimentally, this equation can be solved by

```
In[97]:= SimplestPriceModel[a_, σ_, p_, T_] :=
           SDESolver[a #2 &, σ #2 &, p, 0, T, 1000]
```

Above a, σ, p are all constants, a yearly average appreciation rate, σ the volatility, and p the initial price of the stock. For example

```
In[98]:= With[{T = 2},
           simple = SimplestPriceModel[.3, .4, 100, T];
           ListPlot[simple, PlotRange → {0, Automatic}]];
```

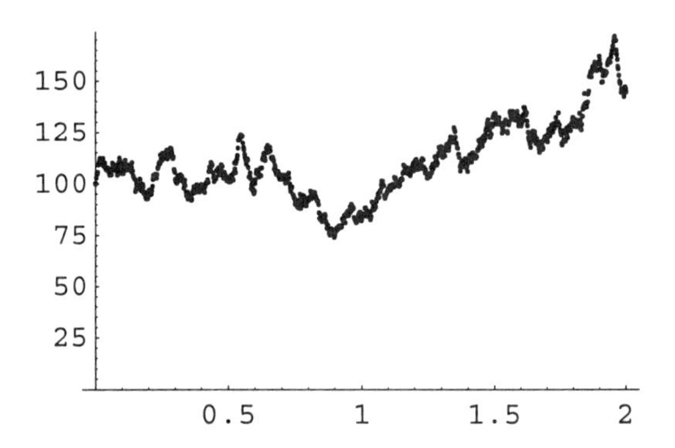

Now we solve the same SDE mathematically. Since the movements of the stock price should be proportional to the current price, it is natural to look at the logarithm of the price. Let

$$z = \log(S) \tag{2.3.19}$$

and using the Itô chain rule, and noticing

$$(dS)^2 = \sigma^2 S^2 \, dt \tag{2.3.20}$$

we derive

$$\begin{aligned} dz = d\log(S) &= \frac{1}{S}\, dS + \frac{1}{2}\,\frac{-1}{S^2}\,(dS)^2 \\ &= a\,dt + \sigma\,dB - \frac{1}{2S^2}\,\sigma^2 S^2\,dt = \left(a - \frac{1}{2}\,\sigma^2\right)dt + \sigma\,dB. \end{aligned}$$

Summarizing

$$dz = \left(a - \frac{1}{2}\,\sigma^2\right)dt + \sigma\,dB, \tag{2.3.21}$$

we conclude that when dealing with the simplest stock price model, we have two alternatives: to consider price $S(t)$ or to consider its logarithm $z(t) = \log(S(t))$. Notice that both have the same σ's (although in the first case σ is the volatility, while in the second it is diffusion), but what may be somewhat surprising, the drift of z is not equal to the appreciation rate of S. This has to be taken into account when going back and forth from one to the other, and in particular when parameters are estimated statistically using one or the other. We will come back to this point when we do the statistical estimation of parameters.

Now, since σ is constant, the above SDE for z, whose right-hand side is independent of z, can be solved explicitly just by applying stochastic integration, yielding:

$$z(t) = t\left(a - \frac{\sigma^2}{2}\right) + \sigma\,(B(t) - B(0)) + z(0) = t\left(a - \frac{\sigma^2}{2}\right) + \sigma\,B(t) + \log(p).$$

This in turn gives the solution of the original stock price SDE by taking the exponential:

$$S(t) = e^{z(t)} = e^{\left(a-\frac{\sigma^2}{2}\right)t+\sigma\,B(t)+\log(p)} = p\,e^{\left(a-\frac{\sigma^2}{2}\right)t+\sigma\,B(t)}. \tag{2.3.22}$$

How useful is this formula? Since on the market stock prices i.e., $S(t)$ are observable, while Brownian motion $B(t)$ is not directly observable, the usefulness of this formula can be questioned. Nevertheless, even though $B(t)$ is not directly observable, useful conclusions can be derived from the above formula. For example, the direct consequence is that $S(t) > 0$ for any time $t < \infty$, provided $p > 0$; the stock price never reaches zero. We shall see later that this fact translates into not having a boundary condition at $S = 0$ for many partial differential equations that will be studied later in this book. Moreover, the above formula can be used to construct confidence bounds for stock price evolution. Indeed, since $B(t) \sim N\left(0, \sqrt{t}\right)$, the confidence bounds for $S(t)$ are given by

$$p\,e^{\left(a-\frac{\sigma^2}{2}\right)t\pm\sigma\,b\sqrt{t}} \tag{2.3.23}$$

for example, for $b = 1, 2, 3$. In particular, the median curve for $S(t)$ is given by

$$p\,e^{\left(a-\frac{\sigma^2}{2}\right)t}. \tag{2.3.24}$$

Notice that the median for $S(t)$ is different (smaller) than the mean value curve for $S(t)$ which is equal to $E(S(t)) = p\,e^{a\,t}$. Indeed, the mean value is computed easily: for example, since

$$\begin{aligned} &d\,S(t) - a\,S(t)\,d\,t = \sigma\,S(t)\,d\,B(t) \\ &S(0) = p \end{aligned}$$

we get, after multiplication by $e^{-a\,t}$ (recall the Itô product rule, and notice furthermore that $d\,e^{-a\,t}\,d\,S(t) = 0$)

$$d\,(e^{-a\,t}\,S(t)) = \sigma\,e^{-a\,t}\,S(t)\,d\,B(t),$$

and after (stochastic) integration

$$e^{-a\,t}\,S(t) - p = \int_0^t \sigma\,e^{-a\,\tau}\,S(\tau)\,d\,B(\tau);$$

taking expectation of both sides we get

$$e^{-a\,t}\,E(S(t)) - p = 0$$

i.e.,

$$E(S(t)) = p\,e^{a\,t} > p\,e^{\left(a-\frac{\sigma^2}{2}\right)t} \tag{2.3.25}$$

if $\sigma \neq 0$. Putting everything together, we get

```
In[99]:= Z[a_, σ_, b_, t_] = y₀ e^((a - σ²/2) t + b σ √t);

In[100]:= y₀ = 70; a = .5; σ = .6; T = 1; Solution[s_] = Table[
      Interpolation[SimplestPriceModel[a, σ, y₀, T],
        InterpolationOrder → 1][s], {100}];
    Pix1 = Plot[Evaluate[Solution[t]], {t, 0, T},
      PlotRange → {0, Z[a, σ, 3, T]},
      PlotStyle → RGBColor[0, 1, 0],
      DisplayFunction → Identity]; Pix2 =
     Plot[Evaluate[Table[Z[a, σ, b, t], {b, -3, 3}]],
      {t, 0, T}, PlotStyle →
       ReplacePart[Table[RGBColor[0, 0, 0], {6}],
        RGBColor[0, 0, 1], 4],
      DisplayFunction → Identity];
    Pix3 = Plot[y₀ e^(a t), {t, 0, T},
      PlotStyle → RGBColor[1, 0, 0],
      DisplayFunction → Identity];

In[101]:= Show[Pix1, Pix2, Pix3,
      DisplayFunction → $DisplayFunction];
```

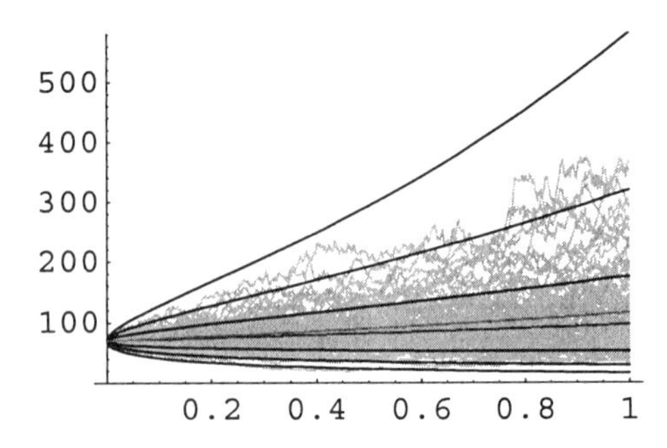

Notice that the blue curve is median, black curves are confidence curves, while red is the average (in black and white red is a bit lighter then blue).

Finally, notice that if $a - \frac{1}{2}\sigma^2 < 0$, all of the confidence curves approach zero when $t \to \infty$. For example if

```
In[102]:= a = .5; σ = 1.6; T = 40;
```

then

```
In[103]:= a - σ^2/2
Out[103]= -0.78
```

and the confidence regions look like

```
In[104]:= Plot[Evaluate[Table[Z[a, σ, b, t], {b, -3, 3}]],
            {t, 0, T}, PlotRange → All, PlotStyle →
             ReplacePart[Table[RGBColor[0, 0, 0], {6}],
              RGBColor[1, 0, 0], 4]];
```

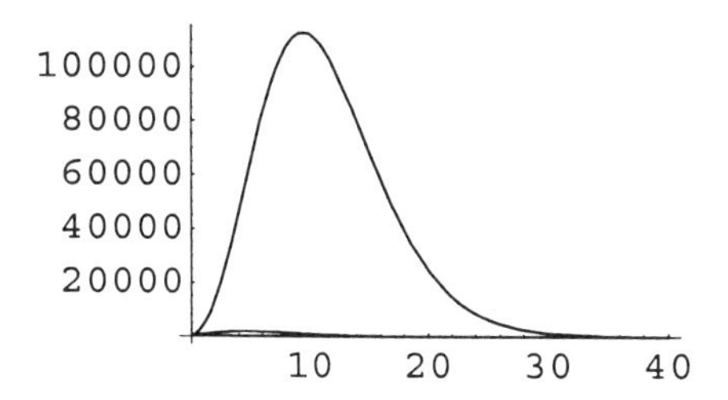

This indicates that if $a - \frac{1}{2}\sigma^2 < 0$, even though, as we have seen earlier, the stock price never reaches zero, it approaches zero: $S(t) \to 0$ when $t \to \infty$. On the other hand, if $a - \frac{1}{2}\sigma^2 > 0$, for example if

```
In[105]:= a = .5; σ = .6; T = 40;
```

then

```
In[106]:= a - σ^2/2
Out[106]= 0.32
```

and the confidence regions look like

```
In[107]:= Plot[Evaluate[Table[Z[a, σ, b, t], {b, -3, 3}]],
            {t, 0, T}, PlotRange → {0, Z[a, σ, -3, T]},
            PlotStyle → ReplacePart[Table[RGBColor[0, 0, 0],
               {6}], RGBColor[1, 0, 0], 4]];
```

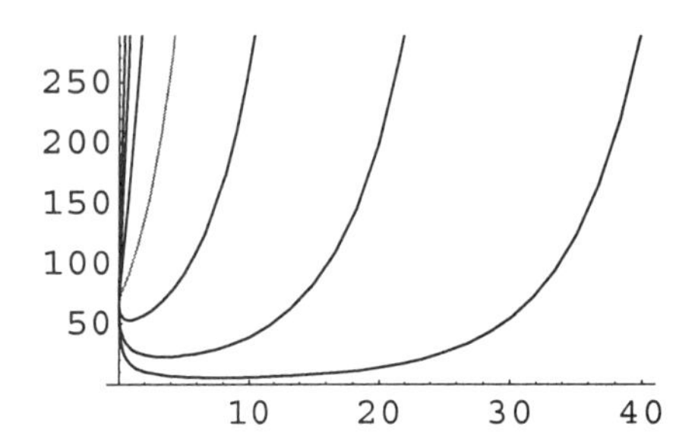

This indicates that if $a - \frac{1}{2}\sigma^2 > 0$, stock price increases to infinity: $S(t) \to \infty$ when $t \to \infty$.

2.4 Multivariable and Symbolic Itô Calculus

2.4.1 Monte–Carlo Solver for m-dimensional SDEs

```
In[108]:= Clear["Global`*"]; Off[General::"spell1"];
```

We develop now a solver for an m-dimensional vector SDE, driven by n independent Brownian motions. Of course, there are many ways to do this. Here we chose to treat time t and space $S = \{S_1, ..., S_2\}$ differently (time is not just yet another variable, as it would be in $\{t, S_1, ..., S_2\}$, but rather we consider $\{t, S\} = \{t, \{S_1, ..., S_2\}\}$). Furthermore, space $S = \{S_1, ..., S_2\}$ is chosen to be a *one-dimensional* array (it is neither a column vector $\{\{S_1\}, ..., \{S_2\}\}$, nor a row vector $\{\{S_1, ..., S_2\}\}$, it is just a vector $\{S_1, ..., S_2\}$, i.e., a one-dimensional array). So, to fix ideas, we are looking at the SDE initial value problem

$$\begin{aligned} dS(t) &= b(t, S(t))\,dt + c(t, S(t)).dB(t) \\ S(t_0) &= s \end{aligned} \tag{2.4.1}$$

where $t \in [t_0, t_1] \subset \mathbb{R}$ is time, $B(t) = \{B_1(t), ..., B_n(t)\}$ with $B_i(t)$ being independent Brownian motions, $b(t, S)$ is an m-vector (one-dimensional array) valued function on $\mathbb{R} \times \mathbb{R}^m$, $c(t, S)$ is an $m \times n$ matrix (two-dimensional array) valued function on $\mathbb{R} \times \mathbb{R}^m$, and $s = \{s_1, ..., s_m\} \in \mathbb{R}^m$ is a (deterministic) initial value. Notice the matrix multiplication "." in the above equation:

```
In[109]:= ? .

    a.b.c or Dot[a, b, c] gives products of vectors, matrices
      and tensors. More...
```

For example, the simplest two-dimensional stock price model (the two-dimensional Log-Normal model) can be written as

```
In[110]:= σ = {{0.4, 0.2, 0.2, 0.2, 0.02, 0.1, 0.02},
            {0.02, 0.4, 0.2, -0.2, 0.2, 0.2, 0.1}};
          a = {0.2, .5}; s = {100, 115};

In[111]:= c[t_, S_] := S σ

In[112]:= b[t_, S_] := S a
```

One should carefully examine the above syntax, since although it is correct *Mathematica*® syntax, it is not, as yet, so commonplace in *Mathematics*. Indeed, what is implied by the above is that the vector simplest price model can be written as

$$\begin{aligned} dS(t) &= S(t)\,a\,dt + S(t)\,\sigma.dB(t) \\ S(t_0) &= s \end{aligned} \tag{2.4.2}$$

where the space " ", i.e., the *Mathematica*® function Times between arrays means componentwise multiplication of arrays (a very useful *Mathematica*® convention). Indeed,

```
In[113]:= {q, w} {e, r}

Out[113]= {e q, r w}
```

while

```
In[114]:= {q, w} {{e, t}, {r, y}}
```

Out[114]= $\begin{pmatrix} e\,q & q\,t \\ r\,w & w\,y \end{pmatrix}$

We now solve the simplest price dynamics model in higher dimension. Let

```
In[115]:= m = Length[σ]; n = Length[Transpose[σ]];
          K = 1000; t0 = 0; t1 = 1; dt = (t1 - t0)/K;
```

Then the above SDE can be solved quickly as

```
In[116]:= << "Statistics`NormalDistribution`";
          SDESolver2[b_, c_, s_, t0_, t1_, K_] :=
            Module[{n, dt, dB, G},
             n = Length[Transpose[σ]]; dt = (t1 - t0)/K;
             dB = Table[Random[NormalDistribution[0, √dt]],
                {K}, {n}]; G[{t_, S_}, dB_] :=
               {t + dt, S + b[t, S] dt + c[t, S].dB};
             FoldList[G, {t0, s}, dB]];
          Sτ = SDESolver2[b, c, s, t0, t1, K];
```

and its realized trajectories can be seen

```
In[119]:= StockExtract[i_, Sτ_] :=
            ({#1[[1]], #1[[2, i]]} &) /@ Sτ;
          (ListPlot[StockExtract[#1, Sτ],
              DisplayFunction → Identity] &) /@ Range[m];
          Show[%, DisplayFunction → $DisplayFunction];
```

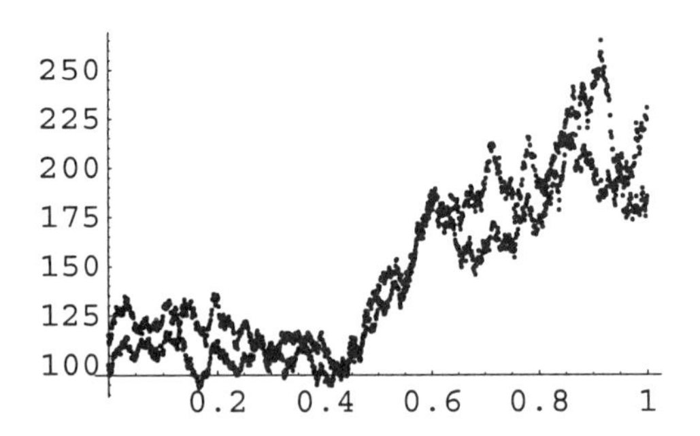

The above function was named SDESolver2 in order to distinguish it from the SDESolver defined before for the non-vector, i.e., scalar data. We rename that one as

```
In[121]:= SDESolver1[b_, c_, s_, t0_, t1_, K_] :=
            Module[{dt, dB, G}, dt = N[(t1 - t0)/K];
             dB = Table[Random[NormalDistribution[0, √dt]],
               {K}]; G[{t_, S_}, dB_] :=
              {t + dt, S + b[t, S] dt + c[t, S] dB};
             FoldList[G, {t0, s}, dB]]
```

and put both definitions into a single definition

```
In[122]:= SDESolver[b_, c_, s_, t0_, t1_, K_] :=
            If[Head[b[t0, s]] === List,
             SDESolver2[b, c, s, t0, t1, K],
             SDESolver1[b, c, s, t0, t1, K]]
```

So let us see how the new function (placed in the package CFMLab`ItoSDEs`) works. Consider data

```
In[123]:= σ = {{0, 0, 0, 0, 0, 0, 0}}; a = {1}; s = {0};
```

```
In[124]:= c[t_, S_] := σ
```

```
In[125]:= b[t_, S_] := a
```

and we get a trivial solution $S(t) = t$, (just to check the veracity of the solver)

```
In[126]:= Sτ = SDESolver[b, c, s, t0, t1, K];
          {τ, S} = Transpose[Sτ];
```

which can be seen as

```
In[128]:= ListPlot[StockExtract[1, Sτ]];
```

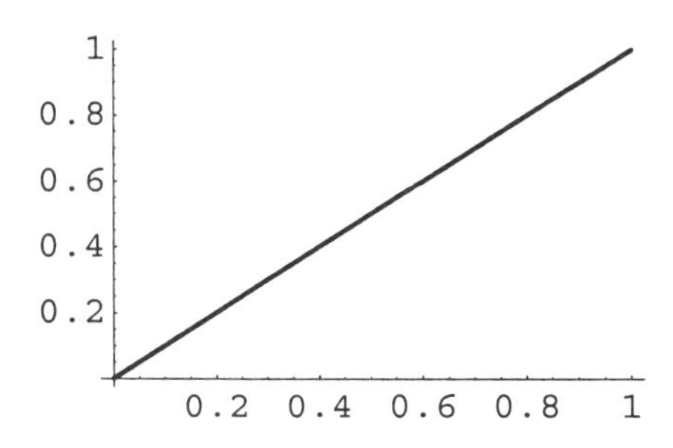

One more example: we generate m independent Brownian motions as a solution of the m-dimensional SDE. To that end, let

```
In[129]:= m = 25; n = m; σ = IdentityMatrix[m];
          a = Table[0, {m}]; s = a;

In[130]:= c[t_, S_] := σ

In[131]:= b[t_, S_] := a
```

and we get the solution

```
In[132]:= Sτ = SDESolver[b, c, s, t0, t1, K];
          {τ, S} = Transpose[Sτ];
```

which can be seen

```
In[134]:= (ListPlot[StockExtract[#1, Sτ], PlotJoined → True,
               DisplayFunction → Identity] &) /@ Range[m];
          Show[%, DisplayFunction → $DisplayFunction];
```

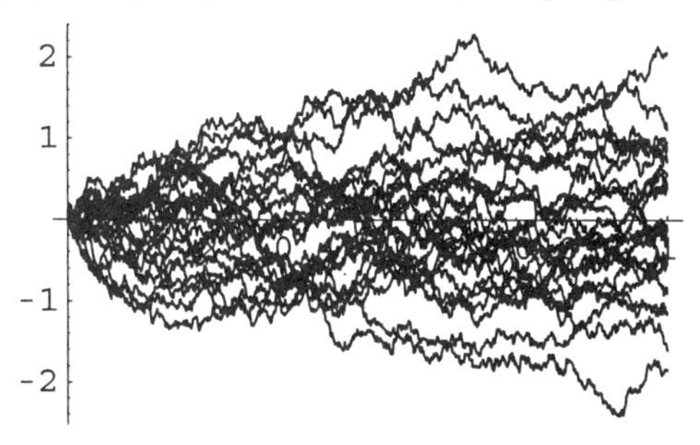

2.4.2 Symbolic *m*-Dimensional Itô Chain Rule

2.4.2.1 Some Calculus and Linear Algebra

```
In[136]:= Clear["Global`*"]
```

Before we introduce the m-dimensional Itô chain rule, we need to recall, and moreover, introduce in a *Mathematica®* context, some notions and facts from multivariable calculus and linear algebra. The gradient of a function $u(x_1, \ldots, x_n)$ of an arbitrary number of variables $V = \{x_1, \ldots, x_n\}$, as studied in calculus, is a vector (one-dimensional array) $\nabla u = \text{grad}\, u = D\, u = D_V(u)$ of all partial derivatives of u, and, in *Mathematica®*, can be defined simply as

```
In[137]:= DV_[u_] := (∂#1 u &) /@ V
```

Indeed, for example, if

```
In[138]:= u[x_, y_] := x^6 + x^2 y^4
```

then

```
In[139]:= D{x,y,z}[u[x, y]]
```

Out[139]= $\{6\,x^5 + 2\,y^4\,x,\ 4\,x^2\,y^3,\ 0\}$

The Hessian of a function $u(x_1, \ldots, x_n)$ is a symmetric matrix, i.e., symmetric two-dimensional array $D^2\,u$ of all second order partial derivatives of function u, and, in *Mathematica®*, can be defined simply as

```
In[140]:= (D^2)V_[u_] := DV[DV[u]]
```

Continuing the above example

```
In[141]:= (D^2){x,y,z}[u[x, y]]
```

Out[141]=
$$\begin{pmatrix} 30\,x^4 + 2\,y^4 & 8\,x\,y^3 & 0 \\ 8\,x\,y^3 & 12\,x^2\,y^2 & 0 \\ 0 & 0 & 0 \end{pmatrix}$$

Furthermore, generally, even though we shall not need this, we can define the k-dimensional (symmetric) array of all k-th order partial derivatives

```
In[142]:= (D^0)V_[u_] := u
          (D^k_)V_[u_] := Nest[DV[#1] &, u, k]
```

in which case, for example,

```
In[144]:= Table[(D^k){x,y,z}[u[x, y]], {k, 0, 4}]
```

Out[144]= $\left\{x^6+y^4 x^2, \{6 x^5+2 y^4 x, 4 x^2 y^3, 0\}, \begin{pmatrix} 30 x^4+2 y^4 & 8 x y^3 & 0 \\ 8 x y^3 & 12 x^2 y^2 & 0 \\ 0 & 0 & 0 \end{pmatrix},\right.$

$$\begin{pmatrix} \{120 x^3, 8 y^3, 0\} & \{8 y^3, 24 x y^2, 0\} & \{0, 0, 0\} \\ \{8 y^3, 24 x y^2, 0\} & \{24 x y^2, 24 x^2 y, 0\} & \{0, 0, 0\} \\ \{0, 0, 0\} & \{0, 0, 0\} & \{0, 0, 0\} \end{pmatrix},$$

$$\left.\begin{pmatrix} \begin{pmatrix} 360 x^2 & 0 & 0 \\ 0 & 24 y^2 & 0 \\ 0 & 0 & 0 \end{pmatrix} & \begin{pmatrix} 0 & 24 y^2 & 0 \\ 24 y^2 & 48 x y & 0 \\ 0 & 0 & 0 \end{pmatrix} & \begin{pmatrix} 0 & 0 & 0 \\ 0 & 0 & 0 \\ 0 & 0 & 0 \end{pmatrix} \\ \begin{pmatrix} 0 & 24 y^2 & 0 \\ 24 y^2 & 48 x y & 0 \\ 0 & 0 & 0 \end{pmatrix} & \begin{pmatrix} 24 y^2 & 48 x y & 0 \\ 48 x y & 24 x^2 & 0 \\ 0 & 0 & 0 \end{pmatrix} & \begin{pmatrix} 0 & 0 & 0 \\ 0 & 0 & 0 \\ 0 & 0 & 0 \end{pmatrix} \\ \begin{pmatrix} 0 & 0 & 0 \\ 0 & 0 & 0 \\ 0 & 0 & 0 \end{pmatrix} & \begin{pmatrix} 0 & 0 & 0 \\ 0 & 0 & 0 \\ 0 & 0 & 0 \end{pmatrix} & \begin{pmatrix} 0 & 0 & 0 \\ 0 & 0 & 0 \\ 0 & 0 & 0 \end{pmatrix} \end{pmatrix}\right\}$$

The information about all directional derivatives of order k of the function u are contained in the (symmetric) array $D^k u$. Indeed, for example, the second derivative in the direction $\left\{\frac{1}{\sqrt{2}}, \frac{1}{\sqrt{2}}, 0\right\}$ of the first derivative in the direction $\{-1, 0, 0\}$ of the function u is equal to

In[145]:= $(D^3)_{\{x,y,z\}}[u[x, y]].\{-1, 0, 0\}.\left\{\frac{1}{\sqrt{2}}, \frac{1}{\sqrt{2}}, 0\right\}.\left\{\frac{1}{\sqrt{2}}, \frac{1}{\sqrt{2}}, 0\right\}$

Out[145]= $\frac{-60\sqrt{2}\, x^3 - 4\sqrt{2}\, y^3}{\sqrt{2}} + \frac{-4\sqrt{2}\, y^3 - 12\sqrt{2}\, x y^2}{\sqrt{2}}$

or

In[146]:= `Simplify[%]`

Out[146]= $-4(15 x^3 + 3 y^2 x + 2 y^3)$

Notice that the "matrix" multiplications above are done from the left to the right in succession, since otherwise it would not be possible to carry on the multiplication (the "matrix" multiplication here is *not* associative). Moreover, the multivariable Taylor series for $u(v)$ at the point w can be computed for example, as

In[147]:=
```
TaylorSeries_n_[u_, V_, v_, w_] :=
```
$$\sum_{k=0}^{n} \frac{\text{Nest}[\#1.(w-v)\ \&,\ (D^k)_V[u]\ /.\ \text{Thread}[V \to v],\ k]}{k!}$$

where the nontrivial new *Mathematica*® function is

In[148]:=
```
? Thread
```

```
Thread[f[args]] "threads" f over any lists that appear
   in args. Thread[f[args], h] threads f over any objects
   with head h that appear in args. Thread[f[args], h,
   n] threads f over objects with head h that appear in
   the first n args. Thread[f[args], h, -n] threads over
   the last n args. Thread[f[args], h, {m, n}] threads
   over arguments m through n. More...
```

Indeed, for example,

In[149]:=
```
Simplify[TaylorSeries6[u[x, y],
    {x, y, z}, {α, β, γ}, {x, y, z}]]
```

Out[149]= $x^2 (x^4 + y^4)$

which is, as it should be, equal to u

In[150]:=
```
Simplify[% == u[x, y]]
```

Out[150]= True

verifying the formula. On the other hand, for example,

In[151]:=
```
Simplify[
  TaylorSeries10[e^-(x^2+y^2), {x, y}, {0, 0}, {x, y}]]
```

Out[151]= $$\frac{1}{120}(-x^{10} - 5(y^2-1)x^8 - 10(y^4-2y^2+2)x^6 - 10(y^6-3y^4+6y^2-6)x^4 - 5(y^8-4y^6+12y^4-24y^2 + 24)x^2 - y^{10} + 5y^8 - 20y^6 + 60y^4 - 120y^2 + 120)$$

We shall also need the following observations from linear algebra. Let A be a symmetric matrix, such as

In[152]:=
```
a_{i_,j_} := a_{j,i} /; i > j
A = Array[a_{#1,#2} &, {3, 3}]
```

Out[153]= $$\begin{pmatrix} a_{1,1} & a_{1,2} & a_{1,3} \\ a_{1,2} & a_{2,2} & a_{2,3} \\ a_{1,3} & a_{2,3} & a_{3,3} \end{pmatrix}$$

and let Σ be a rectangular matrix, such as

```
In[154]:= Σ = Array[σ#1,#2 &, {3, 5}]
```

$$Out[154]= \begin{pmatrix} \sigma_{1,1} & \sigma_{1,2} & \sigma_{1,3} & \sigma_{1,4} & \sigma_{1,5} \\ \sigma_{2,1} & \sigma_{2,2} & \sigma_{2,3} & \sigma_{2,4} & \sigma_{2,5} \\ \sigma_{3,1} & \sigma_{3,2} & \sigma_{3,3} & \sigma_{3,4} & \sigma_{3,5} \end{pmatrix}$$

Then $\Sigma.\Sigma^T$ is symmetric and

```
In[155]:= Table[Sum[σi,k σj,k, {k, 1, 5}], {i, 1, 3}, {j, 1, 3}] ==
            Σ.Transpose[Σ]
Out[155]= True
```

Moreover,

```
In[156]:= Simplify[Tr[A.Σ.Transpose[Σ]] ==
             Plus @@ Flatten[A Σ.Transpose[Σ]] ==
             Tr[Σ.Transpose[Σ].A]]
Out[156]= True
```

where

```
In[157]:= ? Tr

      Tr[list] finds the trace of the matrix or tensor list.
         Tr[list, f] finds a generalized trace, combining terms
         with f instead of Plus. Tr[list, f, n] goes down to
         level n in list. More...
```

and

```
In[158]:= ? Apply

      Apply[f, expr] or f @@ expr replaces the head of expr
         by f. Apply[f, expr, levelspec] replaces heads in
         parts of expr specified by levelspec. More...
```

2.4.2.2 Derivation of the m-Dimensional Itô Chain Rule

Now we are ready to consider the SDE as before

$$dS(t) = b(t, S(t))\, dt + c(t, S(t)).dB(t) \tag{2.4.3}$$

where $S(t) = \{S_1(t), \ldots, S_m(t)\}$, $B(t) = \{B_1(t), \ldots, B_n(t)\}$ with $B_i(t)$ being independent Brownian motions, $b(t, S)$ is an m-vector (one-dimensional array) valued function on $\mathbb{R} \times \mathbb{R}^m$, $c(t, S)$ is an $m \times n$ matrix (two-dimensional array) valued function on $\mathbb{R} \times \mathbb{R}^m$. Denote also by $c_i = c_i(t, S)$ the rows (as one-dimensional arrays) of the matrix $c(t, S)$, and with $c_{i,j}$ the entries of the matrix c, or which is the same, of vectors c_i. Then, since $(dt)^2 = dt\, dB_i = dB_i\, dB_j = 0$, $i \neq j$, and $dB_i\, dB_i = dt$, we have

$$dS_i(t)\, dS_j(t) = (c_i.dB)\,(c_j.dB) = \sum_{k=1}^{n} c_{i,k}\, c_{j,k}\, dt = (c.c^T)_{i,j}\, dt. \tag{2.4.4}$$

Therefore, if g is a twice continuously differentiable function (this assumption can be relaxed somewhat), then

$$\begin{aligned} dg(t, S(t)) &= g_t(t, S(t))\, dt + D_S(g(t, S(t))).dS(t) \\ &+ \frac{1}{2} D^2{}_S(g(t, S(t))).dS(t).dS(t) \end{aligned} \tag{2.4.5}$$

where, again matrix multiplication is carried out, as in the Taylor series above from left to right. This, by the way might be enough to remember; the rest can be computed in examples. Nevertheless, we proceed in the usual way the multivariable Itô formula is written:

$$\begin{aligned} dg(t, S(t)) &= g_t(t, S(t))\, dt + D_S(g(t, S(t))).dS(t) \\ &+ \frac{1}{2} \sum_{i=1}^{m} \sum_{j=1}^{m} \frac{\partial^2 g(t, S(t))}{\partial S_i \partial S_j} (c.c^T)_{i,j}\, dt = g_t\,(t, S\,(t))dt \\ &+ D_S(g(t, S(t))).(b(t, S(t))\, dt + c(t, S(t)).dB(t)) \\ &+ \frac{1}{2} \operatorname{Tr}(c(t, S(t)).c(t, S(t))^T.D^2{}_S(g(t, S(t))))\, dt. \end{aligned} \tag{2.4.6}$$

Summarizing, the Itô multidimensional chain rule can be written either as

$$\begin{aligned} dg(t, S(t)) = \Big(&g_t(t, S(t)) + D_S(g(t, S(t))).b(t, S(t))\, dt \\ &+ \frac{1}{2} \operatorname{Tr}(c(t, S(t)).c(t, S(t))^T.D^2{}_S(g(t, S(t))))\Big)\, dt \\ + D_S(&g(t, S(t))).c(t, S(t)).dB(t) \end{aligned}$$

or

$$\begin{aligned} dg(t, S(t)) = \Big(&g_t(t, S(t)) + D_S(g(t, S(t))).b(t, S(t))\, dt \\ &+ \frac{1}{2} \text{Plus @@ Flatten}(c(t, S(t)).c(t, S(t))^T\, D^2{}_S(g(t, S(t))))\Big)\, dt \\ + D_S(&g(t, S(t))).c(t, S(t)).dB(t). \end{aligned}$$

We shall refer to the parts of the above formula(s). The scalar-valued function

$$g_t(t,\,S(t)) + D_S(g(t,\,S(t))).b(t,\,S(t))$$
$$+\frac{1}{2}\,\mathrm{Tr}(c(t,\,S(t)).c(t,\,S(t))^T.D^2{}_S(g(t,\,S(t))))$$

will be referred to as ItoDrift, while the vector-valued function

$$D_S(g(t,\,S(t))).c(t,\,S(t))$$

will be referred to as ItoDiffusion.

2.4.2.3 Implementation of the *m*-Dimensional Itô Chain Rule

It is interesting, and it might be quite useful, to implement the symbolic Itô chain rule, i.e., to implement the formulas for ItoDrift and ItoDiffusion symbolically. Having already introduced the gradient D and the Hessian D^2, we simply define

```
In[159]:= ItoDrift[g_, {b_, c_}, {t_, S_}] := ∂t g + DS[g].b +
          1/2 Plus @@ Flatten[c.Transpose[c] (D^2)S[g]]
```

and

```
In[160]:= ItoDiffusion[g_, c_, {t_, S_}] := DS[g].c
```

For example, consider a process $\{S(t)\}$, such that

$$dS(t) = b(t,\,S(t))\,dt + c(t,\,S(t)).dB(t)$$

with

```
In[161]:= b = {0.2 Sin[2 π t] S1, -0.3 Sin[2 π t] S2,
             0.4 Sin[2 π t] S3, -0.5 Sin[2 π t] S4};
```

and

```
In[162]:= c = {{0.02 S1, 0, 0, 0, 0, 0},
             {0, 0.02 S2, 0, 0, 0, 0}, {0, 0, 0.02 S3, 0, 0, 0},
             {0, 0, 0, 0.02 S4, 0, 0}};
```

On the top of the process $\{S(t)\}$ consider function

```
In[163]:= g = t (e^(S3^3) + Sin[S1^2 + S2^2]) / ((t + 1) ArcTan[S4]);
```

i.e., consider a process $\{G(t)\} = \{g(t,\,S(t))\}$. Then compute (for no particular reason) the drift and the diffusion of the process $\{G(t)\}$ as functions of $(t,\,S)$:

```
In[164]:= ItoDrift[g, {b, c}, {t, {S1, S2, S3, S4}}]
```

Out[164]=
$$\frac{1.2\, e^{S_3^3}\, t \sin(2\pi t)\, S_3^3}{(t+1)\tan^{-1}(S_4)} + \frac{0.4\, t \cos(S_1^2+S_2^2)\sin(2\pi t)\, S_1^2}{(t+1)\tan^{-1}(S_4)}$$
$$- \frac{0.6\, t\cos(S_1^2+S_2^2)\sin(2\pi t)\, S_2^2}{(t+1)\tan^{-1}(S_4)} + \frac{\sin(S_1^2+S_2^2)+e^{S_3^3}}{(t+1)\tan^{-1}(S_4)}$$
$$- \frac{t\left(\sin(S_1^2+S_2^2)+e^{S_3^3}\right)}{(t+1)^2\tan^{-1}(S_4)}$$
$$+ \frac{1}{2}\left(0.0004\left(\frac{2\, t\cos(S_1^2+S_2^2)}{(t+1)\tan^{-1}(S_4)} - \frac{4\, t\sin(S_1^2+S_2^2)\, S_1^2}{(t+1)\tan^{-1}(S_4)}\right) S_1^2\right.$$
$$+ 0.0004\, S_2^2\left(\frac{2\, t\cos(S_1^2+S_2^2)}{(t+1)\tan^{-1}(S_4)} - \frac{4\, t\sin(S_1^2+S_2^2)\, S_2^2}{(t+1)\tan^{-1}(S_4)}\right)$$
$$+ 0.0004\, S_3^2\left(\frac{9\, e^{S_3^3}\, t\, S_3^4}{(t+1)\tan^{-1}(S_4)} + \frac{6\, e^{S_3^3}\, t\, S_3}{(t+1)\tan^{-1}(S_4)}\right)$$
$$+ 0.0004\, S_4^2\left(\frac{2\, t\, S_4\left(\sin(S_1^2+S_2^2)+e^{S_3^3}\right)}{(t+1)\tan^{-1}(S_4)^2\,(S_4^2+1)^2}\right.$$
$$\left.\left. + \frac{2\, t\left(\sin(S_1^2+S_2^2)+e^{S_3^3}\right)}{(t+1)\tan^{-1}(S_4)^3\,(S_4^2+1)^2}\right)\right)$$
$$+ \frac{0.5\, t\sin(2\pi t)\left(\sin(S_1^2+S_2^2)+e^{S_3^3}\right) S_4}{(t+1)\tan^{-1}(S_4)^2\,(S_4^2+1)}$$

and

```
In[165]:= ItoDiffusion[g, c, {t, {S1, S2, S3, S4}}]
```

Out[165]=
$$\left\{\frac{0.04\, t\cos(S_1^2+S_2^2)\, S_1^2}{(t+1)\tan^{-1}(S_4)}, \frac{0.04\, t\cos(S_1^2+S_2^2)\, S_2^2}{(t+1)\tan^{-1}(S_4)}, \frac{0.06\, e^{S_3^3}\, t\, S_3^3}{(t+1)\tan^{-1}(S_4)},\right.$$
$$\left. - \frac{0.02\, t\left(\sin(S_1^2+S_2^2)+e^{S_3^3}\right) S_4}{(t+1)\tan^{-1}(S_4)^2\,(S_4^2+1)}, 0, 0\right\}$$

This was, of course, a totally artificial example. The following is not. Consider the one-dimensional stock price dynamics:

```
In[166]:= b = {a S}; c = {{σ S}};
```

i.e.,

$$dS(t) = a\, S(t)\, dt + \sigma\, S(t)\, dB(t)$$

where a is the appreciation rate and σ is the volatility. Now, consider a function of time t and stock price S, which, as we shall see in the next chapter, will have a fundamental importance (this is one of the Black–Scholes formulas)

In[167]:=
$$g = \frac{1}{2}\left(e^{r\,(t-T)}\,k\,\text{Erfc}\left[\frac{2\,\text{Log}[\frac{S}{k}] - (t-T)\,(2\,r-\sigma^2)}{2\,\sqrt{2}\,\sqrt{(T-t)\,\sigma^2}}\right] - S\,\text{Erfc}\left[\frac{2\,\text{Log}[\frac{S}{k}] - (t-T)\,(\sigma^2+2\,r)}{2\,\sqrt{2}\,\sqrt{(T-t)\,\sigma^2}}\right]\right);$$

We compute the components of the Itô differential $dg(t, S(t))$:

```
In[168]:= ItoDrift[g, {b, c}, {t, {S}}] // Simplify
```

Out[168]=
$$\frac{1}{2}\left(e^{r\,(t-T)}\,k\,r\,\text{erfc}\left(\frac{2\log\left(\frac{S}{k}\right) - (t-T)\,(2\,r-\sigma^2)}{2\,\sqrt{2}\,\sqrt{(T-t)\,\sigma^2}}\right) - a\,S\,\text{erfc}\left(\frac{2\log\left(\frac{S}{k}\right) - (t-T)\,(\sigma^2+2\,r)}{2\,\sqrt{2}\,\sqrt{(T-t)\,\sigma^2}}\right)\right)$$

and

```
In[169]:= ItoDiffusion[g, c, {t, {S}}]
```

Out[169]=
$$\left\{\frac{1}{2}\,S\,\sigma\left(-\frac{e^{r\,(t-T)-\frac{\left(2\log\left(\frac{S}{k}\right)-(t-T)\,(2\,r-\sigma^2)\right)^2}{8\,(T-t)\,\sigma^2}}\sqrt{\frac{2}{\pi}}\,k}{S\,\sqrt{(T-t)\,\sigma^2}} - \text{erfc}\left(\frac{2\log\left(\frac{S}{k}\right) - (t-T)\,(\sigma^2+2\,r)}{2\,\sqrt{2}\,\sqrt{(T-t)\,\sigma^2}}\right) + \frac{e^{-\frac{\left(2\log\left(\frac{S}{k}\right)-(t-T)\,(\sigma^2+2\,r)\right)^2}{8\,(T-t)\,\sigma^2}}\sqrt{\frac{2}{\pi}}}{\sqrt{(T-t)\,\sigma^2}}\right)\right\}$$

Those two functions determine the evolution of process $\{g(t, S(t))\}$: even though the stochastic dynamics of $S(t)$ is quite simple, the dynamics of $g(t, S(t))$ is a bit more complicated, and it would look much more complicated if above we did not use function Simplify.

Finally, it may be worrisome whether these calculations are actually correct (maybe we missed some of the terms in the implementation). So, consider an example, simple enough to be verified by hand: the same SDE as in the last example but now g is equal to

```
In[170]:= g = t + S^2;
```

Then, computing by hand first

$$d\,g(t, S(t)) = g_t(t, S(t))\,d\,t + g_S(t, S(t))\,d\,S(t) + \frac{1}{2}\,g_{S,S}(t, S(t))\,(d\,S(t))^2$$
$$= d\,t + 2\,S(t)\,d\,S(t) + (d\,S(t))^2$$
$$= (\sigma^2\,S(t)^2 + 2\,a\,S(t)^2 + 1)\,d\,t + 2\,S(t)^2\,\sigma\,d\,B(t).$$

On the other hand

```
In[171]:= ItoDrift[g, {b, c}, {t, {S}}]
```

Out[171]= $\sigma^2\,S^2 + 2\,a\,S^2 + 1$

and

```
In[172]:= ItoDiffusion[g, c, {t, {S}}]
```

Out[172]= $\{2\,S^2\,\sigma\}$

which checks out. Functions ItoDrift and ItoDiffusion are stored in the package CFMLab`ItoSDEs` for further reference.

2.4.2.4 Itô Product Rule Revisited

We have already discussed the Itô product rule:

$$d(X(t)\,Y(t)) = X(t)\,d\,Y(t) + Y(t)\,d\,X(t) + d\,X(t)\,d\,Y(t).$$

Here we show that the product rule is a consequence of the 2-Dimensional Itô chain rule. Indeed, consider

```
In[173]:= f[X_, Y_] := X Y
```

Then, since

```
In[174]:= D_{X,Y}[f[X, Y]]
```

Out[174]= $\{Y, X\}$

and

```
In[175]:= (D^2)_{X,Y}[f[X, Y]]
```

Out[175]= $\begin{pmatrix} 0 & 1 \\ 1 & 0 \end{pmatrix}$

we have, from the 2-dimensional Itô chain rule

$$
\begin{aligned}
d(X\,Y) &= d(f(X, Y)) = D_{\{X,Y\}}[f(X, Y)].\{dX, dY\} \\
&+ \frac{1}{2} D^2{}_{\{X,Y\}}[f[X, Y]].\{dX, dY\}.\{dX, dY\} \\
&= \{Y, X\}.\{dX, dY\} + \frac{1}{2}\begin{pmatrix} 0 & 1 \\ 1 & 0 \end{pmatrix}.\{dX, dY\}.\{dX, dY\} \\
&= Y\,dX + X\,dY + dX\,dY
\end{aligned}
\tag{2.4.7}
$$

where again the matrix multiplication is performed from left to right.

2.4.3 The Simplest Price Model: m-Stocks

2.4.3.1 Derivation

```
In[176]:= Clear["Global`*"]
```

The simplest vector SDE for modeling stock-price evolution in higher dimensions can be written as

$$\frac{dS(t)}{S(t)} = a\,dt + \sigma.dB(t) \tag{2.4.8}$$

where $S(t)$ (the stock prices at time t) is an n-vector valued process, a (the appreciation rate) is a constant n-vector, σ (the volatility matrix) is a constant $n \times m$ matrix, and $B(t)$ is m-vector valued Brownian motion (i.e., components are independent scalar valued Brownian motions). The above SDE makes sense if we adopt the *Mathematica*® convention (which we do): if a and b are n-vectors (arrays), then $\frac{a}{b}$ is defined componentwise. Indeed,

```
In[177]:= With[{a = {a1, a2}, b = {b1, b2}}, a/b]
```

Out[177]= $\left\{\frac{a_1}{b_1}, \frac{a_2}{b_2}\right\}$

We have already emphasized how vectors can be multiplied componentwise. If both conventions are adopted, then the above vector SDE is equivalent to

$$dS(t) = S(t)\,(a\,dt + \sigma.dB(t)). \tag{2.4.9}$$

Is it possible to mimic the derivation from the one-dimensional case to solve the present SDE? Let us first solve a simpler problem of finding not $S(t)$ but rather its expectation $E_{s,S(s)}\,S(t)$. To that end, a bit formally, by means of taking the expectation of the above equation, we get

$$d\,E_{s,S(s)}\,S(t) = a\,E_{s,S(s)}\,S(t)\,dt$$

which is just a (decoupled) ODE. It is solved immediately by (a is vector, *Mathematica*® convention: Exponential is listable, i.e., $e^{\{a_1,a_2\}} := \{e^{a_1}, e^{a_2}\}$)

$$E_{s,S(s)}\, S(t) = S(s)\, e^{a\,(t-s)}$$

or

$$E_{s,S(s)}\, \frac{S(t)}{S(s)} = e^{a\,(t-s)}.$$

Now we solve the SDE (again, Log is listable)

$$z = \log(S) \tag{2.4.10}$$

using the Itô chain rule, and noticing that

$$\left(\frac{dS}{S}\right)^2 = (a\,dt + \sigma.dB(t))^2 = (\sigma.dB(t))^2 = \mathrm{Diag}(\,\sigma.\sigma^T)\,dt \tag{2.4.11}$$

where Diag(A) is by definition equal to the vector whose components are equal to the diagonal entries of the matrix A, i.e.,

```
In[178]:= Diag[A_] := Table[A[[i, i]], {i, 1, Length[A]}]
```

We derive

$$\begin{aligned} dz = d\log(S) &= \frac{dS}{S} - \frac{1}{2}\left(\frac{dS}{S}\right)^2 \\ &= a\,dt + \sigma.dB(t) - \frac{1}{2}\,\mathrm{Diag}(\,\sigma.\sigma^T)\,dt \\ &= \left(a - \frac{1}{2}\,\mathrm{Diag}(\,\sigma.\sigma^T)\right)dt + \sigma.dB(t). \end{aligned} \tag{2.4.12}$$

Therefore

$$\begin{aligned} z(t) &= (t-s)\left(a - \frac{1}{2}\,\mathrm{Diag}(\,\sigma.\sigma^T)\right) + \sigma.(B(t) - B(s)) + z(s) \\ &= (t-s)\left(a - \frac{1}{2}\,\mathrm{Diag}(\,\sigma.\sigma^T)\right) + \sigma.(B(t) - B(s)) + \log(S(s)). \end{aligned}$$

This in turn gives the solution of the original stock price SDE by taking the (listable) exponential (and using the above convention for multiplication of same size vectors):

$$\begin{aligned} S(t) = e^{z(t)} &= e^{(t-s)\left(a-\frac{1}{2}\,\mathrm{Diag}\left(\sigma.\sigma^T\right)\right)+\sigma.(B(t)-B(s))+\log(S(s))} \\ &= S(s)\, e^{(t-s)\left(a-\frac{1}{2}\,\mathrm{Diag}\left(\sigma.\sigma^T\right)\right)+\sigma.(B(t)-B(s))} \end{aligned}$$

or

$$\frac{S(t)}{S(s)} = e^{(t-s)\left(a-\frac{1}{2}\,\mathrm{Diag}\left(\sigma.\sigma^T\right)\right)+\sigma.(B(t)-B(s))}. \tag{2.4.13}$$

So, the simplest price dynamics equation is solved in the vector case as well. Notice that for any t (consider s fixed)

$$(t-s)\left(a-\frac{1}{2}\,\text{Diag}(\sigma.\sigma^T)\right)+\sigma.(B(t)-B(s))$$

is a multinormal random variable, with mean value (vector)

$$\begin{aligned}&E\left((t-s)\left(a-\frac{1}{2}\,\text{Diag}(\sigma.\sigma^T)\right)+\sigma.(B(t)-B(s))\right)\\&=(t-s)\left(a-\frac{1}{2}\,\text{Diag}(\sigma.\sigma^T)\right)\end{aligned} \tag{2.4.14}$$

and covariance matrix

$$E\,(\sigma.(B(t)-B(s))).(\sigma.(B(t)-B(s)))^T=\sigma.\sigma^T\,(t-s). \tag{2.4.15}$$

So, we can write

$$\frac{S(t)}{S(s)}=e^Z \tag{2.4.16}$$

where

$$Z\sim N_n\left((t-s)\left(a-\frac{1}{2}\,\text{Diag}(\sigma.\sigma^T)\right),\ \sigma.\sigma^T\,(t-s)\right) \tag{2.4.17}$$

where $N_n(A, B)$ stands for n-dimensional multinormal distribution with mean A (n-vector) and covariance B ($n\times n$ nonnegative definite symmetric matrix).

2.4.3.2 Example; Using Multinormal Distribution

We shall need the package

```
In[179]:= << "Statistics`MultinormalDistribution`"
```

and in particular

```
In[180]:= ? MultinormalDistribution

     MultinormalDistribution[mu, sigma] represents the
        multivariate normal (Gaussian) distribution with mean
        vector mu and covariance matrix sigma. For a p-variate
        random vector to be distributed MultinormalDistribution[
        mu, sigma], mu must be a p-variate vector, and sigma
        must be a p x p symmetric  positive definite matrix.
       More...
```

Notice a bit of inconsistency in the way data enters: in the multinormal distribution (co)variance is entered, while in the normal distribution, as we have seen earlier, instead of variance σ^2, the standard deviation σ is entered.

Consider the two-dimensional (simplest) stock price evolution

$$\frac{dS(\tau)}{S(\tau)} = a\,d\tau + \sigma.dB(\tau) \tag{2.4.18}$$

for $s < \tau < t$, where

```
In[181]:= s = 0; t = 1;
```

and where appreciation rate vector is equal to

```
In[182]:= a = {0.6, 0.3};
```

while (the volatility) matrix σ is, for example, a 2×5 matrix

```
In[183]:= σ =
        {{0.5, 0.1, 0, -0.1, 0.2}, {0.1, 0.3, 0, 0, 0.2}};
```

Although, σ is a rectangular matrix, its action is, for all practical purposes, determined by the 2×2 matrix

```
In[184]:= σ.Transpose[σ]
```

$$\text{Out[184]=} \begin{pmatrix} 0.31 & 0.12 \\ 0.12 & 0.14 \end{pmatrix}$$

For example, the individual stock volatilities are

```
In[185]:= √Diag[σ.Transpose[σ]]

Out[185]= {0.556776, 0.374166}
```

while the same can be computed from σ as

```
In[186]:= (√Plus @@ #1² &) /@ σ

Out[186]= {0.556776, 0.374166}
```

As shown above, the solution of problem (2.4.18) is equal to $S(t)$, given by (2.4.16) and (2.4.17).

Compute the covariance matrix of the multinormal random vector Z from (2.4.14):

```
In[187]:= Σ = σ.Transpose[σ] (t - s)
```

$$\text{Out[187]=} \begin{pmatrix} 0.31 & 0.12 \\ 0.12 & 0.14 \end{pmatrix}$$

as well as its mean value

```
In[188]:= μ = (t - s) (a - 1/2 Diag[σ.Transpose[σ]])

Out[188]= {0.445, 0.23}
```

Define its distribution

```
In[189]:= MD = MultinormalDistribution[μ, Σ];
```

and the probability density function:

```
In[190]:= FF[z1_, z2_] = PDF[MD, {z1, z2}]
```

$$\text{Out[190]= } 0.93459\, e^{\frac{1}{2}(-(z1-0.445)(4.82759(z1-0.445)-4.13793(z2-0.23))-(10.6897(z2-0.23)-4.13793(z1-0.445))(z2-0.23))}$$

One can plot the density

```
In[191]:= Plot3D[FF[z1, z2], {z1, -1, 2},
           {z2, -2, 2}, PlotRange → All, Mesh → False,
           PlotPoints → 100, AxesLabel → {"z1", "z2", ""}];
```

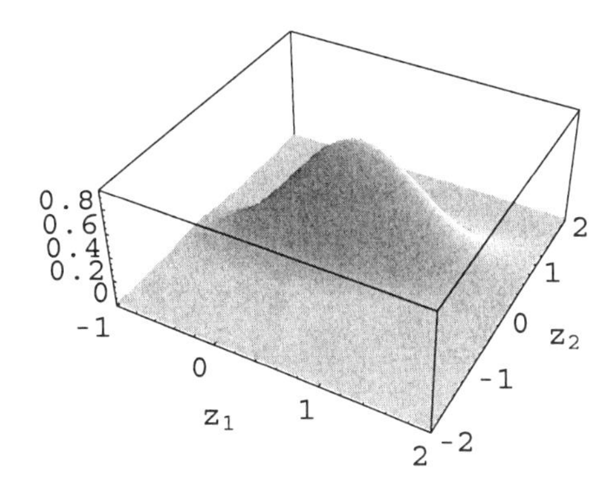

but even more interesting and precise would be to see the ellipsoid quantiles, i.e., the natural regions in the outcome space with prescribed probabilities:

```
In[192]:= ?EllipsoidQuantile

    EllipsoidQuantile[{{x11, ..., x1p}, ..., {xn1, ...,
       xnp}}, q] gives the locus of the qth quantile of the
       p-variate data, where the data have been ordered using
       ellipsoids centered on Mean[{{x11, ..., x1p}, ...,
       {xn1, ..., xnp}}]. The fraction of the data lying
       inside of this locus is q. EllipsoidQuantile[distribution,
       q] gives the ellipsoid centered on Mean[distribution]
       containing a fraction q of the specified distribution.
       EllipsoidQuantile[distribution, q] gives the ellipsoid
       centered on Mean[distribution] containing a fraction
       q of the specified distribution. More...
```

For example, if the prescribed probability is

```
In[193]:= p = .9;
```

Then ellipsoid quantile returns

```
In[194]:= eq = EllipsoidQuantile[MD, p]
```

Out[194]= $\text{Ellipsoid}\left(\{0.445, 0.23\}, \{1.30896, 0.599127\}, \begin{pmatrix} 0.888262 & 0.459338 \\ -0.459338 & 0.888262 \end{pmatrix}\right)$

where

```
In[195]:= ?Ellipsoid
```

```
Ellipsoid[{x1, ..., xp}, {r1, ..., rp}, {d1, ..., dp}]
  represents a p-dimensional ellipsoid centered at the
  point {x1, ..., xp}, where the ith semi-axis has radius
  ri and lies in direction di.  Ellipsoid[{x1, ..., xp},
  {r1, ..., rp}, IdentityMatrix[p]] simplifies to Ellipsoid[
  {x1, ..., xp}, {r1, ..., rp}], an ellipsoid aligned
  with the coordinate axes.
```

Now consider the pieces of the Ellipsoid return above. The first one

```
In[196]:= eq[[1]]
```

Out[196]= {0.445, 0.23}

is the mean vector of the multinormal distribution, i.e., the center of the ellipsoid. The second one

```
In[197]:= eq[[2]]
```

Out[197]= {1.30896, 0.599127}

is the vector of radiuses of the semi-axes. The third one

```
In[198]:= eq[[3]]
```

Out[198]= $\begin{pmatrix} 0.888262 & 0.459338 \\ -0.459338 & 0.888262 \end{pmatrix}$

is the matrix whose rows are axes directions (unit vectors). To understand this we also import

```
In[199]:= << "Graphics`Arrow`"
```

from where we shall use function

```
In[200]:= ?Arrow
```

```
Arrow[start, finish, (opts)] is a graphics primitive
   representing an arrow starting at start and ending
   at finish. More...
```

So, the ellipsoid, together with the axes vectors, looks like

```
In[201]:= (Show[Graphics[eq], Graphics[{#1, Point[
              eq⟦1⟧ + eq⟦2⟧⟦1⟧ eq⟦3⟧⟦1⟧]}], Graphics[
            {#1, Point[eq⟦1⟧ - eq⟦2⟧⟦1⟧ eq⟦3⟧⟦1⟧]}],
           Graphics[{#1, Point[
              eq⟦1⟧ + eq⟦2⟧⟦2⟧ eq⟦3⟧⟦2⟧]}], Graphics[
            {#1, Point[eq⟦1⟧ - eq⟦2⟧⟦2⟧ eq⟦3⟧⟦2⟧]}],
           Graphics[{#1, Point[eq⟦1⟧]}],
           (Graphics[{RGBColor[
                Sequence @@ Append[Abs[#1], 1]],
               Arrow[eq⟦1⟧, eq⟦1⟧ + #1]}] &) /@ eq⟦3⟧,
           Frame → True, AxesLabel → {"z1", "z2"},
           Axes → True, AspectRatio → Automatic] &)[
        AbsolutePointSize[4]];
```

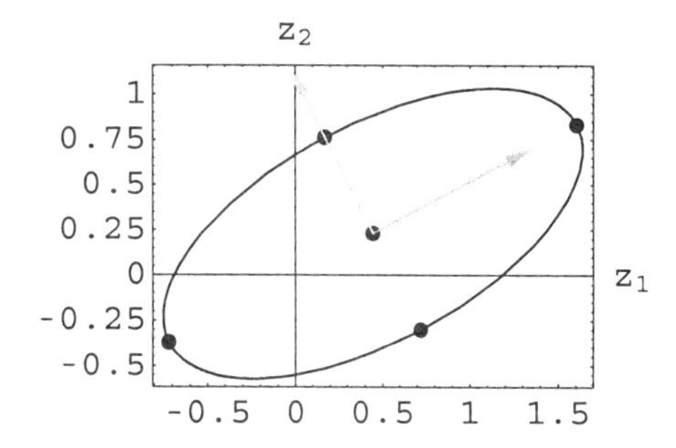

This ellipsoid can be translated into an ellipsoid centered at the origin by means of the transformation

$$\{y_1, y_2\} = Y = Z - \text{eq}[[1]] = \{z_1, z_2\} - \text{eq}[[1]]$$

and furthermore, since the axes directions are already orthogonal, rotated so that the direction vectors are in the directions of coordinate axes:

$$\{x_1, x_2\} = X = Y.\text{eq}[\![3]\!]^{-1} = \{y_1, y_2\}.\text{eq}[\![3]\!]^{-1} = (\{z_1, z_2\} - \text{eq}[[1]]).\text{eq}[\![3]\!]^{-1}$$

The transformed ellipsoid, together with axes vectors, looks like

```
In[202]:= (Show[Graphics[Ellipsoid[{0, 0},
             {eq⟦2⟧⟦1⟧, eq⟦2⟧⟦2⟧}]], Graphics[{#1,
             Point[eq⟦2⟧⟦1⟧ eq⟦3⟧⟦1⟧.Inverse[eq⟦3⟧]]}],
           Graphics[{#1, Point[-eq⟦2⟧⟦1⟧ eq⟦3⟧⟦1⟧.
                Inverse[eq⟦3⟧]]}], Graphics[{#1,
             Point[eq⟦2⟧⟦2⟧ eq⟦3⟧⟦2⟧.Inverse[eq⟦3⟧]]}],
           Graphics[{#1, Point[-eq⟦2⟧⟦2⟧
               eq⟦3⟧⟦2⟧.Inverse[eq⟦3⟧]]}],
```

```
    Graphics[{#1, Point[{0, 0}]}],
    (Graphics[{RGBColor[
          Sequence @@ Append[Abs[#1], 1]],
         Arrow[{0, 0}, #1.Inverse[eq[[3]]]]}] &) /@
     eq[[3]], Frame → True, Axes → True,
    AspectRatio → Automatic] &)[
 AbsolutePointSize[
  4]];
```

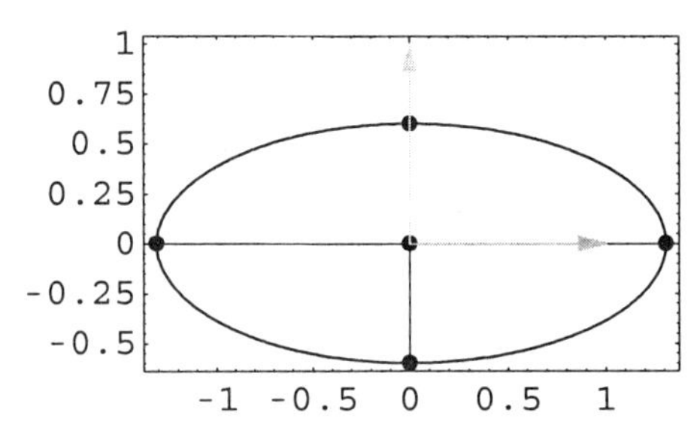

and has an equation

$$\frac{z_1{}^2}{a^2} + \frac{z_2{}^2}{b^2} = 1$$

or, more precisely,

$$\frac{z_1{}^2}{\text{eq}[[2, 1]]^2} + \frac{z_2{}^2}{\text{eq}[[2, 2]]^2} = 1$$

This implies that if

```
In[203]:= qq[p_] := EllipsoidQuantile[MD, p]
```

then the equation for the original ellipsoid can be written as

```
In[204]:= EllipsoidEquation[p_] := Plus @@
              (({z1, z2} - qq[p][[1]]).Inverse[qq[p][[3]]])^2
              ----------------------------------------------- == 1
                             qq[p][[2]]^2
```

For example,

```
In[205]:= EllipsoidEquation[p]
```

Out[205]= $0.583644\,(0.888262\,(z_1 - 0.445) + 0.459338\,(z_2 - 0.23))^2 + 2.78588\,(0.888262\,(z_2 - 0.23) - 0.459338\,(z_1 - 0.445))^2 == 1$

Now let the prescribed probabilities be

```
In[206]:= PrescribedProbabilities = Range[.11, .99, .11]
```

Out[206]= {0.11, 0.22, 0.33, 0.44, 0.55, 0.66, 0.77, 0.88, 0.99}

Then the ellipsoid quantiles for $Z = \{z_1, z_2\}$, together with some random realizations of Z, look like

```
In[207]:= Quant =
           Show[Table[Graphics[Point[Random[MD]]], {500}],
            (Graphics[EllipsoidQuantile[MD, #1]] &) /@
             PrescribedProbabilities, Frame → True,
            FrameLabel → {z1, z2}, AspectRatio → Automatic,
            RotateLabel → False, Axes → True];
```

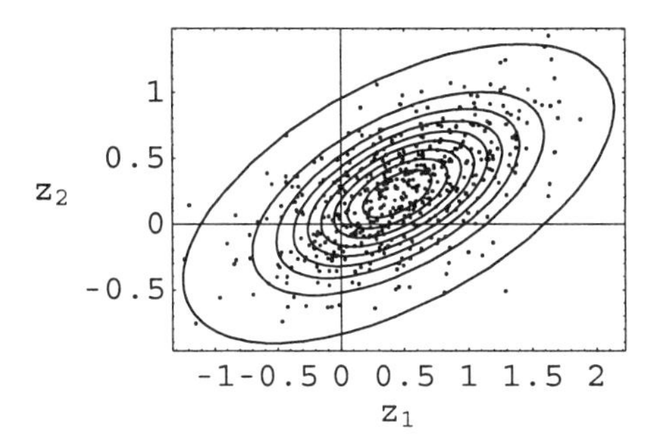

We first check whether our construction of the Ellipsoid Equations is correct. To that end we shall need a package

```
In[208]:= Needs["Graphics`ImplicitPlot`"]
```

Then, checking,

```
In[209]:= ImplicitPlot[(EllipsoidEquation[#1] &) /@
             PrescribedProbabilities /.
            {z1 → z[1], z2 → z[2]}, {z[1], -1.5, 2.5},
           Frame → True, FrameLabel → {z1, z2},
           AspectRatio → Automatic,
           RotateLabel → False, Axes → True];
```

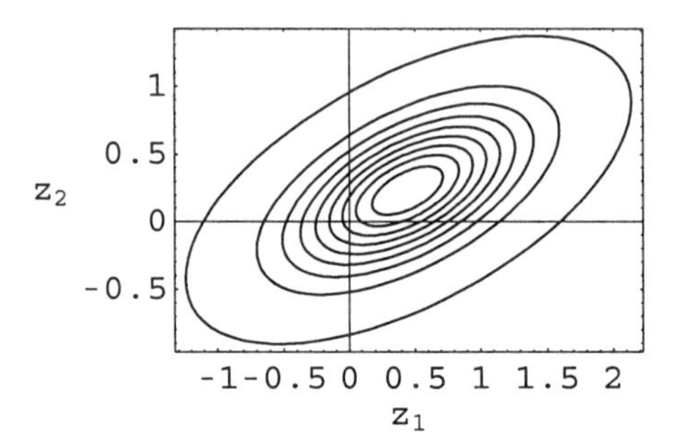

which looks good. We are now in the position to compute and draw confidence regions for

$$\frac{S(t)}{S(s)} = e^{(t-s)\left(a-\frac{1}{2}\operatorname{Diag}\left(\sigma.\sigma^T\right)\right)+\sigma.(B(t)-B(s))} = e^Z$$

Indeed, define $w_i = \frac{S_i(t)}{S_i(s)}$. Then, since $Z = \log\left(\frac{S(t)}{S(s)}\right)$, i.e., $z_i = \log\left(\frac{S_i(t)}{S_i(s)}\right) = \log(w_i)$, we get an equation for the quantile for $\frac{S(t)}{S(s)}$ simply by

```
In[210]:= ZQuantileEquation[p_] :=
          EllipsoidEquation[p] /. z_i_ → Log[w[i]]
```

For example,

```
In[211]:= ZQuantileEquation[.99]
```

$$\text{Out[211]=}\ 0.291822\,(0.888262\,(\log(w(1)) - 0.445) + 0.459338\,(\log(w(2)) - 0.23))^2 + 1.39294\,(0.888262\,(\log(w(2)) - 0.23) - 0.459338\,(\log(w(1)) - 0.445))^2 == 1$$

So, let

```
In[212]:= IP = ImplicitPlot[(ZQuantileEquation[#1] &) /@
              PrescribedProbabilities,
            {w[1], 0.1, 26}, AspectRatio → Automatic];
```

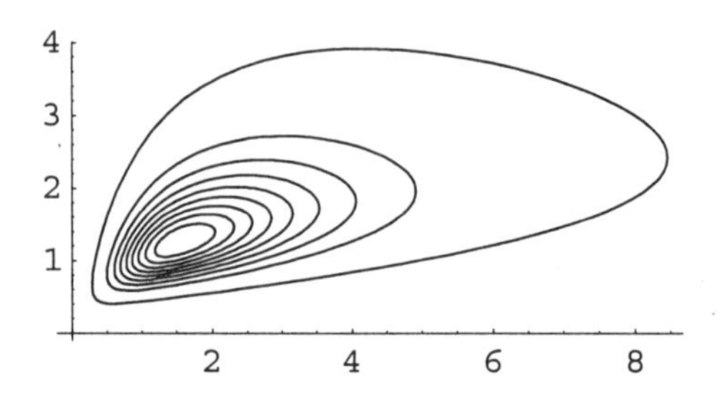

and we also superimpose its expected value $E_{s,S(s)} \frac{S(t)}{S(s)} = e^{a(t-s)}$

In[213]:= `EV = ℯ^(a (t-s))`

Out[213]= {1.82212, 1.34986}

as well as some realizations of the random variable $\frac{S(t)}{S(s)}$, yielding

In[214]:=
```
IPEV =
   Show[Table[Graphics[Point[ℯ^Random[MD]]], {500}],
    Graphics[{AbsolutePointSize[4],
      RGBColor[1, 0, 0], Point[EV]}],
    IP, PlotRange → All, Frame → True,
    DisplayFunction → $DisplayFunction,
    AspectRatio → Automatic, PlotLabel → S[t]/S[s]];
```

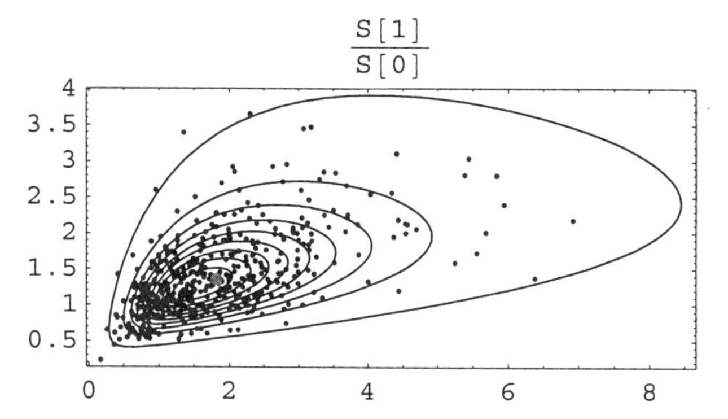

Unfortunately, the above experiments also indicate that the simplest price model, i.e., the Log-Normal stock-price dynamics model is not very descriptive: all kind of stock-price evolutions are quite probable. This is of course much better then using a model which *is* descriptive but wrong. Also this simple model will allow for very far-reaching conclusions to be developed in subsequent chapters. We shall also develop a theory, mostly in Chapter 8, that does not rely on the simplicity of the Log-Normal stock-price dynamics model.

2.5 Relationship Between SDEs and PDEs

Consider the SDE system

$$dS(t) = b(t, S(t))\, dt + c(t, S(t)).dB(t) \tag{2.5.1}$$

where $S(t) = \{S_1(t), \ldots, S_m(t)\}$, $B(t) = \{B_1(t), \ldots, B_n(t)\}$ with $B_i(t)$ being independent Brownian motions, $b(t, S)$ is an m-vector (one-dimensional array) valued function on $\mathbb{R} \times \mathbb{R}^m$, $c(t, S)$ is an $m \times n$ matrix (two-dimensional array) valued function on $\mathbb{R} \times \mathbb{R}^m$. Denote also by $c_i = c_i(t, S)$ the rows (as one-dimensional arrays) of the matrix $c(t, S)$, and with $c_{i,j}$ the entries of the matrix c, or which is the same, of vectors c_i.

We shall consider the evolution of the process S until it exits from some prescribed domain $\Omega \subset \mathbb{R}^m$, or due to the continuity of the trajectories, until the process hits the boundary $\partial\Omega$ (this hitting time is random stopping time that we shall denote θ), or until some fixed future time denoted T, whichever happens earlier, i.e., until $\mathrm{Min}[\theta, T] = \theta \wedge T$. In the case $\Omega = \mathbb{R}^m$, then $\partial\Omega = \varnothing$ and $\theta \wedge T = T$ is deterministic. Also we denote $Q_T = (t_0, T) \times \Omega$.

As derived above, the Itô chain rule holds:

$$\begin{aligned} d g(t, S(t)) &= \Big(g_t(t, S(t)) + D_S(g(t, S(t))).b(t, S(t))\, dt \\ &\quad + \frac{1}{2}\,\mathrm{Tr}(c(t, S(t)).c(t, S(t))^T.D^2{}_S(g(t, S(t))))\Big)\, dt \\ &\quad + D_S(g(t, S(t))).c(t, S(t)).dB(t) \\ &= (g_t(t, S(t)) + \mathcal{A}(t)\, g(t, S(t)))\, dt + D_S(g(t, S(t))).c(t, S(t)).dB(t) \end{aligned}$$

where obviously a linear differential operator of second order $\mathcal{A}(t)$ is defined to be $\mathcal{A}(t) : g \mapsto \mathcal{A}(t)\, g$, where

$$(\mathcal{A}(t)\, g)\,(S) = \frac{1}{2}\,\mathrm{Tr}(c(t, S).c(t, S)^T.D^2{}_S(g(S))) + b(t, S).D_S(g(S))$$

Also if we define a backward evolution operator $\mathcal{B} = \frac{\partial}{\partial t} + \mathcal{A}(t)$, then the Itô chain rule can be written as

$$d g(t, S(t)) = \mathcal{B}\, g(t, S(t))\, dt + D_S(g(t, S(t))).c(t, S(t)).dB(t)$$

Integrating between τ and $\theta \wedge T > \tau$

$$\begin{aligned} g(\theta \wedge T, S(\theta \wedge T)) - g(\tau, S(\tau)) &= \int_\tau^{\theta \wedge T} d g(t, S(t)) \\ &= \int_\tau^{\theta \wedge T} \mathcal{B}\, g(t, S(t))\, dt + \int_\tau^{\theta \wedge T} D_S(g(t, S(t))).c(t, S(t)).dB(t). \end{aligned}$$

Furthermore, for $x \in \Omega$ and $\tau < T$, taking the conditional expectation $E_{\tau,x}$ of both sides, and since $E_{\tau,x} \int_\tau^{\theta \wedge T} D_S(g(t, S(t))).c(t, S(t)).dB(t) = 0$, we get

$$\begin{aligned} E_{\tau,x}\, g(\theta \wedge T, S(\theta \wedge T)) - g(\tau, x) &= E_{\tau,x}(g(\theta \wedge T, S(\theta \wedge T)) - g(\tau, S(\tau))) \\ &= E_{\tau,x} \int_\tau^{\theta \wedge T} \mathcal{B}\, g(t, S(t))\, dt = -E_{\tau,x} \int_\tau^{\theta \wedge T} f(t, S(t))\, dt \end{aligned}$$

provided g is a solution of the backward evolution problem

$$\mathcal{B}\, g(t, S) = -f(t, S) \tag{2.5.2}$$

for $\tau \le t < T$ and $S \in \Omega$. If additionally the boundary condition is fulfilled

$$g = \psi \tag{2.5.3}$$

on $\partial_{\mathcal{B}} Q_T$, for some given function $\psi(t, S)$, the following formula holds:

$$g(\tau, x) = E_{\tau,x}\left(\psi(\theta \wedge T, S(\theta \wedge T)) + \int_{\tau}^{\theta \wedge T} f(t, S(t))\, dt\right). \tag{2.5.4}$$

What is meant by $\partial_{\mathcal{B}} Q_T$? This will depend on the properties of the operator $\mathcal{B}$, i.e., on the properties of the family of operators $\{\mathcal{A}(t)\}_{t<T}$, and further down, on the property of the matrix function $C(t) = c(t, S).c(t, S)^T$. Obviously, $C(t)$ is always non-negative definite. If additionally, $C(t)$ is uniformly positive definite, then $\partial_{\mathcal{B}} Q_T$ is simply the backward parabolic boundary of Q_T, i.e., $\partial_{\mathcal{B}} Q_T = (t_0, T) \times \partial\Omega \bigcup T \times \Omega$. In general, $\partial_{\mathcal{B}} Q_T$ is the subset of the backward parabolic boundary of Q_T that can be reached by a process $\{t, S(t)\}$ starting from the inside of Q_T. More about this in Chapter 8.

So, the solution of the PDE (2.5.2)–(2.5.3) has the probabilistic representation (2.5.4), and vice-versa. In other words, formulas (2.5.2)–(2.5.4) connect probability and and PDEs, or more precisely, continuous Markov processes and elliptic, parabolic, and as we are going to see later even hypoelliptic PDEs. Why is it important to connect probability and PDEs? With the exception of Chapter 4, much of the rest of this book is based on this connection. So instead of offering a short answer, we shall give a rather long, and I believe, a convincing one.

3 European Style Stock Options

Symbolic Solutions of Black–Scholes Partial Differential Equations and their Extensions

3.1 What Are Stock Options?

We have seen in Chapters 1 and 2 how different it is to invest in risk-free assets than in stocks. Going down the line of available financial instruments chosen for consideration in this book, we come to the study of stock options. There exists a further fundamental difference between stocks and options, and consequently, even much more of a difference between risk-free investing and options trading. The fundamental difference between stocks and options is that options have quite complicated dynamics, are much more volatile, and moreover, have a limited life time. As a matter of fact, options usually have a very limited life time. Options are available with expirations of up to eight months on over 1700 stocks, and for over 200 stocks as far in the future as three years; nevertheless, the most liquid options are those which expire sooner rather than later, and consequently the old fashioned buy and hold strategy obviously is almost *never an option*. On the other hand, although individual options by themselves are extremely volatile, i.e., risky, the price movements of different kind of options on the same underlying stock, as well as of the price movements of the underlying stock, are almost perfectly correlated, positively or negatively. That fact alone can be exploited for constructing various trading strategies that have the objective of actually reducing or even, at least theoretically, eliminating the risk involved in investing rather than increasing the upside potential aggressively.

So, what are the stock options?

Consider a particular stock, and to fix ideas, suppose that the price dynamics is simply:

$$\begin{aligned} dS(t) &= a\, S(t)\, dt + \sigma\, S(t)\, dB(t) \\ S(0) &= S_0. \end{aligned}$$

An option is a contract between two parties: buyer and seller, or which is the same, between holder and writer. We shall first describe a *call option*. A buyer of a *European* call option, who thereby becomes a holder (owner) of the option, is buying for a fee the right from the owner of the (underlying) stock, i.e., from the writer (seller) of the (*covered*) call, to buy the stock, if he/she chooses, at the prescribed price, the *strike price* (k), and at the prescribed future date, the *expiry* or *expiration date* (T). This contract can be, and usually is, frequently resold on the option market up until the expiration date. If, at the expiration date T, the holder decides to buy the stock from the writer, then the option is *exercised*. Of course, that decision is going to be based solely on whether or not the market price of the underlying stock is higher than the strike price k. Indeed, in order to exercise an option, the option holder has to buy the underlying stock from a writer, and only after that transaction is completed (and paid for), can the option holder sell the same stock on the stock market, pocketing the difference. So, at the expiration date the payoff to the holder of an option is equal to zero if the price of the stock is less than or equal to the strike price (i.e., the option is not exercised), or, in the opposite case, it is equal, modulo transaction costs that are ignored throughout most (but not all) of this book, to the difference between the stock and the strike price. Therefore, the plot of the payoff to the holder of the call option at the expiration date looks like

```
In[1]:= Clear["Global`*"]
```

```
In[2]:= k = 80; Plot[Max[0, S - k], {S, 0, 2 k},
          AxesLabel → {S, ""}, PlotStyle →
           {AbsoluteThickness[3], RGBColor[1, 0, 0]}];
```

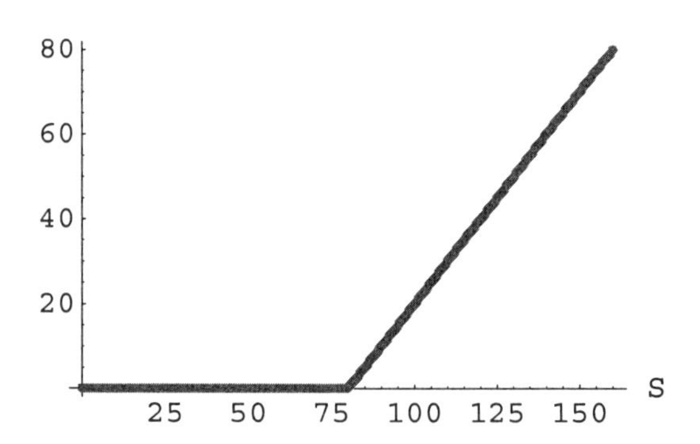

(Here and elsewhere, we try to follow the convention, even though this is not possible to see in the print, that red color refers to calls, blue color refers to puts).

There are variations to such contracts. For example, the writer does not necessarily have to own a stock in order to write an option. Although this might be risky, and assuming one has a good credit rating, one can write un-covered, or *naked calls* on the underlying stock that he/she does not own. The obligation that is assumed is the same. Also, an *American* call option is same as the European, except in regard to the time of possible exercise—an American option can be exercised at *any* time before the expiration, and not only at the expiration date as in the case of European options. On the option market one actually trades most often American options only. The popularity of European options is due to their mathematical simplicity, when compared to American options, and to the fact that the computational results are often very similar. We shall study in great details both kinds.

On the other hand, the buyer of a *put option*, who thereby becomes a holder (owner) is buying the right from the writer (seller) of the put for a fee to *sell* the underlying stock if he/she chooses to at the prescribed price, the strike price, at (in the case of European style options, or possibly before in the case of American style options) the prescribed future date, the expiration date. In order to exercise the put option, the option owner has to buy the underlying stock on the stock market (unless he/she already owns the stock), and only after that transaction is completed can he/she sell the stock to the writer of the option pocketing the difference. So at the expiry date the payoff to the holder of the option is equal to zero if the price of the stock is greater than or equal to the strike price, or, in the opposite case, it is equal to the difference between the strike and the stock price. The plot of the payoff at the expiry date for a put option then looks like the following:

```
In[3]:= Plot[Max[0, k - S], {S, 0, 2 k},
          AxesLabel → {S, ""}, PlotStyle →
           {AbsoluteThickness[3], RGBColor[0, 0, 1]}];
```

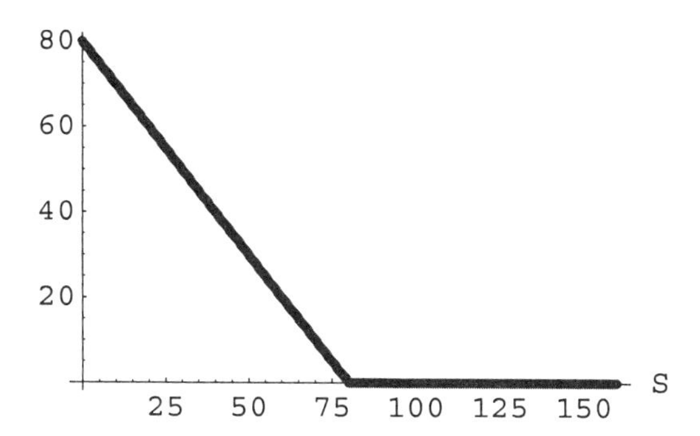

To gain some intuition about what happens to the fortunes of the holder of an option consider the following calculation. Recall the SDESolver:

```
In[4]:= << "CFMLab`ItoSDEs`"
```

Suppose that expiration date is

```
In[5]:= T = 3/12;
```

i.e., 3 months from now. Suppose that the strike price is, as above,

```
In[6]:= k

Out[6]= 80
```

Also choose the underlying stock average appreciation rate, stock volatility, the present stock price, and the interest rate to be

```
In[7]:= a = .5; σ = .5; S0 = 75; r = .05;
```

Finally compute

```
In[8]:= K = 30;
```

random stock price trajectories:

```
In[9]:= Experiments = Table[
          SDESolver[a #2 &, σ #2 &, S0, 0, T, 1000], {K}];
```

We compute the payoffs at the time of expiration

```
In[10]:= CallPayoffs = (Max[#1[[-1, 2]] - k, 0] &) /@ Experiments

Out[10]= {5.73541, 16.3561, 39.3257, 0, 0, 0, 25.2707, 9.28551, 0, 0, 0,
          0, 0, 0, 0, 0, 0, 0, 0, 7.33951, 0, 0, 16.0792, 0, 0, 0, 0, 83.5078,
          11.0598, 0}
```

for a call, and

```
In[11]:= PutPayoffs = (Max[k - #1[[-1, 2]], 0] &) /@ Experiments

Out[11]= {0, 0, 0, 27.8458, 2.56773, 11.953, 0, 0, 34.4392, 16.0753,
          4.47443, 7.47636, 12.4898, 12.6156, 8.49877, 7.22054,
          15.3662, 0.87802, 24.4698, 0, 10.5856, 2.98396, 0, 12.5647,
          4.67862, 14.9923, 16.9333, 0, 0, 4.42666}
```

for a put. (Option trades are done for option contracts; one contract is for 100 options so that if only one contract is traded above, then these numbers should be multiplied by 100.) The first observation, in either case, is how very different the outcomes are. Also, compute the average payoffs. To that end define the average function

```
In[12]:= Av[List_] := Plus @@ List / Length[List]
```

Then

```
In[13]:= Av[CallPayoffs]

Out[13]= 7.13199

In[14]:= Av[PutPayoffs]

Out[14]= 8.45119
```

The practical question arises: What would be the *fair price* of these options before the experiment starts? A reasonable candidate for such fair prices seems to be the *present value* of the above averages (possibly over much larger samples)? Those, at least over these small samples, by the way, would have been equal to

```
In[15]:= Av[CallPayoffs] e^(-r T)
Out[15]= 7.0434
```

and

```
In[16]:= Av[PutPayoffs] e^(-r T)
Out[16]= 8.34621
```

The correct answer, a bit different from this one and a lot more complete, provided by Black and Scholes, widely accepted and celebrated by practitioners and theoreticians alike, is discussed in details in the present chapter. Some extensions will be discussed as well.

3.2 Black–Scholes PDE and Hedging

3.2.1 Derivation of the Black–Scholes PDE

The price of options is determined on the option market. They cost as much as anybody is willing to pay for them, or as little as anybody is willing to sell them for. Nevertheless, for many reasons, some of which will be discussed in detail later in this book, it is useful to have an "objective" way, and preferably a *formula*, for relating the price of an option to the time left until the expiration, and to the price and some of the other parameters characterizing the performance of the underlying stock, or even the economy. This is provided by the famous Black–Scholes formula. The Black–Scholes formula is derived as a solution of the Black–Scholes partial differential equation (BSPDE), which in its own right needs to be derived. Contrary to the popular derivation of the Black–Scholes PDE (see, e.g., [31,58]), we shall be careful in our derivation. We shall provide here a mathematically completely satisfactory and yet elementary derivation. Furthermore, in the subsequent section this derivation is going to be implemented so that all becomes completely transparent.

Consider an underlying stock, whose stock price obeys an SDE

$$\begin{aligned} dS(t) &= S(t)\ ((a(t, S(t)) - D_0(t, S(t)))\ dt + \sigma(t, S(t))\, dB(t)) \\ S(0) &= S_0. \end{aligned} \tag{3.2.1}$$

Here $a(t, S)$ is the appreciation rate, $\sigma(t, S)$ is the volatility, and $D_0(t, S)$ is the (yearly) dividend payment rate. In regard to the dividend payment rate in the above equation, for example, if the dividend is assumed to be 2% per year then the cash dividend payment of $\$\, S(t)\, .02\, dt$ reduces the value of the stock; indeed if a company pays $\$\, S(t)\, .02\, dt = \$1$ for each stock outstanding, and let's say there are one million stocks outstanding, this means that the company paid \$1M from its cash holdings, and therefore the objective value of the company has been reduced by \$1M, and consequently each stock should lose on the stock market the value of \$1. We shall derive the Black–Scholes PDE for the fair price of an option on that underlying stock by considering a *put* option. Our choice was based on the fact that in the case of a put, the hedging portfolio consists of long positions, and long positions seem to be easier to grasp than short ones.

Suppose an investor has a brokerage account which consists of a small portfolio consisting of a single put option and a yet undetermined number of underlying stocks in addition to the cash account. The investor has to decide how many stocks to hold at each time t. So the portfolio consists of a put option worth $V(t, S(t))$, and of a long position of $-\Delta(t) = -\Delta(t, S(t)) > 0$ underlying stocks worth $-\Delta(t, S(t))\, S(t) > 0$. This portfolio is accompanied by a cash account with balance $C(t)$, from which, and where to, a cash stream is going in and out to account for all hedging transactions (in addition to interest). Function $\Delta(t, S)$ is called a *hedging strategy*, or hedging rule. The cash value of the whole position, i.e., the brokerage account balance, is equal to

$$X(t) = V(t, S(t)) - \Delta(t, S(t))\, S(t) + C(t). \tag{3.2.2}$$

Usually when the Black–Scholes PDE is derived (see, e.g., [31,58]), $C(t)$ is ignored, even though, as we shall see soon very explicitly, understanding its dynamics and influence is crucial for understanding the derivation. So, we start there. We postulate that

$$dC(t) = r(t)\, C(t)\, dt - \Delta(t, S(t))\; S(t)\, D_0(t, S(t))\, dt + S(t + dt)\; d\Delta(t) \tag{3.2.3}$$

where $r(t)\, C(t)\, dt$ is the interest received or paid during the time interval dt on the balance of $C(t)$, $-\Delta(t, S(t))\; S(t)\, D_0(t, S(t))\, dt$ is the dividend received during the time interval dt, while, and this is very important,

$$S(t + dt)\, d\Delta(t) = S(t + dt)\, (\Delta(t + dt) - \Delta(t)) \tag{3.2.4}$$

is the cost or pay-off from buying or selling the stock, as dictated by the chosen hedging strategy Δ (notice carefully that the trading decision $d\Delta(t) = \Delta(t + dt) - \Delta(t)$ could be made only at the time $t + dt$, and not before, and therefore the stock price $S(t + dt)$ is used rather than $S(t)$; the latter choice would lead to a wrong conclusion, since in the stochastic calculus they are not equivalent).

Now consider the dynamics of the brokerage account balance. We compute the Itô differential of $X(t)$, relying on the Itô product rule (contrary to e.g. [31,58]):

$$\begin{aligned}
dX(t) &= dV(t, S(t)) - d(\Delta(t, S(t))\, S(t)) + dC(t) \\
&= dV(t, S(t)) - \Delta(t, S(t))\, dS(t) - S(t + dt)\, d\Delta(t) + dC(t) \\
&= dV(t, S(t)) - \Delta(t, S(t))\, dS(t) - S(t + dt)\, d\Delta(t) \\
&\quad + r(t)\, C(t)\, dt - D_0(t, S(t))\, dt + S(t + dt)\, d\Delta(t) \\
&= dV(t, S(t)) - \Delta(t, S(t))\, dS(t) + r(t)\, C(t)\, dt - D_0(t, S(t))\, dt
\end{aligned} \tag{3.2.5}$$

and, applying the Itô chain rule

$$
\begin{aligned}
&dX(t) = V_t(t, S(t))\,dt + V_S(t, S(t))\,dS(t) \\
&+\frac{1}{2} V_{S,S}(t, S(t))\,(dS(t))^2 - \Delta(t, S(t))\,dS(t) \\
&+r(t)\,C(t)\,dt - \Delta(t, S(t))\; S(t)\,D_0(t, S(t))\,dt.
\end{aligned}
\tag{3.2.6}
$$

This is true for any hedging strategy $\Delta(t, S)$ and any option pricing formula $V(t, S)$ observed by the market. Is it possible to design a riskless hedging strategy? Where is the risk? Looking into the expression for $dX(t)$, the most obvious place to start is $dS(t)$ since it contains $dB(t)$, the source of all the randomness. One may even think that by eliminating $dB(t)$ from the above expression for $dX(t)$, the risk is eliminated altogether. This is false. Indeed, eliminating $dB(t)$ is a good place to start, but this is not the end of the risk elimination; the risk elimination is going to be achieved in two steps, as we shall see shortly. To start with, we eliminate $dB(t)$ by eliminating $dS(t)$, which is very simple, by choosing the hedging strategy

$$
\Delta(t, S) = V_S(t, S) = \frac{\partial V(S(t), t)}{\partial S}.
\tag{3.2.7}
$$

Consequently

$$
\begin{aligned}
&dX(t) = V_t(t, S(t))\,dt + \frac{1}{2} V_{S,S}(t, S(t))\,(dS(t))^2 \\
&+r(t)\,C(t)\,dt - \Delta(t, S(t))\; S(t)\,D_0(t, S(t))\,dt \\
&= V_t(t, S(t))\,dt + \frac{1}{2} V_{S,S}(t, S(t))\,S(t)^2\,\sigma(t, S(t))^2\,dt \\
&+r(t)\,C(t)\,dt - V_S(t, S(t))\; S(t)\,D_0(t, S(t))\,dt
\end{aligned}
\tag{3.2.8}
$$

which holds for any option pricing formula $V(t, S)$ observed by the market. We notice that the risk, although vastly reduced, is still present, i.e., if no special requirements on the option pricing formula $V(t, S)$ are imposed, the randomness still exists, since in the above expression for $dX(t)$ we still have $S(t)$ which is a random process. So, in order to completely eliminate the randomness, we require that $X(t)$ satisfy an (deterministic) ODE:

$$
dX(t) = r(t)\,X(t)\,dt.
\tag{3.2.9}
$$

This ODE is chosen not just because it is the simplest one, but also because it is the most natural one: if randomness is to be eliminated, then the profit should be no more and no less than the one gained in the bank. This finally yields

$$
\begin{aligned}
&V_t(t, S(t))\,dt + \frac{1}{2} V_{S,S}(t, S(t))\,S(t)^2\,\sigma(t, S(t))^2\,dt \\
&+r(t)\,C(t)\,dt - V_S(t, S(t))\; S(t)\,D_0(t, S(t))\,dt \\
&= dX(t) = r(t)\,X(t)\,dt \\
&= r(t)\,(V(t, S(t)) - \Delta(t, S(t))\,S(t) + C(t))\,dt
\end{aligned}
\tag{3.2.10}
$$

or, after cancelling $r(t)\, C(t)\, dt$,

$$\begin{aligned} &V_t(t, S(t))\, dt + \frac{1}{2}\, V_{S,S}(t, S(t))\, S(t)^2\, \sigma(t, S(t))^2\, dt \\ &- V_S\,(t, S(t))\; S(t)\, D_0(t, S(t))\, dt \\ &= r(t)\,(V(t, S(t)) - V_S(t, S(t))\, S(t))\, dt \end{aligned} \tag{3.2.11}$$

and finally

$$\begin{aligned} &V_t(t, S) + \frac{1}{2}\, V_{S,S}(t, S)\, S^2\, \sigma(t, S)^2 \\ &+ (r(t) - D_0(t, S))\, V_S(t, S)\, S - r(t)\, V(t, S) = 0 \end{aligned} \tag{3.2.12}$$

which is the (famous) Black–Scholes partial differential equation (BSPDE), which holds for any underlying stock price $S > 0$ and any time $t < T$, where T is the option expiration time. Also, obviously, the terminal condition

$$V(T, S) = \mathrm{Max}(0, k - S) \tag{3.2.13}$$

in the case of puts (k is the strike price) and

$$V(T, S) = \mathrm{Max}(0, S - k) \tag{3.2.14}$$

in the case of calls must hold as well.

Looking into PDE theory one recognizes that yes, in either case, put or call, there exists such a function, and moreover such a function is uniquely determined from the above set of conditions. Therefore there exists a unique option pricing formula that allows for risk elimination in a fair, i.e., interest-rate-generating, fashion.

We make some initial observations about the PDE just derived. The above *linear* PDE is of *backward–parabolic* type, and it degenerates when $S \to 0$, which in particular implies the fact that no boundary condition is needed (or possible to impose) at $S = 0$. Recall an argument in Chapter 2 amounting to the fact that the price process $S(t)$, under reasonable assumptions on the coefficients in the equation (3.2.1), never hits $S = 0$. These two facts are closely related. Summarizing, only one condition in addition to the PDE itself is needed and it is the *terminal condition* at $t = T$. The terminal condition is the only place where the theory for calls and puts differ.

The problem of finding the fair price of an option is reduced to the problem of solving a terminal value problem for the backward–parabolic PDE. Finally, notice at first a somewhat surprising property of the BSPDE: the appreciation rate of the underlying stock does not appear in it, and accordingly the derived fair option price does not depend on the appreciation rate a of the underlying stock, but it depends very much on the stock's volatility σ. This is actually the way it should be. Shortly, the investor's perceptions about the future performance of the underlying stock, i.e., investor's perceptions about a stock's appreciation rate a, vary considerably from one investor to the other. Moreover, one could even argue that precisely because of the existence of those variations, most of the trades are executed: the seller views the market as if it is in a decline, while the opposite is true for the buyer. Additionally, as known in the statistics and as it is going to be seen in Chapter 4 below, statistical estimates of appreciation rates are not efficient. On the other hand, market volatility, and in particular volatility of the particular underlying stock is much easier to observe, i.e., simple statistical estimates are very effective in measuring it, and consequently there is a much higher level of agreement among the traders regarding the value of the current volatility σ. It is quite natural then that the option prices that traders agree upon do not depend on the appreciation rate of the underlying stock, and that, on the other hand, they have to depend on the volatility, since otherwise it would be very easy to make money based only on stock price random fluctuations.

3.2.2 Black–Scholes Hedging Implemented

3.2.2.1 Put Hedging Implemented

```
In[17]:= Clear["Global`*"];
```

```
In[18]:= Off[General::"spell1"]
```

The derivation of the Black–Scholes PDE presented above was the correction of the prominent and yet often criticized derivation. The criticism was based on the lack of applying of the Itô, or even classical, product rule when differentiating the product $\Delta(t, S(t))\,S(t)$. The additional observation was made that hedging strategy $\Delta(t, S) = V_S(t, S)$, by itself, contrary to what is explicitly stated in many well-known texts does not eliminate the randomness, i.e., risk completely, but instead the risk elimination is achieved in two steps. This correction of the derivation of the Black–Scholes PDE was discovered in stages, and every step of the way the following kind of examples was a confident guide. So, let

```
In[19]:= << "Statistics`NormalDistribution`"
```

```
In[20]:= T = 2/12; K = 1000; dt = T/K; a = -1; σ = .4; r = .05;
         k = 100; S0 = 100; Off[General::"spell"]
```

where T is the expiration time, K is the number of time subintervals, or in practical terms, the number of portfolio adjustments to be made during the life of the option; a is the appreciation rate of the underlying stock, σ is the volatility, r is the interest rate; k is the option strike price (in this section we consider the put, while in the subsequent one we shall consider a call), and S_0 is the initial stock price. We shall assume that no dividend is paid, i.e., $D_0 = 0$. We define the list of Brownian increments (we do not evaluate them since, we shall perform several experiments at the same time below)

```
In[21]:= dB :=
           Table[Random[NormalDistribution[0, √dt]], {K}]
```

The next is the iterative function to be used to generate the stock price trajectories

```
In[22]:= G[{t_, S_}, db_] := {dt + t, a dt S + db σ S + S}
```

```
In[23]:= SolList[dB_] := FoldList[G, {0, S0}, dB]
```

The stock prices list will be called SList:

```
In[24]:= SList[dB_] := Transpose[SolList[dB]][[2]]
```

and the corresponding times are:

```
In[25]:= TList = Range[0, T, dt];
```

We shall try different hedging strategies $\Delta(t, S)$ below. But no matter what the particular hedging strategy, it is going to be applied on top of the underlying stock price trajectory $(\Delta(t, S(t)))$ as

```
In[26]:= ΔList[dB_] := N[Δ /@ SolList[dB]]
```

Similarly, the option price is calculated on top of the stock price trajectory $(V(t, S(t)))$:

```
In[27]:= VList[dB_] := N[V /@ SolList[dB]]
```

as well as the cash value of the stock investment $(-\Delta(t, S(t))\, S(t))$:

```
In[28]:= MDSList[dB_] := N[(-Δ[#1] #1[[2]] &) /@ SolList[dB]]
```

The next is $d\Delta(t, S(t))$, i.e., the necessary adjustment in the number of stocks needed in the portfolio:

```
In[29]:= dΔList[dB_] :=
           Drop[ΔList[dB], 1] - Drop[ΔList[dB], -1]
```

As explained above, this adjustment yields *cash transactions* $d\Delta(t, S(t))\, S(t + dt)$:

```
In[30]:= dΔSList[dB_] := dΔList[dB] Drop[SList[dB], 1]
```

Those cash transactions accumulate in the cash account in addition to the interest:

```
In[31]:= CList[InitialBalance_, dB_] := FoldList[
            dt r #1 + #1 + #2 &, InitialBalance, dΔSList[dB]]
```

Finally, the total brokerage account balance $X(t) = V(t, S(t)) - \Delta(t, S(t))\, S(t) + C(t)$ is going to be computed as

```
In[32]:= XList[InitialBalance_, dB_] :=
           CList[InitialBalance, dB] + MDSList[dB] + VList[dB]
```

Now we are ready to perform some computations. First we show that if an arbitrary pricing formula is chosen, the hedging strategy did not eliminate all of the randomness. For example, let

```
In[33]:= V[{t_, S_}] = 50 (Sin[S/10] Sin[t] + 1);
```

and let the corresponding hedging be given by

```
In[34]:= Δ[{t_, S_}] = ∂S V[{t, S}]
```

Out[34]= $5\cos\left(\frac{S}{10}\right)\sin(t)$

We compute 5 realizations. Even though quite a bit of randomness is eliminated it still exists:

```
In[35]:= Show[(ListPlot[Transpose[{TList, XList[0, #1]}],
              PlotJoined → True, DisplayFunction →
               Identity] &) /@ Table[dB, {5}],
         DisplayFunction → $DisplayFunction,
         PlotRange → All];
```

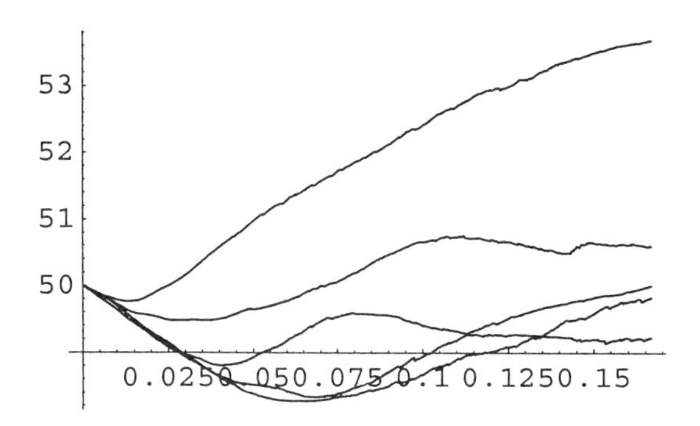

The reason is that all that was achieved by hedging $\Delta(t, S) = V_S(t, S)$ is

$$dX(t) = V_t(t, S(t))\,dt + \frac{1}{2}\,V_{S,S}(t, S(t))\,S(t)^2\,\sigma(t, S(t))^2\,dt + r\,C(t)\,dt$$

and since $S(t)$ is random, so is the balance $X(t)$. Now, by requesting that the balance $X(t)$ be non-random, and moreover that it grow exponentially as money in a bank, i.e., requesting $dX(t) = r\,X(t)\,dt$, we deduce, posteriori, that the option pricing formula cannot be an arbitrary function such as the one above, but rather it is a unique solution of the Black–Scholes PDE. In later sections we shall find this explicit unique solution (VP in the case of puts, and VC in the case of calls), and it will be placed into the package CFMLab`BlackScholes`. We shall use these functions here:

```
In[36]:= << "CFMLab`BlackScholes`"
```

```
In[37]:= V[{t_, S_}] :=
           If[t < T, VP[t, S, T, k, r, σ], Max[k - S, 0]];
```

yielding in particular the hedging strategy

```
In[38]:= Δ[{t_, S_}] = ∂S V[{t, S}]
```

Out[38]= $\text{If}\left[t < \frac{1}{6}, \text{VP}^{(0,1,0,0,0,0)}(t, S, T, k, r, \sigma), -\text{Max}^{(1,0)}[k - S, 0]\right]$

Now we repeat the above experiments, and this time we arrive at the (almost) deterministic evolution of $X(t)$. The leftover randomness can be further reduced by increasing K, and when $K \to \infty$, the randomness disappears:

```
In[39]:= bp =
    Plot[(V[{0, S0}] + (-Δ[#1] #1[[2]] &)[{0, S0}]) e^(r t),
     {t, 0, T},
     PlotStyle → {RGBColor[1, 0, 0], Thickness[.015]},
     DisplayFunction → Identity]; Show[bp,
    (ListPlot[Transpose[{TList, XList[0, #1]}],
        PlotJoined → True, DisplayFunction →
         Identity] &) /@Table[dB, {5}],
    DisplayFunction → $DisplayFunction,
    PlotRange → All];
```

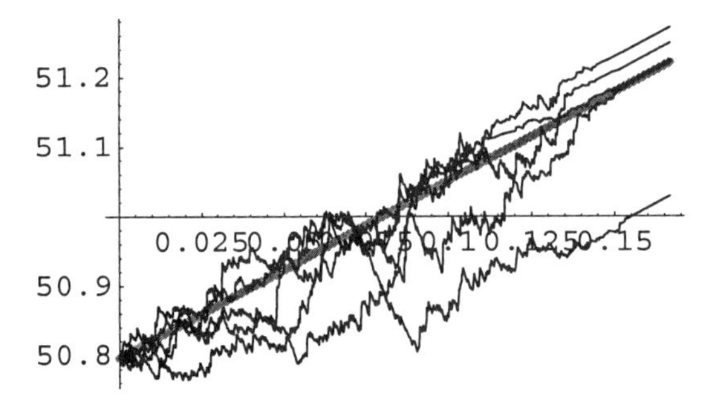

We concentrate now on a single random trajectory, but showcasing all the different quantities that amount to the evolution of $X(t)$. So let the randomization be fixed:

```
In[40]:= db = dB;
```

then compute the stock price trajectory $S(t)$:

```
In[41]:= StockPricePlot =
     (ListPlot[SolList[#1], PlotJoined → True,
         PlotRange → All,
         PlotStyle → RGBColor[0, 1, 0]] &)[db];
```

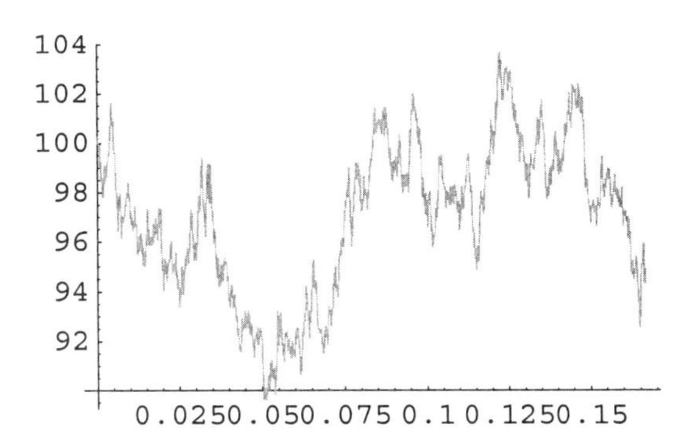

the cash value of the (put) option investment $V(t, S(t))$:

```
In[42]:= OptionInvestmentPlot =
           (ListPlot[Transpose[{TList, VList[#1]}],
              PlotJoined → True, PlotRange → All,
              PlotStyle → {RGBColor[0, 0, 1]}] &)[db];
```

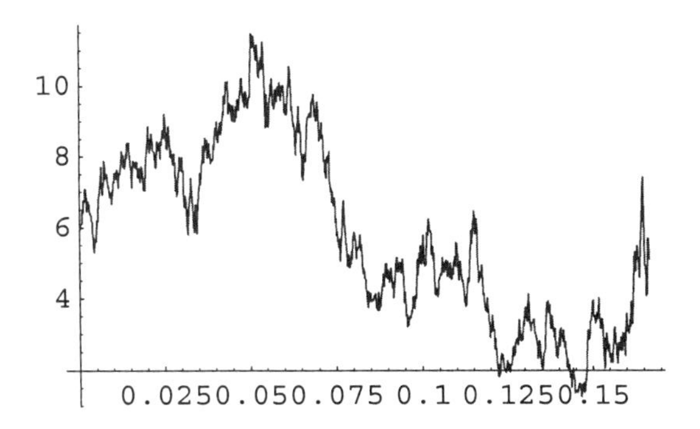

and the cash value of the stock investment $-\Delta(t, S(t))\, S(t)$:

```
In[43]:= StockInvestmentPlot =
           (ListPlot[Transpose[{TList, MDSList[#1]}],
              PlotJoined → True, PlotRange → All,
              PlotStyle → {RGBColor[0, 1/2, 0]}] &)[db];
```

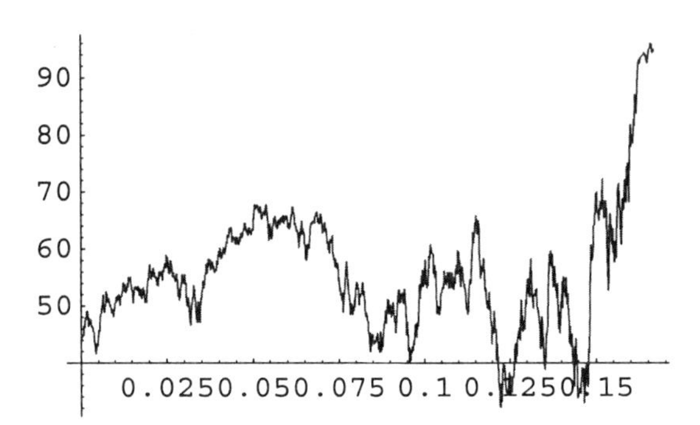

Next, compute all the hedging cash transactions $d\Delta(t, S(t))\,S(t + dt)$ (notice that since $-\Delta(t, S(t))$ is the number of stocks held at time t, $+d\Delta(t, S(t))\,S(t + dt)$ is the cash *influx* into the cash account, i.e., $+d\Delta(t, S(t))\,S(t + dt) > 0$ means a *sale* of the stock was executed at time $t + dt$):

```
In[44]:= AllTransactions =
          (ListPlot[Transpose[{Drop[TList, 1],
                dΔSList[#1]}], PlotRange → All,
              PlotStyle → RGBColor[.5, .5, .5]] &) [db];
```

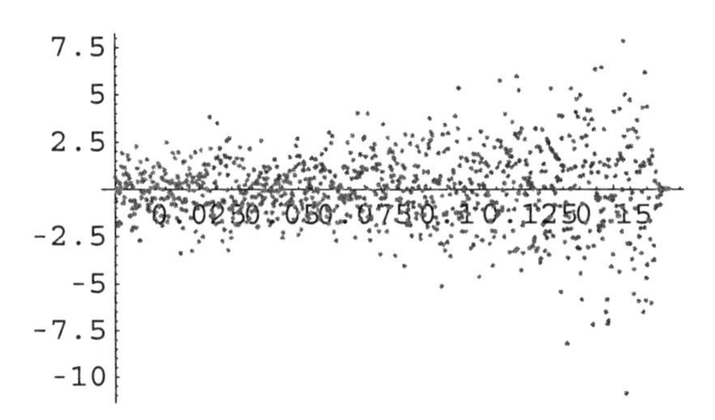

Notice how many $K \to \infty$ transactions are needed, and how buys and sales almost perpetually alternate. All these transactions accumulate into the cash account balance (in addition to the interest):

```
In[45]:= CashAccountPlot =
         (ListPlot[Transpose[{TList, CList[0, #1]}],
              PlotJoined → True, PlotRange → All,
              PlotStyle → {Thickness[.01],
                 RGBColor[0, 1/2, 0]}] &) [db];
```

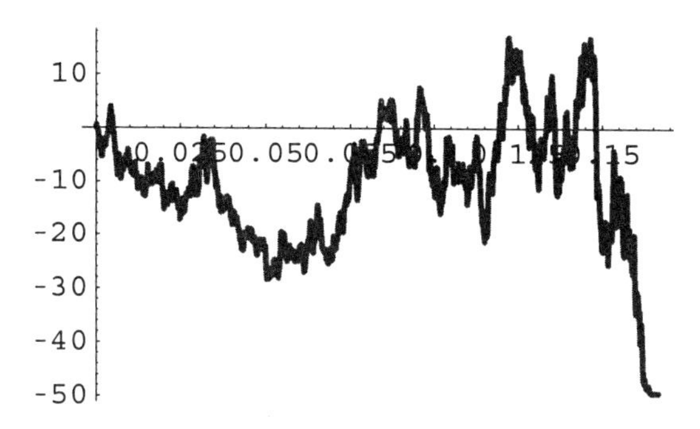

and finally, the total brokerage account balance $X(t)$ is (almost deterministic and) equal to

```
In[46]:= BrokerageAccountPlot =
           (ListPlot[Transpose[{TList, XList[0, #1]}],
               PlotJoined → True, PlotStyle →
                Thickness[.01], PlotRange → All] &)[db];
```

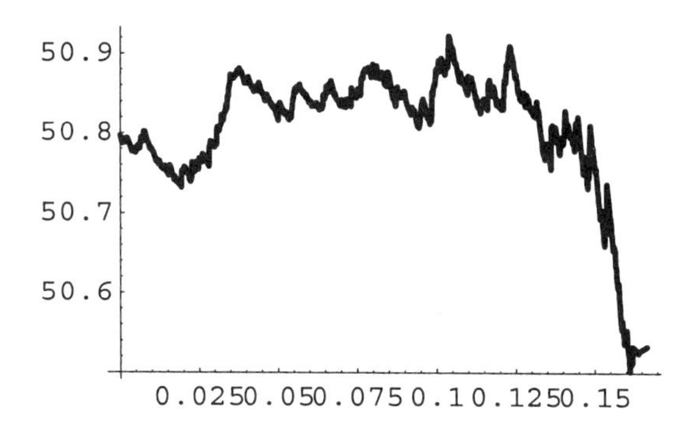

Putting them all together we get

```
In[47]:= TheCompletePutPicture =
          Show[StockPricePlot, OptionInvestmentPlot,
           StockInvestmentPlot, AllTransactions,
           CashAccountPlot, BrokerageAccountPlot];
```

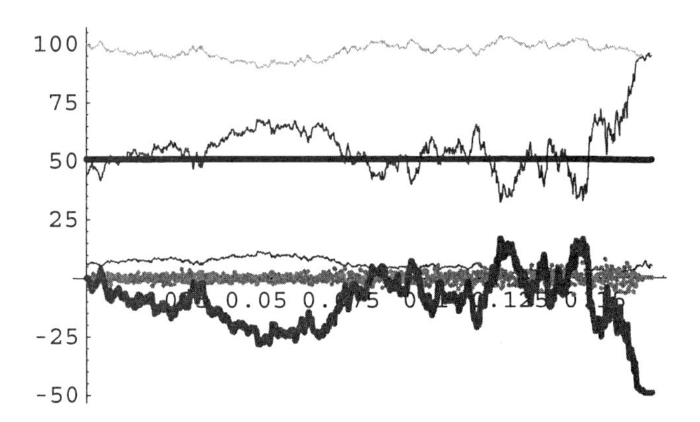

The picture is self-explanatory. The reader should pause to understand the dynamics and interaction of various components involved, and also experiment with various market conditions (bull/bear market, etc.).

3.2.2.2 Call Hedging Implemented

What is different if instead of a put option we consider a call on the same underlying. We shall use the same data as in the case of a put. As in the case of a put, we have the call Black–Scholes formula (to be derived shortly):

```
In[48]:= V[{t_, S_}] :=
           If[t < T, VC[t, S, T, k, r, σ], Max[S - k, 0]];
```

yielding the hedging strategy

```
In[49]:= Δ[{t_, S_}] = ∂S V[{t, S}]
```

Out[49]= $\text{If}\left[t < \frac{1}{6}, \text{VC}^{(0,1,0,0,0,0)}(t, S, T, k, r, \sigma), \text{Max}^{(1,0)}[S - k, 0]\right]$

As above we compute several trajectories for $X(t)$:

```
In[50]:= bp =
          Plot[(V[{0, S0}] + (-Δ[#1] #1[[2]] &)[{0, S0}]) e^(r t),
           {t, 0, T},
           PlotStyle → {RGBColor[1, 0, 0], Thickness[.015]},
           DisplayFunction → Identity]; Show[bp,
          (ListPlot[Transpose[{TList, XList[0, #1]}],
              PlotJoined → True, DisplayFunction →
               Identity] &) /@Table[dB, {5}],
          DisplayFunction → $DisplayFunction,
          PlotRange → All];
```

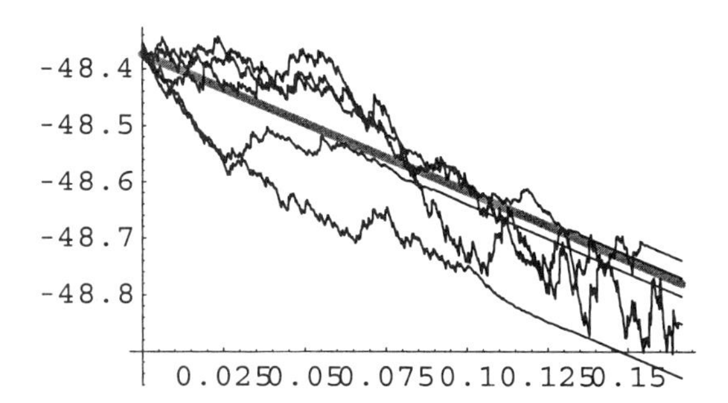

Notice at first glance the surprising difference compared to the put-hedging in the total brokerage balance evolution: it is decreasing! The reason is, that since the cash account starts at zero, and since the call option is hedged by the short position in the underlying stock, the initial brokerage balance is negative, it is a debt, and therefore the interest rate works against the account owner—the debt is increasing. We concentrate now on a single random trajectory, generated by the same randomization as in the case of a put above; also we introduce some initial cash holding so that we do not get into debt as above:

```
In[51]:= InitialCashHolding = 70;
```

Compute first the stock price evolution:

```
In[52]:= StockPricePlot =
           (ListPlot[SolList[#1], PlotJoined → True,
              PlotRange → All,
              PlotStyle → RGBColor[0, 1, 0]] &) [db];
```

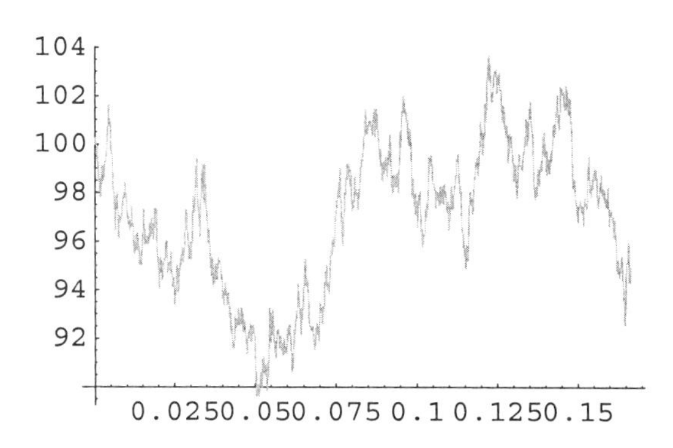

Next, compute the evolution of the cash value of the (call) option investment $V(t, S(t))$:

```
In[53]:= OptionInvestmentPlot =
           (ListPlot[Transpose[{TList, VList[#1]}],
               PlotJoined → True, PlotRange → All,
               PlotStyle → {RGBColor[1, 0, 0]}] &)[db];
```

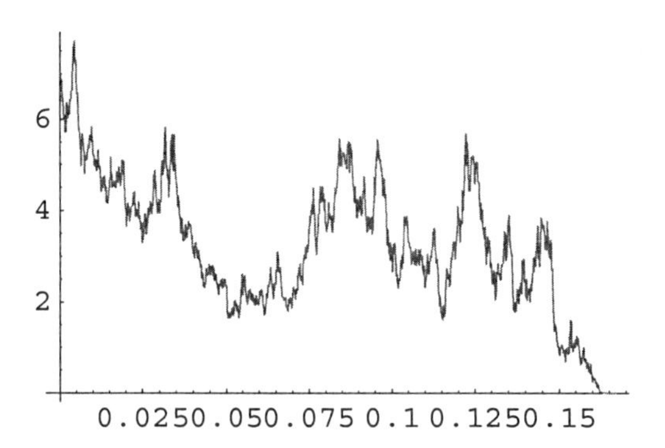

and the cash value of the stock investment $-\Delta(t, S(t))\, S(t)$:

```
In[54]:= StockInvestmentPlot =
           (ListPlot[Transpose[{TList, MDSList[#1]}],
               PlotJoined → True, PlotRange → All,
               PlotStyle → {RGBColor[0, 1/2, 0]}] &)[db];
```

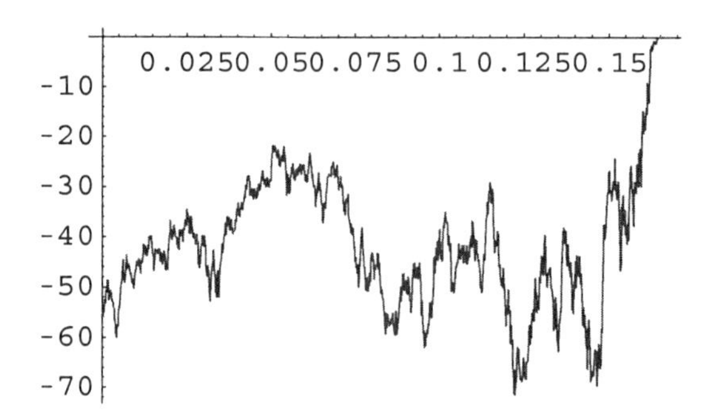

Next, compute all the hedging cash transactions $d\Delta(t, S(t))\, S(t + dt)$:

```
In[55]:= AllTransactions =
           (ListPlot[Transpose[{Drop[TList, 1],
                dΔSList[#1]}], PlotRange → All,
               PlotStyle → RGBColor[.5, .5, .5]] &)[db];
```

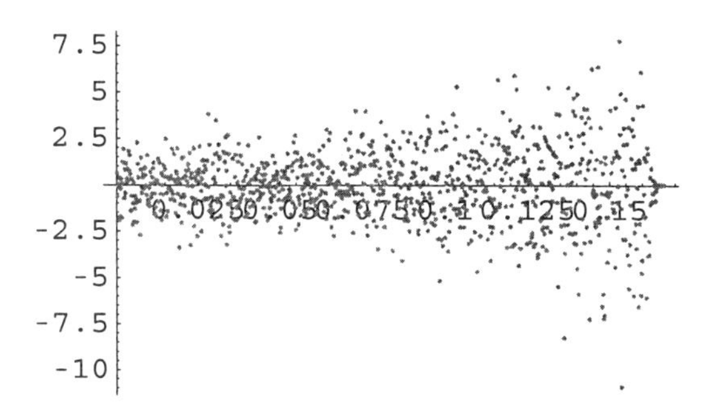

that accumulate into the cash account (in addition to the interest):

```
In[56]:= CashAccountPlot = (ListPlot[Transpose[
            {TList, CList[InitialCashHolding, #1]}],
           PlotJoined → True, PlotRange → All,
           PlotStyle → {Thickness[.01],
             RGBColor[0, 1/2, 0]}] &) [db];
```

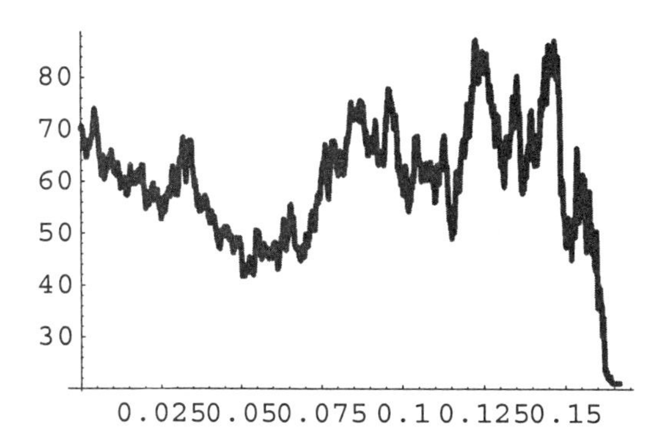

and finally, the total (almost deterministic) brokerage account balance $X(t)$

```
In[57]:= BrokerageAccountPlot =
        (ListPlot[Transpose[{TList, XList[
              InitialCashHolding, #1]}], PlotJoined →
           True, PlotStyle → Thickness[.01],
          PlotRange → All] &) [db];
```

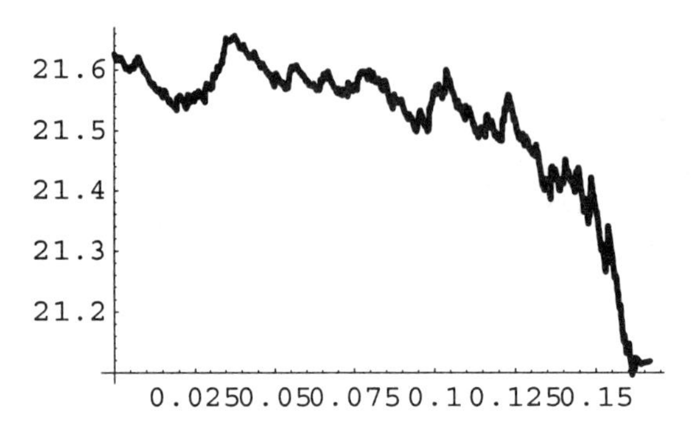

Putting them all together we get

```
In[58]:= TheCompleteCallPicture =
           Show[StockPricePlot, OptionInvestmentPlot,
            StockInvestmentPlot, AllTransactions,
            CashAccountPlot, BrokerageAccountPlot];
```

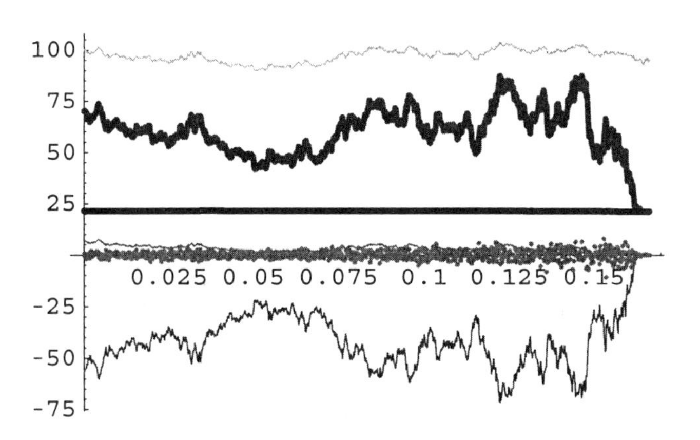

Many scenarios are possible, and can all be discovered through experiments. This completes the derivation, as well as the verification of the derivation of the Black–Scholes PDE.

3.2.3 Probabilistic Interpretation of the Solution of the Black–Scholes PDE

3.2.3.1 Derivation

Consider an underlying stock, and suppose that the price dynamics is the simplest one:

$$
\begin{aligned}
&dS(t) = a\,S(t)\,dt + \sigma\,S(t)\,dB(t)\\
&S(0) = S_0.
\end{aligned}
\tag{3.2.15}
$$

Also, for simplicity we shall assume that no dividend is paid ($D_0 = 0$). The BSPDE reduces to

$$
\begin{aligned}
&\frac{\partial V(t,S)}{\partial t} + \frac{1}{2} S^2 \frac{\partial^2 V(t,S)}{\partial S^2}\sigma^2 + r\,S\,\frac{\partial V(t,S)}{\partial S} - r\,V(t,S) = 0\\
&V(T,S) = \phi(S)
\end{aligned}
\tag{3.2.16}
$$

for $S > 0$, $t < T$. Here

$$
\phi(S) = \mathrm{Max}(S - k, 0) \tag{3.2.17}
$$

in the case of calls, and

$$
\phi(S) = \mathrm{Max}(k - S, 0) \tag{3.2.18}
$$

in the case of puts. Notice that no boundary condition is imposed for the reason already explained above.

Consider now an auxiliary SDE

$$
\begin{aligned}
&dZ(t) = r\,Z(t)\,dt + \sigma\,Z(t)\,dB(t)\\
&Z(0) = z.
\end{aligned}
\tag{3.2.19}
$$

Notice that this is *not* the same SDE as the one governing the evolution of the stock price $S(t)$. The difference is in the average appreciation rate, which is this time taken to be equal to the interest rate r.

Consider $W(t, Z(t)) = e^{r(T-t)}\,V(t, Z(t))$. By the Itô product rule

$$
\begin{aligned}
&dW(t,Z(t)) = V(t,Z(t))\,d e^{r(T-t)} + e^{r(T-t)}\,dV(t,Z(t))\\
&+d e^{r(T-t)}\,dV(t,Z(t)) = -r\,e^{r(T-t)}\,V(t,Z(t)) + e^{r(T-t)}\,dV(t,Z(t))\\
&= -r\,e^{r(T-t)}\,V(t,Z(t)) + e^{r(T-t)}\Big(V_t(t,Z(t))\,dt + V_Z(t,Z(t))\,dZ(t)\\
&\qquad + \frac{1}{2}V_{ZZ}(t,Z(t))\,(dZ(t))^2\Big) = -r\,e^{r(T-t)}\,V(t,Z(t))\\
&+e^{r(T-t)}\Big(V_t(t,Z(t))\,dt + V_Z(t,Z(t))\,(r\,Z(t)\,dt + \sigma\,Z(t)\,dB(t))\\
&\qquad + \frac{1}{2}V_{ZZ}(t,Z(t))\,\sigma^2\,Z(t)^2\,dt\Big) = e^{r(T-t)}\Big(V_t(t,Z(t))\\
&\qquad + V_Z(t,Z(t))\,r\,Z(t) + \frac{1}{2}V_{ZZ}(t,Z(t))\,\sigma^2\,Z(t)^2\Big)\,dt\\
&+e^{r(T-t)}\,V_Z(t,Z(t))\,\sigma\,Z(t)\,dB(t) = e^{r(T-t)}\,V_Z(t,Z(t))\,\sigma\,Z(t)\,dB(t)
\end{aligned}
$$

Integrating in t we get

$$W(T, Z(T)) - W(t, Z(t)) = \int_t^T e^{r(T-\tau)} V_Z(\tau, Z(\tau))\, \sigma\, Z(\tau)\, dB(\tau).$$

Taking the conditional expectation (with respect to $(t, Z(t))$) of both sides, we get

$$E_{t,Z(t)}(W(T, Z(T)) - W(t, Z(t))) = 0,$$

i.e., since $E_{t,Z(t)}\, W(t, Z(t)) = W(t, Z(t))$,

$$\begin{aligned} W(t, Z) &= E_{t,Z}\, W(T, Z(T)) = E_{t,Z}\, e^{r(T-T)}\, V(T, Z(T)) \\ &= E_{t,Z}\, \phi(Z(T)). \end{aligned} \tag{3.2.20}$$

Therefore,

$$V(t, S) = e^{-r(T-t)}\, W(t, S) = e^{-r(T-t)}\, E_{t,S}\, \phi(Z(T)) \tag{3.2.21}$$

i.e., the Black–Scholes fair price of an option $V(t, S)$ is equal to the *present value (i.e., the value discounted by the interest rate r) of the conditional expectation of the payoff* $\phi(Z(T))$. The crucial thing to observe is that it is $Z(t)$ under ϕ, and not $S(t)$, and that the dynamics of $Z(t)$ differ from the dynamics of $S(t)$ by the fact that no matter what the appreciation rate of the stock price $S(t)$ is, the appreciation rate of $Z(t)$ is equal to the bank interest rate r.

One final observation in regard to a possible misunderstanding of the "paradox" that the appreciation rate is not present in the option pricing formula, or equivalently, that there is a need for this artificial price dynamics: one can make a comparison with stocks; for example if the stock price dynamics is given by

$$\begin{aligned} &dS(t) = a\, S(t)\, dt + \sigma\, S(t)\, dB(t) \\ &S(0) = p, \end{aligned}$$

then *today*'s price of that stock is not equal to the present value of some future price of that stock, for example $e^{-rT}\, E_{0,p}\, S(T)$ for some future time T, but rather it is equal to the present price p; therefore the appreciation rate a does not affect the current stock price in an explicit manner.

The above can be used to compute the Black–Scholes fair price experimentally. Of course, this is a very inefficient way of doing this, and we will soon see how to do it explicitly.

3.2.3.2 An Experiment

```
In[59]:= Clear["Global`*"];
         << "CFMLab`ItoSDEs`"
```

Suppose data

```
In[61]:= T = 3/12; k = 80; σ = .5; r = .05; s0 = 75;
```

Perform

```
In[62]:= K = 50;
```

random experiments:

```
In[63]:= Experiments = Table[
             SDESolver[r #2 &, σ #2 &, s0, 0, T, 1000], {K}];
```

The payoffs are

```
In[64]:= CallPayoffs = (Max[#1[[-1, 2]] - k, 0] &) /@ Experiments
```

```
Out[64]= {0, 8.48917, 2.04882, 0, 0, 22.2884, 0.993161, 0, 0, 0, 28.3202,
          0, 0, 0, 13.383, 0, 0, 0, 0, 9.28865, 0, 0, 0, 0, 0, 0, 7.84614,
          48.3128, 0, 0, 0, 18.7489, 21.3319, 0, 46.9793, 10.581, 0, 0,
          4.28266, 4.35241, 5.21846, 19.5578, 0, 12.5382, 0.660331,
          0, 14.8615, 12.4422, 0, 12.8775}
```

in the case of a call, and

```
In[65]:= PutPayoffs = (Max[k - #1[[-1, 2]], 0] &) /@ Experiments
```

```
Out[65]= {44.8644, 0, 0, 14.377, 9.8522, 0, 0, 10.9006, 12.4526, 7.14031,
          0, 31.4345, 19.3885, 22.2661, 0, 17.1228, 17.5629, 17.094,
          33.4142, 0, 11.9367, 6.58108, 0.674495, 9.86173, 1.84584,
          5.27651, 0, 0, 18.2913, 16.9431, 15.2133, 0, 0, 15.262, 0, 0,
          15.4607, 17.1563, 0, 0, 0, 0, 11.3245, 0, 0, 10.791, 0, 0,
          15.2028, 0}
```

in the case of a put (with same strike price).

Then the Black–Scholes fair price of the call option is (estimated to be; recall Av is average)

```
In[66]:= Av[CallPayoffs] e^(-r T)
```

```
Out[66]= 6.42721
```

in the case of call, and

```
In[67]:= Av[PutPayoffs] e^(-r T)
```

```
Out[67]= 8.48707
```

in the case of puts.

By the way, the exact value is computed, using the Black–Scholes formulas, available in

```
In[68]:= << "CFMLab`BlackScholes`"
```

as

```
In[69]:= VC[0, s0, T, k, r, σ]
Out[69]= 5.82133
```

in the case of a call, and

```
In[70]:= VP[0, s0, T, k, r, σ]
Out[70]= 9.82755
```

in the case of a put. The reader is encouraged to experiment with different values for K (the number of trajectories calculated). Indeed, one can see that quite a high number K is needed in order to get an estimate for the fair call and put option prices that approximate reasonably well the above exact values, reconfirming the utility of having explicit solutions.

The above Monte–Carlo calculation can be improved significantly if analytic knowledge, which is available in this simple situation, is utilized. Namely, as derived in Chapter 2 the simplest price dynamics SDE

$$dZ(t) = r\,Z(t)\,dt + \sigma\,Z(t)\,dB(t)$$
$$Z(0) = z$$

has an explicit solution

$$Z(T) = Z(t)\;e^{(T-t)\left(r-\frac{\sigma^2}{2}\right)+\sigma\,B(T-t)} \tag{3.2.22}$$

and therefore

$$V(t, S) = e^{-r(T-t)}\,E_{t,S}\,\phi(Z(T)) = e^{-r(T-t)}\,E_{t,S}\,\phi\left(Z(t)\,e^{(T-t)\left(r-\frac{\sigma^2}{2}\right)+\sigma\,B(T-t)}\right)$$

$$= e^{-r(T-t)}\,E\,\phi\left(S\,e^{(T-t)\left(r-\frac{\sigma^2}{2}\right)}\,e^{\sigma\,B(T-t)}\right)$$

where

```
In[71]:= φ1[S_] := Max[S - k, 0]
```

in the case of calls, and

```
In[72]:= φ2[S_] := Max[k - S, 0]
```

in the case of puts. The last formula can be implemented as

```
In[73]:= Off[General::"spell1"];
         Timing[Module[{x, tr, y, φy}, x = s0 e^(T (r - σ^2/2));
           tr = Table[Random[NormalDistribution[0, 1]],
             {100000}]; y = e^(σ √T tr) x; φy = φ1 /@ y; Av[φy] e^(-r T)]]
```

Out[73]= {6.65 Second, 5.85194}

in the case of calls, and

```
In[74]:= Timing[
   Module[{x, tr, y, ϕy}, x = s0 ⅇ^(T (r - σ^2/2)); tr = Table[
      Random[NormalDistribution[0, 1]], {100000}];
    y = ⅇ^(σ √T tr) x; ϕy = ϕ2 /@ y; Av[ϕy] ⅇ^(-r T)]]
```

Out[74]= {6.92 Second, 9.8692}

in the case of puts. Indeed, the improvement is significant. On the other hand, one can see that even if 100000 simulations were performed the error is still there. This suggests that if the above or some other analytic simplifications are not possible, the Monte–Carlo simulation cannot be applied very efficiently and accurately at the same time.

3.3 Solving Black–Scholes PDE Symbolically

3.3.1 Heat PDE

3.3.1.1 Probabilistic Derivation of the Solution of Heat PDE

```
In[75]:= Clear["Global`*"]
```

The closest relative of the Black–Scholes PDE, and at the same time much simpler, is the Heat PDE. The Black–Scholes is backward (the terminal condition is imposed). Heat equation has two variants, forward and backward, with the forward one being more common.

The forward problem, i.e., the *initial* value problem, for the one-dimensional heat equation in the whole x-space is

$$\frac{\partial^2 u(\tau, x)}{\partial x^2} - \frac{\partial u(\tau, x)}{\partial \tau} = 0 \tag{3.3.1}$$

$$u(0, x) = \phi(x) \tag{3.3.2}$$

for any $-\infty < x < \infty$, $\tau > 0$, and where $\phi(x)$ is a given initial value. Notice there are no boundary conditions since there are no boundaries; the domain is infinite in both directions. The only condition is a very generous growth condition at infinity. We come back to this issue below when we discuss uniqueness. We shall show that the unique solution of the heat equation is given by

$$u(t, x) = \frac{1}{\sqrt{4 \pi t}} \int_{-\infty}^{\infty} e^{-\frac{(\xi - x)^2}{4 t}} \phi(\xi) \, d\xi \tag{3.3.3}$$

for any $t > 0$.

There are other ways to derive this formula (see [17]). Nevertheless, taking into account the spirit of this book, we shall use Normal Distribution and Itô calculus (to that end we shall assume smoothness of the initial data; in the general case when ϕ is not smooth the above formula still holds, and the proof can be obtained by the smooth approximation of the initial value, and by passing to the limit). Consider a random process such that

$$d X(t) = \sqrt{2}\, d B(t)$$
$$X(0) = x$$

i.e.,

$$X^{0,x}(t) = X^x(t) = x + \sqrt{2}\, B(t) \tag{3.3.4}$$

where $B(t)$ is a Brownian motion. Then, since $B(t) \sim N(0, \sqrt{t})$, for each t

$$X^x(t) \sim N(x, \sqrt{2t})$$

and therefore, since

```
In[76]:= << "Statistics`NormalDistribution`"

In[77]:= PDF[NormalDistribution[x, √2 t], ξ]
```

$$\text{Out[77]=} \quad \frac{e^{-\frac{(\xi-x)^2}{4t}}}{2\sqrt{\pi}\sqrt{t}}$$

we have

$$E_{0,x}\, \phi(X(t)) = E\, \phi(X^x(t)) = \frac{1}{\sqrt{4\pi t}} \int_{-\infty}^{\infty} e^{-\frac{(\xi-x)^2}{4t}}\, \phi(\xi)\, d\xi \tag{3.3.5}$$

which is the right-hand side of (3.3.3). On the other hand, using the Itô chain rule,

$$d\phi(X^x(t)) = \phi_x(X^x(t))\, d X^x(t) + \frac{1}{2}\, \phi_{x,x}(X^x(t))\, (d X^x(t))^2$$
$$= \sqrt{2}\, \phi_x(X^x(t))\, d B(t) + \phi_{x,x}(X^x(t))\, d t.$$

This implies, after integration in time,

$$\phi(X^x(t)) - \phi(X^x(s)) = \int_s^t \sqrt{2}\, \phi_x(X^x(w))\, d B(w) + \int_s^t \phi_{x,x}(X^x(w))\, d w.$$

Taking the expectation, and noticing that (using the change of variable $u = \xi - x$)

$$(E\, \phi(X^x(t)))_{x,x} = \left(\frac{1}{\sqrt{4\pi t}} \int_{-\infty}^{\infty} e^{-\frac{(\xi-x)^2}{4t}}\, \phi(\xi)\, d\xi \right)_{x,x}$$

$$= \left(\frac{1}{\sqrt{4 \pi t}} \int_{-\infty}^{\infty} e^{-\frac{u^2}{4t}} \phi(u + x) \, du \right)_{x,x}$$

$$= \frac{1}{\sqrt{4 \pi t}} \int_{-\infty}^{\infty} e^{-\frac{u^2}{4t}} \phi_{x,x}(u + x) \, du$$

$$= \frac{1}{\sqrt{4 \pi t}} \int_{-\infty}^{\infty} e^{-\frac{(\xi - x)^2}{4t}} \phi_{x,x}(\xi) \, d\xi = E \, \phi_{x,x}(X^x(t))$$

we derive

$$E \, \phi(X^x(t)) - E \, \phi(X^x(s)) = E \int_s^t \phi_{x,x}(X^x(w)) \, dw$$

$$= \int_s^t E \, \phi_{x,x}(X^x(w)) \, dw = \int_s^t (E \, \phi(X^x(w)))_{x,x} \, dw.$$

Dividing by $(t - s)$ and sending the difference to zero

$$\frac{\partial (E \, \phi(X^x(t)))}{\partial t} = \lim_{s \to t} \frac{E \, \phi(X^x(t)) - E \, \phi(X^x(s))}{t - s}$$

$$= \lim_{s \to t} \left(\frac{1}{t - s} \int_s^t (E \, \phi(X^x(w)))_{x,x} \, dw \right) = (E \, \phi(X^x(t)))_{x,x}$$

from which we conclude that

$$u(t, x) = E \, \phi(X^x(t)) \tag{3.3.6}$$

solves the above *initial value problem* for the forward Heat PDE. Taking into account (3.3.5), we conclude that (3.3.3) solves the equation (3.3.1) and (3.3.2), i.e., we have constructed an "explicit" solution of the heat equation. We also notice that

$$\begin{aligned} v(t, x) &= E_{t,x} \, \phi(X(T)) = E \, \phi(X^{t,x}(T)) \\ &= \frac{1}{\sqrt{4 \pi (T - t)}} \int_{-\infty}^{\infty} e^{-\frac{(\xi - x)^2}{4(T-t)}} \phi(\xi) \, d\xi \end{aligned} \tag{3.3.7}$$

solves the *terminal value problem* for the backward Heat PDE

$$\frac{\partial^2 u(\tau, x)}{\partial x^2} + \frac{\partial u(\tau, x)}{\partial \tau} = 0 \tag{3.3.8}$$

for $t < T$, and

$$u(T, x) = \phi(x). \tag{3.3.9}$$

Indeed, by the Itô chain rule, if u is the solution of the above PDE,

$$\begin{aligned} du(t, X(t)) &= u_t(t, X(t))\,dt + u_x(t, X(t))\,dX(t) + \frac{1}{2}u_{x,x}(t, X(t))\,(dX(t))^2 \\ &= u_t(t, X(t))\,dt + u_x(t, X(t))\sqrt{2}\,dB(t) + u_{x,x}(t, X(t))\,dt \\ &= u_x(t, X(t))\sqrt{2}\,dB(t) \end{aligned}$$

and therefore

$$u(T, X(T)) - u(t, X(t)) = \int_t^T u_x(w, X(w))\sqrt{2}\,dB(w).$$

Integrating in time, using the terminal condition, and taking the expectation, we get

$$\begin{aligned} &E_{t,x}\,\phi(X(T)) - u(t, x) = E_{t,x}(u(T, X(T)) - u(t, X(t))) \\ &= E_{t,x}\left(\int_t^T u_x(w, X(w))\sqrt{2}\,dB(w)\right) = 0. \end{aligned}$$

So, we have shown that if

$$dX(t) = \sqrt{2}\,dB(t)$$

then

$$u(t, x) = E\,\phi(X^{0,x}(t))$$

is a solution of the forward Heat equation, and that if v is a solution of the backward Heat equation with ϕ as a terminal condition, then

$$v(t, x) = E\,\phi(X^{t,x}(T)).$$

Actually, in both cases, those two functions respectively are the unique solutions.

3.3.1.2 Solving Heat PDE: Examples of Explicit Solutions

In[78]:=
```
Off[Unique::"usym"]
```

This section can be skipped if you are in a hurry. We have seen that (3.3.3) solves the forward heat equation. Here are some examples when the solution integral can be actually computed explicitly:

In[79]:=
```
1/(2 Sqrt[π t]) Integrate[e^(-(ξ-x)^2/(4 t)) e^ξ,
  {ξ, -∞, ∞}, Assumptions → t > 0]
```

Out[79]= e^{t+x}

In[80]:=
```
1/(2 Sqrt[π t]) Integrate[e^(-(ξ-x)^2/(4 t)) e^(-5 ξ),
  {ξ, -∞, ∞}, Assumptions → t > 0]
```

Out[80]= e^{25t-5x}

In[81]:= $\frac{1}{2\sqrt{\pi t}}$ `Integrate`$\left[e^{-\frac{(\xi-x)^2}{4t}}\ (\xi^{25}+\xi^4),\right.$
$\{\xi, -\infty, \infty\}$, `Assumptions` $\rightarrow$ `{Im[x] == 0, t > 0}`$\left.\right]$

Out[81]= $x^{25} + 21252000\,t^3\,x^{19} + 1817046000\,t^4\,x^{17} + 98847302400\,t^5\,x^{15} + 3459655584000\,t^6\,x^{13} + 77100895872000\,t^7\,x^{11} + 1060137318240000\,t^8\,x^9 + 8481098545920000\,t^9\,x^7 + 35620613892864000\,t^{10}\,x^5 + x^4 + 64764752532480000\,t^{11}\,x^3 + 32382376266240000\,t^{12}\,x + 12\,t^2\,(12650\,x^{21}+1) + 12\,t\,(50\,x^{23}+x^2)$

In[82]:= $\frac{1}{2\sqrt{\pi t}}$ `Integrate`$\left[e^{-\frac{(\xi-x)^2}{4t}}\ e^{-\xi^2},\right.$
$\{\xi, -\infty, \infty\}$, `Assumptions` $\rightarrow$ `{Im[x] == 0, t > 0}`$\left.\right]$

Out[82]= $$\frac{e^{-\frac{x^2}{4t+1}}}{\sqrt{4+\frac{1}{t}}\ \sqrt{t}}$$

In[83]:= `sol =` $\frac{1}{2\sqrt{\pi t}}$ `Integrate`$\left[e^{-\frac{(\xi-x)^2}{4t}}\ e^{\xi^2},\ \{\xi, -\infty, \infty\},\right.$
`Assumptions` $\rightarrow \left\{\text{Im[x] == 0},\ \frac{1}{4} > t > 0\right\}\left.\right]$

Out[83]= $$\frac{e^{\frac{x^2}{1-4t}}}{\sqrt{\frac{1}{t}-4}\ \sqrt{t}}$$

The last example is very interesting: finite time blow-up for the (linear) Heat PDE. Heuristic reasoning for the finite time blow-up is that the initial condition e^{x^2} is so hot at $x \to \pm\infty$ that the heat collapses towards the middle, into the blow-up. Also, it is interesting to note that the uniqueness proof given below fails in this example. Nevertheless, uniqueness does hold in an appropriate class of functions (see the proof of that fact that actually utilizes the above function in [17]). The blow-up occurs at the time $t = \frac{1}{4}$, for any $-\infty < x < \infty$ (infinite speed of propagation). Here is the plot of the solution, truncated at height 20:

In[84]:= `Plot3D[sol, {x, -.3, .3}, {t, 0.0005,` $\frac{1}{4}$ `- 0.0005},`
`PlotRange` $\rightarrow$ `{1, 20}, PlotPoints` $\rightarrow$ `60,`
`Mesh` $\rightarrow$ `False, AxesLabel` $\rightarrow$ `{x, t, ""}];`

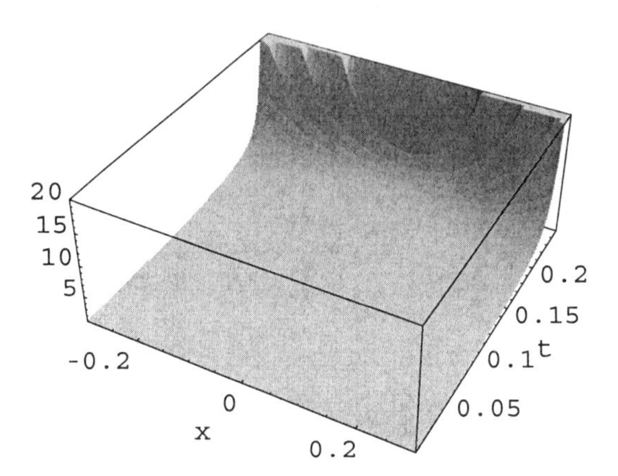

We make some comments about the properties of these solutions and their initial conditions: many of them grow rapidly when $x \to \pm\infty$. The growth rate of some of them is exponential such as in the case of $u(x, t) = e^{t+x}$. We remark that the exponential growth at infinity is going to be observed also in the case we are particularly interested in: the heat equation to be derived below from the Black–Scholes PDE. By the way the last example has even more rapid growth: the solution $u(x, t) = e^{\frac{x^2}{1-4t}} \Big/ \left(\sqrt{\frac{1}{t}} - 4\sqrt{t}\right)$ has a growth rate $e^{c x^2}$. This is going to be relevant when we discuss the question of uniqueness of the solution of the Heat PDE.

3.3.1.3 Uniqueness for Heat PDE with Exponential Growth at Infinity

We have computed some solutions of the Heat PDE above, and we shall compute two more, when we solve the Black–Scholes PDE below, but the natural question arises as to whether those that are (or to be) computed, are the right ones. If there is only one solution, or at least only one in a reasonably large class of functions, then those computed, being the only ones, are the correct ones. So we address the issue of uniqueness now.

Since the Heat equation is linear the uniqueness is addressed by considering the difference of two possible solutions $w = u_1 - u_2$. Since both u_1 and u_2 solve the same equation and the same initial condition, the difference w satisfies

$$\begin{aligned} &w_t(t, x) = w_{x,x}(t, x),\ t > 0 \\ &w(0, x) = 0. \end{aligned} \tag{3.3.10}$$

So, proving the uniqueness is equivalent to proving that w has to be identically equal to zero. Proving that claim will depend very much on the class of functions we consider. Ideally, we would like to allow all the functions. Since this is not possible, the considered class of functions should be the largest possible, or at least large enough. We have computed some explicit solutions above that can guide us in guessing what would be reasonable classes of functions to consider. If, on the other hand, the objective is to have a very simple argument, then we have to assume a lot. Indeed, define the "energy"

$$E(t) = \int_{-\infty}^{\infty} w(t, x)^2 \, dx. \tag{3.3.11}$$

Obviously, $E(0) = 0$, $E(t) \geq 0$ for $t > 0$. Also

$$E'(t) = 2 \int_{-\infty}^{\infty} w(t, x) \; w_t(t, x) \; dx$$

$$= 2 \int_{-\infty}^{\infty} w(t, x) \, w_{x,x}(t, x) \, dx = -2 \int_{-\infty}^{\infty} (w_x(t, x))^2 \; dx$$

integrating by parts, provided

$$w(t, x) \, w_x(t, x) \to 0 \tag{3.3.12}$$

when $x \to \pm\infty$. Assuming (3.3.12), i.e., restricting the uniqueness discussion on the class of functions for which (3.3.12) holds, we conclude that $E'(t) \leq 0$. Together with previously observed properties of $E(t)$, it is easy to conclude that $E(t) = 0$ for every $t \geq 0$, and therefore w is identically equal to zero.

On the other hand, many of the above computed solutions do not satisfy the above condition; think for example of $u(x, t) = e^{t+x}$. More importantly, the equation we are particularly interested in, the Black–Scholes PDE, will yield a Heat PDE whose solution will exhibit similar exponential behavior. Therefore we need an additional argument.

Consider a class of functions such that, for some $\mu > 0$,

$$\int_{-\infty}^{\infty} e^{-\mu |x|} \; (w(t, x)^2 + w_x(t, x)^2) \, dx < \infty \tag{3.3.13}$$

for any $t > 0$. Consider for some $\lambda > 0$, and any $t \geq 0$, the "discounted energy" function

$$E(t) = e^{-\lambda t} \int_{-\infty}^{\infty} e^{-\mu |x|} \, w(t, x)^2 \, dx \geq 0.$$

Then

$$\frac{dE(t)}{dt} = \frac{d}{dt} \left(e^{-\lambda t} \int_{-\infty}^{\infty} e^{-\mu |x|} \, w(t, x)^2 \, dx \right)$$

$$= e^{-\lambda t} \left(-\lambda \int_{-\infty}^{\infty} e^{-\mu |x|} \, w(t, x)^2 \, dx + 2 \int_{-\infty}^{\infty} e^{-\mu |x|} \, w(t, x) \, w_t(t, x) \, dx \right)$$

$$= e^{-\lambda t} \left(-\lambda \int_{-\infty}^{\infty} e^{-\mu |x|} \, w(t, x)^2 \, dx + 2 \int_{-\infty}^{\infty} e^{-\mu |x|} \, w(t, x) \, w_{x,x}(t, x) \, dx \right)$$

$$= e^{-\lambda t}\left(-\lambda \int_{-\infty}^{\infty} e^{-\mu|x|}\, w(t, x)^2\, dx + 2 \int_{-\infty}^{\infty} (e^{-\mu|x|})_x\, w(t, x)\, w_x(t, x)\, dx\right.$$
$$\left. - 2 \int_{-\infty}^{\infty} e^{-\mu|x|}\, w_x(t, x)^2\, dx\right) \leq 0$$

if λ is large enough. Indeed,

$$\left| \int_{-\infty}^{\infty} (e^{-\mu|x|})_x\, w(t, x)\, w_x(t, x)\, dx \right| \leq \mu \int_{-\infty}^{\infty} e^{-\mu|x|} \left| w(t, x)\, w_x(t, x) \right| dx$$
$$\leq \mu \left(\int_{-\infty}^{\infty} e^{-\mu|x|}\, w(t, x)^2\, dx \right)^{1/2} \left(\int_{-\infty}^{\infty} e^{-\mu|x|}\, w_x(t, x)^2\, dx \right)^{1/2}.$$

So, $E(t) \geq 0$, $E'(t) \leq 0$, $E(0) = 0$. We conclude $E(t) = 0$, for all t's, and therefore $w(t, x) = 0$.

3.3.1.4 Monte–Carlo Simulation Solutions of PDEs

```
In[85]:= Clear["Global`*"]
```

This section is optional. Recall solutions of the forward and backward Heat PDE

$$u(t, x) = E\, \phi(X^{0,x}(t))$$

$$v(t, x) = E\, \phi(X^{t,x}(T))$$

respectively. Can we construct these functions experimentally?

Somewhat surprisingly, there is no symmetry between u and v. The reason is that the same process appears in both formulas; moreover $X(t)$ is a stochastic process, and there is no time reversibility. The function u seems *somewhat* easier to construct than the function v. Indeed, $X^{0,x}(t) = X^x(t) = x + \sqrt{2}\, B(t)$ is the process starting at time zero at the position x, and the single trajectory can be traced over different times while for $X^{t,x}(T)$ we need to construct processes for different starting times t.

So, let's construct u. Consider, for example, the initial condition

```
In[86]:= ϕ[x_] := e^x
```

The exact solution is available, and as computed already in the previous section, it is equal to

```
In[87]:= u[t_, x_] := e^(t+x)
```

Now let us use the Monte–Carlo simulation method, and compare it with the exact solution. We shall compute $u(t, x) = E\, \phi(X^{0,x}(t)) \approx \mathrm{Av}[\phi(X^{0,x}(t))]$. Let the time interval be $(0, T)$, let the number of t-subinterval be denoted by K, and let the number of randomizations be L:

```
In[88]:= T = 2; K = 10; L = 3000;
```

We shall need SDESolver:

```
In[89]:= << "CFMLab`ItoSDEs`"
```

Compute L realizations of $\sqrt{2}\,B(t)$

```
In[90]:= SolList =
           Table[SDESolver[0 &, √2 &, 0, 0, T, K], {L}];
```

and for any x, average all realizations of $\phi(X^{0,x}(t)) = \phi(x + \sqrt{2}\,B(t))$. For the sake of efficiency, it is important to keep the generation of $\sqrt{2}\,B(t)$ and averaging separate if more than one x is considered:

```
In[91]:= Function[x,
             Sol = Interpolation[Transpose[{Transpose[
                   SolList[[1]]][[1]], Av /@ φ /@ Transpose[
                    Transpose[Transpose[# + {0, x} & /@ # & /@
                        SolList, {3, 2, 1}][[2]]]]}],
               InterpolationOrder → 1]; Plot[{u[t, x],
               Sol[t]}, {t, 0, T},
              PlotRange → {0, Automatic},
              PlotStyle → {AbsoluteDashing[{5, 0}],
                AbsoluteDashing[{5, 5}]}]] /@ {-100, 0};
```

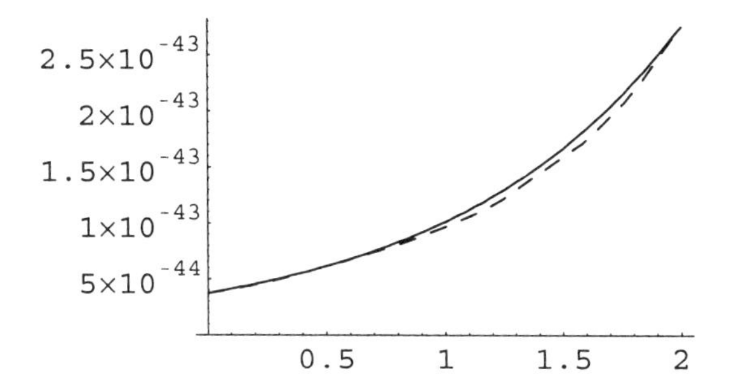

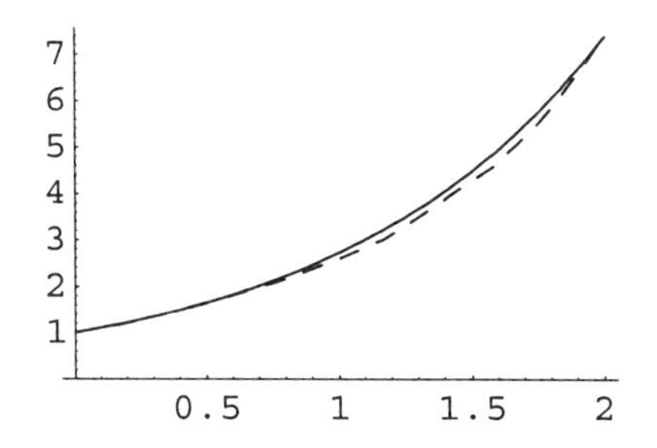

The good property of this method is that it can compute the solution of the PDE for a single x, and, which implies in particular that *no truncation of space is needed for problems in infinite domains.* Sometimes this can be very important since it is usually difficult to find a natural and accurate boundary condition for the truncated domain. In a way, the trajectories of the underlying stochastic process spread around x, and determine the size of the "truncation of the infinite space". A possible application is to use this method for determining boundary value for the truncated problem, and then to use more efficient, numerical PDE methods to fill in the solution in between.

3.3.2 Reduction of Black–Scholes to Heat PDE

In[92]:= Clear["Global`*"]

Finally, we shall derive now an explicit solution of the Black–Scholes PDE. Ideally we would like to find an explicit solution of the equation (3.2.12). Unfortunately this is not possible—the stock price dependency of the equation cannot be handled in an explicit/symbolic fashion. We first consider a constant coefficient case (later in this chapter we shall extend the solution to the time-dependent case; and much later, in Chapter 5, we shall extend the solution, using numerical methods, to the stock price dependent case, as well):

In[93]:= `BSPDE =` ∂_t `V[t, S] +` $\frac{1}{2}$ `S`^{2} $\partial_{\{S,2\}}$ `V[t, S]` σ^2 `+`
`a S` ∂_S `V[t, S] - b V[t, S] == 0`

Out[93]= $\frac{1}{2} S^2 V^{(0,2)}(t, S)\,\sigma^2 - b\, V(t, S) + a\, S\, V^{(0,1)}(t, S) + V^{(1,0)}(t, S)$
$== 0$

(obviously, for $a = r - D_0$, $b = r$ we recover the Black–Scholes PDE) together with the terminal condition

In[94]:= `TC = V[T, S] == Max[0, S - k];`

in the case of calls, and

In[95]:= `TCput = V[T, S] == Max[0, k - S];`

in the case of puts.

The strategy is to apply a sequence of transformations that will transform the above PDE into the Heat PDE. Of course, such a sequence of transformations will deform the terminal condition into an appropriate initial condition for the Heat PDE. Once the Heat PDE is solved for that transformed initial condition, one has to go back to recover the solution of the original BSPDE.

The first set of substitutions is

```
In[96]:= x[S_] := Log[S/k];
         τ[t_] := 1/2 (T - t) σ^2;
```

together with

```
In[98]:= Sub1[t_, S_] := k v[τ[t], x[S]]
```

Under these substitutions, the above BSPDE becomes

```
In[99]:= BSPDE2 = Simplify[
           BSPDE /. V → Sub1 /. {1/2 (T - t) σ^2 → τ, Log[S/k] → x}]
```

Out[99]= $\frac{1}{2}k\left(\left(v^{(1,0)}(\tau,x)-v^{(0,2)}(\tau,x)\right)\sigma^2+2\,b\,v(\tau,x)+\left(\sigma^2-2\,a\right)v^{(0,1)}(\tau,x)\right)==0$

What happens to the terminal condition? It becomes an initial condition:

```
In[100]:= TC2 = TC /. V → Sub1 /.
            {1/2 (T - t) σ^2 → τ, Log[S/k] → x, S → k e^x}
```

Out[100]= $k\,v(0,x)==\text{Max}(0,\,e^x\,k-k)$

in the case of calls, and

```
In[101]:= Off[General::"spell1"];
          TC2put = TCput /. V → Sub1 /.
            {1/2 (T - t) σ^2 → τ, Log[S/k] → x, S → k e^x}
```

Out[101]= $k\,v(0,x)==\text{Max}(0,\,k-e^x\,k)$

in the case of puts. The BSPDE2 is simpler than the BSPDE for the obvious reason that it is a constant coefficient PDE. Furthermore, looking closely, the Heat PDE starts to appear. Nevertheless, to have a true Heat PDE, we need to dispose of some of the terms in BSPDE2. To that end, define a new function $u(\tau, x)$ via formula

```
In[102]:= v[τ_, x_] := u[τ, x] e^(x α+β τ)
```

with α and β yet to be determined. This yields

```
In[103]:= BSPDE3 = FullSimplify[BSPDE2]
```

Out[103]= $e^{x\,\alpha+\beta\,\tau}\,k\left(\left(u^{(1,0)}(\tau,x)-u^{(0,2)}(\tau,x)\right)\sigma^2+\left(\left(-\alpha^2+\alpha+\beta\right)\sigma^2+2\,b-2\,a\,\alpha\right)u(\tau,x)+\left(-2\,\alpha\,\sigma^2+\sigma^2-2\,a\right)u^{(0,1)}(\tau,x)\right)==0$

We shall need the property of the Max function

```
In[104]:= MaxProperty = {a_Max[0, b_] → Max[0, a b] /; a ≥ 0};
```

It implies the following form of the initial condition

```
In[105]:= TC3 = Reduce[TC2, u[0, x]][[1, 1]] /. MaxProperty
```

Out[105]= $u(0, x) == \left(\text{Max}\left(0, \frac{e^{-x\alpha}(e^x k - k)}{k}\right) /; \frac{e^{-x\alpha}}{k} \geq 0\right)$

for calls, and

```
In[106]:= TC3put =
            Reduce[TC2put, u[0, x]][[1, 1]] /. MaxProperty
```

Out[106]= $u(0, x) == \left(\text{Max}\left(0, \frac{e^{-x\alpha}(k - e^x k)}{k}\right) /; \frac{e^{-x\alpha}}{k} \geq 0\right)$

for puts. Can we find α and β so that BSPDE3 is reduced to the Heat PDE? BSPDE3 will be reduced to the Heat PDE provided α and β are found such that

```
In[107]:= Coefficient[BSPDE3[[1]], u[τ, x]]
```

Out[107]= $e^{x\alpha+\beta\tau} k((-\alpha^2 + \alpha + \beta)\sigma^2 + 2b - 2a\alpha)$

and

```
In[108]:= Coefficient[BSPDE3[[1]], ∂x u[τ, x]]
```

Out[108]= $e^{x\alpha+\beta\tau} k(-2\alpha\sigma^2 + \sigma^2 - 2a)$

are both equal to zero. Therefore, solve the system of equations

```
In[109]:= αβSub = Simplify[
             Solve[{Coefficient[BSPDE3[[1]], ∂x u[τ, x]][[3]] ==
                 0, Coefficient[BSPDE3[[1]], u[τ, x]][[3]] ==
                 0}, {α, β}][[1]]]
```

Out[109]= $\left\{\beta \to -\frac{a^2}{\sigma^4} + \frac{a}{\sigma^2} - \frac{2b}{\sigma^2} - \frac{1}{4}, \alpha \to \frac{1}{2} - \frac{a}{\sigma^2}\right\}$

Choosing such α and β, we get the transformed PDE

```
In[110]:= BSPDE4 = Simplify[BSPDE3 /. αβSub]
```

Out[110]= $e^{x\left(\frac{1}{2}-\frac{a}{\sigma^2}\right)+\left(-\frac{a^2}{\sigma^4}+\frac{a}{\sigma^2}-\frac{2b}{\sigma^2}-\frac{1}{4}\right)\tau} k\sigma^2 (u^{(0,2)}(\tau, x) - u^{(1,0)}(\tau, x)) == 0$

and transformed initial conditions:

```
In[111]:= TC4 = TC3[[1]] == Simplify[TC3[[2, 1]]] /. αβSub
```

Out[111]= $u(0, x) == \mathrm{Max}\left(0, e^{-x\left(\frac{1}{2}-\frac{a}{\sigma^2}\right)}(-1+e^x)\right)$

for calls, and

```
In[112]:= TC4put =
           TC3put⟦1⟧ == Simplify[TC3put⟦2, 1⟧] /. αβSub
```

Out[112]= $u(0, x) == \mathrm{Max}\left(0, -e^{-x\left(\frac{1}{2}-\frac{a}{\sigma^2}\right)}(-1+e^x)\right)$

for puts. It is interesting to take a closer look at the initial condition. Define

```
In[113]:= tc[a_, σ_, x_] = TC4⟦2⟧
```

Out[113]= $\mathrm{Max}\left(0, e^{-x\left(\frac{1}{2}-\frac{a}{\sigma^2}\right)}(-1+e^x)\right)$

and

```
In[114]:= tcput[a_, σ_, x_] = TC4put⟦2⟧
```

Out[114]= $\mathrm{Max}\left(0, -e^{-x\left(\frac{1}{2}-\frac{a}{\sigma^2}\right)}(-1+e^x)\right)$

What is interesting to notice is that both initial conditions, for calls as well as puts, have *exponential growth*:

```
In[115]:= Plot[tc[.05, .5, x], {x, -3, 3}, PlotStyle →
              {AbsoluteThickness[3], RGBColor[1, 0, 0]}];
```

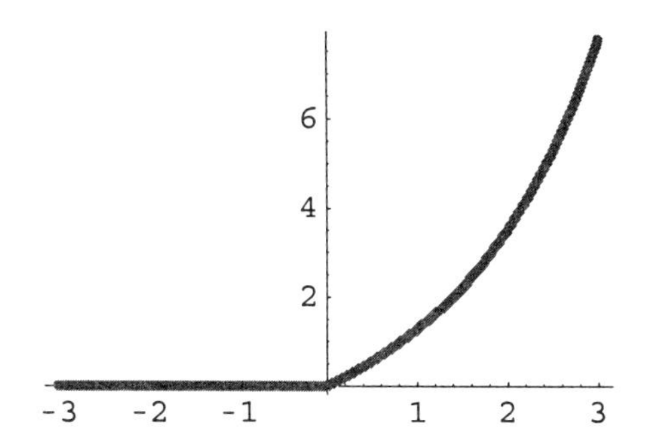


```
In[116]:= Plot[tcput[.05, .5, x], {x, -10, 10}, PlotStyle →
              {AbsoluteThickness[3], RGBColor[0, 0, 1]}];
```

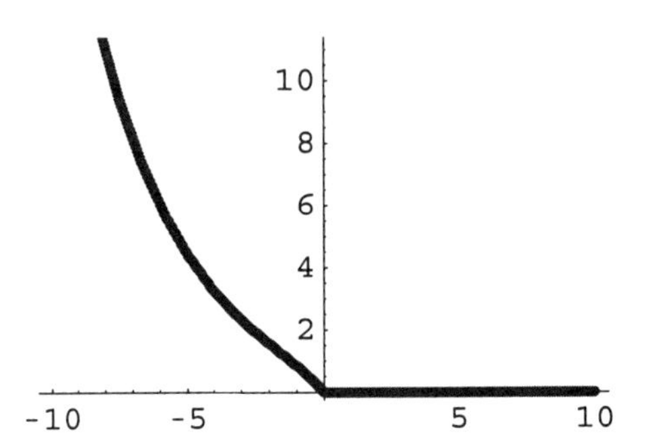

Therefore, the solvability and uniqueness of the Heat PDE has to be understood, as it has been above, in the class of such functions.

The BSPDE4 is obviously equivalent to

In[117]:= BSPDE5 = BSPDE4 [[1, 4]] == 0

Out[117]= $u^{(0,2)}(\tau, x) - u^{(1,0)}(\tau, x) == 0$

which is nothing but a Heat PDE.

3.3.3 Solution of Heat PDE; Back to Black–Scholes PDE

As derived above, the forward Heat PDE has the unique solution given by (3.3.3). Recall the initial condition for calls

In[118]:= TC4

Out[118]= $u(0, x) == \text{Max}\left(0, e^{-x\left(\frac{1}{2}-\frac{a}{\sigma^2}\right)}(-1 + e^x)\right)$

and for puts

In[119]:= TC4put

Out[119]= $u(0, x) == \text{Max}\left(0, -e^{-x\left(\frac{1}{2}-\frac{a}{\sigma^2}\right)}(-1 + e^x)\right)$

and extract

In[120]:= ψ[x_] = TC4[[2, 2]]

Out[120]= $e^{-x\left(\frac{1}{2}-\frac{a}{\sigma^2}\right)}(-1 + e^x)$

and

In[121]:= ψput[x_] = TC4put[[2, 2]]

Out[121]= $-e^{-x\left(\frac{1}{2}-\frac{a}{\sigma^2}\right)}(-1 + e^x)$

Then compute corresponding solutions of the Heat PDE (notice the truncation of the region of integration)

In[122]:= BSSolu[τ_, x_] = $\frac{1}{2\sqrt{\pi\tau}}$ Integrate[$e^{-\frac{(\xi-x)^2}{4\tau}}$ ψ[ξ], {ξ, 0, ∞}, Assumptions → τ > 0]

Out[122]= $$\frac{1}{2} e^{-\frac{(\sigma^2-2a)(2x\sigma^2-\tau\sigma^2+2a\tau)}{4\sigma^4}} \left(-\text{erf}\left(\frac{x\sigma^2-\tau\sigma^2+2a\tau}{2\sigma^2\sqrt{\tau}}\right) + e^{x+\frac{2a\tau}{\sigma^2}} + e^{x+\frac{2a\tau}{\sigma^2}}\,\text{erf}\left(\frac{x\sigma^2+\tau\sigma^2+2a\tau}{2\sigma^2\sqrt{\tau}}\right) - 1\right)$$

which looks like, for some fixed time (notice again the exponential growth, recall the issue of uniqueness, and notice that the parameter b does not appear here):

```
In[123]:= Plot[BSSolu[τ, x] /. {a → .05, σ → .5, τ → .5},
    {x, -2, 2}, PlotStyle →
     {AbsoluteThickness[3], RGBColor[1, 0, 0]}];
```

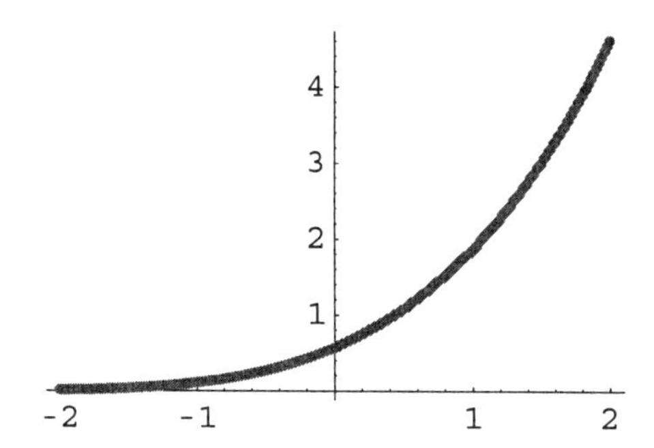

and in the case of puts

In[124]:= BSSoluPut[τ_, x_] = $\frac{1}{2\sqrt{\pi\tau}}$ Integrate[$e^{-\frac{(\xi-x)^2}{4\tau}}$ ψput[ξ], {ξ, -∞, 0}, Assumptions → τ > 0]

Out[124]= $$-\frac{1}{2} e^{-\frac{(\sigma^2-2a)(2x\sigma^2-\tau\sigma^2+2a\tau)}{4\sigma^4}} \left(\text{erf}\left(\frac{x\sigma^2-\tau\sigma^2+2a\tau}{2\sigma^2\sqrt{\tau}}\right) + e^{x+\frac{2a\tau}{\sigma^2}} - e^{x+\frac{2a\tau}{\sigma^2}}\,\text{erf}\left(\frac{x\sigma^2+\tau\sigma^2+2a\tau}{2\sigma^2\sqrt{\tau}}\right) - 1\right)$$

which looks like this:

```
In[125]:= Plot[BSSoluPut[τ, x] /. {a → .05, σ → .5, τ → .5},
            {x, -2, 2}, PlotStyle →
             {AbsoluteThickness[3], RGBColor[0, 0, 1]}];
```

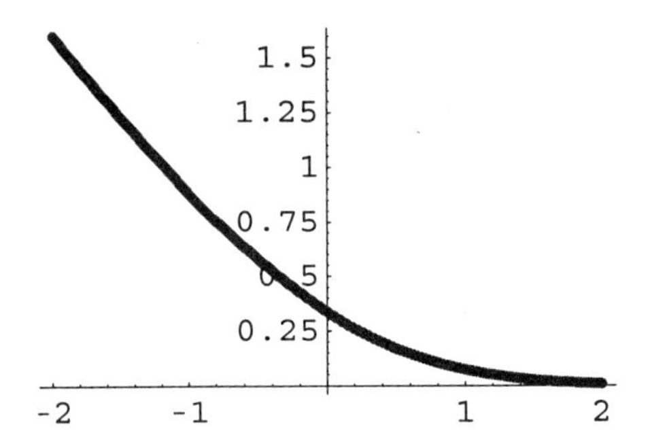

Once the transformed Heat PDE is solved, we can trace our steps back to the original problem. Define,

```
In[126]:= v[τ_, x_] = BSSolu[τ, x] e^(x α+β τ) /. αβSub
```

Out[126]=
$$\frac{1}{2}\, e^{x\left(\frac{1}{2}-\frac{a}{\sigma^2}\right)+\left(-\frac{a^2}{\sigma^4}+\frac{a}{\sigma^2}-\frac{2b}{\sigma^2}-\frac{1}{4}\right)\tau-\frac{(\sigma^2-2a)(2x\sigma^2-\tau\sigma^2+2a\tau)}{4\sigma^4}} \left(-\operatorname{erf}\left(\frac{x\sigma^2-\tau\sigma^2+2a\tau}{2\sigma^2\sqrt{\tau}}\right)+e^{x+\frac{2a\tau}{\sigma^2}} + e^{x+\frac{2a\tau}{\sigma^2}}\operatorname{erf}\left(\frac{x\sigma^2+\tau\sigma^2+2a\tau}{2\sigma^2\sqrt{\tau}}\right)-1\right)$$

and, finally, the solution of the BSPDE in the case of calls, i.e., the call Black–Scholes formula, is

```
In[127]:= VC[t_, S_, T_, k_, a_, b_, σ_] =
            FullSimplify[k v[τ[t], x[S]]]
```

Out[127]=
$$\frac{1}{2}\, e^{b(t-T)}\left(e^{a(T-t)}\, S\left(\operatorname{erf}\left(\frac{2\log\left(\frac{S}{k}\right)-(t-T)(\sigma^2+2a)}{2\sqrt{2}\sqrt{(T-t)\sigma^2}}\right)+1\right) + k\left(\operatorname{erfc}\left(\frac{2\log\left(\frac{S}{k}\right)-(t-T)(2a-\sigma^2)}{2\sqrt{2}\sqrt{(T-t)\sigma^2}}\right)-2\right)\right)$$

The important special case is when no dividends are paid, i.e., when $D_0 = 0$, i.e., when $a = b = r$ the interest rate. In that case we get the simplest form of the **call Black–Scholes formula:**

```
In[128]:= VC[t_, S_, T_, k_, r_, σ_] =
            FullSimplify[VC[t, S, T, k, r, r, σ]]
```

Out[128]=
$$\frac{1}{2}\left(\operatorname{erf}\left(\frac{2\log\left(\frac{S}{k}\right)-(t-T)(\sigma^2+2r)}{2\sqrt{2}\sqrt{(T-t)\sigma^2}}\right)S+S+e^{r(t-T)}k\left(\operatorname{erfc}\left(\frac{2\log\left(\frac{S}{k}\right)-(t-T)(2r-\sigma^2)}{2\sqrt{2}\sqrt{(T-t)\sigma^2}}\right)-2\right)\right)$$

which, for some fixed time, looks as it should:

```
In[129]:= Plot[VC[0, S, .01, 120, .05, .5],
            {S, 80, 170}, PlotRange → All, PlotStyle →
             {AbsoluteThickness[3], RGBColor[1, 0, 0]}];
```

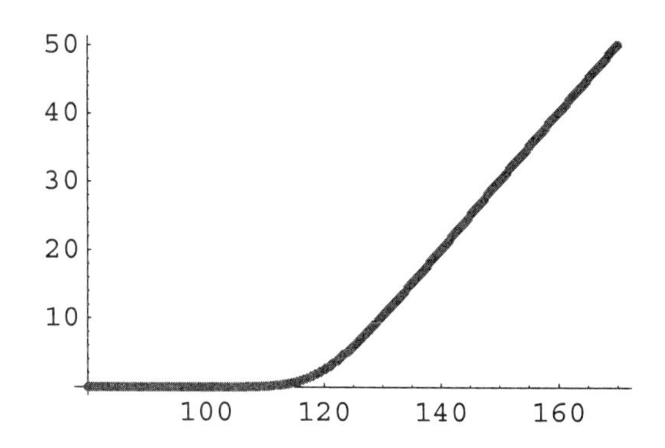

In the case of puts, on the other hand, we have

```
In[130]:= v[τ_, x_] = BSSoluPut[τ, x] e^(x α+β τ) /. αβSub
```

Out[130]=
$$-\frac{1}{2}e^{x\left(\frac{1}{2}-\frac{a}{\sigma^2}\right)+\left(-\frac{a^2}{\sigma^4}+\frac{a}{\sigma^2}-\frac{2b}{\sigma^2}-\frac{1}{4}\right)\tau-\frac{(\sigma^2-2a)(2x\sigma^2-\tau\sigma^2+2a\tau)}{4\sigma^4}}\left(\operatorname{erf}\left(\frac{x\sigma^2-\tau\sigma^2+2a\tau}{2\sigma^2\sqrt{\tau}}\right)+e^{x+\frac{2a\tau}{\sigma^2}}-e^{x+\frac{2a\tau}{\sigma^2}}\operatorname{erf}\left(\frac{x\sigma^2+\tau\sigma^2+2a\tau}{2\sigma^2\sqrt{\tau}}\right)-1\right)$$

```
In[131]:= VP[t_, S_, T_, k_, a_, b_, σ_] =
            FullSimplify[k v[τ[t], x[S]]]
```

Out[131]= $\frac{1}{2}\, e^{b\,(t-T)} \left(k\, \text{erfc}\left(\frac{2\log\left(\frac{S}{k}\right) - (t-T)\,(2\,a - \sigma^2)}{2\sqrt{2}\,\sqrt{(T-t)\,\sigma^2}} \right) - e^{a\,(T-t)}\, S\, \text{erfc}\left(\frac{2\log\left(\frac{S}{k}\right) - (t-T)\,(\sigma^2 + 2\,a)}{2\sqrt{2}\,\sqrt{(T-t)\,\sigma^2}} \right) \right)$

which, again, in the simplest case when $a = b = r$ becomes the **put Black–Scholes formula**:

```
In[132]:= VP[t_, S_, T_, k_, r_, σ_] =
            FullSimplify[VP[t, S, T, k, r, r, σ]]
```

Out[132]= $\frac{1}{2} \left(e^{r\,(t-T)}\, k\, \text{erfc}\left(\frac{2\log\left(\frac{S}{k}\right) - (t-T)\,(2\,r - \sigma^2)}{2\sqrt{2}\,\sqrt{(T-t)\,\sigma^2}} \right) - S\, \text{erfc}\left(\frac{2\log\left(\frac{S}{k}\right) - (t-T)\,(\sigma^2 + 2\,r)}{2\sqrt{2}\,\sqrt{(T-t)\,\sigma^2}} \right) \right)$

and looks like this:

```
In[133]:= Plot[VP[0, S, .01, 120, .05, .5],
             {S, 80, 170}, PlotRange → All, PlotStyle →
              {AbsoluteThickness[3], RGBColor[0, 0, 1]}];
```

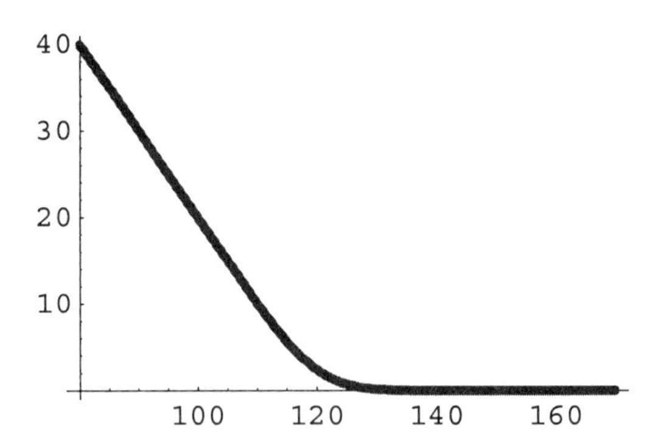

This concludes the derivation of the Black–Scholes formula. We store these four formulas in the package CFMLab`BlackScholes` for future reference.

We notice in the above formulas function

$$\text{erf}(z) = \frac{2}{\sqrt{\pi}} \int_0^z e^{-t^2}\, \text{dt}$$

which is a built-in *Mathematica*® function, called the error function, as well as $\mathrm{erfc}(z) = 1 - \mathrm{erf}(z)$, called the complementary error function. The functions look like

```
In[134]:= Plot[{Erf[x], Erfc[x]},
            {x, -3, 3}, Ticks → {{-2, 2}, Automatic},
            PlotStyle → {Dashing[{0}], Dashing[{0.01}]}];
```

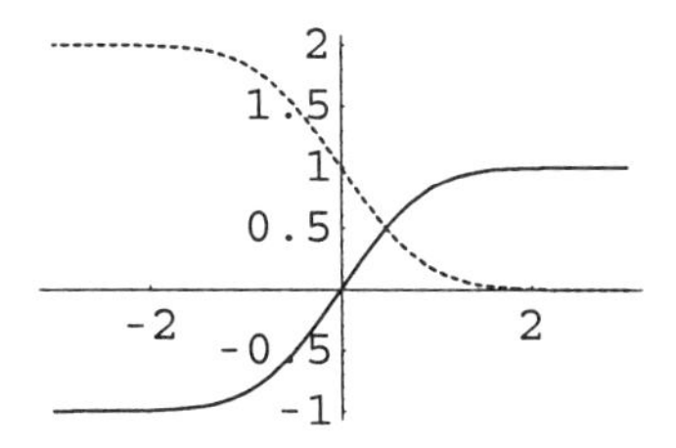

Depending on the point of view, there are differences among many arguments of the above two functions. Although not always (see Chapters 5 and 6 on implied volatility), most of the time, the *variables* will be current date (time) and current underlying stock price (i.e., t and S), while the rest of the arguments are going to be considered as parameters (T, k, a, b, σ). So, define call and put values as:

```
In[135]:= CV[t_, S_] := VC[t, S, T, k, a, b, σ]
```

```
In[136]:= PV[t_, S_] := VP[t, S, T, k, a, b, σ]
```

Let us *verify*, by means of substitution, that the above derivation is correct, i.e., that the two functions that were derived indeed *do* satisfy the BSPDE:

```
In[137]:= BSPDE /. V → CV
```

Out[137]=
$$\frac{1}{4} e^{b(t-T)} S^2 \left(k \left(\left(\frac{e^{-\frac{\left(2\log\left(\frac{S}{k}\right)-(t-T)(2a-\sigma^2)\right)^2}{8(T-t)\sigma^2}} \left(2\log\left(\frac{S}{k}\right) - (t-T)(2a-\sigma^2)\right)}{\sqrt{2\pi}\, S^2 (T-t)\sigma^2 \sqrt{(T-t)\sigma^2}} \right. \right. \right.$$

$$\left. + \frac{e^{-\frac{\left(2\log\left(\frac{S}{k}\right)-(t-T)(2a-\sigma^2)\right)^2}{8(T-t)\sigma^2}} \sqrt{\frac{2}{\pi}}}{S^2 \sqrt{(T-t)\sigma^2}} \right)$$

$$+ e^{a(T-t)} S \left(- \left(e^{-\frac{\left(2\log\left(\frac{S}{k}\right)-(t-T)(\sigma^2+2a)\right)^2}{8(T-t)\sigma^2}} \left(2\log\left(\frac{S}{k}\right) - (t-T)(\sigma^2+2a)\right) \right) \right/$$

$$\left(\sqrt{2\pi}\, S^2 (T-t)\sigma^2 \sqrt{(T-t)\sigma^2}\right) - \frac{e^{-\frac{\left(2\log\left(\frac{S}{k}\right)-(t-T)\left(\sigma^2+2a\right)\right)^2}{8(T-t)\sigma^2}} \sqrt{\frac{2}{\pi}}}{S^2\sqrt{(T-t)\sigma^2}}\Bigg)$$

$$+ \frac{2\, e^{a(T-t)-\frac{\left(2\log\left(\frac{S}{k}\right)-(t-T)\left(\sigma^2+2a\right)\right)^2}{8(T-t)\sigma^2}} \sqrt{\frac{2}{\pi}}}{S\sqrt{(T-t)\sigma^2}}\Bigg)\sigma^2$$

$$+\frac{1}{2}\, a\, e^{b(t-T)}\, S\left(-\frac{e^{-\frac{\left(2\log\left(\frac{S}{k}\right)-(t-T)\left(2a-\sigma^2\right)\right)^2}{8(T-t)\sigma^2}} \sqrt{\frac{2}{\pi}}\, k}{S\sqrt{(T-t)\sigma^2}} + e^{a(T-t)}\left(\operatorname{erf}\left(\frac{2\log\left(\frac{S}{k}\right)-(t-T)\left(\sigma^2+2a\right)}{2\sqrt{2}\sqrt{(T-t)\sigma^2}}\right)+1\right)\right.$$

$$\left. + \frac{e^{a(T-t)-\frac{\left(2\log\left(\frac{S}{k}\right)-(t-T)\left(\sigma^2+2a\right)\right)^2}{8(T-t)\sigma^2}} \sqrt{\frac{2}{\pi}}}{\sqrt{(T-t)\sigma^2}}\right) + \frac{1}{2}\, e^{b(t-T)}\Bigg(-a\, e^{a(T-t)}\, S$$

$$\left(\operatorname{erf}\left(\frac{2\log\left(\frac{S}{k}\right)-(t-T)\left(\sigma^2+2a\right)}{2\sqrt{2}\sqrt{(T-t)\sigma^2}}\right)+1\right) - \frac{1}{\sqrt{\pi}}\Bigg(2\, e^{-\frac{\left(2\log\left(\frac{S}{k}\right)-(t-T)\left(2a-\sigma^2\right)\right)^2}{8(T-t)\sigma^2}}\, k$$

$$\left(\frac{\left(2\log\left(\frac{S}{k}\right)-(t-T)\left(2a-\sigma^2\right)\right)\sigma^2}{4\sqrt{2}\left((T-t)\sigma^2\right)^{3/2}} + \frac{\sigma^2-2a}{2\sqrt{2}\sqrt{(T-t)\sigma^2}}\right)\Bigg) + \frac{1}{\sqrt{\pi}}$$

$$\Bigg(2\, e^{a(T-t)-\frac{\left(2\log\left(\frac{S}{k}\right)-(t-T)\left(\sigma^2+2a\right)\right)^2}{8(T-t)\sigma^2}}\, S\left(\frac{\left(2\log\left(\frac{S}{k}\right)-(t-T)\left(\sigma^2+2a\right)\right)\sigma^2}{4\sqrt{2}\left((T-t)\sigma^2\right)^{3/2}}\right.$$

$$\left.\left. + \frac{-\sigma^2-2a}{2\sqrt{2}\sqrt{(T-t)\sigma^2}}\right)\Bigg)\Bigg) == 0$$

Mathematica® needs to be told to simplify the answer:

In[138]:= `Simplify[%]`

Out[138]= True

we see that CV(t, S) *is* the solution of the BSPDE. Let us check the PV(t, S):

In[139]:= `Simplify[BSPDE /. V → PV]`

Out[139]= True

The put value is a solution as well. It helps understanding to visualize these functions, especially plotted against the corresponding payoff profiles. For example,

```
In[140]:= With[{k = 40, T = 6/12},
    (Plot[{VC[#1, S, T, k, .07, .5], VP[#1, S, T, k,
        .07, .5], Max[0, S - k], Max[0, k - S]},
      {S, 0, 2 k}, PlotRange → {0, k}, AxesLabel →
       {"S", ""}, PlotStyle → {RGBColor[1, 0, 0],
        RGBColor[0, 0, 1]}] &) /@ Range[0, T, T/40]];
```

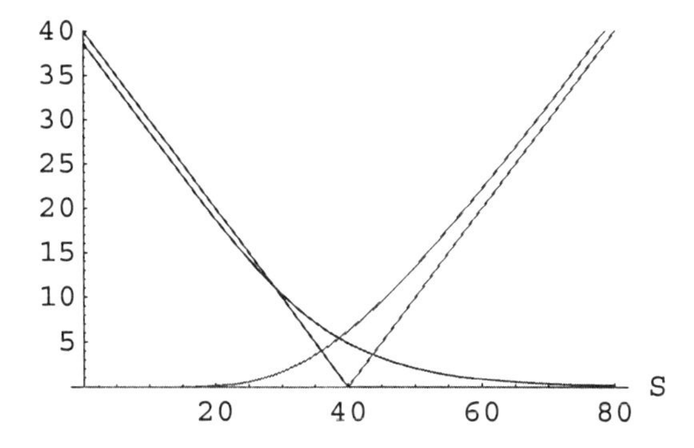

(Double-clicking on the above picture (in its electronic form) one can start the animation to see the dynamics as t varies between $t = 0$ and $t = T$.) Here we emphasize again that due to the degeneracy of the PDE no boundary condition is *imposed*, nor is it possible to impose, at $S = 0$ (also at $S = \infty$ but that is less surprising). It happens though, as a consequence of all of the other conditions, that

$$\lim_{S \to 0} \mathrm{VC}(t, S, T, k, r, \sigma) = 0$$

$$\lim_{S \to 0} \mathrm{VP}(t, S, T, k, r, \sigma) = e^{r(t-T)}\, k.$$

Also, notice that in the case of a call, the solution is *above* the payoff profile, while in the case of a put, the solution crosses *under* the payoff profile for stock prices that are significantly lower than the strike price. This means that in the case of *European* options the value of a put may be lower than the difference of the strike and stock price, i.e., lower than the option's payoff. In order to have a similar phenomenon in the case of calls, one has to consider dividends. Indeed,

```
In[141]:= With[{k = 40, T = 6/12},
    (Plot[{VC[#1, S, T, k, 0, .07, .5], VP[#1, S, T,
        k, 0, .07, .5], Max[0, S - k], Max[0, k - S]},
```

```
{S, 0, 2 k}, PlotRange → {0, k}, AxesLabel →
 {"S", ""}, PlotStyle → {RGBColor[1, 0, 0],
  RGBColor[0, 0, 1]}] &) /@Range[0, T, T/40]];
```

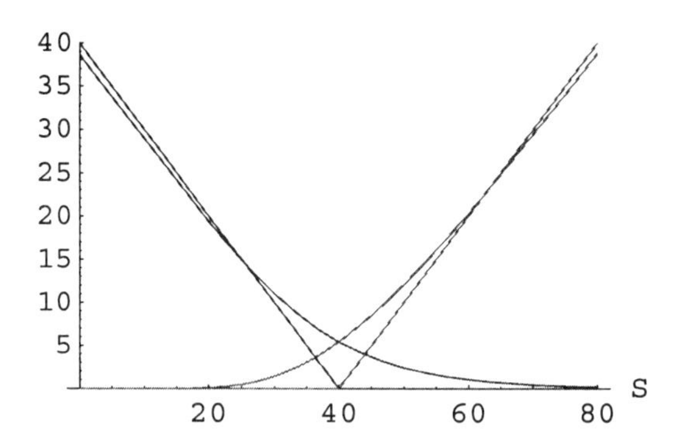

So, in the case of a put option as well as in the case of a call option if dividends are paid, the value of the *European* options crosses under the option payoff due to the fact that, in the European option case, the option holder has to wait in order to exercise the option (or sell it on the option market). In the case of *American* options, like the ones that are actually traded on the option market, one does not have to wait with the exercise, and therefore the option value should never cross under the payoff. This is how obstacle problems come into the play, as we shall see very explicitly, and in great detail in Chapter 6.

3.3.4 Black–Scholes Formulas

3.3.4.1 Options Volatility

```
In[142]:= Clear["Global`*"]
```

The following calculation may help understanding the relative dynamics of stock price vs. option price: moderate stock price volatility yields very significant volatility of the option price. We shall need the SDESolver; below we shall also need ItoDiffusion and ItoDrift; they can be all downloaded from

```
In[143]:= << "CFMLab`ItoSDEs`"
```

For example, let the time horizon T, the number of time subintervals K, the underlying stock appreciation rate a and volatility σ be

```
In[144]:= T = 4/12; K = 1000; a = .36; σ = .3;
```

Compute the stock price evolution

```
In[145]:= SPE = SDESolver[#2 a &, #2 σ &, 100, 0, T, K];
```

We shall also need VC and VP, i.e., the call and put Black–Scholes pricing formulas:

```
In[146]:= << "CFMLab`BlackScholes`"
          r = .05; k = 100;
```

Then compute the call option price evolution

```
In[148]:= COPE = ({#1[[1]], VC[#1[[1]], #1[[2]], T, k, r, σ]} &) /@
            Drop[SPE, -1];
```

as well as the put option price evolution

```
In[149]:= POPE = ({#1[[1]], VP[#1[[1]], #1[[2]], T, k, r, σ]} &) /@
            Drop[SPE, -1];
```

Also, compute their plots:

```
In[150]:= LPS = ListPlot[SPE, PlotRange → {0, Automatic},
             PlotStyle → RGBColor[0, 1/2, 0],
             PlotJoined → True, DisplayFunction → Identity];
          LPCO = ListPlot[COPE, PlotRange → {0, Automatic},
             PlotStyle → RGBColor[1, 0, 0],
             PlotJoined → True, DisplayFunction → Identity];
          LPPO = ListPlot[POPE, PlotRange → {0, Automatic},
             PlotStyle → RGBColor[0, 0, 1],
             PlotJoined → True, DisplayFunction → Identity];
          Show[LPS, LPCO, LPPO, DisplayFunction →
             $DisplayFunction];
```

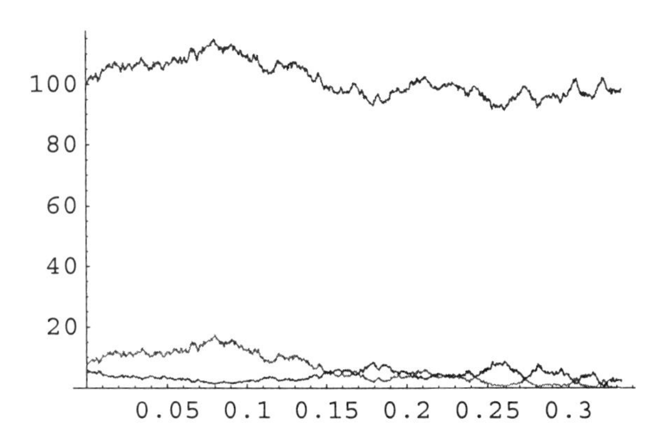

Maybe even more instructive is to track the *relative* prices evolution, i.e., we divide the price of each instrument by its initial price:

```
In[154]:= RSPE = ({#1[[1]], #1[[2]]/SPE[[1, 2]]} &) /@ SPE;
          RCOPE = ({#1[[1]], #1[[2]]/COPE[[1, 2]]} &) /@ COPE;
          RPOPE = ({#1[[1]], #1[[2]]/POPE[[1, 2]]} &) /@ POPE;
          LPRS = ListPlot[RSPE, PlotRange → {0, Automatic},
             PlotStyle → RGBColor[0, 1/2, 0],
             PlotJoined → True, DisplayFunction → Identity];
          LPRCO = ListPlot[RCOPE, PlotRange → {0, Automatic},
             PlotStyle → RGBColor[1, 0, 0],
             PlotJoined → True, DisplayFunction → Identity];
          LPRPO = ListPlot[RPOPE, PlotRange → {0, Automatic},
             PlotStyle → RGBColor[0, 0, 1],
             PlotJoined → True, DisplayFunction → Identity];
          Show[LPRS, LPRCO, LPRPO,
            DisplayFunction → $DisplayFunction];
```

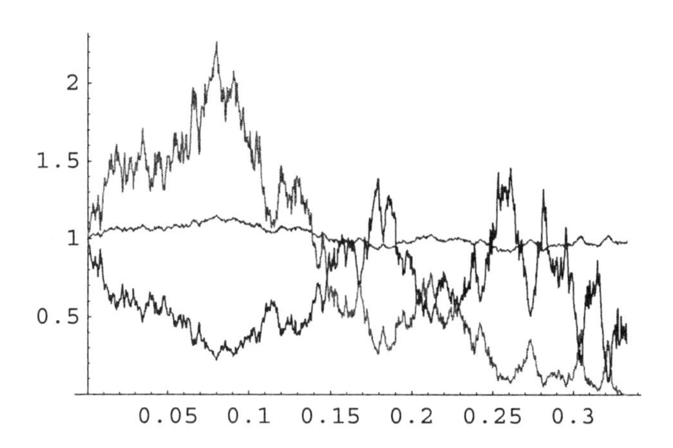

One can realize by means of such experiments how much more volatile, i.e., how much more risky, it is to own options as opposed to owning the underlying stock. The same can be examined more precisely, although not as transparently: indeed, using the symbolic Itô formulas from Chapter 2 (also residing in the package CFMLab`ItoSDEs`), we can plot the call option volatility, as a function of the future time (almost until the expiration of the option), and future possible stock prices:

```
In[161]:= Plot3D[Evaluate[
             ItoDiffusion[VC[t, S, T, k, r, σ], {{σ S}},
               {t, {S}}][[1]] / VC[t, S, T, k, r, σ]],
```

```
{t, 0, 4 T/5}, {S, 3 k/4, 3 k/2}, PlotRange → All,
Mesh → False, PlotPoints → 50,
AxesLabel → {t, S, "σ Call"}];
```

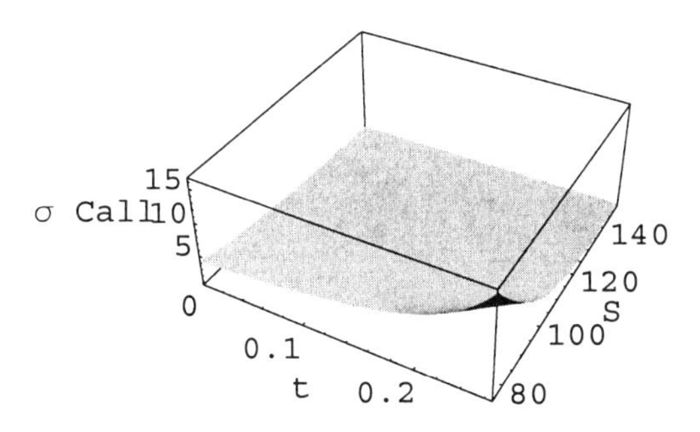

We can see how big the volatility is, i.e., the risk involved. The motivation for taking such a risk, unless it is hedged somehow, could only be possibly justified by a large option appreciation rate (for the given data, and in particular, for the given appreciation rate of the underlying):

```
In[162]:= Plot3D[Evaluate[
           ItoDrift[VC[t, S, T, k, r, σ], {{a S}, {{σ S}}},
             {t, {S}}][[1]] / VC[t, S, T, k, r, σ]],
         {t, 0, 4 T/5}, {S, 3 k/4, 3 k/2}, PlotRange → All,
         Mesh → False, PlotPoints → 50,
         AxesLabel → {t, S, "a Call"}];
```

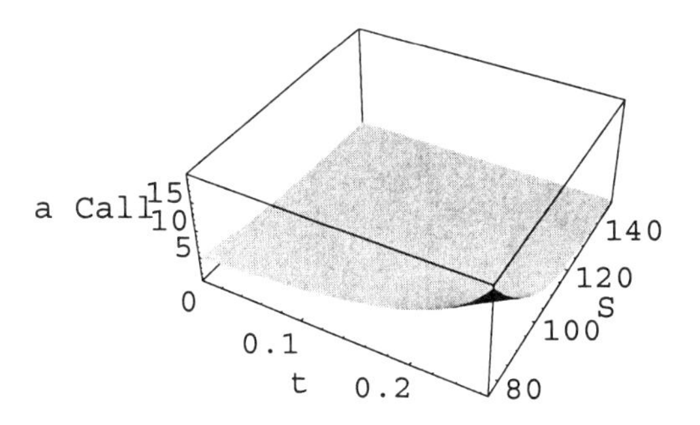

The reader is encouraged to do a similar calculation for the put options. Such quantification of the risk and of the up-side potential is a necessary prerequisite for addressing more advanced option-trading questions such as designing optimal hedging portfolios in accord with the investors risk tolerance/preference (see Section 8.2.4.2 below)

3.3.4.2 Put–Call Parity

Consider a simple portfolio: one stock, one put, and short one call (or sell a covered call). The value of such a portfolio at time t is equal to

$$\Pi(t, S) = S - \mathrm{CV}(t, S) + \mathrm{PV}(t, S).$$

What is the value of the portfolio at the expiration time $t = T$?

$$\Pi(T, S) = S - \mathrm{CV}(T, S) + \mathrm{PV}(T, S) = S + \mathrm{Max}(0, k - S) - \mathrm{Max}(0, S - k)$$
$$= S - \mathrm{Min}(0, S - k) - \mathrm{Max}(0, S - k) = S - (S - k) = k.$$

So the value of the portfolio at the time $t = T$ is not random. Therefore the value of the portfolio at any prior time should not be random as well, and should be equal to the present value of the terminal value:

$$\Pi(t, S) = \Pi(T, S)\, e^{-r(T-t)} = k\, e^{-r(T-t)}.$$

Therefore

$$S - \mathrm{CV}(t, S) + \mathrm{PV}(t, S) = k\, e^{-r(T-t)}$$

or, the so called put–call parity holds:

$$\mathrm{CV}(t, S) - \mathrm{PV}(t, S) = S - k\, e^{-r(T-t)}.$$

This is true only in the case of European options and not in the case of American options.

3.3.4.3 Sensitivity Analysis

Delta (δ)

In the derivation of the Black–Scholes PDE above, the crucial quantity was $\Delta(t, S)$, the Black–Scholes hedging strategy. Additionally, when owning an option, the holder is observing and anticipating movements of the price of the underlying stock, and wishes to be able to anticipate the corresponding movements in the value of the held option. As it is well known from calculus, for each dollar change in the price of the underlying stock, there will be $\Delta(t, S) = \frac{\partial V(t,S)}{\partial S}$ change in the value of the stock option. In other words, *Delta is an option's sensitivity to stock price movements*.

Here is how δ's look like (in the case of a call and put option respectively)

```
In[163]:= Plot[Evaluate[∂S VC[1/12, S, 1/6, 100, .05, .5]],
            {S, 0, 200}, AxesLabel → {S, δ}];
```

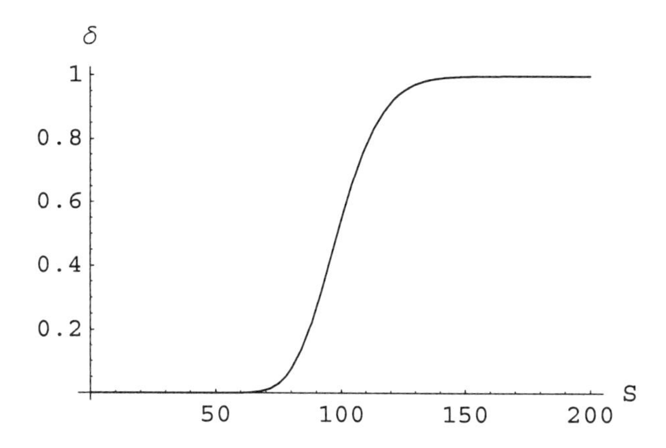


```
In[164]:= Plot[Evaluate[∂S VP[1/12, S, 1/6, 100, .05, .5]],
            {S, 0, 200}, AxesLabel → {S, δ}];
```

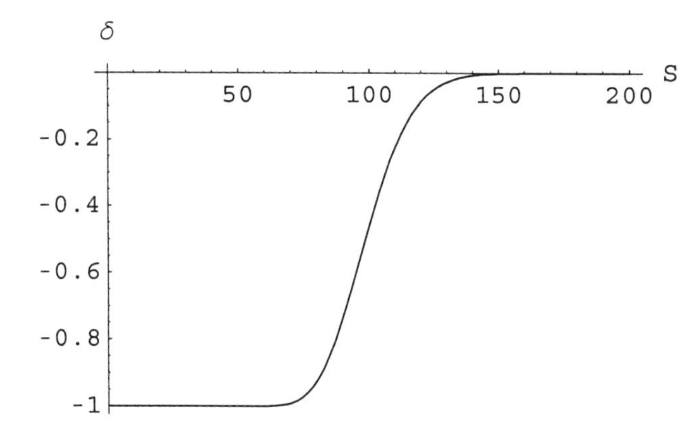

Gamma (γ)

A derivative of delta, gamma measures the change in delta with a small change in stock price:

```
In[165]:= Plot[Evaluate[∂{S,2} VC[1/12, S, 1/6, 100, 0.05, 0.5]],
            {S, 0, 200}, AxesLabel → {S, γ}, PlotRange → All];
```

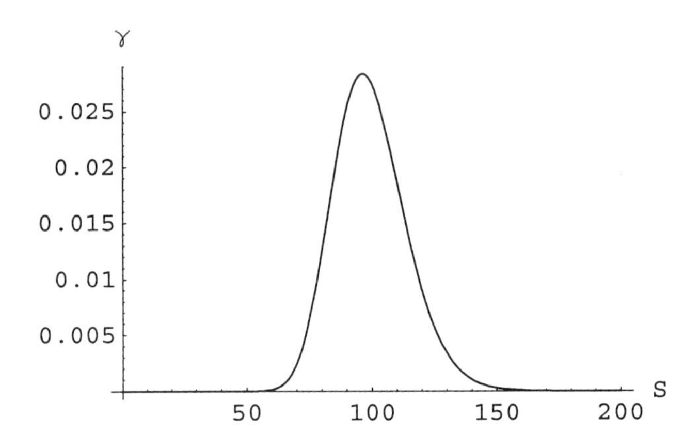

and an almost identical picture in the case of a put.

Rho (ρ)

Rho is an option's sensitivity to changes in the interest rate r:

```
In[166]:= Clear[r];
          Plot[Evaluate[∂r VC[1/12, 100, 1/6, 100, r, 0.5]],
           {r, 0.02, .09}, AxesLabel → {r, ρ},
           PlotRange → All];
```

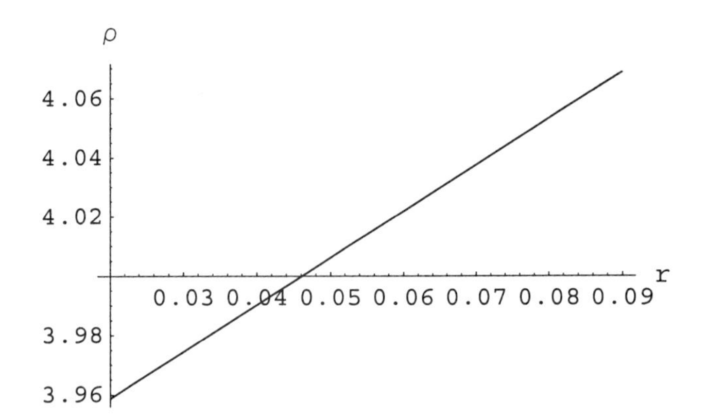


```
In[167]:= Plot[Evaluate[∂r VP[1/12, 100, 1/6, 100, r, 0.5]],
            {r, 0.02, .09}, AxesLabel → {r, ρ},
            PlotRange → All];
```

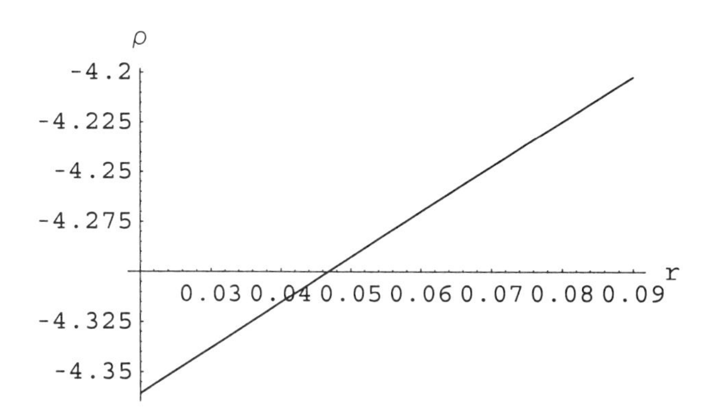

Theta (θ)

Theta is an option's sensitivity with respect to the time

```
In[168]:= Plot[Evaluate[∂t VC[t, 100, 1/6, 100, 0.05, 0.5]],
    {t, 0, .9/6}, AxesLabel → {t, θ}, PlotRange → All];
```

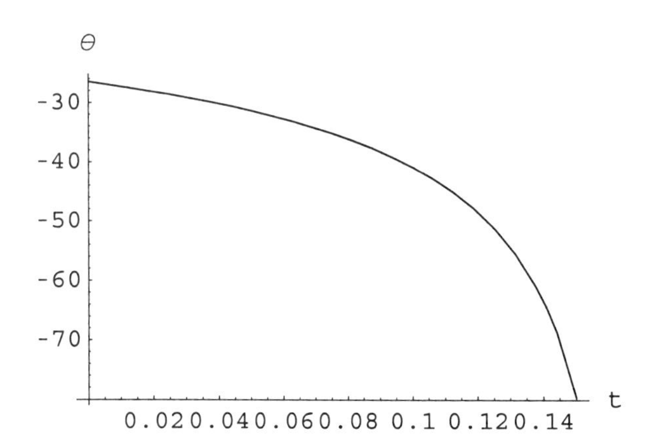

and an almost identical picture in the case of a put.

Vega

Vega is an option's sensitivity with respect to the underlying's volatility

```
In[169]:= Clear[σ]
```

In[170]:= Plot$\left[\text{Evaluate}\left[\partial_\sigma \text{VC}\left[\frac{1}{12}, 100, \frac{1}{6}, 100, 0.05, \sigma\right]\right],\right.$
{σ, 0.1, 1}, AxesLabel → {σ, ""},
PlotRange → All$\left.\right]$;

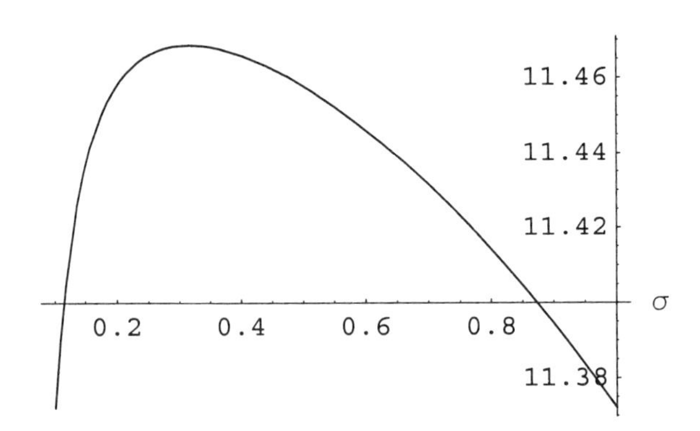

and an almost identical picture in the case of a put.

3.4 Generalized Black–Scholes Formulas: Time-Dependent Data

In[171]:= Clear["Global`*"]

We shall need the Black–Scholes option pricing formulas: VC and VP:

In[172]:= << "CFMLab`BlackScholes`"

Consider now the generalized Black–Scholes partial differential equation

In[173]:= GeneralBSPDE = ∂_t V[t, S] + $\frac{1}{2}$ S^2 $\partial_{\{S,2\}}$ V[t, S] σ[t]^2 +
S a[t] ∂_S V[t, S] - b[t] V[t, S] == 0

Out[173]= $\frac{1}{2} S^2 V^{(0,2)}(t, S)\,\sigma(t)^2 - b(t)\,V(t, S) + S\,a(t)\,V^{(0,1)}(t, S) + V^{(1,0)}(t, S) == 0$

for $t < T$ and $S > 0$, together with terminal conditions

In[174]:= TermCondCall = V[T, S] == Max[S - k, 0]

Out[174]= $V(T, S) == \text{Max}(0, S - k)$

in the case of calls, and

In[175]:= TermCondPut = V[T, S] == Max[k - S, 0]

Out[175]= $V(T, S) == \text{Max}(0, k - S)$

in the case of puts. Again, no boundary condition is imposed, nor is it possible to impose on the boundary $S = 0$ or for $S \to \infty$. The difference between this problem and the one solved before is that all the coefficients a, b and the volatility σ are *time-dependent* $a = a(t)$, $b = b(t)$, $\sigma = \sigma(t)$. The idea is to use the formulas for constant data, with some sort of time-average values. Namely, consider the

In[176]:= `Substitution =`

$$\left\{a \to \frac{\int_t^T a[\tau]\, d\tau}{T - t},\ b \to \frac{\int_t^T b[\tau]\, d\tau}{T - t},\ \sigma^2 \to \frac{\int_t^T \sigma[\tau]^2\, d\tau}{T - t}\right\};$$

and then define

In[177]:=
```
GeneralCall[t_, S_, T_, k_, a_, b_, σ_] =
  FullSimplify[
   VC[t, S, T, k, a, b, σ] /. Substitution]
```

Out[177]=
$$\frac{1}{2} e^{-\int_t^T b(\tau)\,d\tau} \left(e^{\int_t^T a(\tau)\,d\tau} S \left(\text{erf}\left(\frac{2 \int_t^T a(\tau)\,d\tau + \int_t^T \sigma(\tau)^2\,d\tau + 2\log\left(\frac{S}{k}\right)}{2\sqrt{2}\sqrt{\int_t^T \sigma(\tau)^2\,d\tau}} \right) + 1 \right) - k\,\text{erfc}\left(\frac{-2 \int_t^T a(\tau)\,d\tau + \int_t^T \sigma(\tau)^2\,d\tau - 2\log\left(\frac{S}{k}\right)}{2\sqrt{2}\sqrt{\int_t^T \sigma(\tau)^2\,d\tau}} \right) \right)$$

and

In[178]:=
```
GeneralPut[t_, S_, T_, k_, a_, b_, σ_] =
  FullSimplify[
   VP[t, S, T, k, a, b, σ] /. Substitution]
```

Out[178]=
$$\frac{1}{2} e^{-\int_t^T b(\tau)\,d\tau} \left(k\,\text{erfc}\left(-\frac{-2 \int_t^T a(\tau)\,d\tau + \int_t^T \sigma(\tau)^2\,d\tau - 2\log\left(\frac{S}{k}\right)}{2\sqrt{2}\sqrt{\int_t^T \sigma(\tau)^2\,d\tau}} \right) - e^{\int_t^T a(\tau)\,d\tau} S\,\text{erfc}\left(\frac{2 \int_t^T a(\tau)\,d\tau + \int_t^T \sigma(\tau)^2\,d\tau + 2\log\left(\frac{S}{k}\right)}{2\sqrt{2}\sqrt{\int_t^T \sigma(\tau)^2\,d\tau}} \right) \right)$$

We store these two formulas in the package CFMLab`BlackScholes` for future reference. We shall verify whether those two candidates are indeed solutions of the generalized Black–Scholes PDE. To that end let

In[179]:=
```
GC[t_, S_] := GeneralCall[t, S, T, k, a, b, σ]
```

and

```
In[180]:= GP[t_, S_] := GeneralPut[t, S, T, k, a, b, σ]
```

So, checking the general call formula

```
In[181]:= Simplify[GeneralBSPDE /. V → GC]
Out[181]= True
```

as well as the general put formula

```
In[182]:= Simplify[GeneralBSPDE /. V → GP]
Out[182]= True
```

An additional check would be to recover the constant coefficient Black–Scholes formula:

```
In[183]:= FullSimplify[
           GeneralCall[t, S, T, k, r &, r &, σ &] ==
            VC[t, S, T, k, r, σ]]
Out[183]= True
```

Now compute some non-constant coefficient Black–Scholes formulas. For example, if volatility is quadratic function of time

```
In[184]:= (σ2 #1^2 + σ1 #1 + σ0 &) [t]
```

Out[184]= $\sigma_2 t^2 + \sigma_1 t + \sigma_0$

if the rate of dividend payments is quadratic function of time

```
In[185]:= (d2 #1^2 + d1 #1 + d0 &) [t]
```

Out[185]= $d_2 t^2 + d_1 t + d_0$

and if the interest rate is quadratic function of time

```
In[186]:= (r2 #1^2 + r1 #1 + r0 &) [t]
```

Out[186]= $r_2 t^2 + r_1 t + r_0$

Then, for example, the call Black–Scholes formulas read as

```
In[187]:= GeneralCall[t, S, k, T,
           r2 #1^2 + r1 #1 - (d2 #1^2 + d1 #1 + d0) + r0 &,
           r2 #1^2 + r1 #1 + r0 &, σ2 #1^2 + σ1 #1 + σ0 &]
```

Out[187]= $\frac{1}{2} e^{-\frac{1}{3} r_2 k^3 - \frac{r_1 k^2}{2} - r_0 k + t r_0 + \frac{t^2 r_1}{2} + \frac{t^3 r_2}{3}}$

$$\Bigg(e^{\frac{1}{3}(r_2-d_2)k^3+\frac{1}{2}(r_1-d_1)k^2+(r_0-d_0)k-t(r_0-d_0)-\frac{1}{2}t^2(r_1-d_1)-\frac{1}{3}t^3(r_2-d_2)}$$

$$S\Bigg(\operatorname{erf}\Bigg(\Big(\frac{1}{5}\sigma_2^2 k^5+\frac{1}{2}\sigma_1\sigma_2 k^4+\frac{1}{3}(\sigma_1^2+2\sigma_0\sigma_2)k^3+\sigma_0\sigma_1 k^2+\sigma_0^2 k-t\sigma_0^2$$
$$-\frac{1}{5}t^5\sigma_2^2+2\log\Big(\frac{S}{T}\Big)+2\Big(\frac{1}{3}(r_2-d_2)k^3+\frac{1}{2}(r_1-d_1)k^2+(r_0-d_0)$$
$$k-t(r_0-d_0)-\frac{1}{2}t^2(r_1-d_1)-\frac{1}{3}t^3(r_2-d_2)\Big)-t^2\sigma_0\sigma_1$$
$$-\frac{1}{2}t^4\sigma_1\sigma_2-\frac{1}{3}t^3(\sigma_1^2+2\sigma_0\sigma_2)\Big)\Big/\Bigg(2\sqrt{2}\sqrt{\Big(\frac{1}{5}\sigma_2^2k^5}$$
$$+\frac{1}{2}\sigma_1\sigma_2k^4+\frac{1}{3}(\sigma_1^2+2\sigma_0\sigma_2)k^3+\sigma_0\sigma_1k^2+\sigma_0^2k-t\sigma_0^2$$
$$-\frac{1}{5}t^5\sigma_2^2-t^2\sigma_0\sigma_1-\frac{1}{2}t^4\sigma_1\sigma_2-\frac{1}{3}t^3(\sigma_1^2+2\sigma_0\sigma_2)\Big)\Bigg)\Bigg)$$
$$+1\Bigg)-T\operatorname{erfc}\Bigg(\Big(\frac{1}{5}\sigma_2^2k^5+\frac{1}{2}\sigma_1\sigma_2k^4+\frac{1}{3}(\sigma_1^2+2\sigma_0\sigma_2)k^3+\sigma_0\sigma_1k^2$$
$$+\sigma_0^2k-t\sigma_0^2-\frac{1}{5}t^5\sigma_2^2-2\log\Big(\frac{S}{T}\Big)-2\Big(\frac{1}{3}(r_2-d_2)k^3+\frac{1}{2}(r_1-d_1)k^2$$
$$+(r_0-d_0)k-t(r_0-d_0)-\frac{1}{2}t^2(r_1-d_1)-\frac{1}{3}t^3(r_2-d_2)\Big)-t^2\sigma_0\sigma_1$$
$$-\frac{1}{2}t^4\sigma_1\sigma_2-\frac{1}{3}t^3(\sigma_1^2+2\sigma_0\sigma_2)\Big)\Big/\Bigg(2\sqrt{2}\sqrt{\Big(\frac{1}{5}\sigma_2^2k^5}$$
$$+\frac{1}{2}\sigma_1\sigma_2k^4+\frac{1}{3}(\sigma_1^2+2\sigma_0\sigma_2)k^3+\sigma_0\sigma_1k^2+\sigma_0^2k-t\sigma_0^2$$
$$-\frac{1}{5}t^5\sigma_2^2-t^2\sigma_0\sigma_1-\frac{1}{2}t^4\sigma_1\sigma_2-\frac{1}{3}t^3(\sigma_1^2+2\sigma_0\sigma_2)\Big)\Bigg)\Bigg)\Bigg)$$

On the other hand, if there are no dividends and if the interest rate is constant while volatility is periodic, say $\sigma(t) := a\left(\frac{4}{3} - \cos(b\,t)\right)$ and while the interest rate is quadratic in time, the put Black–Scholes formulas read as

```
In[188]:= FullSimplify[GeneralPut[t, S,
              k, T, r &, r &, a (4/3 - Cos[b #1]) &]]
```

Out[188]= $\frac{1}{2}\Bigg(e^{r(t-k)}\,T\operatorname{erfc}\Bigg(\Big(3\,(32\sin(b\,k)-3\sin(2\,b\,k)-32\sin(b\,t)$

$$+3\sin(2\,b\,t))\,a^2-2\,b\,(41\,a^2-36\,r)\,(k-t)+72\,b\log\Big(\frac{S}{T}\Big)\Big)\Big/$$
$$\Bigg(12\sqrt{2}\,b\sqrt{\Big(\frac{1}{b}\,(a^2\,(82\,b\,(k-t)-96\sin(b\,k)}$$
$$+9\sin(2\,b\,k)+96\sin(b\,t)-9\sin(2\,b\,t)))\Big)\Bigg)\Bigg)$$
$$-S\operatorname{erfc}\Big(\Big(3\,(-32\sin(b\,k)+3\sin(2\,b\,k)+32\sin(b\,t)-3\sin(2\,b\,t))$$

$$a^2 + 2\,b\,(41\,a^2 + 36\,r)\,(k - t) + 72\,b\log\left(\frac{S}{T}\right)\Big) \Big/$$
$$\left(12\sqrt{2}\,b\,\sqrt{\left(\frac{1}{b}\,(a^2\,(82\,b\,(k - t) - 96\sin(b\,k)\right.}\right.$$
$$\left.\left.\left. + 9\sin(2\,b\,k) + 96\sin(b\,t) - 9\sin(2\,b\,t)))\right)\right)\right)$$

Unfortunately (or fortunately, for numerical analysts for example), the S dependence of data (e.g. volatility), i.e., data dependence on the underlying stock price, cannot be handled this way, explicit solutions do not exist, and only numerical solutions of the associated partial differential equations are possible. We shall study them in detail in Chapter 5.

4 Stock Market Statistics

Stock Market Data Manipulation and Statistical Analysis

4.1 Remarks

All the mathematics presented so far and all the mathematics to be presented later can have importance in investing and trading only if the basic models that are used to describe the market dynamics are correct to a significant degree, *and* if the *parameters* of those models can be *estimated* with a significant degree of precision, using the available market data as well as possibly some other more or less ad hoc hunches and/or privileged insights. This chapter is about the latter; it is about parameter estimation for stock price models. The next chapter will be devoted to another kind of data and another kind of parameter estimation—one that concerns options price models.

We shall see that some of the crucial parameters of stock market models can be estimated easily and precisely from stock market data, while others, just as crucial, can be estimated only with precision that is quite poor, considering that estimates cannot be based on many years of data (even if such data is available for some companies, market conditions change). Luckily, statistics theory provides a natural framework for mixing information that can be deduced from the statistical data only with the information that can be derived from other sources. Such *prior* information, provided it is not erroneous, can reduce significantly the uncertainty about the final estimates of the parameters that are necessary to put the theory and methods of computational financial mathematics into investing and trading practices. Moreover, the quantitative measure of uncertainty that still remains after the conclusion of the statistical analysis is provided. In addition to estimates of the crucial parameters, a quantitative measure of uncertainty about those estimates will be input as well, and will be quite influential in some of the investment portfolio decision making models to be developed in Chapter 7.

4.2 Stock Market Data Import and Manipulation

```
In[1]:= Clear["Global`*"]
```

Stock market data can be downloaded from various web sites into text files somewhere in the computer. Data used in the present chapter was kindly provided by OptionMetrics, LLC. This section has to be executed at least once since additional data files are created to be used later.

Suppose market data resides in the directory:

```
In[2]:= SetDirectory[ToFileName[{$TopDirectory, "AddOns",
          "Applications", "CFMLab", "MarketData"}]]
```

Out[2]= C:\Program Files\Wolfram Research\Mathematica\4.1\AddOns\Applications\CFMLab\MarketData

One can see some of the (data) files residing in the present working directory:

```
In[3]:= FileNames["*OptionMetrics*"]
```

Out[3]= {OptionMetricsData1.txt, OptionMetricsData2.txt}

The stock price file we shall be working with in this chapter looks like this:

```
In[4]:= !! "OptionMetricsData1.txt"
```

From In[4]:=

```
        QQQ   YHOO AMZN GE     PG
1/2/98        66.25  59.5   74     80.9375
1/5/98        62.9375       57     75.3125        82.5625
1/6/98        64     58.0625       74.3125        82.3125
1/7/98        63.8125       57.375 74.9375        81.4375
1/8/98        64.25  55.375 74.25  80.5

12/21/01      39.48  16.92  10     41.349998      80.639999
12/24/01      39.32  16.67  9.8299999     41.130001      80.260002
12/26/01      39.689999     17.51  11.1   40.549999      80.5
12/27/01      40.009998     17.77  10.6   40.950001      80.150002
12/28/01      40.330002     18.299999     10.9   40.73  79.510002
12/31/01      38.91  17.74  10.82  40.080002      79.129997
```


So the file contains historical price-data for 5 securities: QQQ, YHOO, AMZN, GE, and PG. There is a small problem to be addressed below in the fact that the QQQ data starts late (QQQ, the Nasdaq 100 tracking security, did not exist before 3/10/99). Our goal is to import data, translate it into more useful format meaningful to *Mathematica®*, and save it in such a format for future use.

In order to input the above file into the present *Mathematica®* session one can proceed as follows. Create first an input stream.

```
In[5]:= InputData = OpenRead["OptionMetricsData1.txt"]
```

Out[5]= InputStream[OptionMetricsData1.txt, 4]

Having seen the format of the file above, we know that the first five words are names of the securities

```
In[6]:= Tickers = Table[Read[InputData, Word], {5}]
```

Out[6]= {QQQ, YHOO, AMZN, GE, PG}

Now we can read in the rest of the content (since the data is not uniform due to QQQ, as explained above, we import each word separately instead of utilizing also the line structure of the file)

```
In[7]:= AllData = Table[Read[InputData, Word], {6000}];
```

Let us see the structure of the AllData:

```
In[8]:= Take[AllData, 20]
```

Out[8]= {1/2/98, 66.25, 59.5, 74, 80.9375, 1/5/98, 62.9375, 57, 75.3125, 82.5625, 1/6/98, 64, 58.0625, 74.3125, 82.3125, 1/7/98, 63.8125, 57.375, 74.9375, 81.4375}

```
In[9]:= Dimensions[AllData]
```

Out[9]= {6000}

It is an one-dimensional array. We can close the input stream right away:

```
In[10]:= Close[InputData];
```

In order to recover the line structure which was lost when we read the file in, we identify the dates, as identifiers of different lines. Dates are identified by the presence of characters "/":

```
In[11]:= TradingDates =
           Select[AllData, StringPosition[#, "/"] ≠ {} &];
```

We identify positions of all the dates, and then use this to take parts of the AllData, forming a two-dimensional array, where the lines, representing different trading days, are now recovered:

```
In[12]:= AllData2 = Module[{x, y},
            x = Position[AllData, #] & /@ TradingDates;
            y = Take[AllData, # - {0, 1}] & /@
              Partition[Flatten[x], 2, 1]; Append[y, Take[
              AllData, {x[[-1, 1, 1]], x[[-1, 1, 1]] + 5}]]];
```

We are now in a position to separate data that contains QQQ and the data that does not:

```
In[13]:= AfterQQQ = Select[AllData2, Length[#] == 6 &];

In[14]:= BeforeQQQ = Select[AllData2, Length[#] == 5 &];
```

We can start extracting the data corresponding to different securities:

```
In[15]:= TextPriceData["QQQ"] :=
          {#[[1]], #[[2]]} & /@ AfterQQQ

In[16]:= TextPriceData[Ticker_] := Module[{p},
           p = Position[Tickers, Ticker][[1, 1]];
           Join[{#[[1]], #[[p]]} & /@ BeforeQQQ,
            {#[[1]], #[[p + 1]]} & /@ AfterQQQ]]
```

These functions can only produce text data. The text data needs to be translated into the expressions meaningful to *Mathematica*®. In doing that we utilize again position of characters "/". For example

```
In[17]:= TextPriceData["QQQ"][[1]]

Out[17]= {3/10/99, 102.125}
```

after defining

```
In[18]:= PriceData[Ticker_] := Function[X,
           {Module[{xx, yy}, xx = Partition[Append[Prepend[
                  #[[1]] & /@ StringPosition[#, "/"], 0],
                 StringLength[#] + 1], 2, 1];
              yy = ToExpression[Function[y,
                  StringTake[#, y + {1, -1}]] /@ xx];
              {If[#[[3]] > 50, #[[3]] + 1900,
                  #[[3]] + 2000], #[[1]],
                 #[[2]], 16, 0, 0} &[yy]] &[
            X[[1]]], ToExpression[X[[2]]]}] /@
           TextPriceData[Ticker]
```

is transformed into

```
In[19]:= PriceData["QQQ"][[1]]

Out[19]= {{1999, 3, 10, 16, 0, 0}, 102.125}
```

where one can recognize the *Mathematica*® way of presenting time (year, month, day, hour, minute, second):

```
In[20]:= Date[]

Out[20]= {2002, 8, 3, 23, 34, 41}
```

Notice that if we were to be too quick above, we would have obtained, for example,

```
In[21]:= ToExpression[TextPriceData["QQQ"][[1]]]
```

Out[21]= $\left\{\frac{1}{330}, 102.125\right\}$

which lost the right meaning for the date of the price recorded.

In Mathematical Finance it is customary to measure time in units of years. In *Mathematica*®, on the other hand, in addition to the previous method, time is measured in units of seconds from the time of the date

```
In[22]:= ToDate[0]
Out[22]= {1900, 1, 1, 0, 0, 0}
```

An average year has

```
In[23]:= YearLength = 1/100 FromDate[{2000, 1, 1, 0, 0, 0}]
Out[23]= 31556736
```

seconds, and therefore (we decided to measure time in years since {2000, 1, 1, 0, 0, 0})

```
In[24]:= YearClock[y_] :=
           N[(FromDate[y] - FromDate[{2000, 1, 1, 0, 0, 0}]) / YearLength]
```

For example the question: what is the decimal time now, can be answered as

```
In[25]:= YearClock[Date[]]
Out[25]= 2.61384
```

We could apply the YearClock to the data, hoping that the data transformation is successfully finished. We try to do that and plot one of the price histories available:

```
In[26]:= ListPlot[{YearClock[#[[1]]], #[[2]]} & /@
           PriceData["AMZN"], PlotJoined → True];
```

350
300
250
200
150
100
50
-2 -1 1 2

which does not look right. The reason is the data is not adjusted for stock *splits*. Splits usually happen when the stock price becomes to expensive for efficient trading. So we need to identify the dates when the different stocks had splits, as well as to identify the split-ratio (we need to consider only 2:1 and 3:1 splits here)

```
In[27]:= ListOfSplits = Function[Y, {Y, Module[{x, y},
           x = {#[[2, 1]] - {0, 0, 0, 16, 0, 0}, 2} &/@
             Select[Partition[PriceData[Y], 2, 1],
              2.5 #[[2, 2]] > #[[1, 2]] > 1.9 #[[2, 2]] &];
           y = {#[[2, 1]] - {0, 0, 0, 16, 0, 0}, 3} &/@
             Select[Partition[PriceData[Y], 2, 1],
              #[[1, 2]] > 2.8 #[[2, 2]] &];
           Union[x, y]]}] /@ Tickers
```

Out[27]=

QQQ	{2000, 3, 20, 0, 0, 0}	2
YHOO	{1998, 8, 3, 0, 0, 0}	2
	{1999, 2, 8, 0, 0, 0}	2
	{2000, 2, 14, 0, 0, 0}	2
AMZN	{1998, 6, 2, 0, 0, 0}	2
	{1999, 1, 5, 0, 0, 0}	3
	{1999, 9, 2, 0, 0, 0}	2
GE	{2000, 5, 8, 0, 0, 0}	3
PG	{}	

or if times of splits are decimalized

```
In[28]:= ListOfSplits2 =
        {#[[1]], {YearClock[#[[1]]], #[[2]]} &/@
           #[[2]]} &/@ListOfSplits
```

Out[28]=

QQQ	0.216296	2
YHOO	−1.41277	2
	−0.895302	2
	0.120469	2
AMZN	−1.58252	2
	−0.988391	3
	−0.331289	2
GE	0.350454	3
PG	{}	

In order to be able to extract the splits-data for particular security, we define a simple function

```
In[29]:= LOS[Ticker_] :=
        ListOfSplits2[[Position[ListOfSplits2,
           Ticker][[1, 1]], 2]]
```

which works for example as

```
In[30]:= LOS["AMZN"]
```

$$\text{Out[30]=} \begin{pmatrix} -1.58252 & 2 \\ -0.988391 & 3 \\ -0.331289 & 2 \end{pmatrix}$$

We can see that AMZN had three splits in the considered time interval, and that two of them where 2:1, and one was 3:1. We are now ready to adjust the price data for splits, by dividing the price of the stock before the particular split by the split ratio, and doing that in succession for all the splits for the considered security. To this end, let

```
In[31]:= F[X_, Y_] :=
          If[#[[1]] < Y[[1]], {#[[1]], #[[2]] / Y[[2]]},
             {#[[1]], #[[2]]}] & /@X

In[32]:= PriceData2[Ticker_] :=
          Fold[F, {YearClock[#[[1]]], #[[2]]} & /@
            PriceData[Ticker], LOS[Ticker]]
```

Let us check

```
In[33]:= plots = ListPlot[PriceData2[#], PlotJoined → True,
                PlotRange → All, DisplayFunction → Identity,
                PlotLabel → ToString[#]] & /@ # & /@
          {{"QQQ"}, {"YHOO", "AMZN"}, {"PG", "GE"}};
         Show[GraphicsArray[plots]];
```

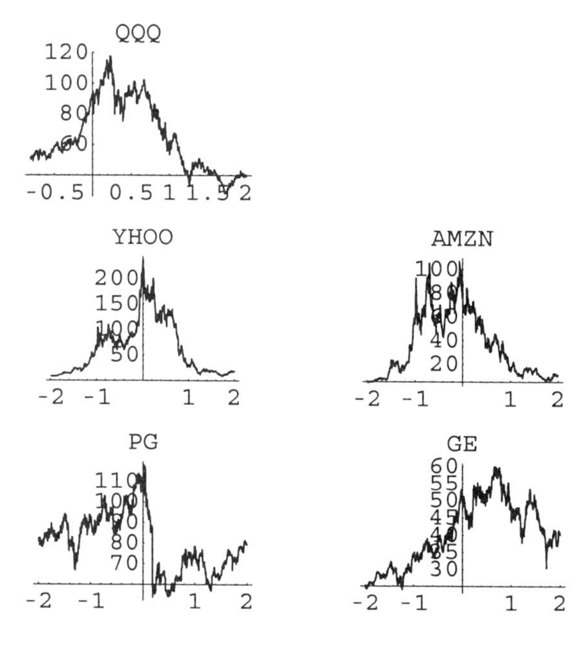

which looks good. There is no need to repeat all what is done in this section every time we need the data. Therefore we store the data in the obtained format as

```
In[34]:= Put[PriceData2[#], ToFileName[
            {$TopDirectory, "AddOns", "Applications",
             "CFMLab", "MarketData"}, StringJoin[
             ToString[#], "PriceData.txt"]]] & /@ Tickers;
```

and prepare the function

```
In[35]:= GetPriceData[Ticker_] :=
          Get[ToFileName[{$TopDirectory, "AddOns",
             "Applications", "CFMLab", "MarketData"},
            StringJoin[ToString[Ticker],
             "PriceData.txt"]]]
```

which will work quickly

```
In[36]:= ListPlot[GetPriceData[amzn], PlotJoined → True];
```

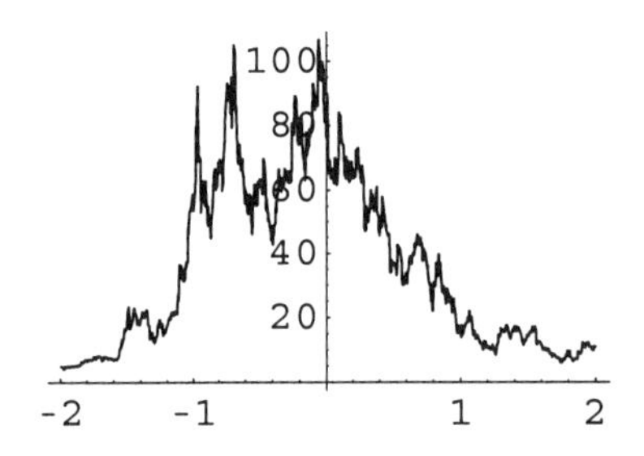

Also we can see the evolution of Nasdaq 100 index, by multiplying the price of QQQ by 40:

```
In[37]:= ListPlot[({#1⟦1⟧, #1⟦2⟧ 40} &) /@ GetPriceData[qqq],
           PlotJoined → True, PlotLabel → "NASDAQ 100"];
```

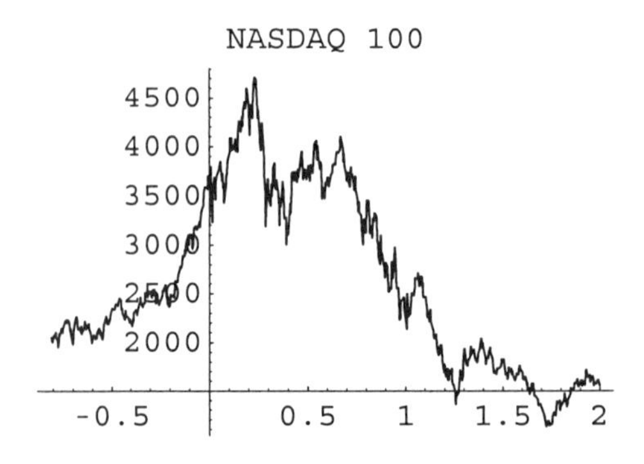

Functions GetPriceData, YearLength, YearClock are stored into the package CFMLab`DataImport` for future reference.

4.3 Volatility Estimates: Scalar Case

4.3.1 First Method

```
In[38]:= Remove["Global`*"]
```

The efficiency of statistical estimates can be assessed by Monte–Carlo experiments. To that end, recall function SDESolver:

```
In[39]:= << CFMLab`ItoSDEs`;
```

Also, we shall need YearLength, YearClock from

```
In[40]:= << CFMLab`DataImport`
```

Let the appreciation rate and volatility be constants:

```
In[41]:= a = .5; σ = .6;
```

and we solve the SDE

$$dS(t) = S(t)\,a\,dt + S(t)\,\sigma\,dB(t). \tag{4.3.1}$$

One trajectory, starting from a year ago and going forward until today, looks like

```
In[42]:= PriceList = SDESolver[a #2 &,
            σ #2 &, 100, YearClock[Date[] - 1],
            YearClock[Date[]], 15000];
         ListPlot[PriceList, PlotJoined → True,
           PlotRange → {0, Max[Transpose[PriceList][[2]]]},
           PlotJoined → True];
```

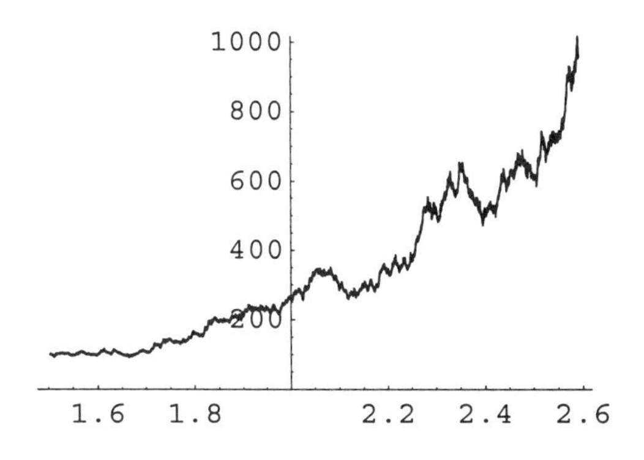

Can we estimate the parameters a and σ from this *single* trajectory?

The natural estimate for the volatility σ is easy. From (4.3.1) we have

$$(dS)^2 = (a\,S\,dt + \sigma\,S\,dB)^2$$
$$= a^2\,S^2\,(dt)^2 + 2\,a\,S\,dt\,\sigma\,S\,dB + \sigma^2\,S^2\,(dB)^2 = \sigma^2\,S^2\,dt$$

and therefore

$$\sigma^2 = \frac{(dS)^2}{S^2\,dt}. \tag{4.3.2}$$

Once σ^2 is estimated, we can also choose σ as either one $\pm\sqrt{\sigma^2}$. For example,

$$\sigma = \sqrt{\frac{(dS)^2}{S^2\,dt}}$$

i.e., volatility is by our choice, or by definition, non-negative. In order to implement this estimate, construct the sample dS, i.e., the list of all differences $dS_i = S_{i+1} - S_i$

```
In[44]:= dS = Drop[Transpose[PriceList][[2]], 1] -
           Drop[Transpose[PriceList][[2]], -1];
```

and the corresponding sample S, i.e., the list of all S_i's (except for the last one)

```
In[45]:= S = Drop[Transpose[PriceList][[2]], -1];
```

We also form the list of time differentials just in case they are not all equal:

```
In[46]:= dt = Drop[Transpose[PriceList][[1]], 1] -
           Drop[Transpose[PriceList][[1]], -1];
```

Then the simple, natural, and quite efficient estimate for the volatility σ is

```
In[47]:= Sqrt[Av[dS^2/(S^2 dt)]]
```

$$\sqrt{\mathrm{Av}\left[\frac{\mathrm{dS}^2}{\mathrm{S}^2\,\mathrm{dt}}\right]}$$

```
Out[47]= 0.601479
```

Since this estimate is going to be used over and over again, we construct a function

```
In[48]:= Σ[PriceList_] := Module[{dS, S, dt},
          dS = Drop[Transpose[PriceList][[2]], 1] -
            Drop[Transpose[PriceList][[2]], -1];
          S = Drop[Transpose[PriceList][[2]], -1];
          dt = Drop[Transpose[PriceList][[1]], 1] - Drop[
              Transpose[PriceList][[1]], -1]; Sqrt[Av[dS^2/(S^2 dt)]]]
```

and store it in the package CFMLab`StockStat`. It works:

```
In[49]:= {Σ[PriceList], σ}
Out[49]= {0.601479, 0.6}
```

What happens when the sample size is smaller, more precisely when every, or every second, or every third, ..., or every hundredth price record is taken into account from the PriceList? This would correspond to the practical situation when not all the realized prices are recorded which indeed is the case (the above real market data, for example, is only for daily closing prices). On the above data we get:

```
In[50]:= ListPlot[
           ({#1, Σ[Take[PriceList, {1, -1, #1}]]} &) /@
            Range[1, 100], PlotJoined → True];
```

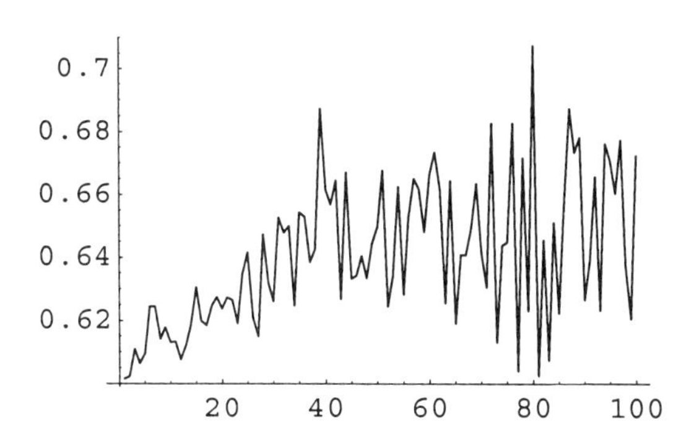

4.3.2 Second Method

On the other hand, if $z = \log(S)$, then, as shown in Chapter 2,

$$dz(t) = \left(a - \frac{1}{2}\,\sigma^2\right)dt + \sigma\, dB(t) \tag{4.3.3}$$

and therefore

$$(dz)^2 = \sigma^2\, dt$$

leading to the estimate

$$\sigma^2 = \frac{(dz)^2}{dt} = \frac{(d\log(S))^2}{dt} \tag{4.3.4}$$

from which we can choose for σ, for example,

$$\sigma = \sqrt{\frac{(dz)^2}{dt}} = \sqrt{\frac{(d\log(S))^2}{dt}}$$

implemented as

```
In[51]:= Σ2[PriceList_] := Module[{dz, dt},
      dz = Drop[Log[Transpose[PriceList][[2]]], 1] -
        Drop[Log[Transpose[PriceList][[2]]], -1];
      dt = Drop[Transpose[PriceList][[1]], 1] -
        Drop[Transpose[PriceList][[1]], -1]; Sqrt[Av[dz^2/dt]]]
```

By the way, the two estimates produce almost identical results:

```
In[52]:= {σ, Σ[PriceList], Σ2[PriceList]}

Out[52]= {0.6, 0.601479, 0.601414}
```

4.3.3 Experiments: Estimating Volatility

One can experiment on many trajectories

```
In[53]:= StandardForm[
      Table[PriceList = SDESolver[a #2 &, σ #2 &, 100,
         YearClock[Date[]] - 1, YearClock[Date[]],
         1000]; {Σ[PriceList], Σ2[PriceList]}, {10}]]

Out[53]//StandardForm=
      {{0.587789, 0.587093}, {0.588116, 0.588621},
       {0.611648, 0.612364}, {0.599763, 0.600182},
       {0.605924, 0.605475}, {0.603242, 0.603758},
       {0.601678, 0.601454}, {0.587722, 0.587526},
       {0.612513, 0.612776}, {0.59206, 0.591405}}
```

always getting very good estimates. *Moreover, even if the time interval is reduced arbitrarily, while sample size is kept constant, the quality of the estimate remains the same.* For example, for sample prices in a *single day:*

```
In[54]:= StandardForm[Table[PriceList =
        SDESolver[a #2 &, σ #2 &, 100, YearClock[Date[]],
         YearClock[Date[] + {0, 0, 1, 0, 0, 0}], 1000];
       {Σ[PriceList], Σ2[PriceList]}, {10}]]

Out[54]//StandardForm=
      {{0.596752, 0.596781}, {0.603674, 0.60362},
       {0.601149, 0.601109}, {0.592854, 0.592921},
       {0.602209, 0.602242}, {0.613371, 0.6134},
       {0.598147, 0.598097}, {0.573013, 0.573035},
       {0.584405, 0.584474}, {0.607307, 0.607328}}
```

It is possible to recover volatility information even during a single day (on a single trajectory). This suggests that it is not essential for statistical estimation purposes that volatility remains constant over a long period of time. This is, unfortunately, *not* the case, as we shall see below if we try to estimate the appreciation rate.

4.3.4 Real Data: Estimating Volatility

Compare the two volatility estimates on the real market data (*assuming the simplest stock price dynamics*):

```
In[55]:= (Σ[GetPriceData[#1]] &) /@ {amzn, yhoo, ge, pg, qqq}
Out[55]= {1.1296, 0.973272, 0.369418, 0.419503, 0.558828}

In[56]:= (Σ2[GetPriceData[#1]] &) /@ {amzn, yhoo, ge, pg, qqq}
Out[56]= {1.1169, 0.968518, 0.368288, 0.439025, 0.556709}
```

It may be interesting to see the "dates", between which the data is taken:

```
In[57]:= (({#1[[1, 1]], #1[[-1, 1]]} &)[GetPriceData[#1]] &) /@
           {amzn, yhoo, ge, pg, qqq}
```

$$
\text{Out[57]=} \begin{pmatrix} -1.99412 & 1.9923 \\ -1.99412 & 1.9923 \\ -1.99412 & 1.9923 \\ -1.99412 & 1.9923 \\ -0.811339 & 1.9923 \end{pmatrix}
$$

4.4 Appreciation Rate Estimates: Scalar Case

4.4.1 An Estimate

In the previous section we saw that without loss of generality we can assume that σ is known. Consider again the logarithm of the stock price $z = \log(S)$, where S solves (4.3.1). Then, if we define

$$\theta = a - \frac{1}{2}\sigma^2 \tag{4.4.1}$$

we get (see 4.3.3)

$$dz(t) = \theta\, dt + \sigma\, dB(t) \tag{4.4.2}$$

and therefore

$$\frac{dz}{dt} = \theta + \frac{\sigma\, dB}{dt} \sim N\left(\theta, \sqrt{\frac{\sigma^2\, dt}{(dt)^2}}\right) = N\left(\theta, \frac{\sigma}{\sqrt{dt}}\right).$$

Now suppose one looks into the trading history of length T (years). The above formula implies a not particularly efficient estimate for θ:

$$\theta_{\text{est}} = \frac{z(T) - z(0)}{T} \sim N\left(\theta, \frac{\sigma}{\sqrt{T}}\right). \tag{4.4.3}$$

Of course, as T becomes larger, the estimate is more and more accurate. Obviously this is not very feasible considering that T is measured in years. One could wish to try to improve this estimate by averaging many $\frac{dz}{dt}$. The disappointing fact is that such averaging does not accomplish anything. Indeed, suppose $dt = \text{const.} = \frac{T}{n}$. Then

$$\frac{1}{n}\sum_{i=1}^{n}\frac{dz_i}{dt} = \frac{1}{n\,dt}\sum_{i=1}^{n} dz_i$$

$$= \frac{1}{n\,dt}\sum_{i=1}^{n}(z(i\,dt) - z((i-1)\,dt)) = \frac{z(T) - z(0)}{T}.$$

So $\theta = a - \frac{1}{2}\sigma^2$ is estimated by $\theta_{\text{est}} = \frac{z(T)-z(0)}{T}$, and this cannot be improved. Consequently a, the appreciation rate, is estimated by

$$a_{\text{est}} = \theta_{\text{est}} + \frac{1}{2}\sigma^2 = \frac{\log(S(T)) - \log(S(0))}{T} + \frac{1}{2}\sigma^2. \tag{4.4.4}$$

Implementing the above estimate

```
In[58]:= AEst[PriceList_] := Module[{σ, T}, σ = Σ[PriceList];
           T = PriceList[[-1, 1]] - PriceList[[1, 1]]; σ^2/2 +
            (Log[PriceList[[-1, 2]]] - Log[PriceList[[1, 2]]])/T]
```

we get, for example,

```
In[59]:= (AEst[GetPriceData[#1]] &) /@
          {amzn, yhoo, ge, pg, qqq}

Out[59]= {0.835593, 0.672532, 0.19404, 0.0835278, 0.0719839}
```

Function AEst is available in the package CFMLab`StockStat`.

4.4.2 Confidence Interval

The above derivation offers a possibility to compute a confidence interval for the estimate of the appreciation rate a. Since

$$a_{\text{est}} - \frac{1}{2}\sigma^2 = \theta_{\text{est}} = \frac{z(T) - z(0)}{T} \sim N\left(\theta, \frac{\sigma}{\sqrt{T}}\right) = N\left(a - \frac{1}{2}\sigma^2, \frac{\sigma}{\sqrt{T}}\right)$$

we get

$$\frac{\log(S(T)) - \log(S(0))}{T} + \frac{1}{2}\sigma^2 = a_{\text{est}} \sim N\left(a, \frac{\sigma}{\sqrt{T}}\right). \tag{4.4.5}$$

Now, this is a bit confusing: a_{est} is the statistics, i.e., a random variable, and the above is saying that this random variable has a distribution with an *unknown* parameter a; so, this implies really a confidence interval for a_{est} in terms of an unknown a; on the other hand, we are looking for a confidence interval for a in terms of the known realization of a_{est}. Little thought would yield a conclusion that there is a symmetry between the two, and that one confidence interval implies the other. The other way to arrive at the same conclusion, which I prefer, can be found later in this chapter when we discuss the Bayesian and frequentists estimates.

So, define

```
In[60]:= << Statistics`NormalDistribution`
         ConfidenceInterval[a_, α_, σ_, T_] :=
          {Quantile[NormalDistribution[a, σ/√T], (1 - α)/2],
           Quantile[NormalDistribution[a, σ/√T], 1 - (1 - α)/2]}
```

and store it in the package CFMLab`StockStat`. For example,

```
In[62]:= ConfidenceInterval[.5, .95, 1, 1]
Out[62]= {-1.45996, 2.45996}
```

Putting all the estimates together we get

```
In[63]:= AΣEst[PriceList_, p_] :=
          Module[{σ, T, aest}, σ = Σ[PriceList];
           T = PriceList[[-1, 1]] - PriceList[[1, 1]]; aest = σ^2/2 +
             (Log[PriceList[[-1, 2]]] - Log[PriceList[[1, 2]]])/T;
           {{aest, ConfidenceInterval[aest, p, σ, T]}, σ}]
```

For example, if we take 90% confidence interval, based on the same data as above, we get:

```
In[64]:= TableForm[({#1, AΣEst[GetPriceData[#1], .9]} &) /@
           {amzn, yhoo, ge, qqq}]
```

Out[64]//TableForm=

amzn	0.835593	−0.0950019 1.76619
	1.1296	
yhoo	0.672532	−0.129276 1.47434
	0.973272	
ge	0.19404	−0.110296 0.498376
	0.369418	
qqq	0.0719839	−0.476981 0.620948
	0.558828	

Function AΣEst is also stored in the package CFMLab\`StockStat\`.

4.4.3 Monte–Carlo Experiments: Estimating the Appreciation Rate

```
In[65]:= << "CFMLab`ItoSDEs`"
```

To gain further insight in the above confidence intervals we perform experiments. Let

```
In[66]:= σ = .5; a = .8;
```

We shall generate random realizations of the simplest stock price model SDE with those parameters, and try to estimate them back using above defined estimates and the confidence interval. We shall do L simulations of SDE, each one with K time steps, and after each simulation we ask the question whether the computed confidence interval contains the true value of a. So, for example

```
In[67]:= With[{L = 50, K = 10000},
          FutureDate = Date[] + {0, 1, 0, 0, 0, 0};
          Experiment = Table[PriceList = SDESolver[a #2 &,
              σ #2 &, 100, YearClock[Date[]], YearClock[
               FutureDate], K]; (#1⟦1⟧ ≤ a ≤ #1⟦2⟧ &)[
             AΣEst[PriceList, .9]⟦1, 2⟧], {L}]]
```

Out[67]= {True, True, True, True, True, True, True, True, True, True,
True, False, True, True, True, False, True, True, True, True,
True, False, False, True, True, False, True, True, True, True,
False, True, True, True, True, True, True, True, True, True,
True, True, True, True, True, False, True, True, True, True}

and the number of those intervals that contain the true appreciation rate a has the frequency

```
In[68]:= N[Length[Select[Experiment, #1 == True &]] / Length[Experiment]]
```

```
Out[68]= 0.86
```

which is not far from the chosen confidence probability (0.9).

4.5 Statistical Experiments: Bayesian and Non-Bayesian

4.5.1 Mathematical Framework for Statistical Experiments and Estimation

```
In[69]:= Clear["Global`*"];
         << "Statistics`"
```

Usually, in probability, one considers a random variable, say X, with values in $\mathcal{X}$ (typically $\mathcal{X} = \mathbb{R}$, or $\mathcal{X} = \mathbb{R}^n$ for some n, and it is called a sample space), and then the problem is to find the probability distribution function $f(x)$ for X. Additionally, the elementary events can also be considered, denoted by $\omega \in \Omega$, so that the particular realizations of the random variable X are $X(\omega)$, and there exists a probability P on the set of suitable subsets of Ω such that

$$P(\{\omega : X(\omega) \in C\}) = \int_C f(x)\, dx$$

for any event $C \subset \mathcal{X}$. Also, sometimes it will be useful to abuse the notation a bit: $\int_C f(x)\, dx := f(C)$.

Now, in statistics, usually one also considers a random variable (called statistics), say X, with values in $\mathcal{X}$ (for example, $X = \frac{\log(S(T)) - \log(S(0))}{T} + \frac{1}{2}\sigma^2$, the appreciation rate estimate in the simplest stock price model) but now, instead of a single ("right") distribution function $f(x)$ for the random variable X, one considers a family of distribution functions $\{f_a(x),\ a \in A\}$, for the same random variable. Set A is called the parameter space. As above, the distribution functions yield a family of probabilities $\{P_a,\ a \in A\}$ on an abstract set of all elementary events Ω.

Since the "right" parameter $a \in A$ is usually not known (except in the self-generating simulation examples), we may just as well forget about it and try to define an intellectual superstructure on top of A and Ω so that a sort of a symmetry between them is achieved, which as we shall see shortly is very useful.

Instead of a space of elementary events Ω, one considers an extended product space $A \times \Omega$ of elementary events. This means that an elementary event now has a representation (a, ω), where the first component is the unknown parameter that characterizes the probability P_a on Ω, or which is the same, the probability distribution $f_a(x)$ of X. Having defined a new space of elementary events $A \times \Omega$, one naturally, in the context of probability and statistics, wishes to determine a probability on it. Can we do that?

Not always, but sometimes it is possible to postulate a *prior probability* on the parameter space A, say probability μ. If that is the case, then it is possible to find the joint probability distribution $\mu \otimes P_a$ for (a, ω), and consequently the statistical experiment can be defined abstractly as a probability space $(A \times \Omega, \mu \otimes P_a)$.

The joint probability distribution $\mu \otimes P_a$ has the representation

$$\mu \otimes P_a = P \otimes \mu_\omega \tag{4.5.1}$$

or which is almost the same

$$\mu \otimes P_a = P \otimes \mu_{X(\omega)} = P \otimes \mu_x \mid_{x=X(\omega)} \tag{4.5.2}$$

where $\mu_{X(\omega)}$ is the conditional distribution for a, given $X(\omega)$, i.e., the *posterior probability* distribution for the parameter $a \in A$. Also, the marginal probability $P(B) = (\mu \otimes P_a)(A \times B)$ for any event $B \subset \Omega$ is called the *predictive probability*. In the case where it is possible to postulate the prior probability μ, the general problem of statistical estimation can be formulated as

$$\text{given } \mu \text{ and } \{P_a\}_{a \in A}, \text{ find } \{\mu_x\}_{x \in \mathcal{X}}. \tag{4.5.3}$$

Once a family $\{\mu_x\}_{x \in \mathcal{X}}$, i.e., function $\mathcal{X} \ni x \mapsto \mu_x$ is computed, then any sort of statistical question about parameter a can be answered (questions such as point estimates, interval estimates, etc.).

As indicated above, and as it is going to be seen below explicitly, sometimes it is not possible to postulate the prior probability μ directly, but instead one has to consider a sequence of approximate statistical experiments $(A \times \Omega, \mu_\sigma \otimes P_a)_\sigma$, and then even though $\lim_{\sigma \to \infty} \mu_\sigma$ fails to be meaningful as a probability distribution, $\lim_{\sigma \to \infty} \mu_{\sigma,x} = \mu_{\infty,x}$, for $x \in \mathcal{X}$, is a legitimate solution of the problem of statistical estimation.

This abstract framework is going to be made clear with the help of examples below.

4.5.2 Uniform Prior

There are two kinds of statistics: frequentist (non-Bayesian) and Bayesian. The frequentist approach is the one that does not assume any preference for possible values of the parameter $a \in A$. To fix ideas, suppose A is a continuum. If A is bounded, one can model such a *non-preference* with the uniform distribution. If A is unbounded, for example $A = (-\infty, \infty)$, then there is a technical difficulty, since unfortunately the uniform distribution for such an A does not exist. But suppose for a moment that A is bounded.

For example, using the abstract notation above, let $\Omega = \mathcal{X} = \{0, 1, 2, 3, \ldots, n\}$, and let $A = (0, 1)$. Furthermore, let $P_a(x) = f_a(x) = \binom{n}{x} a^x (1-a)^{n-x}$, for $x \in \Omega = \mathcal{X}$. Thus X has a binomial distribution with an unknown parameter $a \in (0, 1)$, and known parameter n. The problem is to make a conclusion about the parameter a based on the realization of X.

For the reader who needs a stock market motivation for such a problem, consider a simple model, where we record whether the market goes up or down during a trading day. We assume independence of those daily outcomes. If n is the number of days, say $n = 26$, the number of trading days in a particular month, and if X is the random variable representing the number of trading days during that month that the market has gone up, then the problem is to make a conclusion about a, the probability that in any given day the market goes up, based on the realization of X.

No preference for the possible values for a is assumed, which means that the (prior) probability distribution for a is uniform, or explicitly, $\mu(a) = 1$ for $0 < a < 1$. This implies that the joint probability distribution function for (a, X) on $A \times \Omega$ is equal to $f_a(x)\,\mu(a) = \binom{n}{x} a^x (1-a)^{n-x}$. We now compute the conditional probability distribution function $\mu_x(a) = \mu_{n,x}(a)$. It is equal to

```
In[71]:= μn_,x_[a_] = (Binomial[n, x] a^x (1 - a)^(n-x)) /
           Integrate[Binomial[n, x] a^x (1 - a)^(n-x), {a, 0, 1},
            Assumptions → Re[n - x] > -1 && Re[x] > -1]
```

Out[71]= $$\frac{(1-a)^{n-x}\, a^x\, \Gamma(n+2)}{\Gamma(n-x+1)\,\Gamma(x+1)}$$

This happens to be the density of a Beta distribution, with parameters α and β, which can be computed as

```
In[72]:= βd = BetaDistribution[α, β] /.
           Solve[{n + 2 == α + β, n - x + 1 == β, x + 1 == α,
              n - x == β - 1, x == α - 1}, {α, β}][[1]]
```

Out[72]= $\text{BetaDistribution}(x+1, n-x+1)$

Indeed, the density is equal to

```
In[73]:= dβn_,x_[a_] = PDF[βd, a]
```

Out[73]= $$\frac{(1-a)^{n-x}\, a^x}{\mathrm{B}(x+1, n-x+1)}$$

which can be checked to be the same as the above computed one. The conditional, or sometimes called posterior, density function $g_{n,x}(a)$ contains all the information available about a. For example, the conditional mean or average is equal to

```
In[74]:= m[n_, x_] =
          FullSimplify[Integrate[a μn,x[a], {a, 0, 1},
            Assumptions → Re[n - x] > -1 && Re[x] > -2]]
```

Out[74]= $\frac{x+1}{n+2}$

or, equivalently

```
In[75]:= Mean[βd]
```

Out[75]= $\frac{x+1}{n+2}$

Also

```
In[76]:= Variance[βd]
```

Out[76]= $\frac{(n-x+1)(x+1)}{(n+2)^2(n+3)}$

The mean $m(n, x)$ looks like a very reasonable point estimate for the parameter a. It is somewhat different (better—see the comparison below) than the one usually associated with the frequentist, i.e., with the non-Bayesian approach: the maximum likelihood estimate:

```
In[77]:= Off[Reduce::"ifun"]; MLm[n_, x_] = a /.
           Solve[Reduce[Numerator[Together[∂a μn,x[a]]] == 0,
               a][[1]], a][[1]]
```

Out[77]= $\frac{x}{n}$

The conditional, i.e., the posterior, distribution yields also precise confidence intervals:

```
In[78]:= q[α_, n_, x_] =
           {Quantile[βd, (1 - α)/2], Quantile[βd, 1 - (1 - α)/2]};
```

For example, the 95% confidence interval for the parameter a, given that $x = 0$ out of $n = 26$, can be computed to be

```
In[79]:= ci = q[.95, 26, 0]
```

Out[79]= {0.000937257, 0.127703}

Plotted together, the density function, 95% confidence interval, the maximum likelihood estimate and the mean look like

```
In[80]:= With[{n = 26},
           (With[{x = #1}, pl = Plot[μn,x[a], {a, 0, 1},
                 AxesLabel → {"a", "PDF"}, DisplayFunction →
                  Identity, PlotRange → All]; s1 =
```

```
Show[pl, Graphics[{AbsolutePointSize[5],
    RGBColor[1, 0, 0], Point[{m[n, x], 0}]}],
  Graphics[{AbsolutePointSize[5], RGBColor[
     1/2, 1, 0], Point[{MLm[n, x], 0}]}],
  (Graphics[{AbsolutePointSize[5],
      RGBColor[0, 0, 1], Point[#1]}] &) /@
   ({#1, 0} &) /@ q[.95, n, x],
  PlotLabel -> "x = " <> ToString[x] <>
    "  n = " <> ToString[n],
  DisplayFunction -> $DisplayFunction,
  PlotRange -> {0, 11}]] &) /@ Range[1, n]];
```

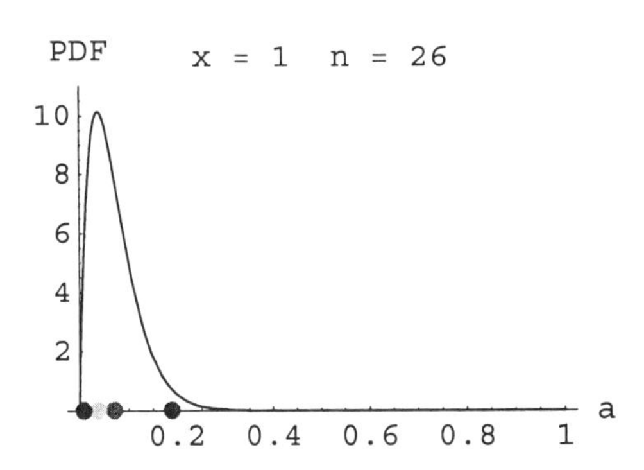

One can see that the maximum likelihood estimate (the second from the left—light green), i.e., the usual frequentist estimate, is skewed. As x increases towards $\frac{n}{2}$, that skewness diminishes and disappears, to increase again as x further increases towards n.

4.5.3 Non-Uniform Prior

Now suppose we do have a preference, i.e., we have a non-uniform prior distribution for a. This is a typical Bayesian framework. For example, suppose the prior distribution has the density

```
In[81]:= bd[a_] = PDF[BetaDistribution[6, 6], a]
```

Out[81]= $2772\,(1-a)^5\,a^5$

which looks like

```
In[82]:= Prior = Plot[bd[a], {a, 0, 1},
           PlotStyle -> RGBColor[0, 2/3, 0]];
```

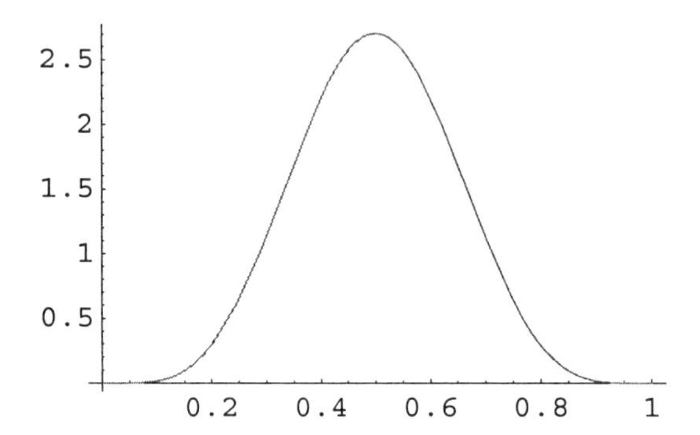

As above, we find the conditional density function for the parameter a:

```
In[83]:= μμn_,x_[a_] =
           (Binomial[n, x] a^x (1 - a)^(n-x) bd[a]) / Integrate[
             Binomial[n, x] a^x (1 - a)^(n-x) bd[a], {a, 0, 1},
             Assumptions → Re[n - x] > -1 && Re[x] > -1]
```

Out[83]= $$\frac{(1-a)^{n-x+5}\, a^{x+5}\, \Gamma(n+12)}{\Gamma(n-x+6)\, \Gamma(x+6)}$$

Then, the conditional mean is equal to

```
In[84]:= m[n_, x_] =
           FullSimplify[Integrate[a μμn,x[a], {a, 0, 1},
             Assumptions → Re[n - x] > -1 && Re[x] > -2]]
```

Out[84]= $$\frac{x+6}{n+12}$$

while the maximum likelihood estimate is this time equal to

```
In[85]:= MLm[n_, x_] = a /.
            Solve[Reduce[Numerator[Together[∂a μμn,x[a]]] ==
                 0, a][[1]], a][[1]] // Simplify
```

Out[85]= $$\frac{x+5}{n+10}$$

Plotted together with the prior density (on the right), "non-Bayesian posterior" density (on the left), and posterior density (in the middle), and together with the non-Bayesian point (maximum likelihood) estimate, these posterior point estimates look as follows (if there were x realizations in n trials, then the parameter a in the governing binomial distribution $\binom{n}{x} a^x (1-a)^{n-x}$ is estimated).

```
In[86]:= With[{n = 26},
    (With[{x = #1}, pl = Plot[{μn,x[a], μμn,x[a]},
        {a, 0, 1}, AxesLabel → {"a", "PDF"},
        DisplayFunction → Identity,
        PlotRange → All]; Show[pl, Prior,
       Graphics[{AbsolutePointSize[5],
         RGBColor[0, 0, 0], Point[{x/n, 0}]}],
       Graphics[{AbsolutePointSize[5],
         RGBColor[1, 0, 0], Point[{m[n, x], 0}]}],
       Graphics[{AbsolutePointSize[5], RGBColor[
          1/2, 1, 0], Point[{MLm[n, x], 0}]}],
       PlotLabel → "x = " <> ToString[x] <>
         "  n = " <> ToString[n],
       DisplayFunction → $DisplayFunction,
       PlotRange → {0, 11}]] &) /@Range[0, n]];
```

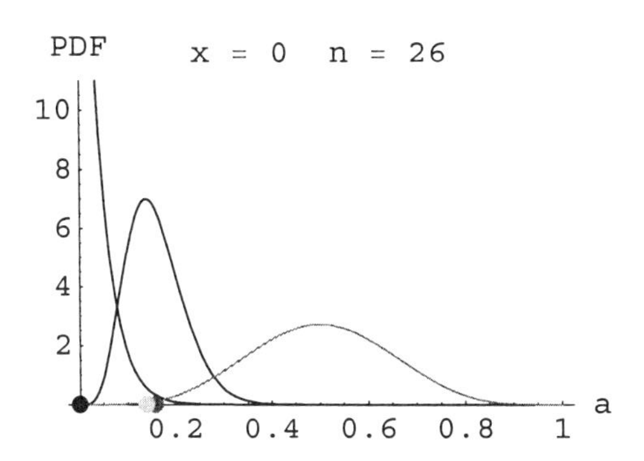

One can see what happens when x runs through its possible values: $0 \le x \le n$, i.e., $\frac{x}{n}$ runs (the first coordinate of the black point in the plot, as well as the frequentist's point estimate for a) through $[0, 1]$.

4.5.4 Unbounded Parameter Space

4.5.4.1 Non-Uniform Prior

Now let the (continuous) parameter space A be unbounded. For example, let $A = (-\infty, \infty)$, and let

```
In[87]:= fa_[x_] = PDF[NormalDistribution[a, σ], x]
```

Out[87]= $\dfrac{e^{-\frac{(x-a)^2}{2\sigma^2}}}{\sqrt{2\pi}\,\sigma}$

for each $a \in A$. Also let the prior distribution for a be normal (even if we wanted to have a uniform distribution here, it would not be possible—it does not exist since A is unbounded; it is a nice exercise to do, with the help of *Mathematica*®, what follows if the prior distribution is uniform on a prescribed bounded interval):

```
In[88]:= h[a_] = PDF[NormalDistribution[b, σp], a]
```

Out[88]= $\dfrac{e^{-\frac{(a-b)^2}{2\sigma_p^2}}}{\sqrt{2\pi}\,\sigma_p}$

This implies that the joint probability distribution function for (a, X) is equal to

```
In[89]:= fa[x] h[a]
```

Out[89]= $\dfrac{e^{-\frac{(a-b)^2}{2\sigma_p^2}-\frac{(x-a)^2}{2\sigma^2}}}{2\pi\sigma\sigma_p}$

Compute now the conditional probability distribution function $g(a \mid x) = g_x(a)$. It is equal to

In[90]:= `gx_[a_] =`

$$\texttt{Simplify}\left[\frac{\texttt{f}_\texttt{a}\texttt{[x] h[a]}}{\int_{-\infty}^{\infty}\texttt{f}_\texttt{a}\texttt{[x] h[a]}\,d\texttt{a}},\ \texttt{Re}\left[-\frac{1}{2\,\sigma_p^2}-\frac{1}{2\,\sigma^2}\right]<0\right]$$

Out[90]= $\dfrac{e^{\frac{1}{2}\left(-\frac{(a-b)^2}{\sigma_p^2}+\frac{(b-x)^2}{\sigma^2+\sigma_p^2}-\frac{(a-x)^2}{\sigma^2}\right)}\sqrt{\frac{1}{\sigma_p^2}+\frac{1}{\sigma^2}}}{\sqrt{2\pi}}$

The conditional mean is equal to

$$\texttt{In[91]:= m[x\_] = FullSimplify}\left[\texttt{Integrate}\left[\texttt{a}\,\texttt{g}_\texttt{x}\texttt{[a]},\ \{\texttt{a}, -\infty, \infty\},\ \texttt{Assumptions} \to \texttt{Re}\left[\frac{1}{\sigma_p^2}+\frac{1}{\sigma^2}\right]>0\right]\right]$$

Out[91]= $\dfrac{b\,\sigma^2 + x\,\sigma_p^2}{\sigma^2+\sigma_p^2}$

while the conditional (posterior) variance is equal to

```
In[92]:= vpost = FullSimplify[Integrate[(a - m[x])^2 gx[a],
            {a, -∞, ∞}, Assumptions → Re[1/σp^2 + 1/σ^2] > 0]]
```

Out[92]= $\frac{\sigma^2 \sigma_p^2}{\sigma^2 + \sigma_p^2}$

Notice that $v_{\text{post}} = \frac{\sigma^2 \sigma_p^2}{\sigma^2 + \sigma_p^2} < \frac{\sigma^2 \sigma_p^2}{\sigma^2} = \sigma_p^2$, and similarly $v_{\text{post}} < \sigma^2$, and therefore $v_{\text{post}} < \text{Min}[\sigma_p^2, \sigma^2]$. Also notice that due to the symmetry, this time the average $m(x)$ is equal to the maximum likelihood estimate:

```
In[93]:= MLm[x_] =
           Simplify[a /. Solve[(∂a gx[a])[[5]] == 0, a][[1]]]
```

Out[93]= $\frac{b\,\sigma^2 + x\,\sigma_p^2}{\sigma^2 + \sigma_p^2}$

Also notice that $g_x(a)$ is the density of a normal distribution with the above parameters:

```
In[94]:= Simplify[
           PDF[NormalDistribution[m[x], √vpost], a] == gx[a],
           {σ > 0, σp > 0}]
```

Out[94]= True

4.5.4.2 "Uniform Prior"

Finally, the "frequentist posterior distribution" can be recovered if "uniform" prior is assumed, i.e., if $\sigma_p \to \infty$:

```
In[95]:= Limit[gx[a], σp → ∞]
```

Out[95]= $\frac{e^{-\frac{(a-x)^2}{2\sigma^2}} \sqrt{\frac{1}{\sigma^2}}}{\sqrt{2\pi}}$

which is the density of a normal distribution with parameters

```
In[96]:= Limit[m[x], σp → ∞]
```

Out[96]= x

and

```
In[97]:= Limit[vpost, σp → ∞]
```

Out[97]= σ^2

This in particular implies the posterior confidence interval for the parameter a with the "uniform prior", i.e. in the "non-Bayesian" approach, to be

```
In[98]:= q[α_, x_, σ_] =
    {Quantile[NormalDistribution[x, σ], (1 - α)/2],
     Quantile[NormalDistribution[x, σ], 1 - (1 - α)/2]};
```

For example, the 95% confidence interval can be computed, as it is well known, to be

```
In[99]:= ci = q[.95, 0, 1]
```

Out[99]= {−1.95996, 1.95996}

4.5.5 Bayesian and Non-Bayesian Appreciation Rate Estimates

We apply the above framework in the context of the appreciation rate estimation in the simplest stock price model:

$$dS(t) = a\,S(t)\,dt + \sigma\,S(t)\,dB$$

As shown before, the random variable

$$X = \frac{\sigma^2}{2} + \frac{\log(S(T)) - \log(S(0))}{T} = \frac{\sigma^2}{2} + \frac{\log\left(\frac{S(T)}{S(0)}\right)}{T}$$

has a normal distribution with known standard deviation $\frac{\sigma}{\sqrt{T}}$ and unknown mean a:

$$X \sim N\left(a, \frac{\sigma}{\sqrt{T}}\right).$$

If no prior preference in regard to a exists, then according to Section 4.5.4.2, the posterior distribution for a, given X, is equal to

$$N\left(X, \frac{\sigma}{\sqrt{T}}\right).$$

On the other hand, if there is the prior preference in regard to a, and say, such a preference is modeled with the normal distribution

$$N(a_p, \sigma_p)$$

then according to Section 4.5.4.1, the posterior distribution for a, given X, is equal to

$$N\left(\frac{a_p\,\sigma^2 + T\,X\,\sigma_p^2}{\sigma^2 + T\,\sigma_p^2},\ \sqrt{\frac{\sigma^2\,\sigma_p^2}{\sigma^2 + T\,\sigma_p^2}}\right)$$

from which point and interval estimates are deduced easily. Here we compare the prior density for the appreciation rate, with the posterior, as well as with the non-Bayesian posterior density. The corresponding point estimates are given too.

```
In[100]:= Off[General::"spell1"];

In[101]:= With[{T = 1, σ = .5, ap = 0, σp = .4},
   (With[{X = #1}, Plot[{PDF[NormalDistribution[ap, σp],
        a], PDF[NormalDistribution[X, σ/√T], a],
       PDF[NormalDistribution[(ap σ^2 + T X σp^2)/(σ^2 + T σp^2),
         Sqrt[(σ^2 σp^2)/(σ^2 + T σp^2)]], a]}, {a, -5, 5},
      PlotLabel → "S[T]/S[0] = " <> ToString[e^(T (X - σ^2/2))],
      AxesLabel → {"a", ""}, PlotRange → All,
      Epilog → {{PointSize[0.03], RGBColor[0, 0, 1],
         Point[{(ap σ^2 + T X σp^2)/(σ^2 + T σp^2), 0}]},
        {PointSize[0.03], RGBColor[1, 0, 0],
         Point[{X, 0}]}}]] &) /@ Range[-4, 4, .1]];
```

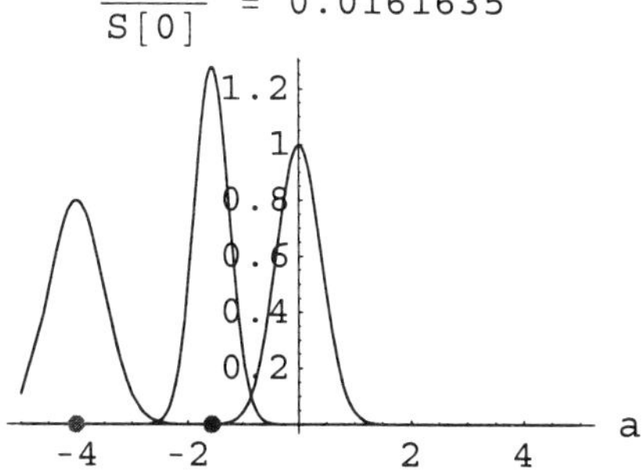

One can see the dependence on the outcome $\frac{S(T)}{S(0)}$, the ratio of the final stock price and the initial one. The reader should also experiment with the parameters T, σ, a_p, σ_p, to gain further insight.

Finally, we transform the formulas for the posterior variance and posterior mean into the forms that will be useful below. For variance we have

$$\frac{\sigma^2\,\sigma_p^2}{\sigma^2+T\,\sigma_p^2}=\frac{1}{\frac{\sigma^2+T\,\sigma_p^2}{\sigma^2\,\sigma_p^2}}=\frac{1}{\frac{1}{\sigma_p^2}+\frac{T}{\sigma^2}}=\left((\sigma_p^2)^{-1}+\left(\frac{\sigma^2}{T}\right)^{-1}\right)^{-1} \tag{4.5.4}$$

while, for the mean, the formula is equal to

$$\begin{aligned}\frac{a_p\,\sigma^2+T\,X\,\sigma_p^2}{\sigma^2+T\,\sigma_p^2}&=\frac{\sigma^2\,\sigma_p^2}{\sigma^2+T\,\sigma_p^2}\;\frac{a_p\,\sigma^2+T\,X\,\sigma_p^2}{\sigma^2\,\sigma_p^2}\\&=\left((\sigma_p^2)^{-1}+\left(\frac{\sigma^2}{T}\right)^{-1}\right)^{-1}\left(\frac{a_p}{\sigma_p^2}+\frac{X}{\frac{\sigma^2}{T}}\right)\\&=\left((\sigma_p^2)^{-1}+\left(\frac{\sigma^2}{T}\right)^{-1}\right)^{-1}\left((\sigma_p^2)^{-1}\,a_p+\left(\frac{\sigma^2}{T}\right)^{-1}X\right).\end{aligned} \tag{4.5.5}$$

The advantage of these formulas is that they suggest what the formulas are in the case of the *vector* basic stock price model, to be discussed next.

4.6 Vector Basic Price Model Statistics

4.6.1 Volatility in the Vector Basic Price Model

Consider m stocks with n sources of randomness. As seen in Chapter 2, the simplest vector SDE for modeling stock price evolution in a higher dimension can be written as

$$\frac{dS(t)}{S(t)}=a\,dt+\sigma.dB(t) \tag{4.6.1}$$

where $S(t)$, a are m-dimensional vectors, $B(t)$ is an n-dimensional vector (Brownian motion), while σ is an $m\times n$ matrix. This is the way we have defined SDEs so far, and the way we shall work in the future. But here it is more convenient to consider $S(t)$ as a column vector, i.e., a two-dimensional array, and the same for a and the Brownian motion $B(t)$. The reason is that the simple estimate for $\sigma.\sigma^T$ is then derived as

$$\left(\frac{dS(t)}{S(t)}\right).\left(\frac{dS(t)}{S(t)}\right)^T = (a\,dt + \sigma.dB(t)).(a\,dt + \sigma.dB(t))^T$$
$$= \sigma.dB(t).(\sigma.dB(t))^T = \sigma.dB(t).(dB(t))^T.\sigma^T = \sigma.\sigma^T\,dt$$

where we have used the fact that since the components of $B(t)$ are *independent* scalar Brownian motions, then $dB(t).(dB(t))^T = I\,dt$, where I is the $m \times m$ identity matrix. This derivation would not make sense if, say $\frac{dS(t)}{S(t)}$ is a one-dimensional array, there would be no transpose. Of course, averaging is needed, as well:

$$(\sigma.\sigma^T)_{\text{est}} = \text{Av}\Big[\frac{\left(\frac{dS(t)}{S(t)}\right).\left(\frac{dS(t)}{S(t)}\right)^T}{dt}\Big]. \tag{4.6.2}$$

Alternatively, the unique solution of the above SDE is derived to be equal by

$$\frac{S(t)}{S(s)} = e^{(t-s)\left(a-\frac{1}{2}\,\text{Diag}(\sigma.\sigma^T)\right)+\sigma.(B(t)-B(s))}$$

which implies

$$\log(S(t)) - \log(S(s)) = (t-s)\left(a - \frac{1}{2}\,\text{Diag}(\sigma.\sigma^T)\right) + \sigma.(B(t) - B(s))$$

and in particular

$$d\log(S(t)) = \left(a - \frac{1}{2}\,\text{Diag}(\sigma.\sigma^T)\right)dt + \sigma.dB(t).$$

This in turn yields

$$(d\log(S(t))).(d\log(S(t)))^T$$
$$= \left(\left(a - \frac{1}{2}\,\text{Diag}(\sigma.\sigma^T)\right)dt + \sigma.dB(t)\right).\left(\left(a - \frac{1}{2}\text{Diag}(\sigma.\sigma^T)\right)dt + \sigma.dB(t)\right)^T$$
$$= \sigma.dB(t).(\sigma.dB(t))^T = \sigma.\sigma^T\,dt$$

and therefore, an alternative estimate for $\sigma.\sigma^T$ is given by

$$(\sigma.\sigma^T)_{\text{est}} = \text{Av}\Big[\frac{(d\log(S(t))).(d\log(S(t)))^T}{dt}\Big]. \tag{4.6.3}$$

We shall see in examples below how efficient these estimates are. Drawing on one-dimensional examples we expect the same high level of accuracy. Also notice that σ is not estimated but rather $\sigma.\sigma^T$. As a matter of fact, σ is not uniquely determined (the same is true in the scalar case $\sigma = \pm\sqrt{\sigma^2}$) but sometimes can be computed as a matrix square root of $\sigma.\sigma^T \geq 0$. In particular, although the original σ may have been a non-square rectangular matrix, $\sigma.\sigma^T$ is a square (possibly degenerate) matrix.

What is the connection between *estimated* individual stock volatilities and the individual volatilities deduced from the *estimated* matrix σ? By the way, individual volatilities can be deduced from σ:

$$\text{IndividualVolatilities} = \sqrt{\text{Diag}(\sigma.\sigma^T)}$$

where $\sqrt{\square}$ is listable. They are the same. So estimating the matrix σ does not improve the knowledge of individual volatilities, which would be difficult anyway, since, as we have seen, volatility estimates are already excellent. Significant improvement exists nevertheless: the matrix σ contains the information about mutual dependencies. We shall see this in an example below.

4.6.2 Non-Bayesian Estimate for the Vector Appreciation Rate

Assume $\sigma.\sigma^T$ is known, or which is much more realistic, that it can be estimated very precisely. As before, SDE (4.6.1) has the unique solution given by

$$\frac{S(t)}{S(s)} = e^{(t-s)\left(a-\frac{1}{2}\text{Diag}(\sigma.\sigma^T)\right)+\sigma.(B(t)-B(s))}$$

which implies

$$\log\left(\frac{S(t)}{S(s)}\right) = (t-s)\left(a - \frac{1}{2}\text{Diag}(\sigma.\sigma^T)\right) + \sigma.(B(t)-B(s))$$

and therefore

$$X = \frac{\log\left(\frac{S(t)}{S(s)}\right)}{t-s} + \frac{1}{2}\text{Diag}(\sigma.\sigma^T) = a + \frac{\sigma.(B(t)-B(s))}{t-s} \tag{4.6.4}$$

is the basic non-Bayesian estimate for the vector appreciation rate a. Indeed,

$$E\,X = E\left(a + \frac{\sigma.(B(t)-B(s))}{t-s}\right) = a.$$

Also, the covariance matrix of X can be computed as

$$E\,X.X^T = E\left(\frac{\sigma.(B(t)-B(s))}{t-s}\right).\left(\frac{\sigma.(B(t)-B(s))}{t-s}\right)^T$$

$$= \frac{\sigma.}{(t-s)^2}\,E\,(B(t)-B(s)).(B(t)-B(s))^T.\sigma^T = \frac{\sigma.I\,(t-s).\sigma^T}{(t-s)^2} = \frac{\sigma.\sigma^T}{t-s}.$$

So, we conclude that the statistics X has a multinormal distribution

$$X \sim N_n\left(a, \frac{\sigma.\sigma^T}{t-s}\right) \tag{4.6.5}$$

or, which is equivalent, as shown in the scalar case above, that the posterior distribution μ_X of a, given X, is equal to

$$\mu_X = N_n\left(X, \frac{\sigma.\sigma^T}{t-s}\right) = N_n\left(\frac{\log(\frac{S(t)}{S(s)})}{t-s} + \frac{1}{2}\,\mathrm{Diag}(\sigma.\sigma^T),\ \frac{\sigma.\sigma^T}{t-s}\right).$$

The denominator $t-s$ in the covariance matrix $\frac{\sigma.\sigma^T}{t-s}$ of the above appreciation rate estimate X shows that the longer the time interval, the more accurate is the appreciation rate estimate. It also shows that in realistic market time frames, the above estimate is not sufficiently efficient, and therefore may need to be improved by some possibly ad hoc insights.

What is the connection between *estimated* individual appreciation rates and the individual appreciation rates deduced from the *estimated* vector a? Again they are the same. The advantage of the vector model over individual studies of the stocks is in the study of their dependencies, which is not contained in appreciation rates, although dependencies do affect the nature and level of our confidence in the estimates of the appreciation rates. This point is discussed in some detail below .

4.6.3 Bayesian Statistical Estimate for the Vector Appreciation Rate

Consider the simplest vector stock price model (4.6.1). For any vector $a \in A = \mathbb{R}^n$, the statistics X, defined in (4.6.4), has a normal $N_n\left(a, \frac{\sigma.\sigma^T}{t-s}\right)$ distribution (with density $f_a(x)$ and probability P_a). Assume the prior distribution $\mu(a)$ for a is normal $N_n(a_p, \sigma_p.\sigma_p{}^T)$, with prior mean vector a_p and prior covariance matrix $\sigma_p.\sigma_p{}^T$. As in the scalar case, the problem is to find the posterior distribution μ_x, $x \in A$. Formulas at the end of Section 4.5.5 extend as it was then announced:

$$\mu_X = N_n\left[\left((\sigma_p.\sigma_p{}^T)^{-1} + \left(\frac{\sigma.\sigma^T}{t-s}\right)^{-1}\right)^{-1} \cdot \left((\sigma_p.\sigma_p{}^T)^{-1}.a_p + \left(\frac{\sigma.\sigma^T}{t-s}\right)^{-1}.X\right),\right.$$

$$\left.\left((\sigma_p.\sigma_p{}^T)^{-1} + \left(\frac{\sigma.\sigma^T}{t-s}\right)^{-1}\right)^{-1}\right] = N_n\left[\left((\sigma_p.\sigma_p{}^T)^{-1} + \left(\frac{\sigma.\sigma^T}{t-s}\right)^{-1}\right)^{-1}\right.$$

$$\cdot\left((\sigma_p.\sigma_p{}^T)^{-1}.a_p + \left(\frac{\sigma.\sigma^T}{t-s}\right)^{-1} \cdot \left(\frac{\log(\frac{S(t)}{S(s)})}{t-s} + \frac{1}{2}\,\mathrm{Diag}(\sigma.\sigma^T)\right)\right),$$

$$\left.\left((\sigma_p.\sigma_p{}^T)^{-1} + \left(\frac{\sigma.\sigma^T}{t-s}\right)^{-1}\right)^{-1}\right].$$

This in particular implies that as a point posterior estimate of a one should take:

$$a_{\text{post}} = \left((\sigma_p.\sigma_p{}^T)^{-1} + \left(\frac{\sigma.\sigma^T}{t-s} \right)^{-1} \right)^{-1}$$
$$.\left((\sigma_p.\sigma_p{}^T)^{-1}.a_p + \left(\frac{\sigma.\sigma^T}{t-s} \right)^{-1} . \left(\frac{\log(S(t)) - \log(S(s))}{t-s} + \frac{1}{2}\,\text{Diag}(\sigma.\sigma^T) \right) \right)$$

and such an estimate then has a measure of indeterminacy, i.e., the confidence region for the true value of a around the point estimate a_{post} is determined via the covariance matrix

$$\left((\sigma_p.\sigma_p{}^T)^{-1} + \left(\frac{\sigma.\sigma^T}{t-s} \right)^{-1} \right)^{-1}.$$

This kind of information about the appreciation rate will be used in Chapter 7 as an input for various portfolio optimization problems. So, not only the appreciation rate estimate, but also the confidence we have about that estimate, which is contained in the covariance matrix $\left((\sigma_p.\sigma_p{}^T)^{-1} + \left(\frac{\sigma.\sigma^T}{t-s} \right)^{-1} \right)^{-1}$ in the case of this Bayesian estimate, or in the covariance matrix $\frac{\sigma.\sigma^T}{t-s}$ in the case of the non-Bayesian estimate, can be used in measuring and optimizing the risk undertaken in the portfolio selection.

4.6.4 Experiments: Statistics for Vector Basic Price Model

4.6.4.1 Experiment set up

We shall need the vector SDE solver:

```
In[102]:= Remove["Global`*"]
          << "CFMLab`ItoSDEs`"
          Off[General::"spell1"]
```

```
In[105]:= << "Statistics`MultinormalDistribution`"
```

Consider the simplest vector stock price model (4.6.1), where

```
In[106]:= σ = {{0.257, -0.188, -0.461, -0.209},
              {0.189, 0.291, 0.32, 0.2}};
```

yielding individual stock volatilities:

```
In[107]:= vol = Transpose[{(√Plus @@ #1² &) /@ σ}]
```

$$\text{Out[107]=} \begin{pmatrix} 0.597992 \\ 0.512642 \end{pmatrix}$$

and the *correlation* matrix

In[108]:= $\frac{\sigma.\text{Transpose}[\sigma]}{\text{vol.Transpose}[\text{vol}]}$

Out[108]= $\begin{pmatrix} 1. & -0.637583 \\ -0.637583 & 1. \end{pmatrix}$

Let

```
In[109]:= ap = {0, .2}; σp = {{1., .01}, {0, .1}}
```

Out[109]= $\begin{pmatrix} 1. & 0.01 \\ 0 & 0.1 \end{pmatrix}$

be the parameters of the prior distribution for the stock's appreciation rate vector a:

```
In[110]:= prior :=
            MultinormalDistribution[ap, σp.Transpose[σp]]
```

So let the *real* appreciation rate be (unknown)

```
In[111]:= a = Random[prior];
```

Let also initial market prices be

```
In[112]:= P0 = {100, 120};
```

We can simulate the stock prices possible future evolutions by

```
In[113]:= s = 0; t = 1; b[t_, S_] := S a; c[t_, S_] := S σ;
          stocksSDE = SDESolver[b, c, P0, s, t, 1500];
```

and visualize them (StockExtract was introduced before in Chapter 2, and placed in the package CFMLab`ItoSDEs`)

```
In[114]:= lp = (ListPlot[StockExtract[#1, stocksSDE],
                PlotStyle → RGBColor[Random[], Random[],
                  Random[]], PlotJoined → True,
                DisplayFunction → Identity] &) /@
            Range[Length[a]]; Show[lp,
           DisplayFunction → $DisplayFunction,
           PlotRange → {0, Automatic}];
```

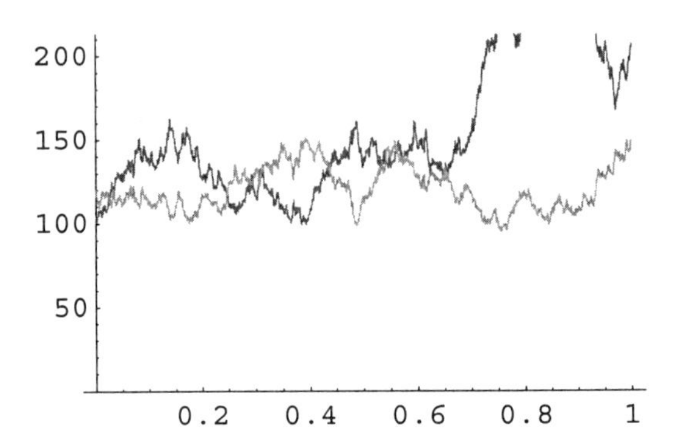

4.6.4.2 Estimate of $\sigma.\sigma^T$

Above, we derived two estimates for $\sigma.\sigma^T$, (4.6.2) and (4.6.3). We implement only the first one. Since $S(t)$ and consequently $dS(t)$, is, by construction, a one-dimensional array, $\left(\frac{dS(t)}{S(t)}\right)^T$ in the above formula does not make sense. So, we have to make a two-dimensional array (for every time t) along the way:

```
In[115]:= preS = Transpose[stocksSDE] [[2]];
          preT = Transpose[stocksSDE] [[1]];
          S = Drop[preS, -1];
          dS = (Transpose[{#1}] &) /@ (Drop[preS, 1] - S);
          dT = Drop[preT, 1] - Drop[preT, -1];
          Av[List_] := Plus @@ List / Length[List]
```

and then

$$\text{In[120]:= Av}\left[\frac{(\#1.\text{Transpose}[\#1]\ \&)\ /@\ \frac{\text{dS}}{\text{S}}}{\text{dT}}\right]$$

$$\text{Out[120]=}\begin{pmatrix} 0.363204 & -0.202346 \\ -0.202346 & 0.269 \end{pmatrix}$$

gives the estimate. We wrap these few lines into a function named StockCovariance and store it in the package CFMLab`StockStat`. Indeed,

```
In[121]:= << "CFMLab`StockStat`"
```

```
In[122]:= StockCovariance[stocksSDE]
```

$$\text{Out[122]=}\begin{pmatrix} 0.363204 & -0.202346 \\ -0.202346 & 0.269 \end{pmatrix}$$

while the true value is

```
In[123]:= σ.Transpose[σ]
```

Out[123]= $\begin{pmatrix} 0.357595 & -0.195455 \\ -0.195455 & 0.262802 \end{pmatrix}$

4.6.4.3 Non-Bayesian Estimate for *a*

It has been shown before that if no prior is taken into account, and if X is given by (4.6.4), then the true value of a has the posterior distribution μ_X

$$\mu_X = N_n\left(X, \frac{\sigma.\sigma^T}{t-s}\right)$$

This of course yields the point estimate

```
In[124]:= GrowthRateEstimate[stocksSDE_] :=
              Log[ stocksSDE[[-1,2]] / stocksSDE[[1,2]] ]
          ----------------------------------------- +
          stocksSDE[[-1, 1]] - stocksSDE[[1, 1]]
          1/2 Diag[StockCovariance[stocksSDE]];
        Diag[Matrix_] := Table[Matrix[[i, i]],
          {i, 1, Length[Matrix]}]
```

which works as

```
In[126]:= GrowthRateEstimate[stocksSDE]
Out[126]= {0.909853, 0.341822}
```

while the true value is

```
In[127]:= a
Out[127]= {1.1184, 0.277419}
```

It seems reasonable to also define

```
In[128]:= GrowthRatePosterior[stocksSDE_] :=
           MultinormalDistribution[
            GrowthRateEstimate[stocksSDE],
                 StockCovariance[stocksSDE]
            -------------------------------------- ]
            stocksSDE[[-1, 1]] - stocksSDE[[1, 1]]
```

As an application, we test whether, for a given confidence probability, the confidence region contains the true value of a (the black point):

```
In[129]:= NoPrior = With[{Probability = .95}, Show[
              Graphics[{RGBColor[1, 0, 0], PointSize[.02],
                Point[GrowthRateEstimate[stocksSDE]]}],
              Graphics[{RGBColor[0, 0, 0],
                PointSize[.03], Point[a]}], Graphics[
```

```
{RGBColor[1, 0, 0], EllipsoidQuantile[
  GrowthRatePosterior[stocksSDE],
  Probability]}], Frame → True]];
```

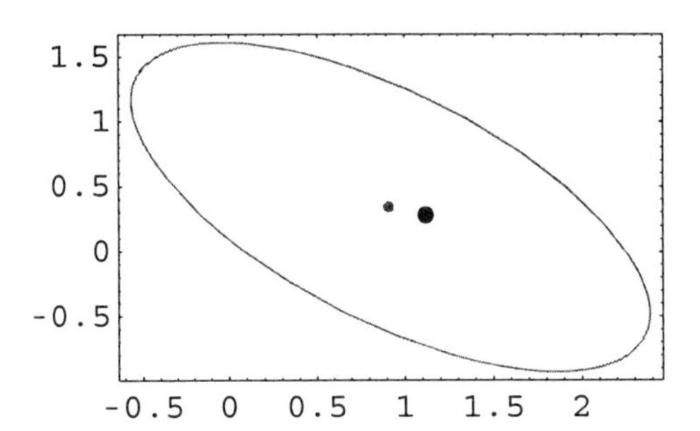

We store functions Diag, GrowthRateEstimate, GrowthRatePosterior into the package CFMLab`StockStat` for future reference.

4.6.4.4 Bayesian Estimate for *a*

Now we take the prior into account. So, as it was derived before, if X is given by (4.6.4), then, this time, the true value of a has the posterior distribution μ_X, given X, where

$$\mu_X = N_n\left[\left((\sigma_p.\sigma_p{}^T)^{-1} + \left(\frac{\sigma.\sigma^T}{t-s}\right)^{-1}\right)^{-1} \cdot \left((\sigma_p.\sigma_p{}^T)^{-1}.a_p + \left(\frac{\sigma.\sigma^T}{t-s}\right)^{-1}.X\right),\right.$$
$$\left.\left((\sigma_p.\sigma_p{}^T)^{-1} + \left(\frac{\sigma.\sigma^T}{t-s}\right)^{-1}\right)^{-1}\right].$$

This of course yields the point estimate

```
In[130]:= GrowthRateEstimateWithPrior[stocksSDE_, ap_, σp_] :=
   Inverse[Inverse[σp.Transpose[σp]] +
      Inverse[StockCovariance[stocksSDE] /
        (stocksSDE[[-1, 1]] - stocksSDE[[1, 1]])]].
    (Inverse[σp.Transpose[σp]].ap +
      Inverse[StockCovariance[stocksSDE] /
        (stocksSDE[[-1, 1]] - stocksSDE[[1, 1]])].
       GrowthRateEstimate[stocksSDE])
```

which works as

```
In[131]:= GrowthRateEstimateWithPrior[stocksSDE, ap, σp]
Out[131]= {0.832229, 0.211912}
```

while the true value is

```
In[132]:= a

Out[132]= {1.1184, 0.277419}
```

Again, it seems reasonable to also define

```
In[133]:= GrowthRatePosteriorWithPrior[stocksSDE_,
            ap_, σp_] := MultinormalDistribution[
            GrowthRateEstimateWithPrior[stocksSDE, ap, σp],
            Inverse[Inverse[σp.Transpose[σp]] +
              Inverse[ StockCovariance[stocksSDE] /
                (stocksSDE[[-1, 1]] - stocksSDE[[1, 1]]) ]]]
```

and we compare the present confidence region (with prior), with the one computed before (without prior), with the same confidence probability. Again, the true value of a is the black point:

```
In[134]:= Off[MultinormalDistribution::"cmsym"]

In[135]:= Off[Experimental`CholeskyDecomposition::"herm"]

In[136]:= With[{Probability = .95}, Show[NoPrior,
             Graphics[{RGBColor[1, 0, 0], PointSize[.02],
               Point[GrowthRateEstimateWithPrior[
                 stocksSDE, ap, σp]]}],
             Graphics[{RGBColor[0, 0, 0],
               PointSize[.03], Point[a]}],
             Graphics[{RGBColor[1, 0, 0], EllipsoidQuantile[
                GrowthRatePosteriorWithPrior[stocksSDE,
                 ap, σp], Probability]}], Frame → True]];
```

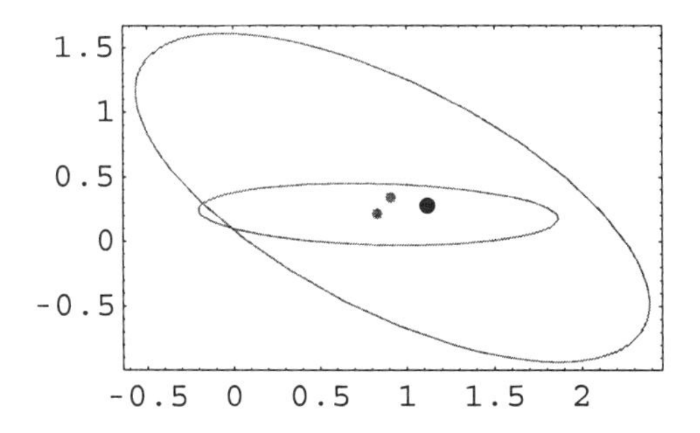

We store these functions into the package CFMLab`StockStat` for future reference as well.

4.6.5 Volatility and Appreciation Rate Estimates for Real Vector Data

4.6.5.1 Data Import and Formatting

```
In[137]:= Clear["Global`*"];
          Off[General::"spell1"]
```

Recall

```
In[139]:= << "CFMLab`DataImport`";
          << "CFMLab`ItoSDEs`";
```

We shall consider

```
In[141]:= Tickers = {ge, pg, amzn, yhoo, qqq};
```

First, import the data

```
In[142]:= StockData = (GetPriceData[#1] &) /@ Tickers;
```

and since not all the data has the same time frame, we have to extract the maximal time interval that is covered in each of the data sets:

```
In[143]:= {s, t} = ({Max[#1[[1]]], Min[#1[[2]]]} &)[
            Transpose[({Min[#1], Max[#1]} &) /@
              (Transpose[#1][[1]] &) /@ StockData]]

Out[143]= {-0.811339, 1.9923}
```

Next we restrict all of the data on that time interval

```
In[144]:= StockData2 =
            (Select[#1, s ≤ #1[[1]] ≤ t &] &) /@ StockData;
```

and then rewrite all of the data so that it looks like the one generated in the Monte–Carlo experiments above:

```
In[145]:= stocksSDE = ({#1[[1, 1]], Transpose[#1][[2]]} &) /@
             Transpose[StockData2];
```

Checking whether everything works, as it should, we see that they look like

```
In[146]:= lp = (ListPlot[StockExtract[#1, stocksSDE],
               PlotJoined → True,
               DisplayFunction → Identity, PlotStyle →
                RGBColor[Random[], 0, Random[]]] &) /@
            Range[Length[Tickers]]; Show[lp,
           DisplayFunction → $DisplayFunction,
           PlotRange → All];
```

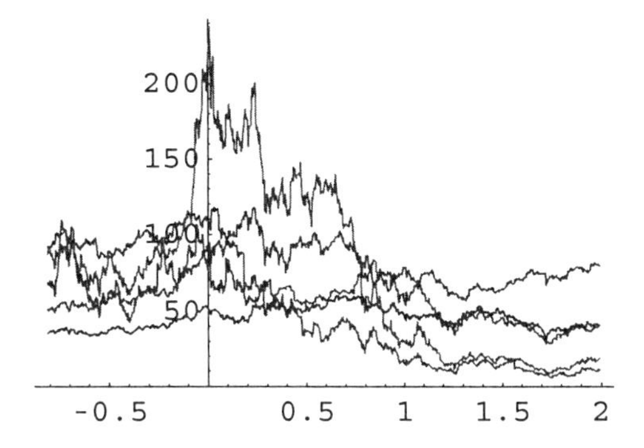

We are ready to apply same set of functions to get all of the estimates.

4.6.5.2 Estimating and Using $\sigma.\sigma^T$

We need

```
In[147]:= << "CFMLab`StockStat`"
```

Then we can quickly compute the covariance matrix $\sigma.\sigma^T$ estimate:

```
In[148]:= SC = StockCovariance[stocksSDE]
```

$$
Out[148]= \begin{pmatrix} 0.15252 & 0.0356272 & 0.108494 & 0.121132 & 0.106561 \\ 0.0356272 & 0.195925 & -0.0121239 & 0.0130063 & 0.000382116 \\ 0.108494 & -0.0121239 & 1.25962 & 0.621459 & 0.317832 \\ 0.121132 & 0.0130063 & 0.621459 & 1.01972 & 0.365724 \\ 0.106561 & 0.000382116 & 0.317832 & 0.365724 & 0.312288 \end{pmatrix}
$$

from which, a possible choice for σ estimate:

```
In[149]:= σest = MatrixPower[SC, 1/2]
```

$$
Out[149]= \begin{pmatrix} 0.365692 & 0.0447186 & 0.0517592 & 0.0640236 & 0.10006 \\ 0.0447186 & 0.440077 & -0.0110283 & 0.0104783 & -0.00505338 \\ 0.0517592 & -0.0110283 & 1.07099 & 0.289878 & 0.160569 \\ 0.0640236 & 0.0104783 & 0.289878 & 0.939534 & 0.220809 \\ 0.10006 & -0.00505338 & 0.160569 & 0.220809 & 0.477192 \end{pmatrix}
$$

It is interesting to know individual volatilities, and mutual correlations. To this end, let

```
In[150]:= Volatility[σ_] := Transpose[{(√Plus @@ #1² &) /@ σ}]
```

and

```
In[151]:= Correlation[σ_] :=
                        σ.Transpose[σ]
          ─────────────────────────────────────────
          Volatility[σ].Transpose[Volatility[σ]]
```

Then, we have the individual volatilities

```
In[152]:= {Tickers, Flatten[Volatility[σest]]}
```

Out[152]=

ge	pg	amzn	yhoo	qqq
0.390539	0.442634	1.12233	1.00981	0.558828

as well as the matrix of mutual correlations

```
In[153]:= Correlation[σest]
```

Out[153]=

1.	0.206098	0.247526	0.307152	0.488265
0.206098	1.	−0.0244049	0.0290984	0.0015448
0.247526	−0.0244049	1.	0.548343	0.506757
0.307152	0.0290984	0.548343	1.	0.64809
0.488265	0.0015448	0.506757	0.64809	1.

from which we can draw the correlation for particular pairs of stocks. If the matrix is much bigger, which indeed is the case in reality, then the following function may be helpful:

```
In[154]:= Corr[Ticker1_, Ticker2_] := Correlation[σest][[
             Position[Tickers, Ticker1][[1, 1]],
             Position[Tickers, Ticker2][[1, 1]]]]
```

which, for example, works like

```
In[155]:= Corr[amzn, yhoo]

Out[155]= 0.548343
```

4.6.5.3 Non-Bayesian Appreciation Rate Estimates

Next, we estimate the appreciation rates, without any prior:

```
In[156]:= Off[General::"spell"];
          aest = GrowthRateEstimate[stocksSDE]

Out[156]= {0.1278, 0.0451945, −0.026116, −0.0454374, 0.0719839}
```

or, we can deduce the full distribution of the vector appreciation rate, without any prior:

```
In[157]:= GRP = GrowthRatePosterior[stocksSDE]
```

```
Out[157]= MultinormalDistribution({0.1278, 0.0451945, −0.026116,
            −0.0454374, 0.0719839}, (≪1≫))
```

implying, in particular that the covariance matrix $S_{\text{est}}.S_{\text{est}}^T$ for the true value of the appreciation rate vector is equal to

```
In[158]:= GRP⟦2⟧
```

Out[158]=

$$\begin{pmatrix} 0.0544009 & 0.0127075 & 0.0386975 & 0.0432052 & 0.038008 \\ 0.0127075 & 0.0698823 & -0.00432434 & 0.00463909 & 0.000136293 \\ 0.0386975 & -0.00432434 & 0.449282 & 0.221662 & 0.113364 \\ 0.0432052 & 0.00463909 & 0.221662 & 0.363713 & 0.130446 \\ 0.038008 & 0.000136293 & 0.113364 & 0.130446 & 0.111387 \end{pmatrix}$$

We can also compute S_{est} as

```
In[159]:= Sest = MatrixPower[%, 1 / 2]
```

Out[159]=

$$\begin{pmatrix} 0.218401 & 0.0267071 & 0.030912 & 0.0382366 & 0.0597584 \\ 0.0267071 & 0.262826 & -0.0065864 & 0.00625792 & -0.00301801 \\ 0.030912 & -0.0065864 & 0.639621 & 0.173123 & 0.0958964 \\ 0.0382366 & 0.00625792 & 0.173123 & 0.561115 & 0.131873 \\ 0.0597584 & -0.00301801 & 0.0958964 & 0.131873 & 0.284992 \end{pmatrix}$$

The ellipsoid quantiles follow again, but now, in 5 dimensions, it is not possible to visualize them. Furthermore, in real applications the dimension of the problem is going to be much higher, in the hundreds. Well, although visualization helps understanding the concepts, it is not important in real applications. Indeed, in Chapter 7, on optimal portfolio rules, we shall use in a crucial way the quantities $\sigma_{\text{est}}.\sigma_{\text{est}}^T$, a_{est}, $S_{\text{est}}.S_{\text{est}}^T$, and for that we don't need to visualize anything.

4.6.5.4 Bayesian Appreciation Rate Estimates

Suppose finally we were smart enough to predict the market decline, and that such a prediction was quantified with a prior

```
In[160]:= ap = Table[-.3, {Length[Tickers]}]

Out[160]= {−0.3, −0.3, −0.3, −0.3, −0.3}
```

and

```
In[161]:= σp[α_] = α IdentityMatrix[Length[Tickers]]
```

Out[161]= $\begin{pmatrix} \alpha & 0 & 0 & 0 & 0 \\ 0 & \alpha & 0 & 0 & 0 \\ 0 & 0 & \alpha & 0 & 0 \\ 0 & 0 & 0 & \alpha & 0 \\ 0 & 0 & 0 & 0 & \alpha \end{pmatrix}$

This would yield a_{est2}

```
In[162]:= aest2 = GrowthRateEstimateWithPrior[
            stocksSDE, ap, σp[1]]
```

Out[162]= {0.0812222, 0.0181061, −0.161997, −0.17696, −0.00689213}

and $S_{est2}.S_{est2}^T$

```
In[163]:= Off[MultinormalDistribution::"cmsym"]
```

```
In[164]:= GrowthRatePosteriorWithPrior[
            stocksSDE, ap, σp[1]][[2]]
```

Out[164]= $\begin{pmatrix} 0.048748 & 0.0112687 & 0.0195551 & 0.0242724 & 0.0276867 \\ 0.0112687 & 0.0651545 & -0.00359178 & 0.0034364 & -0.000307702 \\ 0.0195551 & -0.00359178 & 0.288107 & 0.109453 & 0.0590999 \\ 0.0242724 & 0.0034364 & 0.109453 & 0.24076 & 0.0771191 \\ 0.0276867 & -0.000307702 & 0.0590999 & 0.0771191 & 0.0841965 \end{pmatrix}$

and then $\sigma_{est}.\sigma_{est}^T$, a_{est2}, $S_{est2}.S_{est2}^T$ would be the crucial quantities to be used in portfolio optimization under the appreciation rate uncertainty, to be discussed in Chapter 7. Finally, we notice explicitly that if $\alpha \to \infty$, then $a_{est2} \to a_{est}$, and $S_{est2} \to S_{est}$, or which is the same but easier to accomplish below $a_{est2} - a_{est} \to 0$, and $S_{est2} - S_{est} \to 0$. Indeed,

```
In[165]:= Chop[
           Limit[GrowthRatePosteriorWithPrior[stocksSDE,
               ap, σp[α]][[1]], α → ∞] - aest]
```

Out[165]= {0, 0, 0, 0, 0}

and

```
In[166]:= Chop[
           MatrixPower[Limit[GrowthRatePosteriorWithPrior[
               stocksSDE, ap, σp[α]][[2]], α → ∞], 1/2] - Sest]
```

$$\text{Out[166]=} \begin{pmatrix} 0 & 0 & 0 & 0 & 0 \\ 0 & 0 & 0 & 0 & 0 \\ 0 & 0 & 0 & 0 & 0 \\ 0 & 0 & 0 & 0 & 0 \\ 0 & 0 & 0 & 0 & 0 \end{pmatrix}$$

4.7 Dynamic Statistics: Filtering of Conditional Gaussian Processes

4.7.1 Conditional Gaussian Filtering

We end this chapter with a few useful facts from the theory of filtering of stochastic processes. Our motivation is two-fold: to compare this sophisticated theory with better known aspects of statistical estimation such as the ones already presented and to state an explicit estimate that is needed in Section 8.3.

Stock price history, used in statistical estimation, is updated continuously, each trading day, hour, and minute. The estimates presented so far are not suited for such a practice since they depend on the whole history, and not just on the new information coupled with the current statistical estimate. So what is needed are the recursive statistical estimates. They are provided by the theory of linear and non-linear filtering of random processes.

Also, as we have seen so far, statistical estimates for the appreciation rate are not very efficient. One might wonder whether more efficient estimates can be produced if a much more sophisticated theory such as the theory of filtering of random processes were employed. Unfortunately, to that question the answer is negative. The much more sophisticated theory to be discussed *very* briefly here can be used, as we shall see, only to reproduce the same estimates (see Section 4.7.2.2) in their recursive form, or which is much more important, to produce statistical estimates for more complicated models (see Section 4.7.3).

The general results on filtering of conditional Gaussian processes are from Liptser and Shiryayev's classical book [42]. *Mathematica*® enables us to see those beautiful results in action, and potentially to use them in trading practice. Consider the partially observable random process $\{\theta, \xi\}$, where

$$\begin{aligned} d\theta(t) &= (a_0(t, \xi) + a_1(t, \xi)\,\theta(t))\,dt + b_1(t, \xi)\,dB_1(t) + b_2(t, \xi)\,dB_2(t) \\ d\xi(t) &= (c_0(t, \xi) + c_1(t, \xi)\,\theta(t))\,dt + d(t, \xi)\,dB_2(t) \end{aligned} \tag{4.7.1}$$

and where only ξ is observable. Both θ and ξ are scalar valued (see [42] for the vector case). One should notice that in all of the coefficients above, the ξ-dependence is quite general (e.g. ξ in the equation is not necessarily equal to $\xi(t)$), but it has to be non-anticipative, i.e., for example $a_0(t, \xi)$ cannot depend on $\xi(s)$ for $s > t$. Nevertheless, the most prominent example of such a system, and probably general enough for many applications, is

$$\begin{aligned} d\theta(t) &= (a_0(t,\xi(t)) + a_1(t,\xi(t))\,\theta(t))\,dt + b_1(t,\xi(t))\,dB_1(t) + b_2(t,\xi(t))dB_2(t) \\ d\xi(t) &= (c_0(t,\xi(t)) + c_1(t,\xi(t))\,\theta(t))\,dt + d(t,\xi(t))\,dB_2(t) \end{aligned} \tag{4.7.2}$$

i.e., the case when ξ-dependence is through $\xi(t)$ only. Therefore, we shall restrict our attention to only such a case. Also, for what follows it will be necessary to assume that the conditional distribution of $\theta(0)$ is normal, or more precisely, that

$$P(\theta(0) \le x \mid \xi(0)) \sim N(m_0, g_0) \tag{4.7.3}$$

where $m_0 = m(0) = E[\theta(0) \mid \xi(0)]$ and $g_0^2 = \gamma(0) = E[(\theta(0) - m(0))^2 \mid \xi(0)]$ are the conditional expectation and the conditional variance, respectively.

The problem is to compute the conditional expectation

$$m(t) = E[\theta(t) \mid \xi(s),\ 0 \le s \le t]. \tag{4.7.4}$$

To this end we need to compute also the conditional variance

$$\gamma(t) = E[(\theta(t) - m(t))^2 \mid \xi(s),\ 0 \le s \le t]. \tag{4.7.5}$$

The fundamental result (for the proof see [42]) is that the conditional variance function $\gamma(t)$ is the unique solution of the ODE

$$\begin{aligned} \gamma'(t) &= b_1(t,\xi(t))^2 - \left(\frac{b_2(t,\xi(t))\,d(t,\xi(t)) + c_1(t,\xi(t))\,\gamma(t)}{d(t,\xi(t))}\right)^2 \\ &\quad + b_2(t,\xi(t)) + 2\,a_1(t,\xi(t))\,\gamma(t) \\ \gamma(0) &= g_0^2. \end{aligned} \tag{4.7.6}$$

Once $\gamma(t)$ is computed, the conditional expectation $m(t)$ can be computed recursively via

$$dm(t) = \alpha_0(t,\xi(t))\,dt + \alpha_1(t,\xi(t))\,m(t)\,dt + \alpha_2(t,\xi(t))\,d\xi(t) \tag{4.7.7}$$

where

$$\begin{aligned} \mathrm{A}(t,\xi) &= \frac{b_2(t,\xi)\,d(t,\xi) + c_1(t,\xi)\,\gamma(t)}{d(t,\xi)^2} \\ \alpha_0(t,\xi) &= a_0(t,\xi) - c_0(t,\xi)\,\mathrm{A}(t,\xi) \\ \alpha_1(t,\xi) &= a_1(t,\xi) - c_1(t,\xi)\,\mathrm{A}(t,\xi) \\ \alpha_2(t,\xi) &= \mathrm{A}(t,\xi). \end{aligned} \tag{4.7.8}$$

4.7.2 Kalman–Bucy Filtering Implemented

4.7.2.1 General Equations

```
In[167]:= Remove["Global`*"]
```

```
In[168]:= Off[General::"spell1"]
```

The special case of the above stated conditional Gaussian filtering is the famous Kalman–Bucy filtering. The Kalman–Bucy filtering can be considered as a special case of the above conditional Gaussian filtering if, in the above notation, the coefficients are chosen to be

```
In[169]:= a0[t_, ξ_] := 0; a1[t_, ξ_] := a[t];
          b1[t_, ξ_] := b[t]; b2[t_, ξ_] := 0; c0[t_, ξ_] := 0;
          c1[t_, ξ_] := A[t]; d[t_, ξ_] := B[t]
```

The general system (4.7.2) becomes a *linear* SDE system:

$$d\theta(t) = a(t)\,\theta(t)\,dt + b(t)\,dB_1(t)$$
$$d\xi(t) = A(t)\,\theta(t)\,dt + B(t)\,dB_2(t).$$

The conditional variance is this time computed by solving the ODE:

```
In[170]:= γ'[t] ==
            -((d[t, ξ[t]] b2[t, ξ[t]] + γ[t] c1[t, ξ[t]]) / d[t, ξ[t]])^2 +
            b1[t, ξ[t]]^2 + 2 γ[t] a1[t, ξ[t]] + b2[t, ξ[t]]
```

Out[170]= $\gamma'(t) == b(t)^2 - \frac{A(t)^2\,\gamma(t)^2}{B(t)^2} + 2\,a(t)\,\gamma(t)$

while the optimal linear filter is given by

$$dm(t) = \alpha_0(t, \xi)\,dt + \alpha_1(t, \xi)\,m(t)\,dt + \alpha_2(t, \xi)\,d\xi(t) \tag{4.7.9}$$

where

```
In[171]:= A[t_, ξ_] = (d[t, ξ] b2[t, ξ] + γ[t] c1[t, ξ]) / d[t, ξ]^2;
          α0[t_, ξ_] = a0[t, ξ] - c0[t, ξ] A[t, ξ]
```

Out[172]= 0

```
In[173]:= α1[t_, ξ_] = a1[t, ξ] - c1[t, ξ] A[t, ξ]
```

Out[173]= $a(t) - \frac{A(t)^2\,\gamma(t)}{B(t)^2}$

```
In[174]:= α2[t_, ξ_] = A[t, ξ]
```

Out[174]= $\frac{A(t)\,\gamma(t)}{B(t)^2}$

and therefore equation (4.7.9) becomes

$$dm(t) = \left(a(t) - \frac{A(t)^2\,\gamma(t)}{B(t)^2}\right) m(t)\,dt + \frac{A(t)\,\gamma(t)}{B(t)^2}\,d\xi(t). \tag{4.7.10}$$

4.7.2.2 Recursive Estimate for the Appreciation Rate

Consider the simplest (scalar) stock price model

$$dS(t) = S(t)\,a\,dt + S(t)\,\sigma\,dB(t)$$

and the logarithm of the stock price $\xi(t) = \log(S(t))$ and its dynamics:

$$d\xi(t) = \left(a - \frac{\sigma^2}{2}\right) dt + \sigma\,dB(t).$$

Let $\theta(t) = \theta(0) = a - \frac{\sigma^2}{2}$. Then

$$\begin{aligned} d\theta(t) &= 0 \\ d\xi(t) &= \theta(t)\,dt + \sigma\,dB_2(t) \end{aligned} \tag{4.7.11}$$

and the estimation of $\theta(t) = a - \frac{\sigma^2}{2}$, and consequently of $a = \theta(t) + \frac{\sigma^2}{2}$ is recast into a Kalman–Bucy filtering problem.

It was said in the opening of Section 4.7 that the filtering theory does not bring anything new in the simplest cases such as this one, except for the recursive form of the estimates. We shall check that claim first. It was shown in Section 4.5.5 that the point estimate for a is equal to

$$E[a \mid \xi(t), 0 \le t \le T] = E[a \mid S(t), 0 \le t \le T] = \frac{a_p\,\sigma^2 + T\,X\,{\sigma_p}^2}{\sigma^2 + T\,{\sigma_p}^2}$$

$$= \frac{a_p\,\sigma^2 + T\left(\frac{\sigma^2}{2} + \frac{\log\left(\frac{S(T)}{S(0)}\right)}{T}\right){\sigma_p}^2}{\sigma^2 + T\,{\sigma_p}^2}$$

$$= \frac{\left(m_0 + \frac{\sigma^2}{2}\right)\sigma^2 + T\left(\frac{\sigma^2}{2} + \frac{\log\left(\frac{S(T)}{S(0)}\right)}{T}\right){\sigma_p}^2}{\sigma^2 + T\,{\sigma_p}^2}$$

$$= \frac{m_0\,\sigma^2 + \log\left(\frac{S(T)}{S(0)}\right){\sigma_p}^2 + \frac{\sigma^2}{2}\left(\sigma^2 + T\,{\sigma_p}^2\right)}{\sigma^2 + T\,{\sigma_p}^2}$$

$$= \frac{m_0\,\sigma^2 + (\xi(T) - \xi(0))\,{\sigma_p}^2}{\sigma^2 + T\,{\sigma_p}^2} + \frac{\sigma^2}{2}$$

where $m_0 = m(0) = E[\theta(t) \mid \xi(0)]$, and where the rest of the notation is from Section 4.5.5, or equivalently

$$E[\theta(t) \mid \xi(t),\ 0 \le t \le T] = E\left[a - \frac{\sigma^2}{2} \,\middle|\, \xi(t),\ 0 \le t \le T\right]$$

$$= E[a \mid \xi(t),\ 0 \le t \le T] - \frac{\sigma^2}{2} \tag{4.7.12}$$

$$= \frac{m_0\, \sigma^2 + (\xi(T) - \xi(0))\, {\sigma_p}^2}{\sigma^2 + T\, {\sigma_p}^2}.$$

On the other hand, the Kalman–Bucy filtering equations reduce to

$$d\,m(t) = \left(a(t) - \frac{A(t)^2\, \gamma(t)}{B(t)^2}\right) m(t)\, d\,t + \frac{A(t)\, \gamma(t)}{B(t)^2}\, d\,\xi(t)$$
$$= -\frac{\gamma(t)}{\sigma^2}\, m(t)\, d\,t + \frac{\gamma(t)}{\sigma^2}\, d\,\xi(t) \tag{4.7.13}$$

for

$$\gamma'(t) = -\frac{\gamma(t)^2}{\sigma^2}$$
$$\gamma(0) = E[(\theta(0) - m(0))^2 \mid \xi(0)] = {\sigma_p}^2. \tag{4.7.14}$$

Solving the ODE (4.7.14) we get

```
In[175]:= DSolve[{γ'[t] == -γ[t]^2/σ^2, γ[0] == σp^2}, γ[t], t]
```

$$\text{Out[175]= } \left\{\left\{\gamma(t) \to \frac{\sigma^2\, \sigma_p^2}{\sigma^2 + t\, \sigma_p^2}\right\}\right\}$$

i.e.,

$$\gamma(t) = \frac{\sigma^2\, \sigma_p^2}{\sigma^2 + t\, \sigma_p^2}. \tag{4.7.15}$$

Going back into the equation (4.7.13), we have

$$d\,m(t) + \frac{\gamma(t)}{\sigma^2}\, m(t)\, d\,t = \frac{\gamma(t)}{\sigma^2}\, d\,\xi(t) \tag{4.7.16}$$

which implies

$$d\left(e^{\int_0^t \frac{\gamma(s)}{\sigma^2}\, ds}\, m(t)\right) = e^{\int_0^t \frac{\gamma(s)}{\sigma^2}\, ds}\, \frac{\gamma(t)}{\sigma^2}\, d\,\xi(t) \tag{4.7.17}$$

and consequently, integrating between 0 and T, and after a rearrangement we get

$$m(T) = e^{-\int_0^T \frac{\gamma(s)}{\sigma^2} ds} m_0 + \int_0^T e^{-\int_t^T \frac{\gamma(s)}{\sigma^2} ds} \frac{\gamma(t)}{\sigma^2} d\xi(t). \tag{4.7.18}$$

Using (4.7.15), we compute in (4.7.18)

```
In[176]:= Simplify[e^(-Integrate[(σ^2 σp^2)/((σ^2 + s σp^2) σ^2), {s, 0, T}])]
```

$$\text{Out[176]=} \frac{\sigma^2}{\sigma^2 + T\sigma_p^2}$$

and

```
In[177]:= Simplify[(e^(-Integrate[(σ^2 σp^2)/((σ^2 + s σp^2) σ^2), {s, t, T}]) (σ^2 σp^2))/((σ^2 + t σp^2) σ^2)]
```

$$\text{Out[177]=} \frac{\sigma_p^2}{\sigma^2 + T\sigma_p^2}$$

which somewhat surprisingly does not depend on t. Using this we rewrite (4.7.18) as

$$\begin{aligned} m(T) &= E[\theta(0) \mid \xi(t), 0 \le t \le T] \\ &= \frac{\sigma^2}{\sigma^2 + T\sigma_p^2} m_0 + \frac{\sigma_p^2}{\sigma^2 + T\sigma_p^2} \int_0^T d\xi(t) \\ &= \frac{\sigma^2}{\sigma^2 + T\sigma_p^2} m_0 + \frac{\sigma_p^2}{\sigma^2 + T\sigma_p^2} (\xi(T) - \xi(0)) \\ &= \frac{m_0\,\sigma^2 + (\xi(T) - \xi(0))\,\sigma_p{}^2}{\sigma^2 + T\,\sigma_p{}^2}. \end{aligned} \tag{4.7.19}$$

Comparing (4.7.19) with (4.7.12) we conclude that indeed the new estimate is noting more than the recursive version of the old one. We also notice that in this particular case the recursiveness is not really an improvement, since the old estimate (4.7.19) in the end did not depend on the whole trajectory $\xi(t)$, $0 \le t \le T$ but rather only on $\xi(0)$ and $\xi(T)$, and therefore (4.7.13) is not simpler then (4.7.19).

4.7.2.3 An Experiment

We shall do Monte–Carlo simulations. So we need the package

```
In[178]:= << "CFMLab`ItoSDEs`"
```

The chosen data is

```
In[179]:= T = 2; K = 2000; dt = T/K; m0 = 0; g0 = .5; ξ0 = Log[50];
          σ = .5; θ0 = Random[NormalDistribution[m0, g0]]
Out[179]= 0.111672
```

The coefficients of the Kalman–Bucy filtering equation (4.7.10) are equal to

```
In[180]:= a[t_] := 0; b[t_] := 0; A[t_] := 1; B[t_] := σ
```

First, we compute the conditional variance function (which is unconditional in the present case, as in any other Kalman–Bucy case):

```
In[181]:= Clear[Γ]; Γ[t_] =
           γ[t] /. DSolve[{γ'[t] == -((d[t, ξ[t]] b2[t, ξ[t]] +
                        γ[t] c1[t, ξ[t]]) / d[t, ξ[t]])^2 +
                  b1[t, ξ[t]]^2 + 2 γ[t] a1[t, ξ[t]] +
                  b2[t, ξ[t]], γ[0] == g0^2}, γ[t], t][[1]]
```

$$\text{Out[181]= } \frac{1.}{4.\,t + 4.}$$

It is instructive to see the standard deviation function as an indication of the efficiency of the estimation over the time interval considered:

```
In[182]:= Plot[√Γ[t], {t, 0, T}, PlotRange → {0, √Γ[0]}];
```

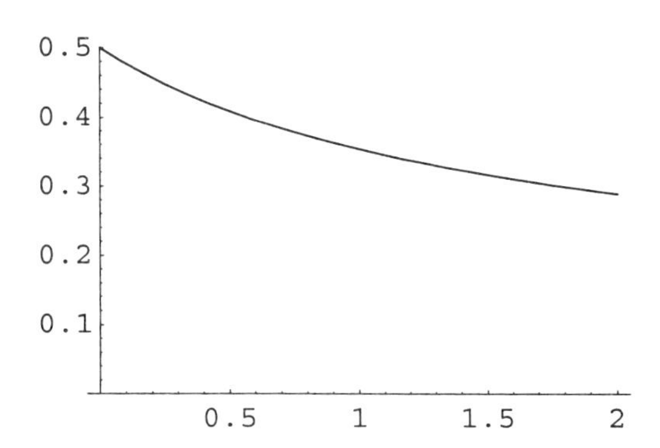

We derive the filter $dm(t) = \alpha_0(t, \xi)\,dt + \alpha_1(t, \xi)\,m(t)\,dt + \alpha_2(t, \xi)\,d\xi(t)$, where

```
In[183]:= A[t_, ξ_] = (d[t, ξ] b2[t, ξ] + Γ[t] c1[t, ξ]) / d[t, ξ]^2;
          α0[t_, ξ_] = a0[t, ξ] - c0[t, ξ] A[t, ξ]
Out[184]= 0

In[185]:= α1[t_, ξ_] = a1[t, ξ] - c1[t, ξ] A[t, ξ]
```

Out[185]= $-\frac{4.}{4.\,t+4.}$

```
In[186]:= α2[t_, ξ_] = A[t, ξ]
```

Out[186]= $\frac{4.}{4.\,t+4.}$

and perform the Monte–Carlo simulation generating a trajectory of the process $\{\theta, \xi\}$:

```
In[187]:= sol =
            SDESolver[{a0[#1, #2⟦2⟧] + #2⟦1⟧ a1[#1, #2⟦2⟧],
               c0[#1, #2⟦2⟧] + #2⟦1⟧ c1[#1, #2⟦2⟧]} &,
             {{b1[#1, #2⟦2⟧], b2[#1, #2⟦2⟧]},
               {0, d[#1, #2⟦2⟧]}} &, {θ0, ξ0}, 0, T, K];
```

The θ-component of the computed trajectory is trivial in this case; nevertheless we compute it explicitly:

```
In[188]:= RealParameter = ({#1⟦1⟧, #1⟦2, 1⟧} &) /@ sol;
          RealParameterFunction =
            Interpolation[RealParameter];
```

The ξ-component of the computed trajectory looks like this:

```
In[190]:= P1 = ListPlot[({#1⟦1⟧, #1⟦2, 2⟧} &) /@ sol,
             PlotStyle → RGBColor[0, 1, 0]];
```

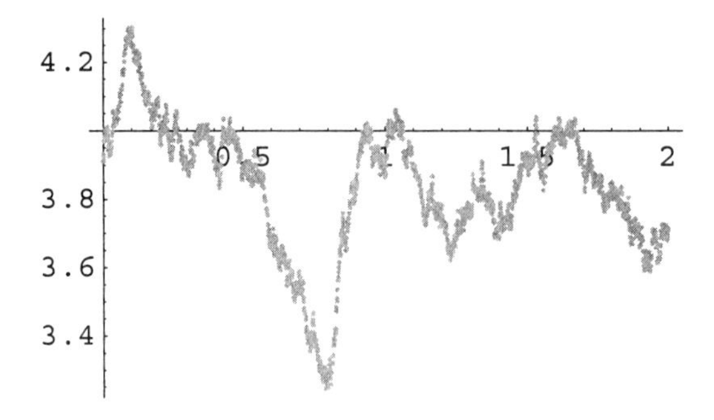

or in the terms of the stock price, i.e., $S(t) = e^{\xi(t)}$ the computed trajectory looks like

```
In[191]:= P2 = ListPlot[({#1⟦1⟧, e^#1⟦2,2⟧} &) /@ sol,
             PlotStyle → RGBColor[0, 1, 0]];
```

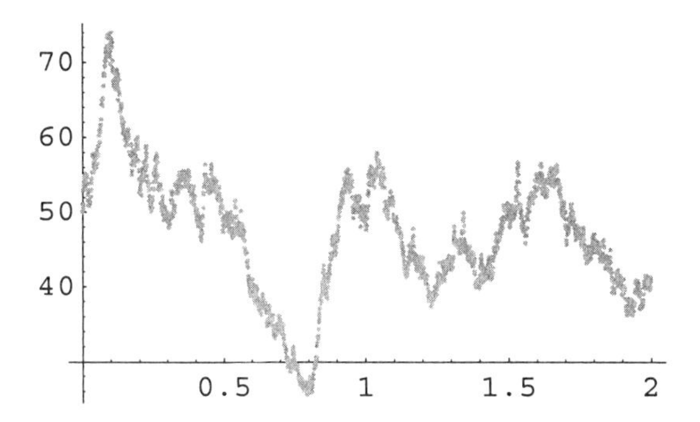

Finally, we do the calculation of the Kalman–Bucy filter on the computed trajectory of $\xi(t)$:

```
In[192]:= ξt = (#1⟦2, 2⟧ &) /@ sol;
          times = (#1⟦1⟧ &) /@ sol;
          dξt = Drop[ξt, 1] - Drop[ξt, -1];
          Estimate = Transpose[
             {times, FoldList[dt α1[#2⟦1⟧, #2⟦2⟧] #1 +
                 #1 + dt α0[#2⟦1⟧, #2⟦2⟧] +
                 #2⟦3⟧ α2[#2⟦1⟧, #2⟦2⟧] &, m0, Transpose[
                {Drop[times, -1], Drop[ξt, -1], dξt}]]}];
          EstimateFunction = Interpolation[Estimate];
```

Putting together the true value of $\theta = \theta(t)$, the best estimate $m(t)$, and the 1, 2, 3-σ confidence regions for θ around $m(t)$, confirms the validity of the theoretical results:

```
In[197]:= Plot[Evaluate[Append[
              Prepend[(EstimateFunction[t] + #1 √Γ[t] &) /@
                {-3, -2, -1, 1, 2, 3}, EstimateFunction[t]],
              RealParameterFunction[t]]],
           {t, 0, T}, PlotStyle → Append[Prepend[
              (RGBColor[0, 1, 0] &) /@ {-3, -2, -1, 1, 2, 3},
              {RGBColor[1, 0, 0], Thickness[.01]}],
             {RGBColor[0, 0, 1], Thickness[.01]}]];
```

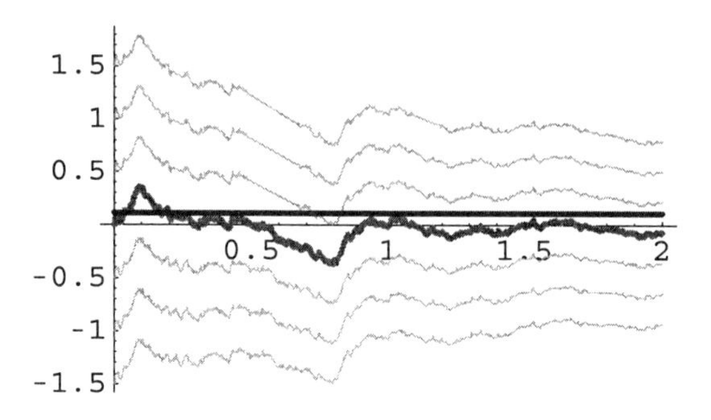

One can experiment with different data to conclude that indeed even over a long periods of time T, unfortunately, the significant indeterminacy remains.

4.7.3 Conditional Gaussian Filtering Implemented: An Example

4.7.3.1 Filtering Equations

```
In[198]:= Clear["Global`*"]
```

Here is an example. Consider the system (that we shall come back to at the very end of this book where we propose it as a model for momentum stock price dynamics—ξ will be the stock price, while θ will be the stock price "*trend*"):

$$\begin{aligned} d\theta(t) &= \left(\frac{2\pi}{p}\right)^2 (e - \xi(t))\, dt \\ d\xi(t) &= \theta(t)\, dt + \sigma\, dB(t). \end{aligned} \tag{4.7.20}$$

So we assume perfect knowledge of parameters e, p, σ, while "trend" $\theta(t)$ is not known and it's evolution needs to be estimated. Using the notation from (4.7.2), the system (4.7.20) corresponds to the coefficients

```
In[199]:= a0[t_, ξ_] := (2π/p)^2 (e - ξ); a1[t_, ξ_] := 0;
          b1[t_, ξ_] := 0; b2[t_, ξ_] := 0;
          c0[t_, ξ_] := 0; c1[t_, ξ_] := 1; d[t_, ξ_] := σ
```

Solving equation (4.7.6) first, we get

```
In[200]:= Remove[g]
```

```
In[201]:= Γ[t_] =
          γ[t] /. DSolve[{γ'[t] == -((d[t, ξ[t]] b2[t, ξ[t]] +
                   γ[t] c1[t, ξ[t]]) / d[t, ξ[t]])^2 +
                b1[t, ξ[t]]^2 + 2 γ[t] a1[t, ξ[t]] +
                b2[t, ξ[t]], γ[0] == g0^2}, γ[t], t][[1]]
```

Out[201]= $\frac{\sigma^2 g_0^2}{\sigma^2 + t g_0^2}$

Therefore

```
In[202]:= A[t_, ξ_] = (b2[t, ξ] d[t, ξ] + c1[t, ξ] Γ[t]) / d[t, ξ]^2
```

Out[202]= $\frac{g_0^2}{\sigma^2 + t g_0^2}$

```
In[203]:= α0[t_, ξ_] = a0[t, ξ] - c0[t, ξ] A[t, ξ]
```

Out[203]= $\frac{4 \pi^2 (e - \xi)}{p^2}$

```
In[204]:= α1[t_, ξ_] = a1[t, ξ] - c1[t, ξ] A[t, ξ]
```

Out[204]= $-\frac{g_0^2}{\sigma^2 + t g_0^2}$

```
In[205]:= α2[t_, ξ_] = A[t, ξ]
```

Out[205]= $\frac{g_0^2}{\sigma^2 + t g_0^2}$

yielding the filtering equations

$$\begin{aligned} d m(t) &= \left(\frac{2\pi}{p}\right)^2 (e - \xi(t))\, d t - \frac{g_0^2}{\sigma^2 + t g_0^2}\, m(t)\, d t + \frac{g_0^2}{\sigma^2 + t g_0^2}\, d\xi(t) \\ &= \left(\frac{2\pi}{p}\right)^2 (e - \xi(t))\, d t + \frac{g_0^2}{\sigma^2 + t g_0^2}\, (d\xi(t) - m(t) d t) \end{aligned} \tag{4.7.21}$$

$m(0) = m_0$

4.7.3.2 Monte–Carlo Simulation

```
In[206]:= Clear["Global`*"]
          << "CFMLab`ItoSDEs`"
```

We now do the same calculation for the specific data:

```
In[208]:= p = 1/4; T = p; K = 2000; dt = T/K;
          g₀ = 200; m₀ = 0; ξ₀ = 1; e = 50; σ = 30;
          θ₀ = Random[NormalDistribution[m₀, g₀]]

Out[208]= -105.584
```

We generate a trajectory of $\{\theta(t), \xi(t)\}$, given by (4.7.20), the corresponding trajectory of $m(t)$ given by (4.7.21), and the confidence regions according to (4.7.6). Plotted are $\theta(t)$ (blue), the estimate $m(t)$ (red), and the confidence regions (green):

```
In[209]:= a₀[t_, ξ_] := (2 π / p)^2 (e - ξ); a₁[t_, ξ_] := 0;
          b₁[t_, ξ_] := 0; b₂[t_, ξ_] := 0;
          c₀[t_, ξ_] := 0; c₁[t_, ξ_] := 1; d[t_, ξ_] := σ
          Clear[Γ]; Γ[t_] =
           γ[t] /. DSolve[{γ'[t] == -((d[t, ξ[t]] b₂[t, ξ[t]] +
                       γ[t] c₁[t, ξ[t]]) / d[t, ξ[t]])^2 +
                  b₁[t, ξ[t]]^2 + 2 γ[t] a₁[t, ξ[t]] +
                  b₂[t, ξ[t]], γ[0] == g₀^2}, γ[t], t][[1]];
          A[t_, ξ_] = (d[t, ξ] b₂[t, ξ] + Γ[t] c₁[t, ξ]) / d[t, ξ]^2;
          α₀[t_, ξ_] = a₀[t, ξ] - c₀[t, ξ] A[t, ξ];
          α₁[t_, ξ_] = a₁[t, ξ] - c₁[t, ξ] A[t, ξ];
          α₂[t_, ξ_] = A[t, ξ];
          sol =
            SDESolver[{a₀[#1, #2[[2]]] + #2[[1]] a₁[#1, #2[[2]]],
               c₀[#1, #2[[2]]] + #2[[1]] c₁[#1, #2[[2]]]} &,
             {{b₁[#1, #2[[2]]], b₂[#1, #2[[2]]]},
               {0, d[#1, #2[[2]]]}} &, {θ₀, ξ₀}, 0, T, K];
          RealParameter = ({#1[[1]], #1[[2, 1]]} &) /@ sol;
          RealParameterFunction =
            Interpolation[RealParameter];
          ξt = (#1[[2, 2]] &) /@ sol;
          times = (#1[[1]] &) /@ sol;
          dξt = Drop[ξt, 1] - Drop[ξt, -1];
          Estimate = Transpose[
             {times, FoldList[dt α₁[#2[[1]], #2[[2]]] #1 +
                  #1 + dt α₀[#2[[1]], #2[[2]]] +
                  #2[[3]] α₂[#2[[1]], #2[[2]]] &, m₀, Transpose[
                 {Drop[times, -1], Drop[ξt, -1], dξt}]]}];
          EstimateFunction = Interpolation[Estimate];
```

```
Plot[Evaluate[Append[
    Prepend[(EstimateFunction[t] + #1 √Γ[t] &) /@
      {-3, -2, -1, 1, 2, 3}, EstimateFunction[t]],
    RealParameterFunction[t]]],
  {t, 0, T}, PlotStyle → Append[Prepend[
      (RGBColor[0, 1, 0] &) /@ {-3, -2, -1, 1, 2, 3},
      {RGBColor[1, 0, 0], Thickness[.01]}],
    {RGBColor[0, 0, 1], Thickness[.01]}]];
```

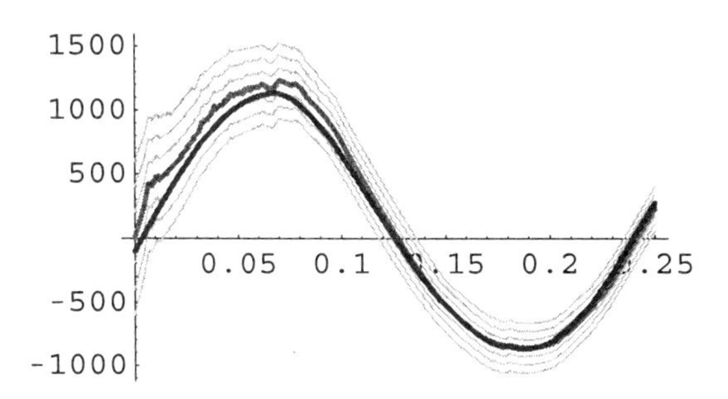

5 Implied Volatility for European Options

Option Market Data vs. Black–Scholes Theory: Numerical PDEs, Optimal Control, and General Implied Volatility

5.1 Remarks

Black–Scholes theory, its explicit formulas presented in Chapter 3, or derived numerical methods for more complicated models to be discussed in the last part of this chapter and the next yield theoretical fair prices for various stock options. On the other hand, the real market prices of stock options, like market prices of anything else, are as high as someone is willing to pay for them, and as low as someone is willing to sell them for. It is a daily routine in options trading practice to reconcile theoretical and observed prices. It was also a matter of recent mathematical research to find new and better, more precise and more efficient methods for such a reconciliation. There are many mathematical methods of various degrees of sophistication to this end. The general rule is that the more precise the methods, the more demanding they are computationally as well.

We shall start our discussion with a very simple routine for finding the constant implied volatility for a single option; then we shall discuss time-dependent implied volatility, utilizing the generalized Black–Scholes formulas, and finally we shall develop a numerical method for solving an optimal control problem for the Dupire partial differential equation. We shall also present a numerical method for solving an optimal control problem for the obstacle problem for the Dupire partial differential equation in the next chapter. These two are the most sophisticated methods presented in this book for solving the problems of finding the implied volatilities for European and American options, respectively.

Which of these methods of various degrees of sophistication and computational difficulty should be employed in trading practice and when? It seems that simple and therefore computationally fast methods have to be employed in the day-to-day calculations involving many underlying stocks. Nevertheless the most sophisticated methods should be utilized here periodically to confirm and calibrate the fast methods, and/or to help compare and choose among many possible alternatives for the fast methods (for example, choosing a class of time-dependent volatilities when employing fast symbolic solutions can be based on the experience developed when preforming full numerical solutions where such restrictive assumptions are not needed). The most sophisticated methods can also be used when the underlying security is extremely significant, so significant that it can gauge the outlook of the whole market (*timing the market*). An example of such an underlying is QQQ (NASDAQ 100) and our computational examples below rely heavily on its data.

Since there is almost a perfect correlation between options and the underlying stock prices, the fact that option price histories are typically quite short, and also since option price dynamics are quite a bit more complicated and therefore statistically more difficult to handle than stock price dynamics, options data are not used for statistical predictions about future moves of option prices. It is used only for calibrating and connecting the Black–Scholes or derived models with real option market data. Once that calibration is performed, along with the statistical analysis of the underlying stock, these two separate inputs and the mathematics that connects them yield a view of the future dynamics of option prices.

So the present problem is: an investor looks at all the options, *different strikes* and *different expiration dates* for an underlying stock at some *fixed time* and consequently at some *fixed underlying stock price*, and based on that data one wishes to calibrate the Black–Scholes model.

5.2 Option Market Data

```
In[1]:= Clear["Global`*"]
```

```
In[2]:= << CFMLab`DataImport`
```

Option market data can be downloaded from various web sites into text files somewhere in the computer. Data used in the present chapter was kindly provided by OptionMetrics, LLC.

Let's, first of all, go where the data is:

```
In[3]:= SetDirectory[ToFileName[{$TopDirectory, "AddOns",
          "Applications", "CFMLab", "MarketData"}]]

Out[3]= C:\Program Files\Wolfram Research\Mathematica\4.1\AddOns\
          Applications\CFMLab\MarketData
```

and see what files there are

```
In[4]:= FileNames["*OptionMetrics*"]

Out[4]= {OptionMetricsData1.txt, OptionMetricsData2.txt}
```

and also see what is the format of the option files that we have to work with:

```
In[5]:= !! "OptionMetricsData2.txt"
```

From In[5]:=

Ticker	Date	Expiration	Strike	Call/Put	BestBid	BestOffer	Volume	OpenInterest
QQQ	10/30/00	11/18/00	64	C	13.375	13.875	3	194
QQQ	10/30/00	11/18/00	64	P	0.625	0.75	67	2243
QQQ	10/30/00	11/18/00	65	C	12.5	13	60	159
QQQ	10/30/00	11/18/00	65	P	0.6875	0.875	47	842
QQQ	10/30/00	11/18/00	66	C	11.75	12.25	6	25
QQQ	10/30/00	1/18/03	90	C	17.5	19	0	642
QQQ	10/30/00	1/18/03	90	P	20.5	22.5	0	243
QQQ	10/30/00	1/18/03	95	C	15.25	17.25	0	295
QQQ	10/30/00	1/18/03	95	P	23.625	25.625	0	82
QQQ	10/30/00	1/18/03	100	C	14.75	15.5	10	1081
QQQ	10/30/00	1/18/03	100	P	26.875	28.875	10	73

One can see that the data looks uniform, and that all the dates have the same format as in Chapter 4, so that we shall be able to use the same function for translating them into the *Mathematica*® format. Let us first open an input stream

In[6]:=
```
InputData = OpenRead["OptionMetricsData2.txt"]
```

Out[6]= InputStream[OptionMetricsData2.txt, 4]

Since the data is uniform, we can utilize the line structure, i.e., input data as a two-dimensional array:

In[7]:=
```
AllData =
   Drop[ReadList[InputData, Table[Word, {9}]], -1];
Close[InputData]
```

Out[8]= OptionMetricsData2.txt

The first element of AllData is the list of definitions of the entries, while the rest have the same format as the second one:

In[9]:=
```
AllData[[-1]]
```

Out[9]= {QQQ, 10/30/00, 1/18/03, 100, P, 26.875, 28.875, 10, 73}

We are again going to need the YearClock function

In[10]:=
```
YearLength := 1/100 FromDate[{2000, 1, 1, 0, 0, 0}]
YearClock[y_] :=
 N[(FromDate[y] - FromDate[{2000, 1, 1, 0, 0, 0}]) / YearLength]
```

$$\text{YearLength} := \frac{1}{100}\,\text{FromDate}[\{2000, 1, 1, 0, 0, 0\}]$$

$$\text{YearClock}[y\_] := N\left[\frac{\text{FromDate}[y] - \text{FromDate}[\{2000, 1, 1, 0, 0, 0\}]}{\text{YearLength}}\right]$$

and similarly as in Chapter 4, we define function

```
In[12]:= F := Module[{xx, yy, zz},
           xx = Partition[Append[Prepend[
               #[[1]] & /@ StringPosition[#, "/"], 0],
              StringLength[#] + 1], 2, 1];
           yy = ToExpression[Function[y,
               StringTake[#, y + {1, -1}]] /@ xx]; zz =
            {If[#[[3]] > 50, #[[3]] + 1900, #[[3]] + 2000],
               #[[1]], #[[2]], 16, 0, 0} &[
             yy]; YearClock[zz]] &
```

which translates the text date into a decimal time. For example,

```
In[13]:= F[AllData[[2, 2]]]

Out[13]= 0.831417
```

We are ready to transform the whole list AllData into

```
In[14]:= AllData2 =
          {#[[1]], #[[2]], F[#[[2]]], #[[3]], F[#[[3]]],
             ToExpression[#[[4]]], #[[5]],
             ToExpression[#[[6]]], ToExpression[#[[7]]],
             ToExpression[#[[8]]], ToExpression[#[[9]]],
             (ToExpression[#[[6]]] + ToExpression[#[[7]]]) / 2
              // N} & /@ Drop[AllData, 1];
```

So, for example

```
In[15]:= AllData[[-1]]

Out[15]= {QQQ, 10/30/00, 1/18/03, 100, P, 26.875, 28.875, 10, 73}
```

is transformed into

```
In[16]:= AllData2[[-1]]

Out[16]= {QQQ, 10/30/00, 0.831417, 1/18/03, 3.04914, 100, P, 26.875,
           28.875, 10, 73, 27.875}
```

Recall

```
In[17]:= AllData[[1]]

Out[17]= {Ticker, Date, Expiration, Strike, Call/Put, BestBid, BestOffer,
           Volume, OpenInterest}
```

One can see that the Date is kept but it is also decimalized, Expiration is kept and decimalized, Strike is transformed from a text into an expression, as well as for the other numerical entries. Finally, one more entry is added: the average of BestBid and BestOffer (Ask) is computed and will be taken as the current option price (this is necessary since, as opposed to usual stocks, options are not liquid enough to be able to take the last realized trade price as the current price). Proceeding, it is important to consider calls and puts separately, since they infer, as we shall see quite explicitly below, different conclusions. Let

```
In[18]:= AllCallData = Select[AllData2, #[[7]] == "C" &];
```

```
In[19]:= AllPutData = Select[AllData2, #[[7]] == "P" &];
```

These lists still contain many unnecessary items so we settle on the following information to be kept:

```
In[20]:= AllCallData2 =
           Transpose[Part[Transpose[AllCallData], #] &/@
             {5, 6, -1, -3}];
```

```
In[21]:= AllPutData2 =
           Transpose[Part[Transpose[AllPutData], #] &/@
             {5, 6, -1, -3}];
```

For example

```
In[22]:= AllCallData2[[101]]

Out[22]= {0.960099, 107, 0.1875, 0}
```

i.e., we kept decimal Expiration, Strike, option price, and the Volume as an indication whether a particular option is liquid, and consequently the option price is relevant enough. The other possibility would be to use the OpenInterest to this end. The Volume seems to work better.

For the analysis that is forthcoming it is also very important to know underlying stock price at the time of this record, i.e., at the time

```
In[23]:= TimeOfRecord =
          Union[Transpose[AllData2][[3]]][[1]]

Out[23]= 0.831417
```

So,

```
In[24]:= UnderlyingStockPrice = Select[GetPriceData[qqq],
            #[[1]] == TimeOfRecord &][[1, 2]]

Out[24]= 76.7656
```

Finally we wrap all of the essential above into a function

```
In[25]:= GetAllData["QQQ", OptionType_,
           Volume_: 300, NumberOfExpirations_: 3] :=
          Module[{InputData, AllData, F, AllData2, AllCallData,
            AllCallData2, AllPutData, AllPutData2, x,
            y, ExpTimes, z}, InputData = InputData =
             OpenRead[ToFileName[{$TopDirectory, "AddOns",
                "Applications", "CFMLab", "MarketData"},
               "OptionMetricsData2.txt"]]; AllData =
            Drop[ReadList[InputData, Table[Word, {9}]], -1];
           Close[InputData];
           F :=
            Module[{xx, yy, zz}, xx = Partition[Append[Prepend[
                  (#1[[1]] &) /@ StringPosition[#1, "/"], 0],
                 StringLength[#1] + 1], 2, 1];
              yy = ToExpression[Function[y,
                  StringTake[#1, y + {1, -1}]] /@ xx]; zz =
               ({If[#1[[3]] > 50, #1[[3]] + 1900, #1[[3]] + 2000],
                   #1[[1]], #1[[2]], 16, 0, 0} &)[
                yy]; YearClock[zz]] &;
           AllData2 = ({#1[[1]], #1[[2]], F[#1[[2]]], #1[[3]],
                F[#1[[3]]], ToExpression[#1[[4]]], #1[[5]],
                ToExpression[#1[[6]]], ToExpression[#1[[7]]],
                ToExpression[#1[[8]]], ToExpression[#1[[9]]],
                N[(1 / 2) * (ToExpression[#1[[6]]] + ToExpression[
                     #1[[7]]])]} &) /@ Drop[AllData, 1];
           AllCallData = Select[AllData2, #1[[7]] == "C" &];
           AllPutData = Select[AllData2, #1[[7]] == "P" &];
           AllCallData2 = Transpose[
             (Transpose[AllCallData][[#1]] &) /@ {5, 6, -1, -3}];
           AllPutData2 = Transpose[
             (Transpose[AllPutData][[#1]] &) /@ {5, 6, -1, -3}];
           x = If[ToString[OptionType] === "Call",
             AllCallData2, AllPutData2];
           y = Select[x, #1[[-1]] ≥ Volume &];
           ExpTimes = Union[Transpose[y][[1]]];
           ExpTimes = Take[ExpTimes,
             Min[NumberOfExpirations, Length[ExpTimes]]];
           z = Function[t, Select[y, #1[[1]] == t &]] /@ ExpTimes;
           {{"QQQ", ToString[OptionType],
             ExpTimes, 0.8314167853101158, 76.765617}, z}]
```

which works as

```
In[26]:= gad = GetAllData["QQQ", "Call", 800]
```

Out[26]= {{QQQ, Call, {0.883437, 0.960099, 1.05593}, 0.831417, 76.7656},

$$\left\{\begin{pmatrix} 0.883437 & 77 & 4.3125 & 1158 \\ 0.883437 & 78 & 3.8125 & 1995 \\ 0.883437 & 79 & 3.375 & 1144 \\ 0.883437 & 80 & 2.96875 & 7208 \\ 0.883437 & 81 & 2.5625 & 3248 \\ 0.883437 & 82 & 2.21875 & 2025 \\ 0.883437 & 84 & 1.625 & 1184 \\ 0.883437 & 85 & 1.375 & 2346 \\ 0.883437 & 87 & 0.9375 & 1783 \\ 0.883437 & 90 & 0.59375 & 1357 \end{pmatrix}, \begin{pmatrix} 0.960099 & 80 & 4.875 & 1121 \\ 0.960099 & 86 & 2.8125 & 2017 \\ 0.960099 & 90 & 1.84375 & 1099 \\ 0.960099 & 100 & 0.53125 & 1267 \end{pmatrix},\right.$$

$$\left.\begin{pmatrix} 1.05593 & 80 & 6.8125 & 1238 \\ 1.05593 & 90 & 3.0625 & 1061 \end{pmatrix}\right\}\}$$

or

In[27]:= `gad = GetAllData["QQQ", "Put"]`

Out[27]= {{QQQ, Put, {0.883437, 0.960099, 1.05593}, 0.831417, 76.7656},

$$\left\{\begin{pmatrix} 0.883437 & 69 & 1.40625 & 450 \\ 0.883437 & 70 & 1.6875 & 1178 \\ 0.883437 & 72 & 2.1875 & 475 \\ 0.883437 & 73 & 2.5 & 382 \\ 0.883437 & 74 & 2.875 & 5552 \\ 0.883437 & 75 & 3.3125 & 4288 \\ 0.883437 & 76 & 3.8125 & 789 \\ 0.883437 & 77 & 4.1875 & 2306 \\ 0.883437 & 78 & 4.6875 & 1560 \\ 0.883437 & 79 & 5.25 & 4661 \\ 0.883437 & 80 & 5.875 & 2184 \\ 0.883437 & 81 & 6.375 & 667 \\ 0.883437 & 83 & 7.8125 & 321 \\ 0.883437 & 84 & 8.5625 & 301 \\ 0.883437 & 85 & 9.3125 & 669 \\ 0.883437 & 87 & 10.9375 & 324 \\ 0.883437 & 90 & 13.5625 & 367 \end{pmatrix}, \begin{pmatrix} 0.960099 & 69 & 2.8125 & 601 \\ 0.960099 & 75 & 4.875 & 367 \\ 0.960099 & 80 & 7.3125 & 773 \end{pmatrix},\right.$$

$$\left.\begin{pmatrix} 1.05593 & 80 & 8.5625 & 553 \end{pmatrix}\right\}\}$$

The return contains the name of the underlying security, the type of options considered, the list of expiration dates, the current (decimal) date, the current underlying price, and then lists of data (expiration, strike, price, open interest) for all sufficiently liquid options (grouped according to the expiration date). Notice that it is more active trading for the options that expire sooner rather than later, as well as for those options with strikes in the neighborhood of the current underlying price.

So, this is the type of data that is considered in the case of stock-options—for fixed (present) time, and the corresponding fixed underlying stock price, a whole array of options, calls and puts, with different strike prices, and different expiration times are considered. This is very different from stock data, where the historical context is of primary interest: the past is used to anticipate the future. With options the present is used to anticipate the future: current option prices with expirations in the future are used to anticipate future performance of the underlying stock.

For future reference, function GetAllData is in the package CFMLab`DataImport`.

5.3 Black–Scholes Theory vs. Market Data: Implied Volatility

5.3.1 Black–Scholes Theory and Market Data

Recall first the Black–Scholes formulas (with constant volatility), i.e., functions VC and VP:

```
In[28]:= << "CFMLab`BlackScholes`"
         << "CFMLab`DataImport`"
         << "CFMLab`NumericalBlackScholes`"
         << "CFMLab`StockStat`"
```

Also, recall how the graph of the call option price, as a function of the underlying security price, looks (for volatility we take the QQQ statistical volatility estimated based on the historical data up until the time the option data is recorded (10/30/00):

```
In[32]:= aod = GetAllData["QQQ", "Call", 200, 4];
```

```
In[33]:= σ = Σ[Select[GetPriceData[qqq],
             #[[1]] ≤ YearClock[{2000, 10, 30, 16, 0, 0}] &]]

Out[33]= 0.494515
```

while the rest of the data is taken from the above imported option data):

```
In[34]:= Plot[{Max[S - aod[[2, 3, 1, 2]], 0], VC[aod[[1, 4]], S,
             aod[[2, 3, 1, 1]], aod[[2, 3, 1, 2]], .04, σ]},
           {S, 50, 120}, PlotRange → All,
           PlotStyle → RGBColor[1, 0, 0],
           AxesLabel → {"S", "V_Call[k]"},
           PlotLabel → "T = " <> ToString[aod[[2, 3, 1, 1]]]];
```

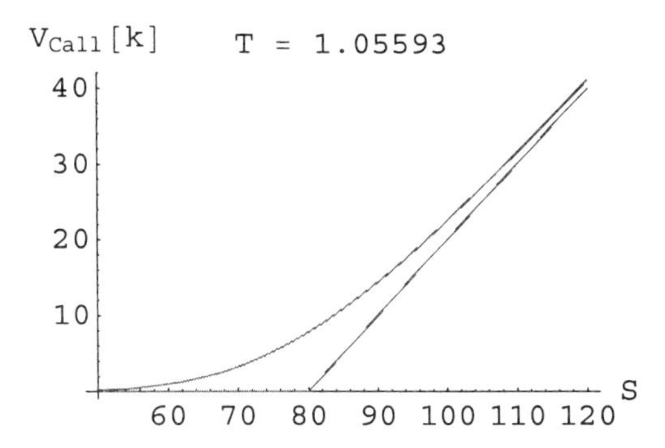

On the other hand, one can fix the stock price, i.e., take the stock price that was observed at the time of the option data record, and vary the *strike* price k, keeping the rest of the data the same to get

```
In[35]:= pl = Plot[{Max[aod⟦1, 5⟧ - k, 0], VC[aod⟦1, 4⟧,
            aod⟦1, 5⟧, aod⟦2, 3, 1, 1⟧, k, .04, σ]},
          {k, 50, 120}, PlotRange → All,
          PlotStyle → RGBColor[1, 0, 0],
          AxesLabel → {"k", "VCall[k]"},
          PlotLabel → "T = " <> ToString[aod⟦2, 3, 1, 1⟧]];
```

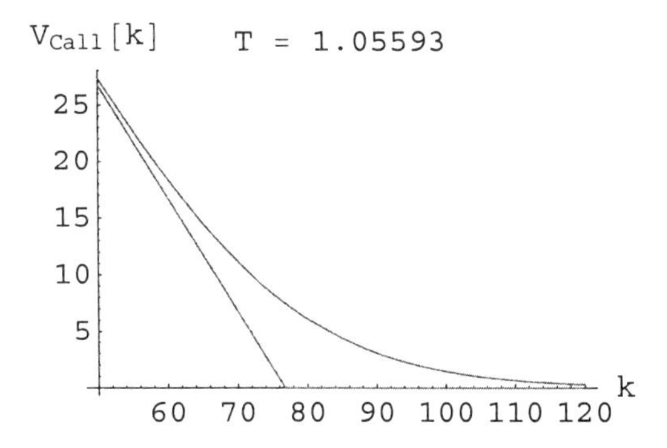

It is very important to notice this "duality". Fully understanding it and the far reaching conclusions that follow are postponed until later in this and the following chapter. The importance of this way of looking at the Black–Scholes formula can be guessed right away. The above option market data looks like:

```
In[36]:= Show[GraphicsArray[
          lp = (ListPlot[Transpose[Take[Transpose[#],
                   {2, 3}]], AxesLabel → {"k", "VC"},
                PlotRange → All, PlotStyle → {RGBColor[
```

```
                   1, 0, 0], AbsolutePointSize[5]},
              DisplayFunction → Identity, PlotLabel →
                "T=" <> ToString[#⟦1, 1⟧]] &) /@
          # & /@ Partition[aod[[2]], 2]],
    DisplayFunction → $DisplayFunction];
```

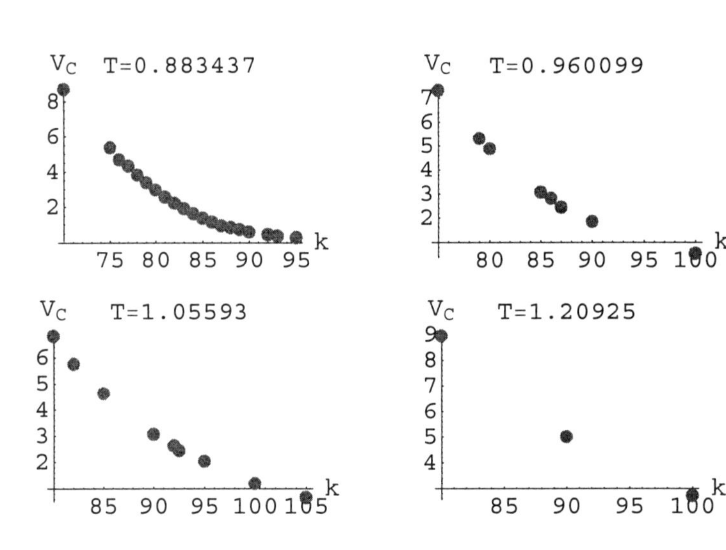

or comparing with the theoretical price:

```
In[37]:= Show[lp⟦2, 1⟧, pl, DisplayFunction →
           $DisplayFunction, PlotRange → All];
```

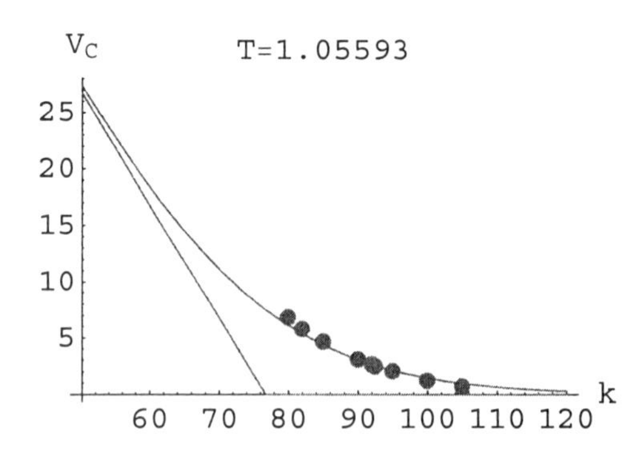

Similarly, in the case of puts we have the theoretical prices

```
In[38]:= plPut = Plot[
           {Max[-aod⟦1, 5⟧ + k, 0], VP[aod⟦1, 4⟧, aod⟦1, 5⟧,
             aod⟦2, 3, 1, 1⟧, k, .04, σ]}, {k, 50, 130},
```

```
    PlotRange → All, PlotStyle → RGBColor[0, 0, 1],
    AxesLabel → {"k", "VPut[k]"},
    PlotLabel → "T = " <> ToString[aod⟦2, 3, 1, 1⟧]];
```

VPut[k] T = 1.05593

50
40
30
20
10
k
80 100 120

as well as the observed ones:

```
In[39]:= aodPut = GetAllData["QQQ", "Put", 100, 4];
    Show[GraphicsArray[
      lp2 = (ListPlot[Transpose[Take[Transpose[#],
              {2, 3}]], AxesLabel → {"k", "VP"},
            PlotRange → All, PlotStyle → {RGBColor[
              0, 0, 1], AbsolutePointSize[5]},
            DisplayFunction → Identity, PlotLabel →
             "T=" <> ToString[#⟦1, 1⟧]] &) /@ # & /@
        Partition[aodPut[[2]], 2]],
     DisplayFunction →
      $DisplayFunction];
```

VP T=0.883437
12
10
8
6
4
k
75 80 85 90

VP T=0.960099
12
10
8
6
4
k
70 75 80 85

VP T=1.05593
11
10
9
8
7
6
k
72747678808284

VP T=1.20925
10
9.5
9
8.5
k
77 78 79 80

or comparing with the theoretical price:

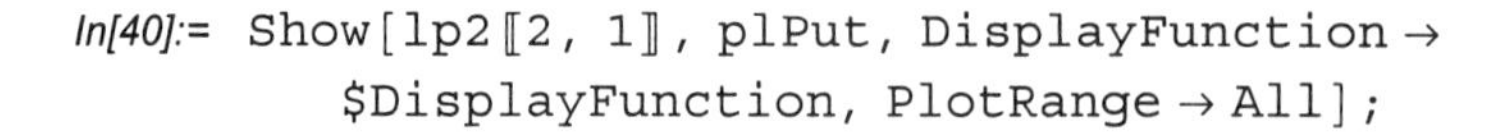

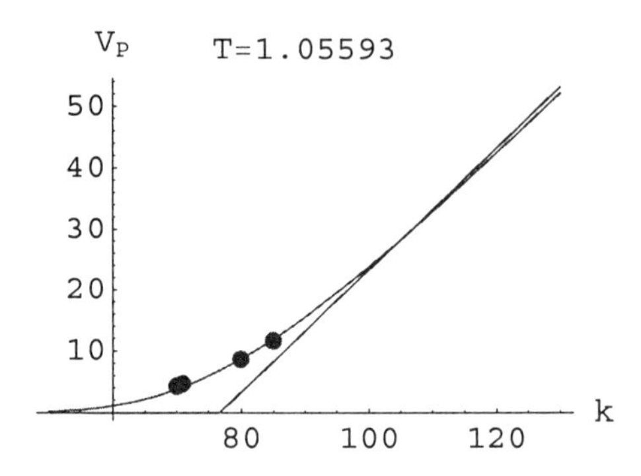

5.3.2 Constant Implied Volatility

5.3.2.1 Single Option Implied Volatility

```
In[41]:= Clear["Global`*"];
         << "CFMLab`BlackScholes`";
         << "CFMLab`DataImport`";
         << "CFMLab`ImpliedVolatility`";
         aod = GetAllData["QQQ", "Call", 200];
         aodPut = GetAllData["QQQ", "Put", 200];
```

We would like to compare the theoretical prices and the observed ones. The theoretical prices, in the simplest case of the Black–Scholes formulas, depend on the current time t, current stock price S, expiration time T, strike price k, the future interest rate r, and the future underlying stock volatility σ. The first four of those six quantities are non-negotiable: they are hard data read out above from the market. The future interest rate r is not known, but it is fairly stable, well understood and anticipated, and therefore assumed known (for example assumed to be equal to the present interest rate, which is known). The future underlying volatility, on the other hand, is not known, and in principle can be quite different from the present or past volatility, which is estimated on the stock market using statistics, as we did in Chapter 4. Indeed, one could as we did above, use the statistical volatility to compute the corresponding fair prices of options. So, in such a framework σ, the volatility, implies V the option prices: $\sigma \to V$.

(The problem with such a framework is that if followed quite often there will be a sizable discrepancy between theoretical option prices and the observed ones. The applicability of the Black–Scholes theory in real trading, as well as the applicability of any mathematical theory in real life applications, depends on its precision. Therefore, one must calibrate the model to make it as precise as possible. This is the first but not the only motivation for the following framework.)

The opposite question is quite useful and consequently quite popular in trading practice. Namely, on the basis of observed (and therefore by definition at least approximately fair) prices of options, one can estimate the *perceived* future underlying volatility. So in such a framework V, the observed option prices imply σ, the volatility: $V \to \sigma$.

The option market prices therefore imply *future* volatility—it is the market consensus about the market future. Oddly enough, the consensus is usually self-fulfilling unless something significant and unpredictable happens in the meantime.

The first question is whether it is possible to find the volatility σ from the observed prices. Plotting the option fair price as a function of volatility σ, while keeping all of the other variables constant (together with the constant c = gad[[2, 4, k, 3]], the observed option price) we get

```
In[47]:= With[{k = 5},
           Plot[{aod⟦2, 3, k, 3⟧, VC[aod⟦1, 4⟧, aod⟦1, 5⟧,
              aod⟦2, 3, k, 1⟧, aod⟦2, 3, k, 2⟧, .04, σ]},
            {σ, .001, 1}, PlotRange → All,
            AxesLabel → {"σ", "V[S]"}]];
```

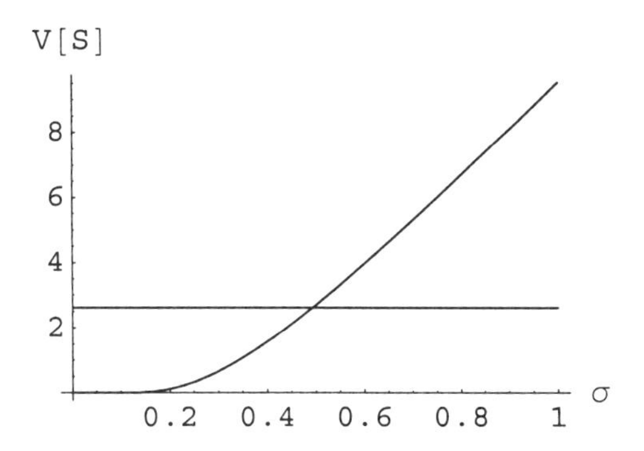

and conclude that, as long as c is not very small, there exists a unique σ such that $V(\sigma) = c$. Such a σ is called implied volatility (for the underlying security, corresponding to the above particular option). It is very easy to compute such (individual) implied volatility:

```
In[48]:= With[{k = 5}, σ /. FindRoot[aod⟦2, 3, k, 3⟧ ==
            VC[aod⟦1, 4⟧, aod⟦1, 5⟧, aod⟦2, 3, k, 1⟧,
             aod⟦2, 3, k, 2⟧, .04, σ], {σ, .8}]]

Out[48]= 0.493975
```

So this is the (future) QQQ volatility implied by a particular current option price. The same can be done for all of the options available:

```
In[49]:= IndividualCallVolatilities[aod_, r_] :=
          Function[y,
            (σ /. FindRoot[{aod⟦2, y, #1, 3⟧ == VC[aod⟦1, 4⟧,
                    aod⟦1, 5⟧, aod⟦2, y, #1, 1⟧, aod⟦2,
                     y, #1, 2⟧, r, σ]}, {σ, .7}] &) /@
             Range[1, Length[aod⟦2, y⟧]]] /@
           Range[1, Length[aod⟦2⟧]]
```

```
In[50]:= AllCallVolatilities =
          IndividualCallVolatilities[aod, .04]
```

```
Out[50]= {{0.659868, 0.630776, 0.606802, 0.622925, 0.615763, 0.612413,
             0.608461, 0.599295, 0.594237, 0.588834, 0.583253, 0.57773,
             0.572602, 0.561706, 0.572869, 0.573456, 0.569491, 0.575613,
             0.578636, 0.587473}, {0.571352, 0.55722, 0.553752, 0.538036,
             0.539418, 0.525758, 0.528295, 0.502992}, {0.544581,
             0.52334, 0.514146, 0.495592, 0.493975, 0.486202, 0.488229,
             0.467498, 0.450057}}
```

in the case of calls, and

```
In[51]:= IndividualPutVolatilities[aod_, r_] :=
          Function[y, (σ /. FindRoot[{aod⟦2, y, #1, 3⟧ ==
                   VP[aod⟦1, 4⟧, aod⟦1, 5⟧, aod⟦2, y, #1, 1⟧,
                    aod⟦2, y, #1, 2⟧, .05, σ]}, {σ, .8}] &) /@
             Range[1, Length[aod⟦2, y⟧]]] /@
           Range[1, Length[aod⟦2⟧]]
```

```
In[52]:= AllPutVolatilities =
          IndividualPutVolatilities[aodPut, .04]
```

```
Out[52]= {{0.642602, 0.645419, 0.64217, 0.623323, 0.615419, 0.612044,
             0.612567, 0.616663, 0.597367, 0.590797, 0.587823, 0.588623,
             0.565364, 0.568674, 0.568014, 0.562166, 0.550323, 0.558434,
             0.562948, 0.563469, 0.559328}, {0.58365, 0.548789, 0.538798,
             0.524177, 0.516324}, {0.538075, 0.498909}}
```

in the case of puts (it is useful to keep them separate as we shall see; also we separate among each one, those options with different times of expiration). They look like

```
In[53]:= AllCallStrikes = ((#1⟦2⟧ &) /@ #1 &) /@ aod⟦2⟧;
         civ = (ListPlot[#1, DisplayFunction → Identity,
                PlotStyle → RGBColor[1, 0, 0], AxesLabel →
                 {"Strike [k]", "Call Implied Volatility"},
```

```
        PlotRange → {0, Automatic},
        PlotJoined → True] &) /@
    Select[Transpose /@ Transpose[{AllCallStrikes,
        AllCallVolatilities}], Length[#1] ≥ 4 &];
Show[civ, DisplayFunction → $DisplayFunction];
```

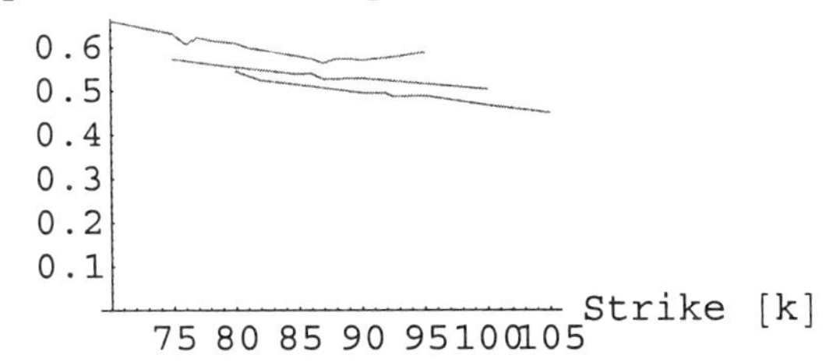

in the case of calls, and

```
In[54]:= AllPutStrikes = ((#1⟦2⟧ &) /@ #1 &) /@ aodPut⟦2⟧;
         piv = (ListPlot[#1, DisplayFunction → Identity,
                 PlotStyle → RGBColor[0, 0, 1], AxesLabel →
                  {"Strike [k]", "Put Implied Volatility"},
                 PlotRange → {0, Automatic},
                 PlotJoined → True] &) /@
             Select[Transpose /@ Transpose[{AllPutStrikes,
                 AllPutVolatilities}], Length[#1] ≥ 4 &];
         Show[piv, DisplayFunction → $DisplayFunction];
```

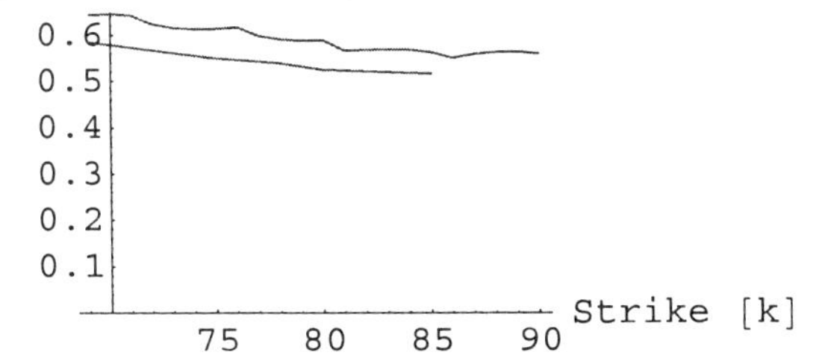

in the case of puts. Also, we store functions IndividualCallVolatilities and IndividualPutVolatilities into the package CFMLab`ImpliedVolatility`.

5.3.2.2 Average Implied Volatility

Putting the above call and put implied volatilities together, we get

```
In[57]:= Show[civ, piv,
           DisplayFunction → $DisplayFunction, AxesLabel →
            {"Strike [k]", "Implied Volatility"}];
```

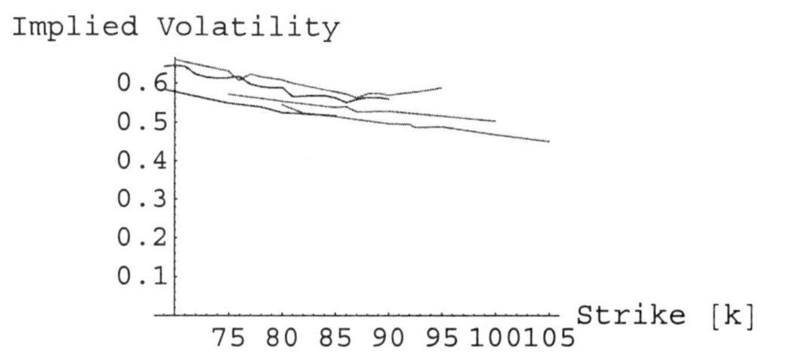

We notice that options with different strikes and different expiration times infer different implied volatilities of the (single) underlying security. To put it more bluntly they *contradict* each other (according to the assumed mathematical model, the Black–Scholes formula, volatility is constant). A trivial way to infer a constant implied volatility from all of the above would be to just take the average:

```
In[58]:= Av[List_] := Plus @@ Flatten[List] / Length[Flatten[List]]

In[59]:= {Av[AllCallVolatilities], Av[AllPutVolatilities]}

Out[59]= {0.55872, 0.577938}
```

Here we make an initial comment as to why it seems to make sense to keep calls and puts separate. It turned out that put-implied volatility was higher than call-implied volatility. Since the higher the volatility, the higher the option, this suggests that the demand for put options was higher than the demand for call options, and therefore, the general market perception was *bearish*, i.e., the stock market was expected to decline.

5.3.2.3 The Least "Square" Constant Implied Volatility

An alternative method to taking the average of individual single-option-implied volatility is to minimize, with respect to volatility, the various kinds of distance between the observed prices and the Black–Scholes theoretical prices. For example, in case of calls, consider the function $f_p(\sigma)$, for p an even integer,

```
In[60]:= Clear[σ];
         fp_[σ_] = (Plus @@ #1 &) [((#1^p &) /@ #1 &) [Flatten[
             Function[y, (aod⟦2, y, #1, 3⟧ - VC[aod⟦1, 4⟧,
                     aod⟦1, 5⟧, aod⟦2, y, #1, 1⟧,
                     aod⟦2, y, #1, 2⟧, 0.05, σ] &) /@
                 Range[1, Length[aod⟦2, y⟧]]] /@
               Range[1, Length[aod⟦2⟧]]]]];
```

So, what $f_p : \sigma \mapsto f_p(\sigma)$ does is as follows: it computes the differences between observed prices and theoretical ones for all the different options considered (on a single underlying), then takes the even power (the reason for that is that absolute value under an odd power would make it difficult for symbolic calculations), and finally sums them all up (while keeping the volatility σ as a variable). The function f_p is very nice. For example, if $p = 2$ or $p = 8$, we have

```
In[61]:= Plot[{f2[σ], f8[σ]},
           {σ, .5, .56}, PlotRange → {0, 10}];
```

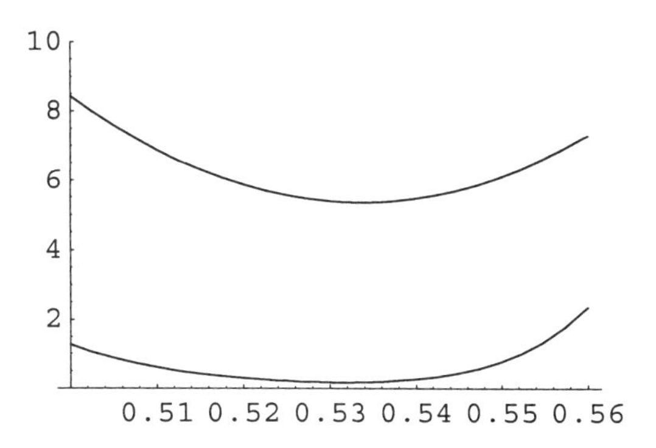

Notice that those two functions have (somewhat) different minimizers (σs for which minimums are taken):

```
In[62]:= (FindMinimum[f#1[σ], {σ, .6}] &) /@ {2, 8}
```

$$\text{Out[62]=} \begin{pmatrix} 5.35836 & \{\sigma \to 0.533708\} \\ 0.175574 & \{\sigma \to 0.531492\} \end{pmatrix}$$

and therefore, not surprisingly, the implied volatility will depend on the way we measure the distance between the observed and the theoretical prices, i.e., on the power p. We shall adopt $p = 2$, as a default value, i.e, we shall adopt the genuine least square method as a default method, and define a function

```
In[63]:= ConstantCallVolatility[aod_, r_] :=
           ConstantCallVolatility[aod, r, 2]
```

```
ConstantCallVolatility[aod_, r_, p_?EvenQ] :=
 Module[{f},
  f = (Plus @@ #1 &)[((#1^p &) /@ #1 &)[Flatten[
      Function[y, (aod[[2, y, #1, 3]] - VC[aod[[1, 4]],
             aod[[1, 5]], aod[[2, y, #1, 1]],
             aod[[2, y, #1, 2]], r, σ] &) /@
         Range[1, Length[aod[[2, y]]]]] /@
       Range[1, Length[aod[[2]]]]]]];
  σ /. FindMinimum[f, {σ, .5}][[2]]]
```

as well as

```
In[65]:= ConstantPutVolatility[aod_, r_] :=
          ConstantPutVolatility[aod, r, 2]
         ConstantPutVolatility[aod_, r_, p_?EvenQ] :=
          Module[{f},
           f = (Plus @@ #1 &)[((#1^p &) /@ #1 &)[Flatten[
               Function[y, (aod[[2, y, #1, 3]] - VP[aod[[1, 4]],
                      aod[[1, 5]], aod[[2, y, #1, 1]],
                      aod[[2, y, #1, 2]], r, σ] &) /@
                  Range[1, Length[aod[[2, y]]]]] /@
                Range[1, Length[aod[[2]]]]]]];
           σ /. FindMinimum[f, {σ, .5}][[2]]]
```

and store them into the package CFMLab`ImpliedVolatility`.

We compute the call and put least square constant implied volatilities

```
In[67]:= {ConstantCallVolatility[aod, .05],
          ConstantPutVolatility[aodPut, .05]}

Out[67]= {0.533708, 0.558524}
```

The first step in improving our estimation of the implied volatility will be in using more elaborate Black–Scholes formulas, formulas that allow for the time dependence of the volatility. This is the subject of the next section. Furthermore, in the last section of this chapter we shall present an elaborate study of the case when implied volatility depends also on the price of the underlying security (or which will turn out to be the same, on the strike price of the option). Such a study will require some substantially different way of thinking about Black–Scholes theory. Indeed, so far, our study of Black–Scholes fair option prices has been based on the study of the Black–Scholes partial differential equation, and therefore implicitly, on the study of fair option prices as a function of the current time and current underlying stock price (t and S), while the rest of the dependencies were considered as parameters (except for a short remark in Section 5.3.1). The option-data-format above (i.e., many strikes, few expiration dates, but a fixed current time and a fixed underlying stock price) suggests that such a study and understanding of fair option prices is not appropriate if the option market data is to be brought into systematic consideration. The discovery of the Dupire partial differential equation, to be discussed below, is filling the void in our understanding of fair option prices from the point of view of incorporating option data, i.e., in our understanding of the dependence of fair option prices on their strikes and expiration dates. A better understanding then yields more precise numerical methods for computing implied volatility.

5.3.3 Time-Dependent Implied Volatility

5.3.3.1 Time-Dependent Implied Volatility: "Symbolic" Solutions

```
In[68]:= aod = GetAllData["QQQ", "Call", 200];
         aodPut = GetAllData["QQQ", "Put", 200];
```

The method presented in this section is based on the symbolic calculations for time-dependent extended Black–Scholes formulas (of Chapter 3), and numerical (*Mathematica*® supplied) minimization functions, such as in the least "square" constant implied volatility method discussed above. So they are half way symbolic. This implies two kinds of restrictions:

(a) to apply *Mathematica*® provided minimization, we need to construct minimization function(al)s (measuring the distance between the observed and theoretical prices) that depend on only several variables, which is equivalent to saying that we need to work with the time-dependent volatilities $\sigma(t)$ that depend on only several parameters (the constant σ was a function that depends on only one such parameter—itself, and consequently $f_2(\sigma)$ was a function of a single variable);

(b) $\sigma(t)$ needs to be simple enough for symbolic calculations to be possible.

Additionally, the class of volatility functions to which we restrict our consideration, needs to be "reasonable" at least from the phenomenological point of view. One way to judge whether it is reasonable or not is to develop other more sophisticated and consequently more computationally demanding methods, as we shall do later, methods that do not impose a priori such strong constraints on the volatility functions that are admissible; and then observe what kind of implied volatility time dependencies are obtained, and as a consequence adjust/calibrate the present method.

Having said this, we adopt piecewise constant time-dependent volatility, with jumps occurring only at the expiration times. The reader may wish to experiment with different kinds of volatility functions. How to construct step functions? For example, let

```
In[70]:= Step[t_, V_, T_] := V⟦1⟧ +
           Plus @@ ((#1⟦2⟧ - #1⟦1⟧ &) /@ Partition[V, 2, 1]
              (UnitStep[t - #1] &) /@ T)
```

which works as

```
In[71]:= Clear[f]; f[x_] = Step[x, {1, 2, 1}, {1, 3}]
```

Out[71]= $-\theta(x-3)+\theta(x-1)+1$

and then

```
In[72]:= Plot[f[x], {x, 0, 4}, PlotRange → {0, 3}];
```

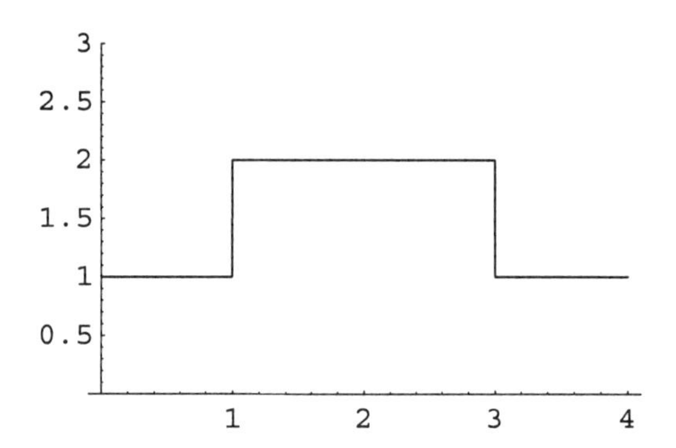

So, V is the set of values that are taken by the step function, while T is the set of times when the jumps occur. Above there are three values {1, 2, 1} and two jump-times {1, 3}, as always: Length[V] = 1 + Length[T]. The values V are actually going to be kept as arbitrary parameters:

```
In[73]:= P[k_] :=
          ToExpression /@ Take[CharacterRange["a", "z"], -k]
```

So, since we have decided to have jumps at the expiration times T, we can define

```
In[74]:= S[k_][t_] := Step[t, P[k], Take[aod⟦1, 3⟧, k - 1]]
```

where k is the number of expiration times. For example, if we restrict our attention to only the first three expiration times, then the volatility function

```
In[75]:= S[3][t]
```

Out[75]= $x + (z - y)\,\theta(t - 0.960099) + (y - x)\,\theta(t - 0.883437)$

has three parameters and two jumps, since at the time of the third expiration the whole process stops. Now that the volatility function is ready, it is time to recall the generalized (time-dependent) Black–Scholes formulas:

```
In[76]:= << "CFMLab`BlackScholes`"
```

Again, we choose as a default method the genuine least square method, while any other even, positive integer power is also allowed. So, consider

```
In[77]:= qq = 3;
         kk = Min[qq, Length[aod⟦2⟧]]

Out[78]= 3
```

expiration dates. Define a pricing formula by means of the above step volatility function

```
In[79]:= w[t_, S_, T_, k_, rr_] =
           GeneralCall[t, S, T, k, rr &, rr &, S[kk]];
```

as well as, for chosen numerical values for p and r, the minimization function(al)

```
In[80]:= With[{p = 2, r = .04},
           f = (Plus @@ #1 &) [((#1^p &) /@ #1 &) [
              Flatten[Function[y, (aod⟦2, y, #1, 3⟧ -
                     w[aod⟦1, 4⟧, aod⟦1, 5⟧, aod⟦2, y,
                       #1, 1⟧, aod⟦2, y, #1, 2⟧, r] &) /@
                  Range[1, Length[aod⟦2, y⟧]]] /@
                Range[1, Length[aod⟦2⟧]]]]]];
```

All that is left to be done is to do the minimizaton:

```
In[81]:= ss[t_] = S[kk][t] /. FindMinimum[Evaluate[f],
             Evaluate[Sequence @@ ({#1, .5} &) /@ P[kk]]]⟦2⟧
```

Out[81]= $-0.0582285\,\theta(t-0.960099)-0.097126\,\theta(t-0.883437)+0.601048$

The solution looks like this:

```
In[82]:= pc = Plot[ss[t], {t, aod⟦1, -2⟧, aod⟦1, 3, 3⟧},
            PlotRange → {0, Automatic},
            PlotStyle → RGBColor[1, 0, 0]];
```

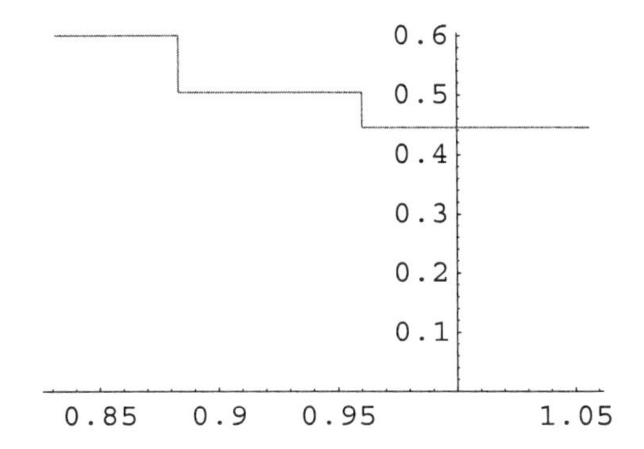

It is time to pack this one into a function TimeDependentCallVolatility, and similarly for the TimeDependentPutVolatility, and place them both in the package CFMLab`ImpliedVolatility`.

5.3.3.2 Market Timing?

Is it possible to use this, and possibly more sophisticated implied volatility methods to be presented later, to draw conclusions about future direction/dynamics of the market (market timing), or at least about the current market "consensus" about the future. For example, using the default least square method we shall compute time-dependent implied volatilities for liquid calls and puts for QQQ. The simple idea is that overpriced puts, i.e., implied volatility for puts is higher than the implied volatility for calls, indicate a "consensus" that the market is expected to decline, and vice-versa.

We compute and draw call and put implied volatilities:

```
In[83]:= aod = GetAllData["QQQ", "Call"];
         aodPut = GetAllData["QQQ", "Put"];
         s1 = TimeDependentCallVolatility[aod, 3, 0.04];
         s2 = TimeDependentPutVolatility[aodPut, 3, 0.05];
```

```
In[86]:= lps1s2 = Plot[{s1[t], s2[t]},
            {t, aod〚1, -2〛, aod〚1, 3, 3〛}, PlotRange →
             {0, Automatic}, PlotLabel → "QQQ", PlotStyle →
             {RGBColor[1, .2, 0], RGBColor[0, 0, .7]}];
```

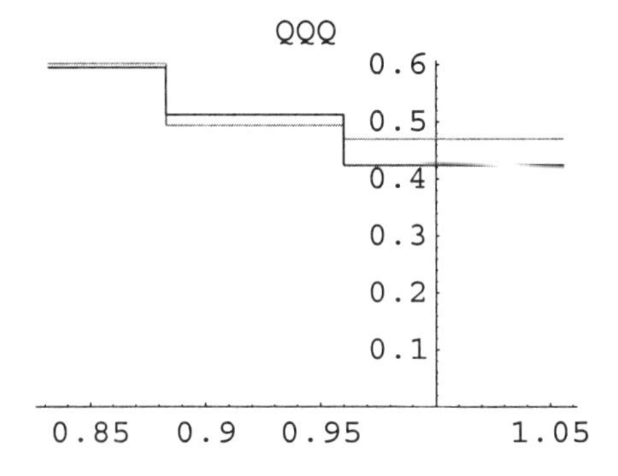

Now let us plot the actual QQQ evolution during the same time period:

```
In[87]:= ListPlot[
           {#[[1]], #[[2]]} & /@ Select[GetPriceData[qqq],
             aod〚1, -2〛 ≤ #[[1]] ≤ aod〚1, 3, 3〛 &],
           PlotJoined → True, PlotStyle →
            {Thickness[.01], RGBColor[0, 1, 0]}];
```

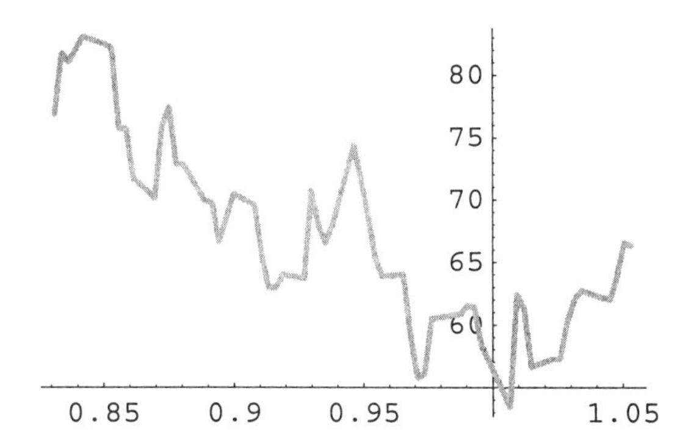

or rescaled and superimposed with the implied volatilities:

```
In[88]:= lpQQQ = ListPlot[{#[[1]], #[[2]] / 100} & /@
           Select[GetPriceData[qqq],
            aod⟦1, -2⟧ ≤ #[[1]] ≤ aod⟦1, 3, 3⟧ &],
          PlotJoined → True, PlotStyle →
           {Thickness[.01], RGBColor[0, 1, 0]},
          DisplayFunction → Identity];
         Show[lps1s2, lpQQQ];
```

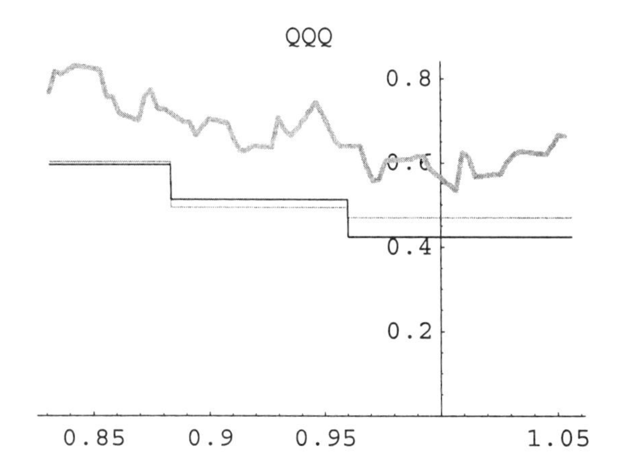

Red (a bit lighter) are the calls, blue (a bit darker) are the puts. The reader can be the judge whether the call and put implied volatilities, their mutual order and difference in size, have predicted the QQQ price evolution or not during the considered time interval. No matter what the conclusion in this instance is, many more experiments are needed.

5.4 Numerical PDEs, Optimal Control, and Implied Volatility

5.4.1 Remarks

Examining what was done above, we can make two observations:

(a) the volatility function depends only on time and not on the price of the underlying stock, i.e., it depends on the expiration time but not on the strike price;

(b) this kind of dependence was prescribed (e.g. piecewise constant, piecewise linear, polynomial) and consequently only a (short) list of numbers was to be determined in the minimization procedure.

In the present study we shall assume neither *(a)* nor *(b)*. The study becomes much more complicated. It will depend on numerical solutions of partial differential equations. So we start there.

5.4.2 Tridiagonal Implicit Finite Difference Solution of Parabolic PDEs

```
In[89]:= Clear["Global`*"]
```

```
In[90]:= << CFMLab`NumericalBlackScholes`
```

We shall present an efficient and precise method for computing approximate solutions of linear one-dimensional parabolic partial differential equations. We shall use and modify this construction also in Chapter 6, when we discuss a solution of the obstacle problem. Consider first a forward parabolic partial differential equation in a bounded domain (interval (x_0, x_1)):

$$\mathcal{P}u = \frac{\partial u(t,x)}{\partial t} - \left(a_2(t,x)\frac{\partial^2 u(t,x)}{\partial x^2} + a_1(t,x)\frac{\partial u(t,x)}{\partial x} + a_0(t,x)\,u(t,x)\right) = f(t,x) \tag{5.4.1}$$

together with the boundary conditions

$$\begin{aligned} u(t, x_0) &= p(t, x_0) \\ u(t, x_1) &= p(t, x_1) \end{aligned} \tag{5.4.2}$$

and the initial condition

$$u(t_0, x) = p(t_0, x) \tag{5.4.3}$$

for $x_0 < x < x_1$ and $t_0 < t < t_1$, where $a_2(t, x) > 0$; no assumption about the sign of $a_1(t, x)$ and $a_0(t, x)$ is made. Initial and boundary conditions can be stated also together as $u = p$ on the *parabolic boundary* $\partial_{\mathcal{P}} Q$ of $Q = (t_0, t_1) \times (x_0, x_1)$.

Implicit solution means that the time interval is divided first into subintervals $(t - dt, t)$, and then that the time derivative is approximated by the finite difference $\frac{\partial u(t,x)}{\partial t} \approx \frac{u(t,x)-u(t-dt,x)}{dt}$; at each time step, after $u(t - dt, x)$ has already been computed $(u(t_0, x) = p(x))$, an elliptic boundary value problem then is

$$
\begin{aligned}
&-\frac{u(t,x) - u(t-dt,x)}{dt} + a_2(t,x)\frac{\partial^2 u(t,x)}{\partial x^2} + a_1(t,x)\frac{\partial u(t,x)}{\partial x} \\
&\qquad + a_0(t,x)\,u(t,x) = -f(t,x) \\
&u(t,x_0) = p(t,x_0) \\
&u(t,x_1) = p(t,x_1)
\end{aligned}
$$

or, more transparently,

$$
\begin{aligned}
&a_2(t,x)\frac{\partial^2 u(t,x)}{\partial x^2} + a_1(t,x)\frac{\partial u(t,x)}{\partial x} + \left(a_0(t,x) - \frac{1}{dt}\right)u(t,x) \\
&= -\frac{1}{dt}\,u(t-dt,x) - f(t,x) \\
&u(t,x_0) = p(t,x_0) \\
&u(t,x_1) = p(t,x_1)
\end{aligned}
$$

solved numerically. Notice that $dt > 0$ needs to be small enough so that $a_0(t, x) - \frac{1}{dt} \leq 0$ (this condition can be relaxed somewhat). Now, in order to solve this elliptic problem one can use finite difference approximations of the derivatives with respect to x: $\frac{\partial u(t,x)}{\partial x} \approx \frac{u(t,x+dx)-u(t,x)}{dx}$ or $\frac{\partial u(t,x)}{\partial x} \approx \frac{u(t,x)-u(t,x-dx)}{dx}$, and $\frac{\partial^2 u(t,x)}{\partial x^2} \approx \frac{u(t,x-dx)-2\,u(t,x)+u(t,dx+x)}{dx^2}$, and thereby the elliptic equation becomes an algebraic linear equation. Indeed, let a_1 be decomposed into its positive and negative parts

$$a_1(t,x) = (a_1(t,x))^+ - (a_1(t,x))^-$$

where $(a_i(t, x))^+ \geq 0$, $i = 1, 2$. The above elliptic equation is discretized into a *system of linear algebraic equations*

$$
\begin{aligned}
&a_2(t,x)\,\frac{u(t,x-dx) - 2\,u(t,x) + u(t,dx+x)}{dx^2} \\
&+(a_1(t,x))^+\,\frac{u(t,x) - u(t,x-dx)}{dx} - (a_1(t,x))^-\,\frac{u(t,x+dx) - u(t,x)}{dx} \\
&+\left(a_0(t,x) - \frac{1}{dt}\right)u(t,x) = -\frac{1}{dt}\,u(t-dt,x) - f(t,x) \\
&u(t,x_0) = p(t,x_0) \\
&u(t,x_1) = p(t,x_1).
\end{aligned}
$$

Moreover, due to the fact that space is only one-dimensional, this is a *tridiagonal* linear system. It is not difficult to write down the above system in the matrix form $A.X = B$. For example, if $dx = \frac{x_1 - x_0}{3}$, then the system matrix A would have been (to reduce the complexity of notation, assume that $x_0 = 0$)

$$A = \begin{pmatrix} 1 & 0 & 0 & 0 \\ -\frac{a_1(t,dx)^+}{dx} + \frac{a_2(t,dx)}{(dx)^2} & \frac{a_1(t,dx)^-}{dx} + \frac{a_1(t,dx)^+}{dx} + a_0(t, dx) - \frac{2\,a_2(t,dx)}{(dx)^2} - \frac{1}{dt} & -\frac{a_1(t,dx)^-}{dx} + \frac{a_2(t,dx)}{(dx)^2} & 0 \\ 0 & -\frac{a_1(t,2\,dx)^+}{dx} + \frac{a_2(t,2\,dx)}{(dx)^2} & \frac{a_1(t,2\,dx)^-}{dx} + \frac{a_1(t,2\,dx)^+}{dx} + a_0(t, 2\,dx) - \frac{2\,a_2(t,2\,dx)}{(dx)^2} - \frac{1}{dt} & -\frac{a_1(t,2\,dx)^-}{dx} + \frac{a_2(t,2\,dx)}{(dx)^2} \\ 0 & 0 & 0 & 1 \end{pmatrix}$$

where X is equal to

$$X = \begin{pmatrix} u(t, 0) \\ u(t, dx) \\ u(t, 2\,dx) \\ u(t, x_1) \end{pmatrix}.$$

The right-hand side vector B would be

$$B = \begin{pmatrix} p(t, 0) \\ -f(t, dx) - \frac{u(t-dt,dx)}{dt} \\ -f(t, 2\,dx) - \frac{u(t-dt,2\,dx)}{dt} \\ p(t, x_1) \end{pmatrix}.$$

It is also then easy to imagine what the linear system would look like in a higher dimension. In practice, of course, dx is going to be much smaller, and consequently the above tridiagonal matrix A and vector B much larger. Since the system matrix is tridiagonal we shall use the standard *Mathematica*® package

```
In[91]:= << "LinearAlgebra`Tridiagonal`"
```

and in particular, we use the function TridiagonalSolve

```
In[92]:= ? TridiagonalSolve

TridiagonalSolve[ a, b, c, r ] solves A . x == r for x,
   where A is a tridiagonal matrix.  The three diagonals
```

```
of A are given by: a11 = b[[1]], a12 = c[[1]], a21 =
a[[1]], a22 = b[[2]], a23 = c[[2]], ... No pivoting
is done, so the algorithm may fail even though a
solution exists. This cannot happen for a symmetric
positive definite matrix. More...
```

Now that we have laid down the ingredients of the solution, it is time to construct the solution itself. Let put data in front (and start with the Heat PDE):

```
In[93]:= a2[t_, s_] := 1; a1[t_, s_] := 0; a0[t_, s_] := 0;
         f[t_, s_] := 0; p[t_, s_] := (s - s0) (s1 - s);
         s0 = 0; s1 = 1; Ns = 50; t0 = 0; t1 = .5; Nt = 50;
```

where Ns and Nt are numbers of subintervals in the space and time variables.

Starting the calculation, we first define space and time grids

```
In[95]:= ds = (s1 - s0)/Ns; ss = N[Range[s0, s1, ds]];
         dt = (t1 - t0)/Nt; tt = N[Range[t0, t1, dt]];
         Arguments = Outer[{#1, #2} &, tt, ss];
```

and then evaluate all the data on the grid

```
In[98]:= PPList = ((p @@ #1 &) /@ #1 &) /@ Arguments;
         A2List = ((a2 @@ #1 &) /@ #1 &) /@ Arguments;
         A1List = ((a1 @@ #1 &) /@ #1 &) /@ Arguments;
         A0List = ((a0 @@ #1 &) /@ #1 &) /@ Arguments;
         RightHandSide = ((f @@ #1 &) /@ #1 &) /@ Arguments;
         A1PlusList = ((Max[#1, 0] &) /@ #1 &) /@ A1List;
         A1MinusList = ((-Min[#1, 0] &) /@ #1 &) /@ A1List;
```

Next, we form three lists in time of vectors that are going to be the first three arguments in TridiagonalSolve:

```
In[105]:= LowerDiagonal = (ReplacePart[#1, 0, -1] &) /@
            (Drop[#1, 1] &) /@ (A2List/ds^2 - A1PlusList/ds);
          Diagonal = (ReplacePart[#1, 1, -1] &) /@
            (ReplacePart[#1, 1, 1] &) /@ (A0List - (2 A2List)/ds^2 +
              A1MinusList/ds + A1PlusList/ds - 1/dt);
          UpperDiagonal = (ReplacePart[#1, 0, 1] &) /@
            (Drop[#1, -1] &) /@ (A2List/ds^2 - A1MinusList/ds);
```

and putting them all together with the right-hand side, as well as with the parabolic boundary data we have

```
In[107]:= AAA = Drop[Transpose[{LowerDiagonal, Diagonal,
              UpperDiagonal, RightHandSide, PPList}], 1];
```

Finally, we define the solver of the elliptic problem above that is going to be applied at each time step:

```
In[108]:= EllipticSolver :=
            TridiagonalSolve[#2⟦1⟧, #2⟦2⟧, #2⟦3⟧,
              ReplacePart[ReplacePart[-#2⟦4⟧ - #1/dt,
                #2⟦5, 1⟧, 1], #2⟦5, -1⟧, -1]] &
```

and perform the calculation, marching in time,

```
In[109]:= ParabolicSolution =
            FoldList[EllipticSolver, PPList⟦1⟧, AAA];
```

Once the solution is computed on the grid, it can be interpolated in a function

```
In[110]:= PSOut = Interpolation[(Append[#1⟦1⟧, #1⟦2⟧] &) /@
              Transpose[{Flatten[Arguments, 1],
                Flatten[ParabolicSolution]}]];
```

and used

```
In[111]:= Plot3D[PSOut[t, x], {t, t0, t1},
            {x, s0, s1}, PlotRange → All, Mesh → False,
            PlotPoints → 50, AxesLabel → {t, x, ""}];
```

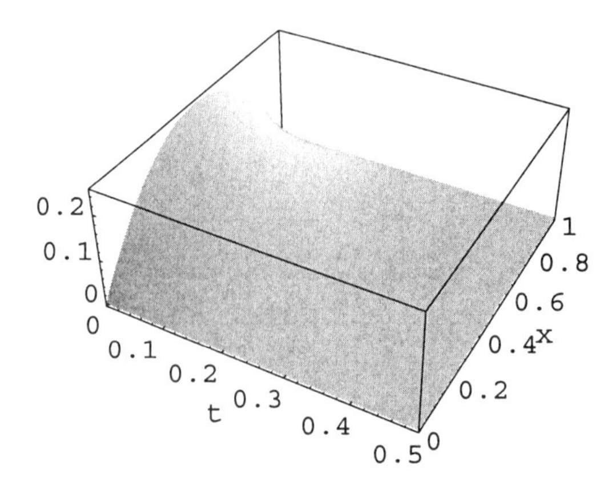

Since this is a solution of the homogeneous Heat PDE, the precision of the above approximate solution can be judged by comparison with the exact solution which can be derived using the method of *separation of variables*:

$$\frac{2}{s_1 - s_0} \sum_{n=1}^{\infty} \left(\int_{s_0}^{s_1} p(t_0, x) \sin\left(\frac{n\pi (x - s_0)}{s_1 - s_0} \right) dx \right) \sin\left(\frac{n\pi (x - s_0)}{s_1 - s_0} \right) e^{-n^2 \pi^2 (t - t_0)}.$$

Unfortunately the above infinite sum does not evaluate, except for the initial value, i.e., except for $t = t_0$:

In[112]:=
$$\frac{1}{\text{s1} - \text{s0}} \left(2 \sum_{\text{n}=1}^{\infty} \left(\int_{\text{s0}}^{\text{s1}} \text{p[t0, x] Sin}\left[\frac{\text{n} \pi (\text{x} - \text{s0})}{\text{s1} - \text{s0}} \right] d\text{x} \right) \text{Sin}\left[\frac{\text{n} \pi (\text{x} - \text{s0})}{\text{s1} - \text{s0}} \right] e^{-\text{n}^2 \pi^2 0} \right)$$

Out[112]=
$$-\frac{2 i (\text{Li}_3(-e^{-i\pi x}) - \text{Li}_3(e^{-i\pi x}) - \text{Li}_3(-e^{i\pi x}) + \text{Li}_3(e^{i\pi x}))}{\pi^3}$$

which looks like

```
In[113]:= Plot[%, {x, -3, 3}];
```

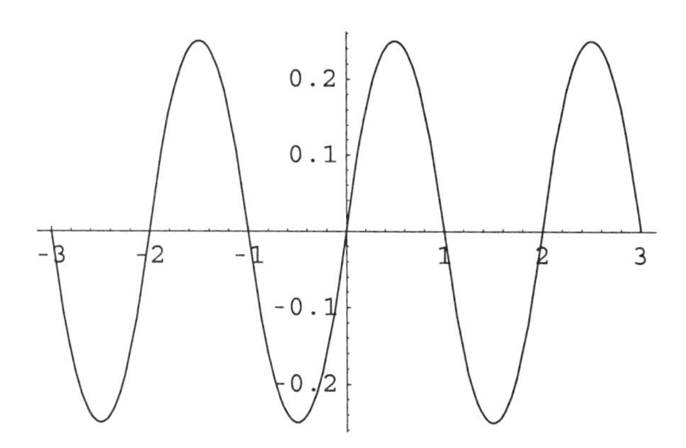

coinciding with the initial value in the domain of consideration (s_0, s_1), and then extended as an odd and periodic function. Nevertheless we can truncate the summation in the above exact solution, and define an approximate solution

In[114]:= FourierSolution[tt_, xx_, K_] := Module[{a},

$$\text{a[n\_]} = \int_{\text{s0}}^{\text{s1}} \text{p[t0, x] Sin}\left[\frac{\text{n} \pi (\text{x} - \text{s0})}{\text{s1} - \text{s0}} \right] d\text{x};$$

$$\text{N}\left[\frac{2 \sum_{\text{n}=1}^{\text{K}} \text{a[n] Sin}\left[\frac{\text{n} \pi (\text{x}-\text{s0})}{\text{s1}-\text{s0}} \right] e^{-\text{n}^2 \pi^2 (\text{t}-\text{t0})}}{\text{s1} - \text{s0}} \text{ /. } \{\text{t} \to \text{tt}, \text{x} \to \text{xx}\} \right]]$$

which can be viewed as

```
In[115]:= Plot3D[Evaluate[FourierSolution[t, s, 100]],
            {t, t0, t1}, {s, s0, s1},
            PlotRange → All, Mesh → False,
            PlotPoints → 50, AxesLabel → {t, x, ""}];
```

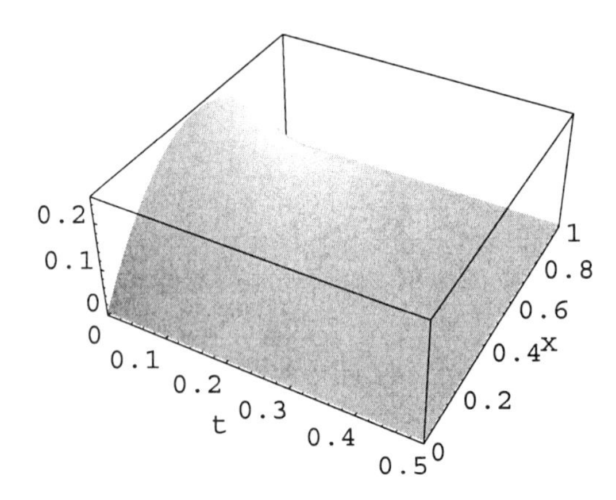

and which can be also computed with arbitrary precision:

```
In[116]:= (FourierSolution[t0 + (t1 - t0)/20,
              s0 + (s1 - s0)/2, #1] &) /@ {1, 3, 5, 7, 10, 100}

Out[116]= {0.201596, 0.200559, 0.200563, 0.200563, 0.200563, 0.200563}
```

while the above finite difference solution yields

```
In[117]:= PSOut[t0 + (t1 - t0)/20, s0 + (s1 - s0)/2]

Out[117]= 0.20206
```

Notice that since the TridiagonalSolve solves linear systems exactly, the error in the above calculation is due only to discretization.

It is time now to pack the above solver into a single module:

```
In[118]:= ForwardParabolicSolver[{a2_, a1_, a0_}, {f_, p_},
            {s0_, s1_, Ns_}, {t0_, t1_, Nt_}, IOrder_ : 2] :=
           Module[{ds, ss, dt, tt, Arguments, PPList,
             A2List, A1List, A0List, RightHandSide,
             A1PlusList, A1MinusList, LowerDiagonal,
             Diagonal, UpperDiagonal, AAA,
             EllipticSolver, ParabolicSolution},
```

```
ds = (s1 - s0)/Ns; ss = N[Range[s0, s1, ds]];
dt = (t1 - t0)/Nt; tt = N[Range[t0, t1, dt]];
Arguments = Outer[{#1, #2} &, tt, ss];
PPList = ((p @@ #1 &) /@ #1 &) /@ Arguments;
A2List = ((a2 @@ #1 &) /@ #1 &) /@ Arguments;
A1List = ((a1 @@ #1 &) /@ #1 &) /@ Arguments;
A0List = ((a0 @@ #1 &) /@ #1 &) /@ Arguments;
RightHandSide = ((f @@ #1 &) /@ #1 &) /@ Arguments;
A1PlusList = ((Max[#1, 0] &) /@ #1 &) /@ A1List;
A1MinusList = ((-Min[#1, 0] &) /@ #1 &) /@ A1List;
LowerDiagonal = (ReplacePart[#1, 0, -1] &) /@
  (Drop[#1, 1] &) /@ (A2List/ds^2 - A1PlusList/ds);
Diagonal = (ReplacePart[#1, 1, -1] &) /@
  (ReplacePart[#1, 1, 1] &) /@
   (A0List - (2 A2List)/ds^2 +
     A1MinusList/ds + A1PlusList/ds - 1/dt);
UpperDiagonal = (ReplacePart[#1, 0, 1] &) /@
  (Drop[#1, -1] &) /@ (A2List/ds^2 - A1MinusList/ds);
AAA = Drop[Transpose[{LowerDiagonal, Diagonal,
     UpperDiagonal, RightHandSide, PPList}], 1];
EllipticSolver := TridiagonalSolve[#2[[1]], #2[[2]],
   #2[[3]], ReplacePart[ReplacePart[-#2[[4]] - #1/dt,
     #2[[5, 1]], 1], #2[[5, -1]], -1]] &;
ParabolicSolution = FoldList[EllipticSolver,
  PPList[[1]], AAA];
Interpolation[(Append[#1[[1]], #1[[2]]] &) /@
  Transpose[{Flatten[Arguments, 1],
    Flatten[ParabolicSolution]}],
 InterpolationOrder -> IOrder]]
```

Above we have checked the precision (and correctness) of the numerical solution when $f(t, x) = 0$. Compute instead (notice that all the coefficients on the right-hand side and the initial condition are pure functions)

```
In[119]:= Timing[sol = ForwardParabolicSolver[{1 &, 0 &, 0 &},
           {1 &, 0 &}, {0, 1, 20}, {0, 1, 20}]]
```

Out[119]= $\{0.82\ \text{Second}, \text{InterpolatingFunction}[\begin{pmatrix} 0. & 1. \\ 0. & 1. \end{pmatrix}, \text{<>}]\}$

The solution converges quickly to the steady state

In[120]:=
```
Plot3D[sol[t, s], {t, 0, 1},
    {s, 0, 1}, PlotRange → All, Mesh → False,
    PlotPoints → 50, AxesLabel → {t, x, ""}];
```

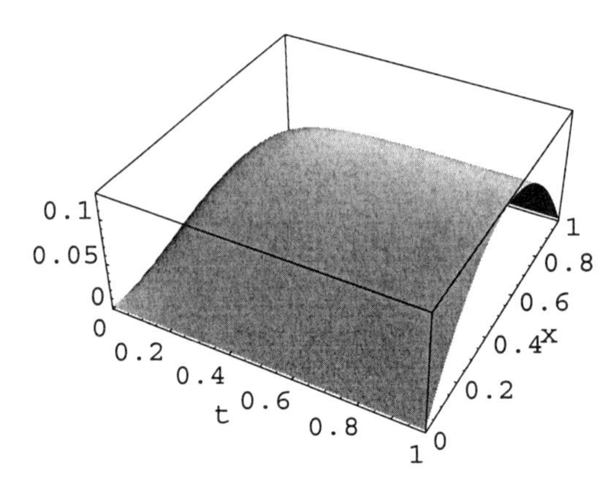

while the steady state is trivial to compute explicitly

In[121]:=
```
z[x_] = y[x] /. DSolve[
        {-y''[x] == 1, y[0] == 0, y[1] == 0}, y[x], x]⟦1⟧
```

Out[121]= $\frac{1}{2}(x - x^2)$

Finally, compare the computed solution and the steady state.

In[122]:=
```
Plot[{sol[1, x], z[x]}, {x, 0, 1}];
```

which looks good. More checking is needed, most importantly when the first order coefficient $a_1 \neq 0$. We postpone this until the next section since a very good example of such problems (backward in time) is the Black–Scholes partial differential equation with its explicit solution.

Consider now the *backward* parabolic partial differential equation in a bounded domain $t_0 < t < t_1$, $x_0 < x < x_1$:

$$\frac{\partial u(t, x)}{\partial t} + \left(a_2(t, x) \frac{\partial^2 u(t, x)}{\partial x^2} + a_1(t, x) \frac{\partial u(t, x)}{\partial x} + a_0(t, x)\, u(t, x) \right) = f(t, x)$$

together with the backward parabolic boundary condition:

$$u(t, x_0) = p(t, x_0)$$
$$u(t, x_1) = p(t, x_1)$$
$$u(t_1, x) = p(t_1, x)$$

where $a_2(t, x) > 0$; no assumption about the sign of $a_1(t, x)$ and $a_0(t, x)$ is made. Notice the change of sign in the partial differential equation (both sides). The solution of the backward problem can be obtained from the solution of the forward problem by the change of time variable:

```
In[123]:= BackwardParabolicSolver[{a2_, a1_, a0_},
            {f_, p_}, {s0_, s1_, Ns_}, {t0_, t1_, Nt_},
            IOrder_ : 2] := Module[{ForwardSol},
            ForwardSol = ForwardParabolicSolver[{a2[t1 - #1, #2] &,
               a1[t1 - #1, #2] &, a0[t1 - #1, #2] &},
              {-f[t1 - #1, #2] &, p[t1 - #1, #2] &}, {s0, s1, Ns},
              {0, t1 - t0, Nt}, IOrder]; ForwardSol[t1 - #1, #2] &]
```

We check the veracity of that claim on an example:

```
In[124]:= BPsol = BackwardParabolicSolver[{1 &, 0 &, 0 &},
              {-1 &, 0 &}, {0, 1, 10}, {1, 2, 10}];
```

which looks good:

```
In[125]:= Plot3D[BPsol[t, s], {t, 1, 2}, {s, 0, 1}, PlotRange → All,
            Mesh → False, PlotPoints → 50, AxesLabel → {t, x, ""}];
```

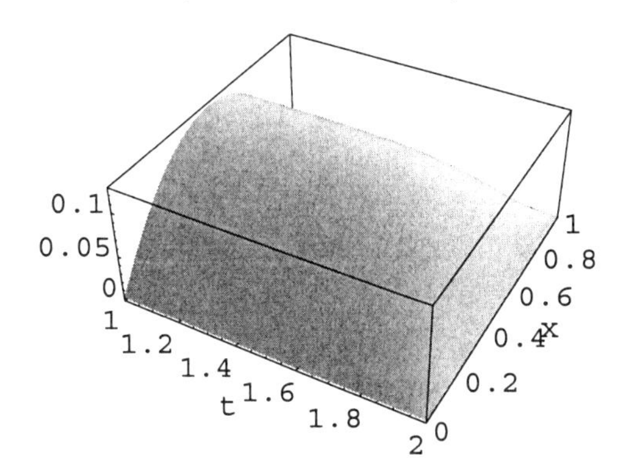

Functions ForwardParabolicSolver and BackwardParabolicSolver are deposited into the package CFMLab`NumericalBlackScholes` to be joined with a few more functions below.

5.4.3 Pricing European Options with Stock Price Dependent Data

When we were deriving the Black–Scholes partial differential equation the starting point was the assumption about the dynamics of the underlying stock:

$$dS(t) = S(t)\,((a(t, S(t)) - D_0(t, S(t)))\,dt + \sigma(t, S(t))\,dB(t))$$
$$S(0) = s_0$$

where $a(t, S)$ is the appreciation rate, $D_0(t, S)$ is the dividend payment rate (dividend payment reduces the value of the stock), and $\sigma(t, S)$ is the volatility. It was then shown that the fair call and put option prices obey, and moreover that they are *the unique solutions of* the partial differential equation

$$\frac{\partial V(t, S)}{\partial t} + \frac{1}{2} S^2 \frac{\partial^2 V(t, S)}{\partial S^2} \sigma(t, S)^2 + (r(t, S) - D_0(t, S))\, S \frac{\partial V(t, S)}{\partial S}$$
$$-r(t, S)\, V(t, S) = 0$$

for $S > 0$ and $t < T$, where T is the expiration time, together with the terminal condition

$$V(T, S) = \mathrm{Max}(0, S - k)$$

in the case of calls, and

$$V(T, S) = \mathrm{Max}(0, k - S)$$

in the case of puts (k is the strike price). Notice that we have allowed here the interest rate $r = r(t, S)$ to be dependent on the underlying price S. If the underlying price is important, such as QQQ which tracks NASDAQ 100, then such a generality may not be just for the sake of generality, but instead may have some practical value: interest rates may indeed be influenced by the performance of the stock market.

As we have seen so far, if further it is assumed that $\sigma(t, S) = \sigma(t)$, as well as $D_0(t, S) = D_0(t)$ and $r(t, S) = r(t)$, explicit solutions are possible. If this is not the case, we are left with the numerical solutions. Considering the high efficiency of the numerical algorithm presented in the previous section, this is not a major inconvenience.

The first issue that arises when attempting to solve the above equation numerically is the issue of truncation of the domain $\{t, t < T\} \times \{S, S > 0\}$. So let $x_0 > 0$, $x_1 < \infty$ and $t_0 < T$ be fixed, and instead of the above equation being considered in $\{t, t < T\} \times \{S, S > 0\}$, we solve the same PDE in $\{t, t_0 < t < T\} \times \{S, x_0 < S < x_1\}$. This implies the necessity of imposing (truncated) boundary conditions on $S = x_0$ and $S = x_1$. Finding natural truncated boundary conditions for solving partial differential equations is sometimes quite difficult. Here though it appears natural to take Black–Scholes (explicit) values, VC in the case of calls, and VP in the case of puts, computed with some kind of averaged data as a good boundary value for the non-constant data problem. Summarizing, we have

```
In[126]:= NumericalCallFairPrice[{σ_, r_, D0_}, {k_, T_},
            {t0_, Nt_}, {s0_, s1_, Ns_}, IOrder_: 2] :=
           Module[{ds, ss, dt, tt, Arguments, σσ, rr},
            ds = (s1 - s0)/Ns; ss = N[Range[s0, s1, ds]];
            dt = (T - t0)/Nt; tt = N[Range[t0, T, dt]];
            Arguments = Outer[{#1, #2} &, tt, ss];
            σσ = Av[(σ @@ #1 &) /@ Flatten[Arguments, 1]];
            rr = Av[(r @@ #1 &) /@ Flatten[Arguments, 1]];
            BackwardParabolicSolver[{1/2 #2^2 σ[#1, #2]^2 &,
              (r[#1, #2] - D0[#1, #2]) #2 &, -r[#1, #2] &}, {0 &,
              If[#1 < T, VC[#1, #2, T, k, rr, σσ], Max[0, #2 - k]] &},
             {s0, s1, Ns}, {t0, T, Nt}, IOrder]]
```

and similarly in the case of puts. Both functions are stored in the package CFMLab`NumericalBlackScholes`. The first computation is to check the veracity and precision of the solver. To this end we use constant data and compare the obtained result with the explicit Black–Scholes solution, with an expectation that no difference will be visible:

```
In[127]:= Off[General::"spell1"]

In[128]:= x0 = 20; x1 = 200; str = 100; σ = .5; r = .05; t0 = 1; T = 1.3;
          ncfp = NumericalCallFairPrice[
             {σ &, r &, 0 &}, {str, T}, {t0, 10}, {x0, x1, 10}];

In[130]:= Plot[{VC[t0, S, T, str, r, σ],
             ncfp[t0, S], Max[0, S - str]}, {S, x0, x1},
            PlotStyle → {RGBColor[1, 0, 0], RGBColor[0, 0, 1],
              RGBColor[0, 0, 0]}, PlotRange → All];
```

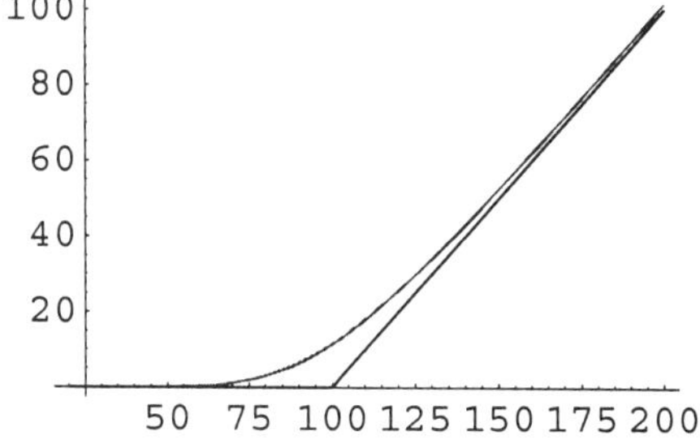

The second computation is to check whether the new solver brings something substantially new. To this end we use stock-price dependent volatility, for example,

In[131]:= $\text{vol}[x\_] := \frac{(x-50)\,(x-150)}{15000} + 0.44827343528098307$

having an "average"

In[132]:= $\text{AverageVolatility} = \sqrt{\frac{\int_{x0}^{x1} \text{vol}[x]^2\, dx}{x1 - x0}}$

Out[132]= 0.5

or

In[133]:= $\text{AverageVolatility2} = \frac{\int_{x0}^{x1} \text{vol}[x]\, dx}{x1 - x0}$

Out[133]= 0.468273

The volatility and the two averages look like

```
In[134]:= Plot[{AverageVolatility,
            AverageVolatility2, vol[x]},
           {x, x0, x1}, PlotRange → {0, Automatic},
           PlotStyle → {RGBColor[1, 0, 0],
             RGBColor[0, 1, 0], RGBColor[0, 0, 0]}];
```

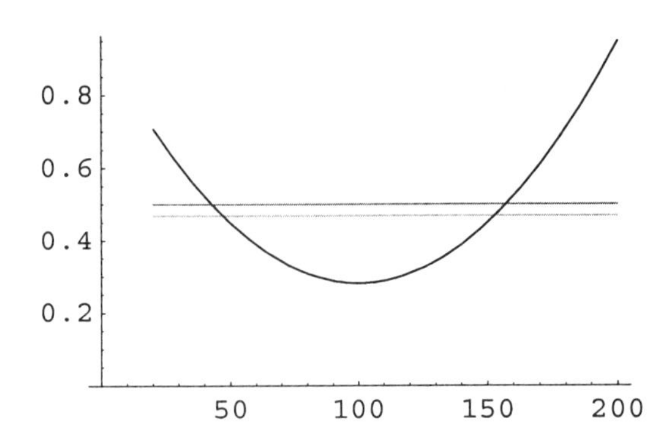

Now compute the numerical call fair price

```
In[135]:= ncfp = NumericalCallFairPrice[{vol[#2] &, r &, 0 &},
            {str, T}, {t0, 10}, {x0, x1, 10}];
```

and compare this with the Black–Scholes fair price for either one of the above "average" volatilities:

```
In[136]:= Plot[{VC[t0, S, T, str, r, AverageVolatility],
            VC[t0, S, T, str, r, AverageVolatility2],
            ncfp[t0, S], Max[0, S - str]},
           {S, x0, x1}, PlotStyle → {RGBColor[1, 0, 0],
             RGBColor[0, 1, 0], RGBColor[0, 0, 0],
             RGBColor[0, 0, 0]}, PlotRange → All];
```

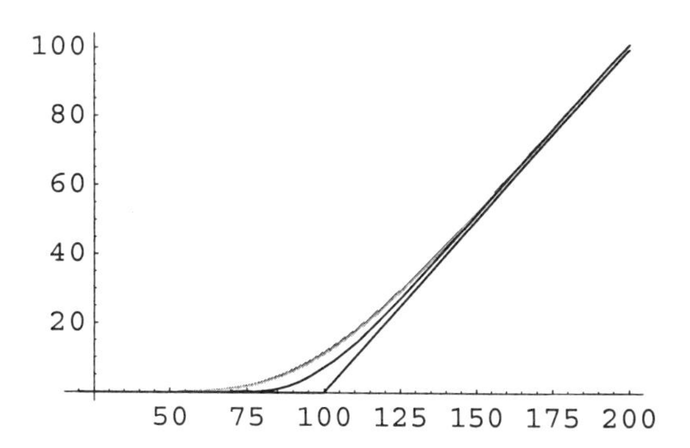

The apparent significant discrepancy between either one of the constant volatility pricing functions and the non-constant volatility pricing function, coupled with the high efficiency and accuracy of the presented numerical algorithm, suggests that it is well worth the effort to use numerical solutions when there exists data dependence on the underlying price.

5.4.4 Dupire Partial Differential Equation

5.4.4.1 Formulation and Verification

```
In[137]:= Remove["Global`*"]
```

When the Black–Scholes partial differential equation

$$\frac{\partial V(t,S)}{\partial t} + \frac{1}{2} S^2 \frac{\partial^2 V(t,S)}{\partial S^2} \sigma(t,S)^2 + (r - D_0) S \frac{\partial V(t,S)}{\partial S} - r V(t,S) = 0 \tag{5.4.4}$$

together with the terminal condition

$$V(T,S) = \text{Max}(0, S - k) \tag{5.4.5}$$

in the case of calls (k is the strike price) and

$$V(T,S) = \text{Max}(0, k - S) \tag{5.4.6}$$

in the case of puts, was solved explicitly, in the constant data case the resulting solution depended on the variables in the above equation, t and S, as well as on other parameters. Furthermore, since it is an explicit solution, in the end there was no real difference between parameters and variables: all of them are considered as variables of the Black–Scholes fair price function. This is of practical importance and has been used for example when we computed implied volatility, both constant and time-dependent.

On the other hand, when there are no explicit solutions, the difference between variables and parameters cannot be ignored: the computation such as the one done in the previous section was for chosen data; one would have to compute the solution for the next set of data, then again for many sets of data, and then to interpolate and to get a solution, for which, in principle, the difference between variables and parameters could be blurred again. Fortunately, there exists a better way.

More precisely, the independent variables in the above equation are t, the current time, and S, the current price of the underlying stock, while the strike price appears as a parameter explicitly in the terminal condition, as well as the expiration time of the considered option T, which appears as the time when the terminal condition is imposed. On the other hand, as seen at the beginning of the present chapter, the option market data comes for frozen times and for a frozen underlying stock price, but for many strikes and several expiration times. Henceforth the Black–Scholes equation is not very useful if one is to use the option market data. Surprisingly, another partial differential equation was discovered by Dupire [8,16]. Namely, the fair option price, as a function of the expiration time T and strike price k for a fixed current time t and a fixed underlying stock price S (they become parameters now), is the unique solution of the *forward* parabolic partial differential equation

$$\begin{aligned}-\frac{\partial V(T,k)}{\partial T}+\frac{1}{2}\,k^2\,\frac{\partial^2 V(T,k)}{\partial k^2}\,\sigma(T,k)^2 \\ -(r-D_0)\,k\,\frac{\partial V(T,k)}{\partial k}-D_0\,V(T,k)=0\end{aligned} \tag{5.4.7}$$

for $T > t_0$, $k > 0$, together with the *initial* condition

$$V(t_0,k)=\mathrm{Max}(0,S-k) \tag{5.4.8}$$

in the case of calls (k is the strike price) and

$$V(t_0,k)=\mathrm{Max}(0,k-S). \tag{5.4.9}$$

Notice that since k is the variable now, the nature of initial conditions inverts as compared to the terminal conditions for the Black–Scholes equation: calls are decreasing with respect to k while puts are increasing. What is quite surprising and extremely useful is that in both partial differential equations the same volatility function σ appears, in the Black–Scholes equation (5.4.4) as a function of time and the stock price $\sigma(t,S)$, and in the Dupire equation (5.4.7) as a function of expiration time and the strike price $\sigma(T,k)$. The usefulness is striking: if we identify $\sigma(T,k)$ using the option market data, we have identified $\sigma(t,S)$ as well.

The mathematical proof of the above fact about the Dupire equation can be found in [8]. Here we confine ourselves only to verify that the solution of the Black–Scholes

PDE is a solution of the Dupire equation in the case when $\sigma(t, S) = \sigma(t)$, i.e., in the case when an explicit solution exists, and therefore verification is possible. So, recall the time-dependent Black–Scholes partial differential equation (for simplicity consider only time dependent volatility)

```
In[138]:= TimeDependentBSPDE =
            ∂t V[t, S] + 1/2 S^2 ∂{S,2} V[t, S] σ[t]^2 +
              r S ∂S V[t, S] - r V[t, S] == 0;
```

and its solutions:

```
In[139]:= << "CFMLab`BlackScholes`"
```

Let first check that indeed this is a solution of the Black–Scholes equation. To this end define

```
In[140]:= BSCall[τ_, S_] = GeneralCall[τ, S, T, k, r &, r &, σ];
```

and plug it into the equation

```
In[141]:= Simplify[TimeDependentBSPDE /. V → BSCall]
Out[141]= True
```

Now define the associated Dupire partial differential equation

```
In[142]:= DupirePDE = -∂T V[T, k] +
              1/2 k^2 ∂{k,2} V[T, k] σ[T]^2 - r k ∂k V[T, k] == 0;
```

and the candidates for solutions:

```
In[143]:= DupireCall[T_, k_] =
             GeneralCall[t, S, T, k, r &, r &, σ];
```

and

```
In[144]:= DupirePut[T_, k_] =
             GeneralPut[t, S, T, k, r &, r &, σ];
```

Indeed they do solve the Dupire PDE:

```
In[145]:= Simplify[DupirePDE /. V → DupireCall]
Out[145]= True
```

in the case of calls, and

```
In[146]:= Simplify[DupirePDE /. V → DupirePut]
Out[146]= True
```

5.4.4.2 Numerical Solution of the Strike Price Dependent Dupire PDE

We now solve numerically the strike price dependent Dupire PDE (5.4.7)–(5.4.9). Considering what was done so far, it is now not difficult to understand the following solution:

```
In[147]:= DupireCallSolution[{σ_, r_}, {S_, t0_},
    {t1_, Nt_}, {s0_, s1_, Ns_}, IOrder_ : 2] :=
   Module[{ds, ss, dt, tt, Arguments, σσ, rr},
    ds = (s1 - s0)/Ns; ss = N[Range[s0, s1, ds]];
    dt = (t1 - t0)/Nt; tt = N[Range[t0, t1, dt]];
    Arguments = Outer[{#1, #2} &, tt, ss];
    σσ = Av[(σ @@ #1 &) /@ Flatten[Arguments, 1]];
    rr = Av[(r @@ #1 &) /@ Flatten[Arguments, 1]];
    ForwardParabolicSolver[{1/2 #2^2 σ[#1, #2]^2 &,
      -r[#1, #2] #2 &, 0 &}, {0 &, If[#1 > t0,
        VC[t0, S, #1, #2, rr, σσ], Max[0, S - #2]] &},
     {s0, s1, Ns}, {t0, t1, Nt}, IOrder]]
```

and similarly in the case of puts. Both functions, the DupireCallSolution and the DupirePutSolution are stored in the package CFMLab`NumericalBlackScholes` for later reference.

5.4.4.3 A Numerical Implied Volatility Problem

The Dupire partial differential equation (5.4.7) characterizes the Black–Scholes fair option prices as a function of expiration time T, and strike price k, exactly those variables that are present in option market prices data.

The implied volatility problem is then the problem of finding an optimal function $\sigma(T, k)$, such that the observed option market prices are as close as possible to the solution of the Dupire partial differential equation, as computed using such a volatility function. This may be considered as a typical problem in the theory of the *optimal control of systems governed by differential equations*. In the next section we present some facts about that theory using some new and fascinating examples of explicit solutions. Then in the following section we come back to the problem of implied volatility, which is a much harder problem, and therefore yields a much less satisfactory, approximate solution. Careful examination of both solutions can yield a better understanding of the practical one.

5.4.5 Optimal Control of Differential Equations

5.4.5.1 Optimal Control of ODEs with Quadratic Cost: Explicit Solutions

```
In[148]:= Clear["Global`*"]
```

How to introduce optimal control of differential equations quickly? The strategy adopted here is to solve the simplest optimal control problem completely. *Mathematica*® helps here by providing some new insights. We showcase both the simplicity and the sophistication of the optimal control theory for systems governed by differential equations. The reader can consult also [12,41,43].

In various branches of optimal control theory the simplest problem is (without being precise about assumptions): given (target) u_d, find f such that if $u = u(f)$ is determined as the unique solution of an abstract, linear equation (called a *state equation*)

$$A\,u = f, \tag{5.4.10}$$

then the real-valued (quadratic) function ($\epsilon > 0$) (*the cost functional*)

$$J(f) = \frac{1}{2}\,(\,|\,u(f) - u_d\,|^2 + \epsilon\,|\,f\,|^2) \tag{5.4.11}$$

is minimized. The first term in the cost functional measures the distance between the state and the target, while the second term measures the cost of applying the control (it also regularizes the problem). It is important to notice that it is not assumed that u_d belongs to the domain of A.

This problem has an explicit solution. First we study how to differentiate the functional J. Let $w = \lim_{\epsilon\to 0} \frac{u(f+h\,\epsilon)-u(f)}{\epsilon}$, the directional derivative of u at f in the (non-unit) direction h. From the state equation we can read the equation characterizing w:

$$A\,w = h. \tag{5.4.12}$$

Now consider the so-called *adjoint equation*

$$A^*\,p = u - u_d \tag{5.4.13}$$

which also admits a unique solution. We can compute the directional derivative of the cost functional:

$$J'[f,\,h] = (u - u_d,\,w) + \epsilon\,(f,\,h) = (A^*\,p,\,w) + \epsilon\,(f,\,h)$$
$$= (\,p,\,A\,w) + \epsilon\,(f,\,h) = (\,p,\,h) + \epsilon\,(f,\,h) = (\,p + \epsilon\,f,\,h)$$

and we conclude that the *gradient* of J is given by

$$\begin{aligned}\nabla J[f] &= p[f] + \epsilon\,f = A^{*-1}[u[f] - u_d] + \epsilon\,f \\ &= A^{*-1}[A^{-1}\,f - u_d] + \epsilon\,f.\end{aligned} \tag{5.4.14}$$

This formula can be used for numerical calculations using the *steepest descent* iteration

$$f_{n+1} = f_n - \rho_n\,\nabla J[f_n] \tag{5.4.15}$$

where $\rho_n > 0$ (if f_n and $\nabla J[f_n]$ are not in the same space, then $\nabla J[f_n]$ also has to be regularized; see the optimal control of the Dupire equation to follow). In many optimal control problems, such as the implied volatility problem to be discussed in the next section, this is pretty much the extent of how much practical insight can be gained. How little this is, and how much more there is to wish for, can be judged from how much will be understood and done in the simplest optimal control problem we are considering now.

Suppose that f is a *minimizer* of the cost functional. This implies $\nabla J[f] = 0$, and therefore $f = \frac{-1}{\epsilon}\, p$, yielding from the above the *optimality system*

$$\begin{aligned} A\,u &= f \\ A^*\,p &= u - u_d \\ f &= \frac{-1}{\epsilon}\,p \end{aligned}$$

or, which is the same

$$\begin{aligned} A\,u &= \frac{-1}{\epsilon}\,p \\ A^*\,p &= u - u_d. \end{aligned} \tag{5.4.16}$$

This system can be solved explicitly. From the second equation

$$p = A^{*-1}\,u - A^{*-1}\,u_d.$$

Therefore going back into the first equation

$$A\,u = \frac{-1}{\epsilon}\,(A^{*-1}\,u - A^{*-1}\,u_d)$$

$$A\,u + \frac{1}{\epsilon}\,A^{*-1}\,u = \frac{1}{\epsilon}\,A^{*-1}\,u_d$$

$$\epsilon\,A^*\,A\,u + u = A^*\,A^{*-1}\,u_d = u_d$$

and finally

$$u = (\epsilon\,A^*\,A + 1)^{-1}\,u_d. \tag{5.4.17}$$

Once u was determined, f can be computed from the state equation

$$f = A\,u = A\,(\epsilon\,A^*\,A + 1)^{-1}\,u_d. \tag{5.4.18}$$

This solution may appear not quite as explicit as announced. On the contrary, examples are below.

Remark: It may be interesting to notice that

$$A^*\,(\epsilon\,A\,A^* + 1)^{-1}\,A^{*-1} = (\epsilon\,A^*\,A + 1)^{-1} \neq (\epsilon\,A\,A^* + 1)^{-1}.$$

The identity is going to be proved below in a more general setting; the inequality can be checked in the example below. Notice also that the trivial case is consistent with the above formula: if $u_d \in \text{dom}(A)$, $\epsilon = 0$, then $f = A\, u_d$.

5.4.5.2 An Example

Many examples can be provided. Let a be a fixed real number. Consider as the state equation an ODE (for $0 < t < 1$):

$$\begin{aligned} &\frac{\partial u(t)}{\partial t} - a\, u(t) = f\,(t) \\ &u(0) = 0 \end{aligned} \tag{5.4.19}$$

where the control f is such that $f \in L^2(0,\, 1) = \left\{f,\ \int_0^1 f(x)^2\, dx < \infty\right\}$. In the present context this yields an abstract equation in $L^2(0,\, 1)$:

$$A\, u = f$$

where A is a linear operator in $L^2(0,\, 1)$ defined by

$$\begin{aligned} &\text{dom}(A) = \{u \in H^1(0,\, 1),\, u(0) = 0\} \\ &(A\, u)\,(t) = \frac{\partial u\,(t)}{\partial t} - a\, u(t) \end{aligned} \tag{5.4.20}$$

or, in the *Mathematica*® meaningful way (we'll take care of the boundary condition separately)

```
In[149]:= A[u_, t_] := ∂t u - a u
```

Above $H^1(0,\, 1) = \left\{u \in L^2(0,\, 1),\ \frac{\partial u}{\partial t} \in L^2(0,\, 1)\right\}$. It is well known from the theory of Sobolev spaces (see, e.g., [17,24]) that if $u \in H^1(0,\, 1)$ then u is continuous on $[0,\, 1]$, and therefore condition $u(0) = 0$ is meaningful. Define also the adjoint operator A^*:

$$\begin{aligned} &\text{dom}(A^*) = \{u \in H^1(0,\, 1),\, u(1) = 0\} \\ &(A^*\, u)\,(t) = -\frac{\partial u(t)}{\partial t} - a\, u(t) \end{aligned} \tag{5.4.21}$$

or, in the *Mathematica*® meaningful way (again, we'll take care of the boundary condition separately)

```
In[150]:= A*[u_, t_] := -a u - ∂t u
```

Maybe (5.4.12) needs a little explaining. The adjoint operator is defined by "integration by parts":

$$(A\, u,\, v) = (u,\, A^*\, v) \tag{5.4.22}$$

which in the present example means

$$(A\,u,\,v) = \int_0^1 \Big(\frac{\partial u(t)}{\partial t} - a\,u(t)\Big)\,v(t)\,dt$$

$$\begin{aligned}
&= \int_0^1 \Big(\frac{\partial u(t)}{\partial t}\,v(t) - a\,u(t)\,v(t)\Big)dt \\
&= \int_0^1 \Big(-u(t)\,\frac{\partial v(t)}{\partial t} - a\,u(t)\,v(t)\Big)dt \\
&= \int_0^1 u(t)\Big(-\frac{\partial v(t)}{\partial t} - a\,v(t)\Big)\,dt = (u,\,A^*\,v)
\end{aligned} \tag{5.4.23}$$

provided $u \in \mathrm{dom}(A)$ and $v \in \mathrm{dom}(A^*)$. The domain $\mathrm{dom}(A^*)$ was chosen in such a way that the integration by parts does not produce a boundary contribution. Let the target function $u_d \in L^2(0,\,1)$ be

```
In[151]:= ud[t_] := 2 - t + 3 t^2
```

(notice that $u_d(0) = 2 \neq 0$ and therefore $u_d \notin \mathrm{dom}(A)$). Also let the cost functional be

$$J_\epsilon(f) = \frac{1}{2}\Big(\int_0^1 (u(t) - u_d(t))^2\,dt + \epsilon \int_0^1 f(t)^2\,dt\Big) \tag{5.4.24}$$

$\epsilon > 0$. From the above we know that the optimal solution is equal to

$$u_{\mathrm{opt}} = (\epsilon\,A^*\,A + 1)^{-1}\,u_d$$

and then the optimal control is

$$f_{\mathrm{opt}} = A\,u_{\mathrm{opt}}.$$

We now compute the two.

Notice the boundary conditions for u_{opt}: $u_{\mathrm{opt}} \in \mathrm{dom}(\epsilon\,A^*\,A + 1)$ if $u_{\mathrm{opt}} \in \mathrm{dom}(A)$ and $A\,u_{\mathrm{opt}} \in \mathrm{dom}(A^*)$, or more explicitly, $u_{\mathrm{opt}}(0) = 0$ and $(A\,u_{\mathrm{opt}})\,(1) = 0$. Also, define the operator $L = \epsilon\,A^*\,A + 1$ by

```
In[152]:= L[u_, t_] := ∈ A*[A[u, t], t] + u
```

and then compute u_{opt}:

```
In[153]:= uopt[t_, a_, ∈_] =
            Simplify[p[t] /. DSolve[{L[p[t], t] == ud[t],
                (A[p[t], t] /. t → 1) == 0,
                p[0] == 0}, p[t], t][[1]]]
```

Out[153]= $$\left(-\sqrt{-\epsilon a^2-1}\,\left(3\left(\epsilon a^2+1\right)t^2-\left(\epsilon a^2+1\right)t+2\left(a^2+3\right)\epsilon+2\right)\cos\left(\frac{\sqrt{-\epsilon a^2-1}}{\sqrt{\epsilon}}\right)\right.$$
$$+2\sqrt{-\epsilon a^2-1}\,\left(\left(a^2+3\right)\epsilon+1\right)\cos\left(\frac{(t-1)\sqrt{-\epsilon a^2-1}}{\sqrt{\epsilon}}\right)$$
$$+\sqrt{\epsilon}\left(a\left(3\left(\epsilon a^2+1\right)t^2-\left(\epsilon a^2+1\right)t+2\left(a^2+3\right)\epsilon+2\right)\sin\left(\frac{\sqrt{-\epsilon a^2-1}}{\sqrt{\epsilon}}\right)\right.$$
$$+2a\left(\left(a^2+3\right)\epsilon+1\right)\sin\left(\frac{(t-1)\sqrt{-\epsilon a^2-1}}{\sqrt{\epsilon}}\right)$$
$$\left.\left.+\left(-4\epsilon a^3+5\epsilon a^2-2(3\epsilon+2)a+5\right)\sin\left(\frac{t\sqrt{-\epsilon a^2-1}}{\sqrt{\epsilon}}\right)\right)\right)\Big/$$
$$\left(\left(\epsilon a^2+1\right)^2\left(a\sqrt{\epsilon}\,\sin\left(\frac{\sqrt{-\epsilon a^2-1}}{\sqrt{\epsilon}}\right)-\sqrt{-\epsilon a^2-1}\,\cos\left(\frac{\sqrt{-\epsilon a^2-1}}{\sqrt{\epsilon}}\right)\right)\right)$$

Once the optimal solution is determined, the corresponding optimal control follows: $f_{\text{opt}} = A\,u_{\text{opt}}$, i.e.,

In[154]:= `fopt[t_, a_, ϵ_] = Simplify[A[uopt[t, a, ϵ], t]]`

Out[154]= $$\left((6t-1)\cos\left(\frac{\sqrt{-\epsilon a^2-1}}{\sqrt{\epsilon}}\right)\left(-\epsilon a^2-1\right)^{3/2}+\sqrt{\epsilon}\right.$$
$$\left(\frac{2a\sqrt{-\epsilon a^2-1}\,\left(\left(a^2+3\right)\epsilon+1\right)\cos\left(\frac{(t-1)\sqrt{-\epsilon a^2-1}}{\sqrt{\epsilon}}\right)}{\sqrt{\epsilon}}\right.$$
$$+\frac{1}{\sqrt{\epsilon}}\left(\sqrt{-\epsilon a^2-1}\,\left(-4\epsilon a^3+5\epsilon a^2-2(3\epsilon+2)a+5\right)\right.$$
$$\left.\cos\left(\frac{t\sqrt{-\epsilon a^2-1}}{\sqrt{\epsilon}}\right)\right)+a(6t-1)\left(\epsilon a^2+1\right)$$
$$\left.\sin\left(\frac{\sqrt{-\epsilon a^2-1}}{\sqrt{\epsilon}}\right)\right)+\frac{2\left(\epsilon a^2+1\right)\left(\left(a^2+3\right)\epsilon+1\right)\sin\left(\frac{(t-1)\sqrt{-\epsilon a^2-1}}{\sqrt{\epsilon}}\right)}{\sqrt{\epsilon}}$$

$$-a\left(-\sqrt{-\epsilon a^2-1}\left(3\left(\epsilon a^2+1\right)t^2-\left(\epsilon a^2+1\right)t+2\left(a^2+3\right)\epsilon+2\right)\right.$$

$$\cos\left(\frac{\sqrt{-\epsilon a^2-1}}{\sqrt{\epsilon}}\right)+2\sqrt{-\epsilon a^2-1}\left(\left(a^2+3\right)\epsilon+1\right)$$

$$\cos\left(\frac{(t-1)\sqrt{-\epsilon a^2-1}}{\sqrt{\epsilon}}\right)+\sqrt{\epsilon}\left(a\left(3\left(\epsilon a^2+1\right)t^2\right.\right.$$

$$-\left(\epsilon a^2+1\right)t+2\left(a^2+3\right)\epsilon+2\right)\sin\left(\frac{\sqrt{-\epsilon a^2-1}}{\sqrt{\epsilon}}\right)$$

$$+2a\left(\left(a^2+3\right)\epsilon+1\right)\sin\left(\frac{(t-1)\sqrt{-\epsilon a^2-1}}{\sqrt{\epsilon}}\right)$$

$$\left.\left.\left.+\left(-4\epsilon a^3+5\epsilon a^2-2(3\epsilon+2)a+5\right)\sin\left(\frac{t\sqrt{-\epsilon a^2-1}}{\sqrt{\epsilon}}\right)\right)\right)\right)\Bigg/$$

$$\left(\left(\epsilon a^2+1\right)^2\left(a\sqrt{\epsilon}\sin\left(\frac{\sqrt{-\epsilon a^2-1}}{\sqrt{\epsilon}}\right)-\sqrt{-\epsilon a^2-1}\cos\left(\frac{\sqrt{-\epsilon a^2-1}}{\sqrt{\epsilon}}\right)\right)\right)$$

If $a = 10$, $\epsilon_1 = 0.004$, $\epsilon_2 = 0.0004$ the corresponding plots of optimal solutions u_1 and u_2, together with the target function u_d are shown:

```
In[155]:= Plot[{ud[t], uopt[t, 10, 0.004],
             uopt[t, 10, 0.0004]}, {t, 0.0001, .9999},
           PlotRange → All, PlotStyle → {RGBColor[0, 0, 0],
             RGBColor[1, 0, 0], RGBColor[0, 0, 1]}];
```


as well as the corresponding optimal controls f_1 and f_2:

```
In[156]:= Plot[{f_opt[t, 10, 0.004], f_opt[t, 10, 0.0004]},
            {t, 0, 1}, PlotRange → All, PlotStyle →
             {RGBColor[1, 0, 0], RGBColor[0, 0, 1]}];
```

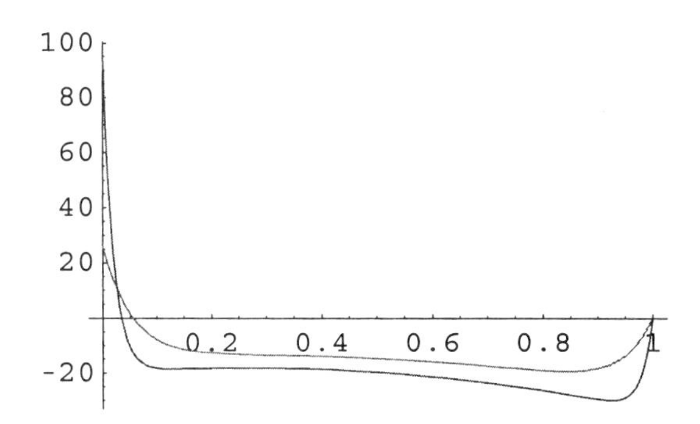

So both trajectories start from zero, quickly reaching optimal trajectories, corresponding to different cost functionals, and giving up active control as time expires.

5.4.5.3 An Extension

It is easy to generalize the above decoupling formula to the case when the state equation is

$$A\,v = f + g \tag{5.4.25}$$

for some given fixed g with the same cost functional. The optimality system in such a case is equal to

$$\begin{aligned} A\,v &= f + g \\ A^*\,p &= v - v_d \\ f &= -\frac{1}{\epsilon}\,p \end{aligned}$$

and tracing the steps of the above derivation, one arrives at the formulas

$$\begin{aligned} v_{\text{opt}} &= (\epsilon\,A^*\,A + 1)^{-1}\,(v_d + \epsilon\,A^*\,g) \\ f_{\text{opt}} &= A\,v_{\text{opt}} - g \end{aligned} \tag{5.4.26}$$

provided $g \in \text{dom}(A^*)$. We can apply this result to solve the above example with a non-zero initial condition. Indeed, the state equation

$$\begin{aligned} &\frac{\partial u(t)}{\partial t} - a\,u(t) = f\,(t) \\ &u(0) = u_0 \end{aligned} \tag{5.4.27}$$

with target u_d can be rewritten for $v(t) = u(t) - w(t)$, where $w(t) = u_0\,(1-t)^2$, i.e.,

```
In[157]:= w[t_] := u0 (1 - t)^2
```

and target $v_d(t) = u_d(t) - w(t)$, as

$$A\,v = f + g$$

where g is equal to

```
In[158]:= g[u0_, t_] = -A[w[t], t]
```

```
Out[158]= a u0 (1 - t)^2 + 2 u0 (1 - t)
```

while v_d is equal to

```
In[159]:= vd[u0_, t_] = ud[t] - w[t]
```

```
Out[159]= -u0 (1 - t)^2 + 3 t^2 - t + 2
```

and where A is (as before) a linear operator in $L^2(0, 1)$ defined by

$$\text{dom}(A) = \{u \in H^1(0, 1), u(0) = 0\}$$

$$(A\,u)(t) = \frac{\partial u(t)}{\partial t} - a\,u(t).$$

The choice of w may seem odd. The quadratic term in $w(t) = u_0\,(1-t)^2$ is present in order to $g \in \text{dom}(A^*)$. So, this time (recall $L = \epsilon\,A^*\,A + 1$)

```
In[160]:= vopt[t_, a_, u0_, ∈_] =
            Simplify[p[t] /. DSolve[{L[p[t], t] == vd[u0, t] +
                  ∈ A*[g[u0, t], t], (A[p[t], t] /. t → 1) ==
                 0, p[0] == 0}, p[t], t][[1]]];
```

while the optimal control is equal to

```
In[161]:= fopt[t_, a_, u0_, ∈_] =
            Simplify[A[vopt[t, a, u0, ∈], t] - g[u0, t]];
```

The optimal solution of the original state equation is then equal to

```
In[162]:= uopt[t_, a_, u0_, ∈_] = vopt[t, a, u0, ∈] + w[t];
```

The optimal solutions of the original state equation look like

```
In[163]:= Plot[{ud[t], uopt[t, 10, 4, .001],
             uopt[t, 10, 0, .001]}, {t, 0.0001, .9999},
            PlotRange → All, PlotStyle → {RGBColor[0, 0, 0],
              RGBColor[1, 0, 0], RGBColor[0, 0, 1]}];
```

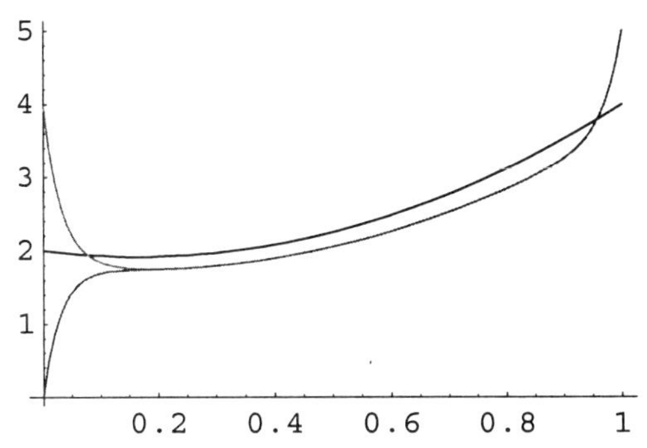

while corresponding optimal controls are

```
In[164]:= Plot[{fopt[t, 10, 4, .001], fopt[t, 10, 0, .001]},
    {t, 0, 1}, PlotRange → All, PlotStyle →
     {RGBColor[1, 0, 0], RGBColor[0, 0, 1]}];
```

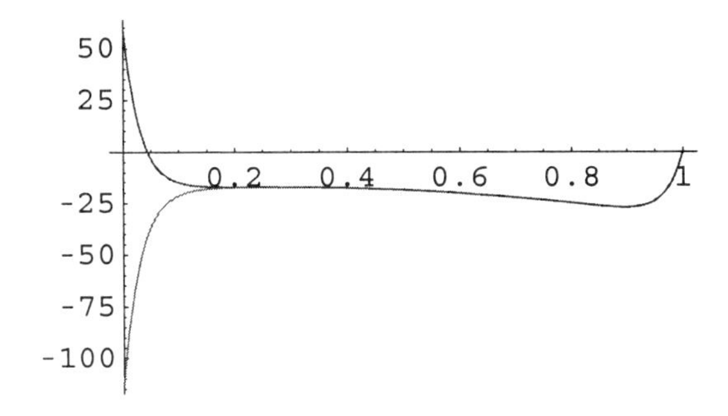

It is interesting to see a plausible result: controls are first used to steer the system from an initial state to the optimal trajectory (for the given cost functional the same one), and afterwards the controls as well as the trajectories are essentially the same.

5.4.5.4 Yet Another Extension

In the previous example we had to employ a trick $w(t) = u_0 (1 - t)^2$ in order to have the condition $g \in \text{dom}(A^*)$ fulfilled. Can we have a theory in the case $g \notin \text{dom}(A^*)$? As before, the optimality system is

$$A\,v = \frac{-1}{\epsilon}\,p + g$$
$$A^*\,p = v - v_d$$

from which we derive

$$A\,v = \frac{-1}{\epsilon}\,(A^{*-1}\,v - A^{*-1}\,v_d) + g$$

$$A\,v + \frac{1}{\epsilon}\,A^{*-1}\,v = \frac{1}{\epsilon}\,A^{*-1}\,v_d + g$$

$$(\epsilon\,A\,A^* + 1)^{-1}\left(A\,v + \frac{1}{\epsilon}\,A^{*-1}\,v\right) = (\epsilon\,A\,A^* + 1)^{-1}\left(\frac{1}{\epsilon}\,A^{*-1}\,v_d + g\right).$$

But

$$(\epsilon\,A\,A^* + 1)^{-1}\left(A\,v + \frac{1}{\epsilon}\,A^{*-1}\,v\right)$$
$$= \frac{1}{\epsilon}\,(\epsilon\,A\,A^* + 1)^{-1}\,(\epsilon\,A\,A^*\,A^{*-1}\,v + A^{*-1}\,v) = \frac{1}{\epsilon}\,A^{*-1}\,v$$

and therefore

$$\frac{1}{\epsilon}\,A^{*-1}\,v = (\epsilon\,A\,A^* + 1)^{-1}\left(\frac{1}{\epsilon}\,A^{*-1}\,v_d + g\right)$$

i.e.,

$$\begin{aligned} v_{\text{opt}} &= A^*\,(\epsilon\,A\,A^* + 1)^{-1}\,(A^{*-1}\,v_d + \epsilon\,g) \\ f_{\text{opt}} &= A\,v_{\text{opt}} - g. \end{aligned} \tag{5.4.28}$$

This formula, although a bit more involved, can be implemented just as successfully as the previous one when $g \in \text{dom}(A^*)$. Indeed, let's solve the above example with a natural choice of w (a constant function, instead of a quadratic function, and consequently the corresponding g will not be a member of $\text{dom}(A^*)$; by the way, it is quite instructive to try to ignore those details above—the result will be wrong). So, let

```
In[165]:= w[t_] := u0;

In[166]:= g[u0_, t_] = -A[w[t], t]; vd[u0_, t_] = ud[t] - w[t];

In[167]:= vopt[t_, a_, u0_, ∈_] = Simplify[
            A*[p[t] /. DSolve[{∈ A[A*[p[t], t], t] + p[t] ==
                (p[t] /. DSolve[{A*[p[t], t] == vd[u0, t],
                      p[1] == 0}, p[t], t][[1]]) +
                ∈ g[u0, t], p[1] == 0, (A*[p[t], t] /.
                 t → 0) == 0}, p[t], t][[1]], t]];

In[168]:= fopt[t_, a_, u0_, ∈_] =
           Simplify[A[vopt[t, a, u0, ∈], t] - g[u0, t]];

In[169]:= uopt[t_, a_, u0_, ∈_] = vopt[t, a, u0, ∈] + w[t];
```

The optimal solutions of the original state equation look as they should:

```
In[170]:= Plot[{ud[t], u_opt[t, 10, 4, .001],
            u_opt[t, 10, 0, .001]}, {t, 0.0001, .9999},
           PlotRange → All, PlotStyle → {RGBColor[0, 0, 0],
             RGBColor[1, 0, 0], RGBColor[0, 0, 1]}];
```

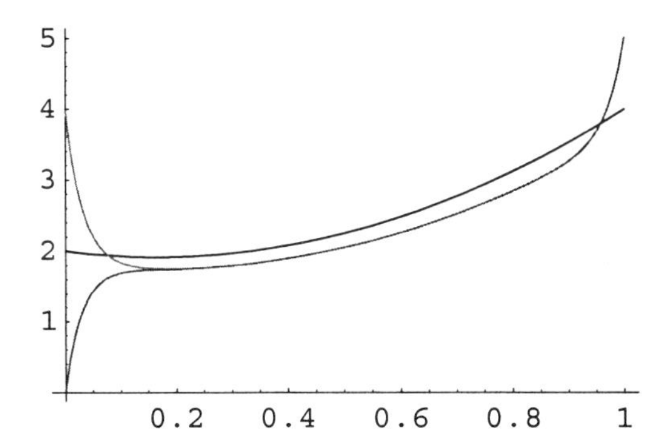


```
In[171]:= Plot[{f_opt[t, 10, 4, .001], f_opt[t, 10, 0, .001]},
           {t, 0, 1}, PlotRange → All, PlotStyle →
            {RGBColor[1, 0, 0], RGBColor[0, 0, 1]}];
```

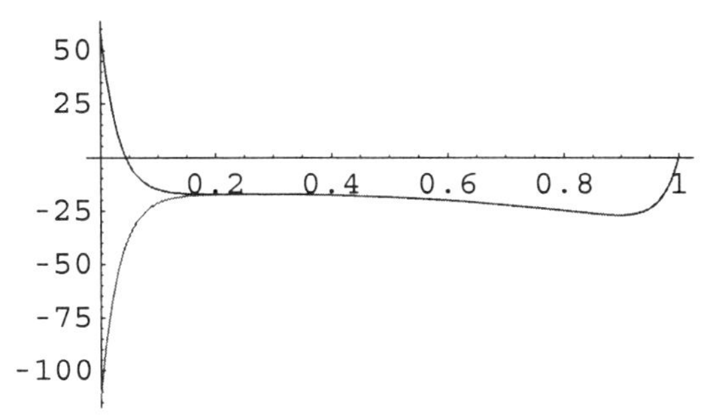

which is the same result as in the previous section but derived using a general method that does *not* assume $g \in \text{dom}(A^*)$.

5.4.6 General Implied Volatility: Optimal Control of Dupire PDEs

Now consider the Dupire partial differential equation (consider calls for example). Let there be no dividend payments, i.e., let for simplicity $D_0 = 0$. The Dupire partial differential equation reads as

$$-\frac{\partial V(T,k)}{\partial T} + \frac{1}{2} k^2 \frac{\partial^2 V(T,k)}{\partial k^2} \sigma(T,k)^2 - r\, k \frac{\partial V(T,k)}{\partial k} = 0$$

for $T > t$, $k > 0$, together with the *initial* condition

$$V(t, k) = \text{Max}(0, S - k).$$

Every function $\sigma = \sigma(T, k) > 0$ yields a different solution of the above equation: $V(\sigma) = V(\sigma)(T, k)$. In particular, if σ is constant, the solution is given by the Black–Scholes formula. As already seen, the theoretical prices obtained in such a manner differ from the market prices, and one searches for the implied volatility to improve the match.

The problem we are going to solve now is a problem of finding implied volatility that is allowed to depend not only on expiration T, but also on the strike prices k, i.e., we are looking for $\sigma(T, k)$ other than the constant statistical volatility which will produce the solution of the Dupire PDE that matches the observed prices in an optimal fashion. We employ the optimal control theory to this end. Furthermore, since there are only a few expiration times to consider and many strikes for each expiration, we consider these two dependencies differently: for each expiration time, using the discrete option price observations, we construct option price functions $v_i(k)$, $i = 1, \ldots, n$ where n is the number of considered expiration times, and then instead of trying to match discrete option price observations, we shall match functions $v_i(k)$, $i = 1, \ldots, n$, thereby reducing somewhat the singularity of the problem.

First truncate the strike price region: consider strikes $0 < k_0 < k < k_1$. Also, let t_0 be the time of the data record, let T_i, $i = 1, \ldots, n$ be the expiration times, and let $t_1 = T_n$. Due to this truncation, we need to impose artificial boundary conditions for $k = k_0$ and for $k = k_1$. So the state equation is

$$-\frac{\partial V(T, k)}{\partial T} + \frac{1}{2} k^2 \frac{\partial^2 V(T, k)}{\partial k^2} \sigma(T, k)^2 - r k \frac{\partial V(T, k)}{\partial k} = 0 \tag{5.4.29}$$

for $t_0 < T < t_1$, $k_0 < k < k_1$, together with the *initial-boundary* condition $V = V_{\text{BS}}$ on the parabolic boundary where V_{BS} is the Black–Scholes statistical volatility solution.

From discrete option price observations we construct the *target* functions $v_i(k)$, $i = 1, \ldots, n$, that the solution of the Dupire PDE is supposed to match. How to measure *distance* between the solution of the Dupire equation and the targets? We adopt the cost functional

$$\begin{aligned} J(\sigma) &= \frac{1}{2} \sum_{i=1}^{n} \int_{k_0}^{k_1} (V(\sigma)(T_i, k) - v_i(k))^2\, q_i(k)\, dk \\ &+ \frac{1}{2} \int_{k_0}^{k_1} \int_{t_0}^{t_1} \left(\epsilon_1 \left(\frac{\partial \sigma(T, k)}{\partial k} \right)^2 + \epsilon_2 \left(\sigma(T, k) - \sigma_{\text{stat}}\right)^2 \right) dT\, dk \end{aligned} \tag{5.4.30}$$

where ϵ_1, ϵ_2, $q_i(k) > 0$ are given and σ_{stat} is some kind of statistical estimate of the constant volatility (or just a guess, or even zero). The first term measures the distance between the Dupire solution and the targets; the second term is necessary to regularize the problem (see the role of $\epsilon > 0$ and its necessity in the previous section in the case of an elementary optimal control problem). The particular form of the regularization term above is very important, and it discloses also the class K of admissible volatilities we are about to consider (we choose not to impose any boundary condition on σ):

$$K = \{\sigma \in L^2((t_0, t_1) \times (k_0, k_1)),\ \sigma(T, k) \geq \sigma_{\min},$$
$$\int_{k_0}^{k_1} \int_{t_0}^{t_1} \left(\left(\frac{\partial \sigma(T, k)}{\partial k} \right)^2 + \sigma(T, k)^2 \right) dT\, dk < \infty \} \tag{5.4.31}$$

for some $\sigma_{\min} > 0$. Of course, the less assumed a priori about admissible volatilities, the better; what is assumed is just minimal smoothness in the strike k-direction, with not even continuity in the time direction; this is consistent with our study of piecewise continuous time-dependent volatilities in the previous section. Also, here we are not concerned with proving mathematically results such as the well posedness of the state equation under such minimal conditions on σ, the existence of an optimal control, etc.; rather our effort is in designing a working numerical minimization algorithm, whose veracity is to be inferred from computational examples. So, the constraints imposed on K are motivated by the numerical procedure below, and whether these constraints also suffice, or have to be accompanied by some less natural constraints to be able to prove accompanying mathematical theorems, is a separate issue. Anyway, the optimal control problem is to find (at least approximately) $\sigma_{\text{opt}} \in K$ such that

$$J(\sigma_{\text{opt}}) = \operatorname*{Min}_{\sigma \in K} J(\sigma). \tag{5.4.32}$$

Motivated by considerations of the optimal control problems in the previous section, we confine ourselves to computing the *gradient* of J, and to using it in the steepest descent minimization algorithm. Of course, it would be very interesting and useful to use *some* of the ideas presented in computing explicit solutions above in the present case. Unfortunately, there are several serious gaps in doing so.

For fixed $\sigma \in K$ and $\overline{\sigma}$ such that $\sigma + \delta\overline{\sigma} \in K$, compute the directional derivative of J in the direction $\overline{\sigma}$:

$$J'(\sigma; \overline{\sigma}) = \lim_{\delta \to 0} \frac{J(\sigma + \delta\overline{\sigma}) - J(\sigma)}{\delta}$$
$$= \sum_{i=1}^{n} \int_{k_0}^{k_1} (V(T_i, k) - v_i(s))\, w(T_i, k)\, q_i(k)\, dk$$
$$+ \int_{t_0}^{t_1} \int_{k_0}^{k_1} \left(\epsilon_1 \frac{\partial \sigma(T, k)}{\partial k} \frac{\partial \overline{\sigma}(T, k)}{\partial k} + \epsilon_2 (\sigma(T, k) - \sigma_{\text{stat}})\, \overline{\sigma}(T, k) \right) dk\, dT$$

where

$$w(T, k) = \lim_{\delta \to 0} \frac{V(\sigma + \delta\overline{\sigma})(T, k) - V(\sigma)(T, k)}{\delta}$$

is the unique solution of the forward equation

$$-\frac{\partial w(T,k)}{\partial T}+\frac{1}{2}k^2\frac{\partial^2 w(T,k)}{\partial k^2}\sigma(T,k)^2-r\,k\frac{\partial w(T,k)}{\partial k}$$
$$=-k^2\frac{\partial^2 V(\sigma)(T,k)}{\partial k^2}\sigma(T,k)\,\overline{\sigma}(T,k)$$

$$w(t_0,k)=0,\ w(T,k_0)=0,\ w(T,k_1)=0.$$

Let $p_i(T,k)$, $1\le i\le n$ be the unique solutions of the *backward* equations

$$\begin{aligned}\frac{\partial p_i(T,k)}{\partial T}&+\frac{1}{2}k^2\frac{\partial^2 p_i(T,k)}{\partial k^2}\sigma(T,k)^2\\&+(2\,\sigma(T,k)^2+2\,k\,\sigma^{(0,1)}(T,k)\,\sigma(T,k)+r)\,k\frac{\partial p_i(T,k)}{\partial k}\\&+\big(\sigma(T,k)^2+k\,(4\,\sigma^{(0,1)}(T,k)+k\,\sigma^{(0,2)}(T,k))\,\sigma(T,k)\\&\quad+k^2\,\sigma^{(0,1)}(T,k)^2+r\big)\,p_i(T,k)=0\end{aligned}\tag{5.4.33}$$

for $t_0<T<T_i$ and $k_0<k<k_1$, with the terminal conditions

$$p_i(T_i,k)=-(V(\sigma)(T_i,k)-v_i(k))\,q_i(k)\tag{5.4.34}$$

and boundary conditions

$$p(T,k_0)=0,\ p(T,k_1)=0.\tag{5.4.35}$$

By means of integration by parts, this terminal boundary value problem is equivalent to its weak formulation:

$$\int_{t_0}^{T_i}\int_{k_0}^{k_1}\left(-\frac{\partial\phi(T,k)}{\partial T}+\frac{1}{2}k^2\frac{\partial^2\phi(T,k)}{\partial k^2}\sigma(T,k)^2-r\,k\frac{\partial\phi(T,k)}{\partial k}\right)p_i(T,k)\,dk\,dT$$
$$=-\int_{k_0}^{k_1}\phi(T_i,k)\,p_i(T,k)\,dk=\int_{k_0}^{k_1}(V(\sigma)(T_i,k)-v_i(k))\,q_i(k)\,\phi(T_i,k)\,dk$$

for any smooth test function ϕ such that, for any k such that $k_0<k<k_1$,

$$\phi(t_0,k)=0$$

and for any T such that $t_0<T<T_i$,

$$\phi(T,k_0)=0,\ \phi(T,k_1)=0.$$

In particular, if $\phi=w$,

$$J'(\sigma;\overline{\sigma})=\sum_{i=1}^{n}\int_{k_0}^{k_1}(V(\sigma)(T_i,k)-v_i(k))\,w(T_i,k)\,q_i(k)\,dk$$

$$+\int_{t_0}^{t_1}\int_{k_0}^{k_1}\left(\epsilon_1\,\frac{\partial\sigma(T,k)}{\partial k}\,\frac{\partial\overline{\sigma}(T,k)}{\partial k}+\epsilon_0\,(\sigma(T,k)-\sigma_{\mathrm{stat}})\,\overline{\sigma}(T,k)\right)dk\,dT$$

$$=\sum_{i=1}^{n}\int_{t_0}^{T_i}\int_{k_0}^{k_1}\left(-\frac{\partial w(T,k)}{\partial T}+\frac{1}{2}\,k^2\,\frac{\partial^2 w(T,k)}{\partial k^2}\,\sigma(T,k)^2-r\,k\,\frac{\partial w(T,k)}{\partial k}\right)$$

$$p_i(T,k)\,dk\,dT+\int_{t_0}^{t_1}\int_{k_0}^{k_1}\left(\epsilon_1\,\frac{\partial\sigma(T,k)}{\partial k}\,\frac{\partial\overline{\sigma}(T,k)}{\partial k}\right.$$

$$\left.+\,\epsilon_0\,(\sigma(T,k)-\sigma_{\mathrm{stat}})\,\overline{\sigma}(T,k)\right)dk\,dT$$

$$=\sum_{i=1}^{n}\int_{t_0}^{T_i}\int_{k_0}^{k_1}\left(-k^2\,\frac{\partial^2 V(\sigma)(T,k)}{\partial k^2}\,\sigma(T,k)\,p_i(T,k)\right)\overline{\sigma}(T,k)\,dk\,dT$$

$$+\int_{t_0}^{t_1}\int_{k_0}^{k_1}\left(-\epsilon_1\,\frac{\partial^2\sigma(T,k)}{\partial k^2}+\epsilon_0\,(\sigma(T,k)-\sigma_{\mathrm{stat}})\right)\overline{\sigma}(T,k)\,dk\,dT$$

$$+\epsilon_1\left(\int_{t_0}^{t_1}\frac{\partial\sigma(T,k_1)}{\partial k}\,\overline{\sigma}(T,k_1)\,dT-\int_{t_0}^{t_1}\frac{\partial\sigma(T,k_0)}{\partial k}\,\overline{\sigma}(T,k_0)\,dT\right)$$

$$=\int_{t_0}^{t_1}\int_{k_0}^{k_1}\left(-k^2\,\frac{\partial^2 V(\sigma)(T,k)}{\partial k^2}\,\sigma(T,k)\sum_{i=1}^{n}(p_i(T,k)\,\chi_{(t_0,T_i)})\right.$$

$$\left.+\left(-\epsilon_1\,\frac{\partial^2\sigma(T,k)}{\partial k^2}+\epsilon_0\,(\sigma(T,k)-\sigma_{\mathrm{stat}})\right)\right)\overline{\sigma}(T,k)\,dk\,dT$$

$$+\epsilon_1\left(\int_{t_0}^{t_1}\frac{\partial\sigma(T,k_1)}{\partial k}\,\overline{\sigma}(T,k_1)\,dT-\int_{t_0}^{t_1}\frac{\partial\sigma(T,k_0)}{\partial k}\,\overline{\sigma}(T,k_0)\,dT\right)$$

where

$$\chi_{(t_0,T_i)}(T)=\begin{cases}1, & t_0<T<T_i\\ 0, & \text{elsewhere}\end{cases}.$$

On the other hand, there exists z such that

$$\nabla J(\sigma)\cdot\overline{\sigma}=\int_{t_0}^{t_1}\int_{k_0}^{k_1}\left(\epsilon_1\,\frac{\partial z(T,k)}{\partial k}\,\frac{\partial\overline{\sigma}(T,k)}{\partial k}+\epsilon_0\,z(T,k)\,\overline{\sigma}(T,k)\right)dk\,dT$$

$$=\int_{t_0}^{t_1}\int_{k_0}^{k_1}\left(-\epsilon_1\,\frac{\partial^2 z(T,k)}{\partial k^2}+\epsilon_0\,z(T,k)\right)\overline{\sigma}(T,k)\,dk\,dT$$

$$+\epsilon_1\left(\int_{t_0}^{t_1}\frac{\partial z(T,k_1)}{\partial k}\,\overline{\sigma}(T,k_1)\,dT-\int_{t_0}^{t_1}\frac{\partial z(T,k_0)}{\partial k}\,\overline{\sigma}(T,k_0)\,dT\right).$$

Since $J'(\sigma; \overline{\sigma}) = \nabla J_{\epsilon}(\sigma) \cdot \overline{\sigma}$ for any admissible $\overline{\sigma}$, we conclude that, for any T, $z(T, \cdot)$ can be computed as a unique solution of the boundary value problem for an ordinary differential equation (regularization equation):

$$-\epsilon_1 \frac{\partial^2 z(T,k)}{\partial k^2} + \epsilon_0 z(T,k) = -k^2 \frac{\partial^2 V(\sigma)(T,k)}{\partial k^2} \sigma(T,k)$$
$$\sum_{i=1}^{n} (p_i(T,k)\, \chi_{(t_0,T_i)}) + \left(-\epsilon_1 \frac{\partial^2 \sigma(T,k)}{\partial k^2} + \epsilon_0 (\sigma(T,k) - \sigma_{\text{stat}})\right) \tag{5.4.36}$$

with boundary conditions

$$\frac{\partial z(T,k_0)}{\partial k} = \frac{\partial \sigma(T,k_0)}{\partial k}$$
$$\frac{\partial z(T,k_1)}{\partial k} = \frac{\partial \sigma(T,k_1)}{\partial k}. \tag{5.4.37}$$

Notice that even though the left-hand side of the above ODE is quite simple and therefore may raise hopes for an explicit solution, since the right-hand side is computed numerically, the ODE has to be solved numerically as well. The single iterate in the steepest descent method is then done in 4 steps:

1) For given $\sigma(T, k)$, $V(\sigma)(T, k)$ is computed via Dupire PDE (5.4.29).

2) Then $p_i(T, k)$'s are computed using $V(\sigma)(T, k)$ in the adjoint equations (5.4.33)–(5.4.35).

3) $V(\sigma)(T, k)$ and all of the $p_i(T, k)$'s are used in the regularization equation (5.4.36)–(5.4.37) computing $z(T, k)$.

4) Finally, $\sigma_{\text{next}}(T, k) = \sigma(T, k) - \rho\, z(T, k)$, for some $\rho > 0$.

The single step is repeated until implied volatility stabilizes sufficiently. Of course, it would have been so much better to have an explicit solution as in the previous section.

Let us check the sign of some crucial quantities in the above calculation. Suppose $\sigma(T, k)$ is a large constant, say $\sigma(T, k) = 1$. This implies that the corresponding option prices $V(\sigma)(T, k)$ are bigger than the observed market prices. From the adjoint equation, and in particular from its terminal value, this in turn implies that $p_i(T, k)$ are all negative. Since also $\frac{\partial^2 V(\sigma)(T,k)}{\partial k^2} > 0$, we conclude that $-k^2 \frac{\partial^2 V(\sigma)(T,k)}{\partial k^2}$ $\sigma(T, k) \sum_{i=1}^{n} (p_i(T, k)\, \chi_{(t_0,T_i)}) > 0$, and consequently $z(T, k) > 0$. This implies that the next iterate $\sigma_{\text{next}}(T, k) = \sigma(T, k) - \rho\, z(T, k) < \sigma(T, k)$ since $\rho > 0$: the next estimate for the implied volatility $\sigma_{\text{next}}(T, k)$ will be lower than the starting large constant. Notice also that ρ needs to be sufficiently small so that $\sigma_{\text{next}}(T, k) = \sigma(T, k) - \rho\, z(T, k) \geq \sigma_{\min} > 0$. The other possibility is to take $\sigma_{\text{next}}(T, k) = \text{Max}(\sigma(T, k) - \rho\, z(T, k), \sigma_{\min})$, or the combination of the two.

5.4.7 Computational Example: Call Implied Volatility for QQQ

5.4.7.1 Data Import

We need functions developed previously

```
In[172]:= Clear["Global`*"]
```

```
In[173]:= << CFMLab`NumericalBlackScholes`
          << CFMLab`DataImport`
```

Next, we import the market data (recall, the liquidity of options is defined in terms of the Volume)

```
In[175]:= CallData = GetAllData["QQQ", "Call", 100];
```

and decide to consider only the first 3 expiration dates (this is defoult for GetAll-Data). The first step is to construct the *target* functions $v_i(k)$, $i = 1, \ldots, n$, that the solution of the Dupire PDE is supposed to match:

```
In[176]:= TradingData = CallData[[2]];
```

```
In[177]:= Observations = {#[[1, 1]],
              Interpolation[{#[[2]], #[[3]]} & /@ #,
               InterpolationOrder → 1]} & /@ TradingData;
```

They look like

```
In[178]:= Plot[#[[2]][k],
             {k, #[[2, 1, 1, 1]], #[[2, 1, 1, 2]]},
             AxesLabel → {"Strike k", "$"},
             PlotLabel → StringJoin["T = ",
               ToString[#[[1]]]]] & /@ Observations;
```

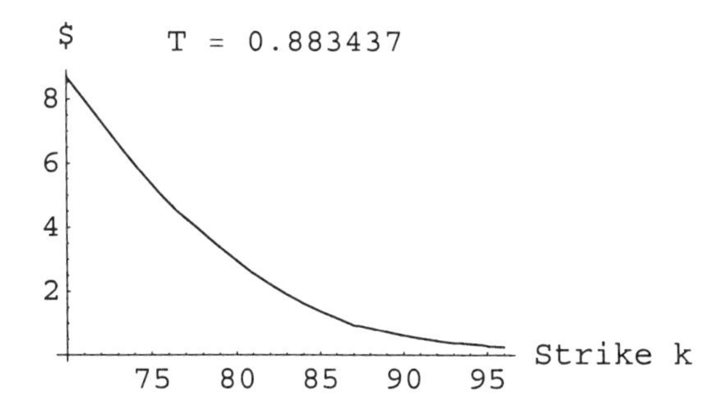

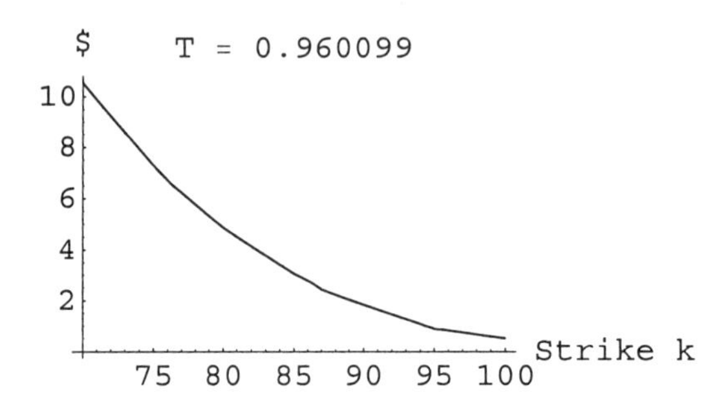

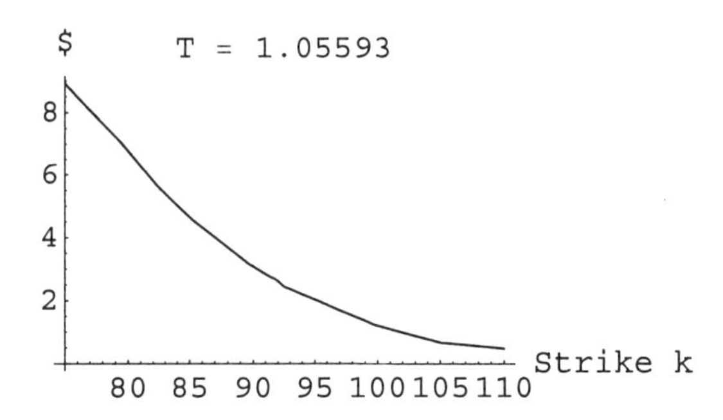

The steepest descent method is initialized by selecting the volatility function $\sigma(T, k) = \sigma_{\text{stat}}$:

```
In[179]:= σstat = Σ[Select[GetPriceData[qqq],
              #[[1]] ≤ YearClock[{2000, 10, 30, 16, 0, 0}] &]]
Out[179]= 0.494515
```

and the computation now starts.

5.4.7.2 Single Iterative Step of the Steepest Descent Method

We choose the computational data, and fix the parameters

```
In[180]:= x0 = 20; x1 = 200; r = .05; Nx = 50; Nt = 50;
          t1 = Transpose[Observations][[1, -1]];
          S = CallData[[1, -1]];
          t0 = CallData[[1, -2]];
```

and the solution of the Dupire PDE can now be computed quite efficiently:

```
In[184]:= Timing[
            DupireCall = DupireCallSolution[{σstat &, r &},
              {S, t0}, {t1, Nt}, {x0, x1, Nx}, {1, 2}]]
```

Out[184]= $\left\{2.14\ \text{Second},\ \text{InterpolatingFunction}\left[\begin{pmatrix} 0.831417 & 1.05593 \\ 20. & 200. \end{pmatrix},\right.\right.$ <>]}

The solution looks like this:

```
In[185]:= P3D = Plot3D[DupireCall[T, k],
              {T, DupireCall[[1, 1, 1]],
               DupireCall[[1, 1, 2]]},
              {k, DupireCall[[1, 2, 1]], DupireCall[[1, 2, 2]]},
              PlotPoints → Nx, Mesh → False,
              ViewPoint -> {-2.788, -0.726, 1.775},
              AxesLabel → {"T", "Strike k", "$"}];
```

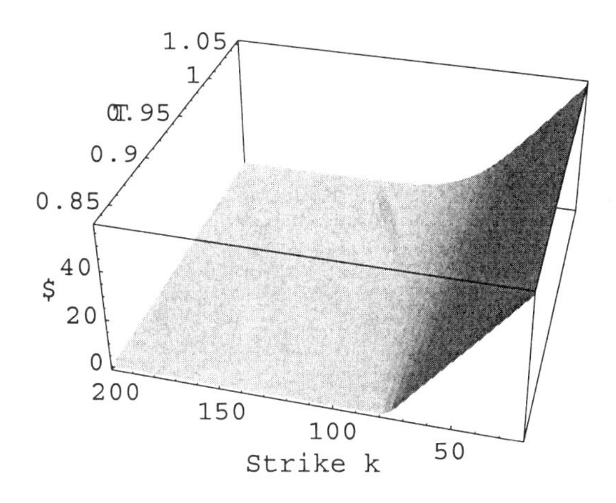

One can see the discrepancy between the theoretical and observed prices:

```
In[186]:= Plot[{#[[2]][k], DupireCall[#[[1]], k]},
              {k, #[[2, 1, 1, 1]], #[[2, 1, 1, 2]]},
              AxesLabel → {"Strike k", "$"}, PlotStyle →
               {RGBColor[0, 0, 0], RGBColor[1, 0, 0]},
              PlotLabel → StringJoin["T = ",
                ToString[#[[1]]]]] & /@ Observations;
```

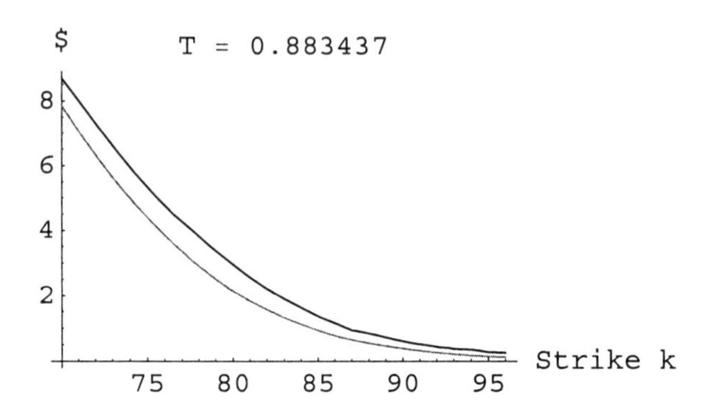

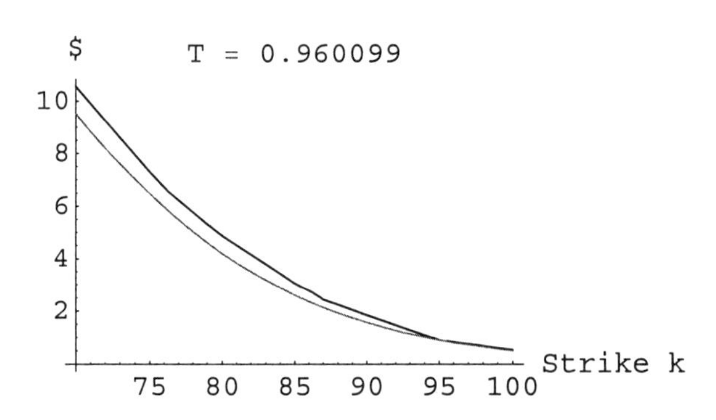

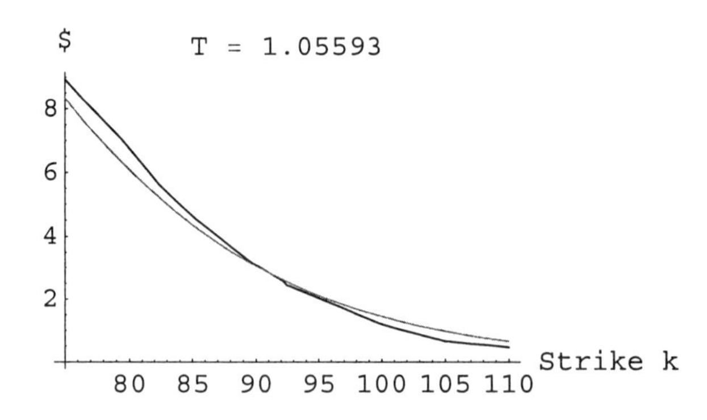

The goal is the make the error as small as possible and identify the corresponding implied volatility.

Proceeding, looking into the adjoint equations we can see that we need to compute the coefficients of the equations and also data on the backward parabolic boundary, i.e., the terminal conditions:

```
In[187]:= CostFunctions =
            If[#[[2, 1, 1, 1]] < s < #[[2, 1, 1, 2]],
              - (DupireCall[#[[1]], s] - #[[2]][s]),
              0] & /@ Observations;
```

They look like

```
In[188]:= Plot[CostFunctions[[#]], {s, 20, 200}] & /@
            {1, 2, 3};
```

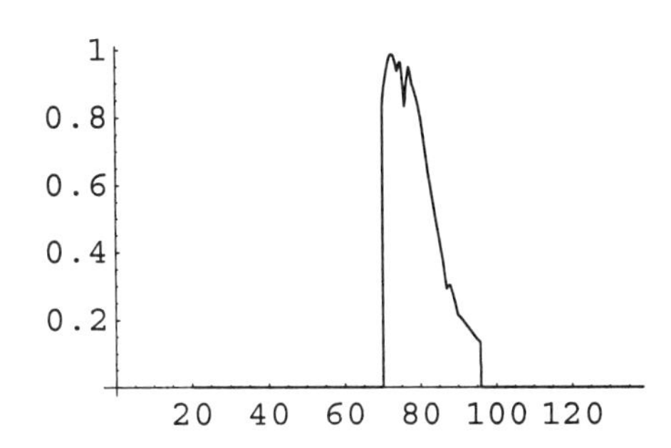

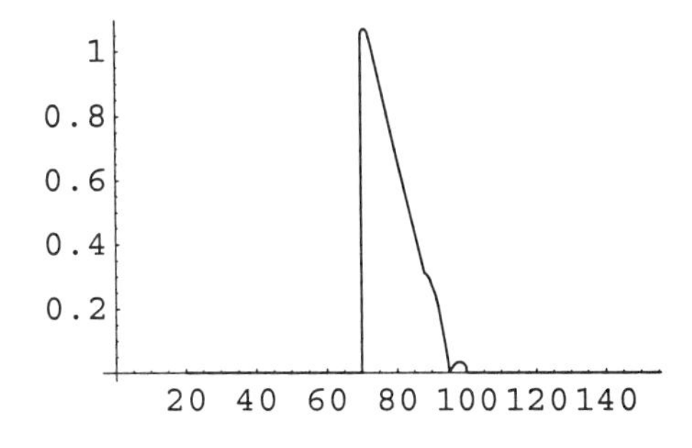

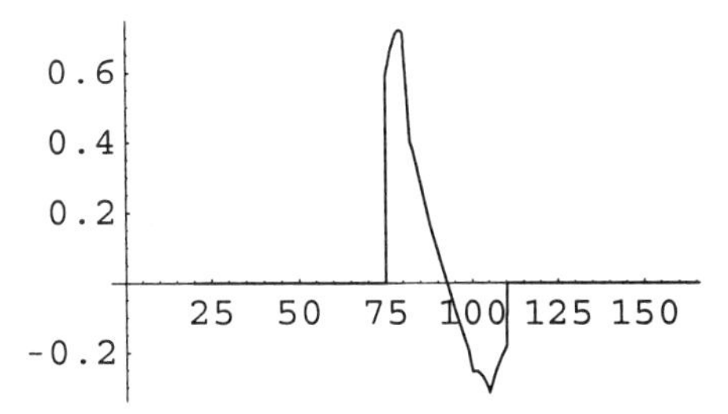

Next we form the coefficients of the adjoint equation

```
In[189]:= σ = σstat &; AA = FunctionInterpolation[
             1/2 #2^2 σ[#1, #2]^2 &[t, s], {t, t0, t1}, {s, x0, x1}]

Out[189]= InterpolatingFunction[( 0.831417   1.05593 ), <>]
                                 ( 20.        200.    )

In[190]:= BB = FunctionInterpolation[
             (2 σ[T, k]^2 + 2 k σ^(0,1)[T, k] σ[T, k] + r) k,
             {T, t0, t1}, {k, x0, x1}]

Out[190]= InterpolatingFunction[( 0.831417   1.05593 ), <>]
                                 ( 20.        200.    )

In[191]:= CC = FunctionInterpolation[
             σ[T, k]^2 + k (4 σ^(0,1)[T, k] + k σ^(0,2)[T, k]) σ[T, k] +
              k^2 σ^(0,1)[T, k]^2 + r, {T, t0, t1}, {k, x0, x1}]

Out[191]= InterpolatingFunction[( 0.831417   1.05593 ), <>]
                                 ( 20.        200.    )
```

and then we solve the adjoint equation

```
In[192]:= AdjointSolver[Target_, t0_, t1_] :=
           BackwardParabolicSolver[{AA, BB, CC},
            {0 &, Target}, {x0, x1, Nx}, {t0, t1, Nt}, {1, 2}]
          ExpiryTimes = Transpose[Observations][[1]];
          Timing[
            AdjointList = Table[AdjointSolver[Function[
                 {t, s}, Evaluate[CostFunctions[[i]]]],
                t0, ExpiryTimes[[i]]],
               {i, 1, Length[ExpiryTimes]}]];
          AdjointList2 = Function[{T, s},
                If[T ≤ #[[2]], #[[1]][T, s], 0]] & /@
              Transpose[{AdjointList, ExpiryTimes}];
          AdjointP = Function[{T, s},
              Sum[AdjointList2[[i]][T, s],
               {i, 1, Length[ExpiryTimes]}]];
```

The solution looks like:

```
In[197]:= Plot3D[AdjointP[T, k], {T, t0, t1}, {k, x0, x1},
            Mesh → False, PlotPoints → 60, PlotRange → All,
            ViewPoint → {-2.622, -1.35, 1.659}];
```

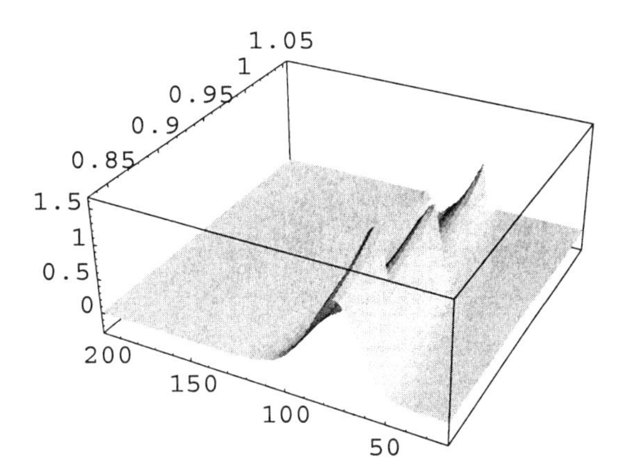

The next is to compute the gradient. To this end we chose the regularization parameters

```
In[198]:= ∈1 = 10000.; ∈2 = 1.;
```

and compute the right-hand side of the regularization ODE:

```
In[199]:= RHS1[T_, k_] = -k^2 D[DupireCall[T, k], {k, 2}]
            σ[T, k] AdjointP[T, k];
```

```
In[200]:= RHS2[T_, k_] =
           -∈1 D[σ[T, k], {k, 2}] + ∈2 (σ[T, k] - σstat);
```

Next we compute the grid for the ODE solvers

```
In[201]:= dx = (x1 - x0)/Nx; dt = (t1 - t0)/Nt;
          xx = Range[x0, x1, dx] // N; tt = Range[t0, t1, dt];
```

For the ODE solver we employ the Tridiagonal solver again:

```
In[202]:= A2List = -∈1 & /@ xx;
          A0List = ∈2 & /@ xx;
          LowerDiagonal = ReplacePart[#, -1, -1] &[
             Drop[#1, 1] &[A2List/dx^2]];
          Diagonal = ReplacePart[#, 1, 1] &[
             ReplacePart[#, 1, -1] &[A0List - 2 A2List/dx^2]];
          UpperDiagonal = ReplacePart[#, -1, 1] &[
             Drop[#, -1] &[A2List/dx^2]];
```

```
Regularization[t_] := TridiagonalSolve[
   LowerDiagonal, Diagonal, UpperDiagonal,
   ReplacePart[ReplacePart[
     Chop[RHS1[t, #] + RHS2[t, #] & /@ xx],
     0, 1], 0, -1]];
Reg = Regularization /@ tt;
```

Once gradient is computed on the grid points, it can be interpolated into a function:

```
In[209]:= Arguments = Outer[{#1, #2} &, tt, xx];
          z = Interpolation[
            Append[#[[1]], #[[2]]] & /@ Transpose[
              {Flatten[Arguments, 1], Flatten[Reg]}],
            InterpolationOrder → 1]
```

Out[209]= InterpolatingFunction[$\begin{pmatrix} 0.831417 & 1.05593 \\ 20. & 200. \end{pmatrix}$, <>]

and viewed

```
In[210]:= Plot3D[z[T, k], {T, t0, t1}, {k, x0, x1},
            Mesh → False, PlotPoints → 60, PlotRange → All,
            ViewPoint → {-2.622, -1.35, 1.659},
            AxesLabel → {"T", "k", ""}];
```

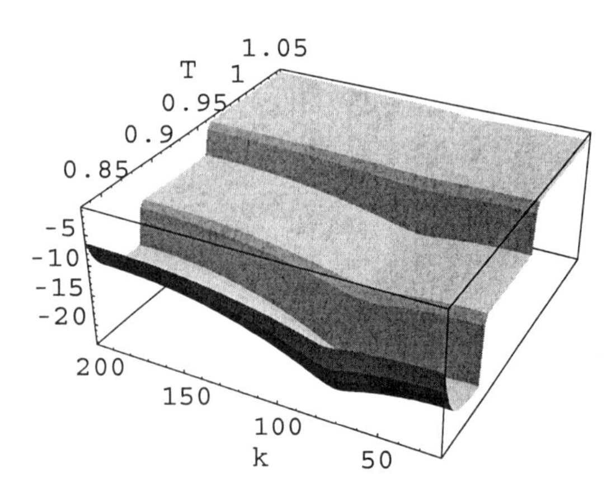

The success of the steepest descent iteration can be measured by the smallness of the gradient $z(T, k)$. Finally, the new volatility $\sigma(T, k)$ is computed by

$$\sigma_{\text{new}}(T, k) = \sigma_{\text{old}}(T, k) - \rho\, z(T, k)$$

for some $\rho > 0$. The choice of a proper ρ is crucial. If ρ is not small enough, convergence *fails*. If, on the other hand, ρ is too small, convergence is too slow. Also, no matter what, σ should always be away from zero. So, let

```
In[211]:= σmin = .05; ρ = .005;
          σnew = FunctionInterpolation[Max[σ[T, k] - ρ z[T, k],
             σmin], {T, t0, t1}, {k, x0, x1}]

Out[211]= InterpolatingFunction[( 0.831417   1.05593 ), <>]
                                 ( 20.        200.    )
```

The new volatility then looks like

```
In[212]:= p2 = Plot3D[σnew[T, k], {T, t0, t1}, {k, x0, x1},
             Mesh → False, PlotPoints → 60, PlotRange → All,
             ViewPoint → {-2.622, -1.35, 1.659},
             AxesLabel → {"T", "k", ""}];
```

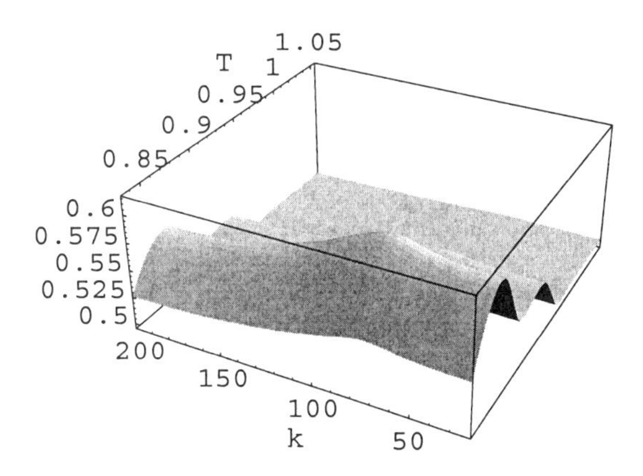

This is not the solution of the implied volatility problem. This is only the first iterate. We need to make many such iterations to arrive at the volatility function that is reasonably close to the optimal one. So, we proceed.

5.4.7.3 Iteration

In order to perform the iteration efficiently, it is necessary to define all of the above operations together as a single function which is then to be iterated. So, let

```
In[213]:= F[σ_, Nx_, Nt_] :=
            (x0 = 20; x1 = 200; r = .05; t1 = Max[ExpiryTimes];
             S = CallData[[1, -1]];
             DupireCall = DupireCallSolution[{σ, r &},
               {S, t0}, {t1, Nt}, {x0, x1, Nx}, {1, 2}];
             CostFunctions = If[#[[2, 1, 1, 1]] < s < #[[2, 1, 1, 2]],
```

```
      - (DupireCall[#[[1]], s] - #[[2]][s]),
      0] & /@ Observations;
  AA = FunctionInterpolation[1/2 k^2 σ[T, k]^2, {T, t0, t1},
    {k, x0, x1}]; BB = FunctionInterpolation[
    (2 σ[T, k]^2 + 2 k σ^(0,1)[T, k] σ[T, k] + r) k, {T, t0, t1},
    {k, x0, x1}]; CC = FunctionInterpolation[
    σ[T, k]^2 + k (4 σ^(0,1)[T, k] + k σ^(0,2)[T, k]) σ[T, k] +
     k^2 σ^(0,1)[T, k]^2 + r, {T, t0, t1}, {k, x0, x1}];
  AdjointList = Table[AdjointSolver[
     Function[{t, s}, Evaluate[CostFunctions[[i]]]],
     t0, ExpiryTimes[[i]]],
    {i, 1, Length[ExpiryTimes]}]; AdjointList2 =
   Function[{T, s}, If[T ≤ #[[2]], #[[1]][T, s], 0]] & /@
    Transpose[{AdjointList, ExpiryTimes}]; AdjointP =
   Function[{T, s}, Sum[AdjointList2[[i]][T, s],
     {i, 1, Length[ExpiryTimes]}]]; ϵ1 = 10000.; ϵ2 = 1.;
RHS1[T_, k_] = -k^2 D[DupireCall[T, k], {k, 2}]
    σ[T, k] AdjointP[T, k]; RHS2[T_, k_] =
   -ϵ1 D[σ[T, k], {k, 2}] + ϵ2 (σ[T, k] - σstat);
  dx = (x1 - x0)/Nx; dt = (t1 - t0)/Nt; xx = Range[x0, x1, dx] // N;
  tt = Range[t0, t1, dt];
A2List = -ϵ1 & /@ xx;
A0List = ϵ2 & /@ xx;
LowerDiagonal =
   ReplacePart[#, -1, -1] &[Drop[#1, 1] &[A2List/dx^2]];
Diagonal = ReplacePart[#, 1, 1] &[
    ReplacePart[#, 1, -1] &[A0List - 2 A2List/dx^2]];
UpperDiagonal = ReplacePart[#, -1, 1] &[
    Drop[#, -1] &[A2List/dx^2]];
Regularization[t_] := TridiagonalSolve[
    LowerDiagonal, Diagonal, UpperDiagonal,
    ReplacePart[ReplacePart[Chop[
       RHS1[t, #] + RHS2[t, #] & /@ xx], 0, 1], 0, -1]];
Reg = Regularization /@ tt;
Arguments = Outer[{#1, #2} &, tt, xx];
z = Interpolation[Append[#[[1]], #[[2]]] & /@
     Transpose[{Flatten[Arguments, 1], Flatten[Reg]}],
    InterpolationOrder → 1];
```

```
σmin = .05; ρ = .005;
FunctionInterpolation[Max[σ[T, k] - ρ z[T, k], σmin],
 {T, t0, t1}, {k, x0, x1}, InterpolationOrder → 1])
```

and perform the first iteration. We choose the number of iterates ad hoc. Of course, those things can be improved.

```
In[214]:= Timing[nl = Nest[F[#, 20, 20] &, σ, 100];]
Out[214]= {451.1 Second, Null}
```

```
In[215]:= Plot3D[nl[T, k], {T, t0, t1}, {k, x0, x1},
           Mesh → False, PlotPoints → 60, PlotRange → All,
           ViewPoint → {-2.622, -1.35, 1.659},
           AxesLabel → {"T", "k", ""}];
```

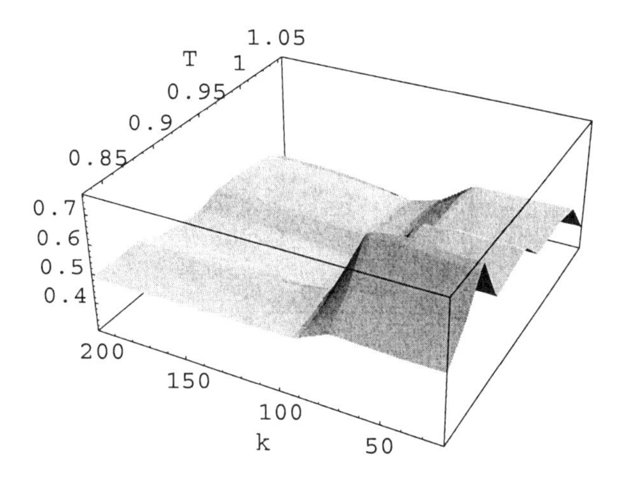

We proceed on a finer grid now

```
In[216]:= Timing[nl2 = Nest[F[#, 50, 50] &, nl, 200];]
Out[216]= {1423.78 Second, Null}
```

arriving at the solution that looks like

```
In[217]:= ImpliedVolatility =
           Plot3D[nl2[T, k], {T, t0, t1}, {k, x0, x1},
            Mesh → False, PlotPoints → 60, PlotRange → All,
            ViewPoint → {-2.622, -1.35, 1.659},
            AxesLabel → {"t", "S", ""}];
```

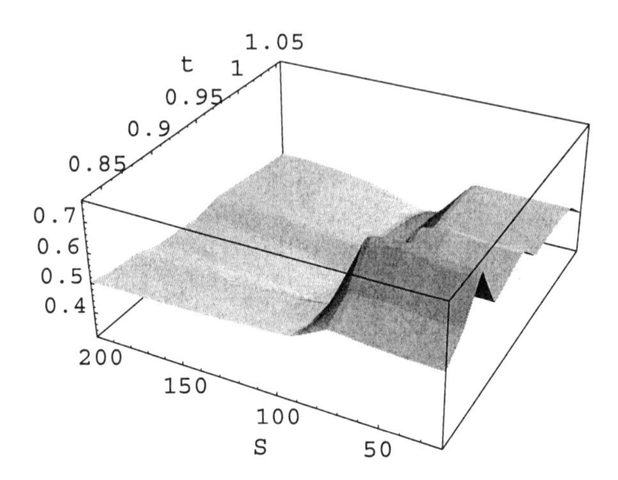

Just in case we would like to use it later, let's save the fruits of our work and heavy computations:

```
In[218]:= nl2 >> "FancyImpliedVolatility"
```

Out[218]= InterpolatingFunction[$\begin{pmatrix} 0.831417 & 1.05593 \\ 20. & 200. \end{pmatrix}$, <>] >>

FancyImpliedVolatility

So, was the minimization successful? How far are the observed prices from the theoretical if new implied volatility is used? The answer looks like this:

```
In[219]:= Plot[{#[[2]][k], DupireCall[#[[1]], k]}, {k,
            #[[2, 1, 1, 1]], #[[2, 1, 1, 2]]}, PlotStyle →
            {RGBColor[0, 0, 0], RGBColor[1, 0, 0]},
           AxesLabel → {"Strike k", "$"},
           PlotLabel → StringJoin["T = ",
             ToString[#[[1]]]]] & /@ Observations;
```

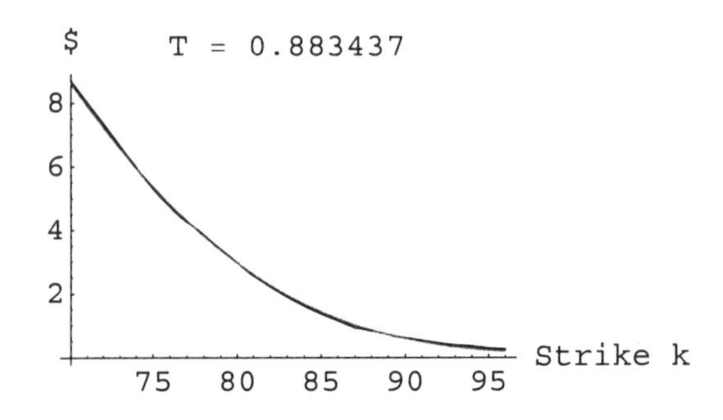

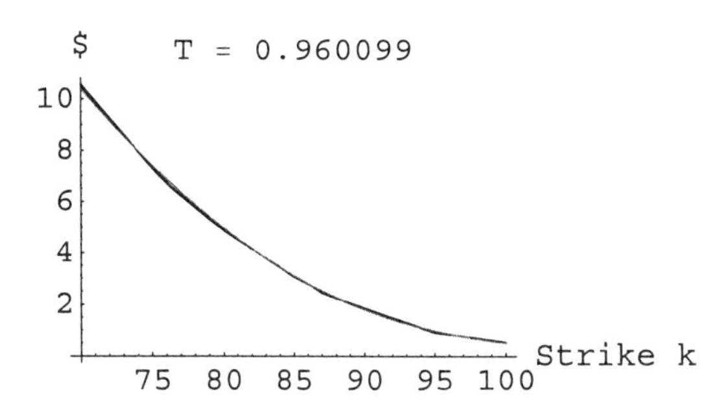

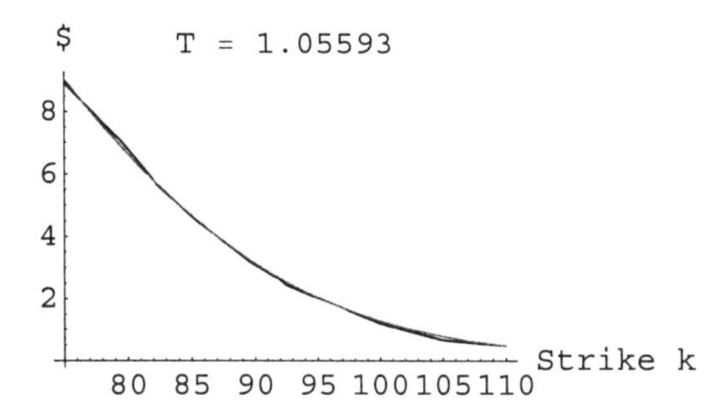

confirming the quality of this complicated but successful, and fully implemented procedure. By the way, the 3D plot of the solution of the Dupire PDE corresponding to the optimal implied volatility looks like this:

```
In[220]:= dc = Plot3D[DupireCall[T, k],
           {T, t0, t1}, {k, x0, x1}, Mesh → False,
           PlotPoints → 130, PlotRange → All,
           ViewPoint → {-2.622, -1.35, 1.659},
           AxesLabel → {"T", "k", ""}];
```

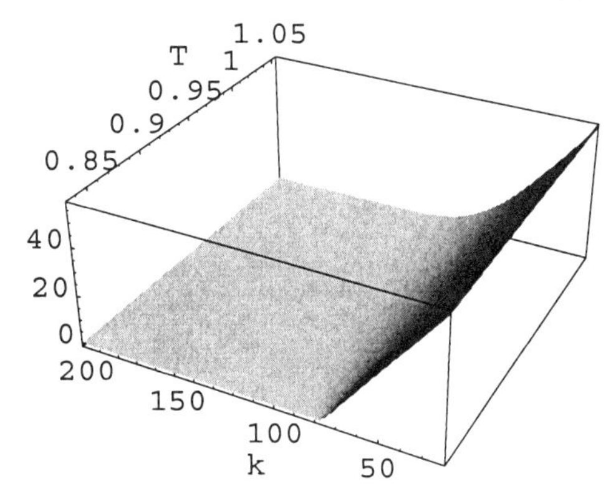

Finally, we can compute the solution of the Black–Scholes PDE (as opposed to Dupire PDE) corresponding to the optimal implied volatility:

```
In[221]:= CallValue = NumericalCallFairPrice[{nl2, r &, 0 &},
            {80, t1}, {t0, 130}, {x0, x1, 130}];
```

The solution looks like this

```
In[222]:= cv = Plot3D[CallValue[T, k],
            {T, t0, t1}, {k, x0, x1}, Mesh → False,
            PlotPoints → 130, PlotRange → All,
            ViewPoint → {-2.622, -1.35, 1.659},
            AxesLabel → {"t", "S", ""}];
```

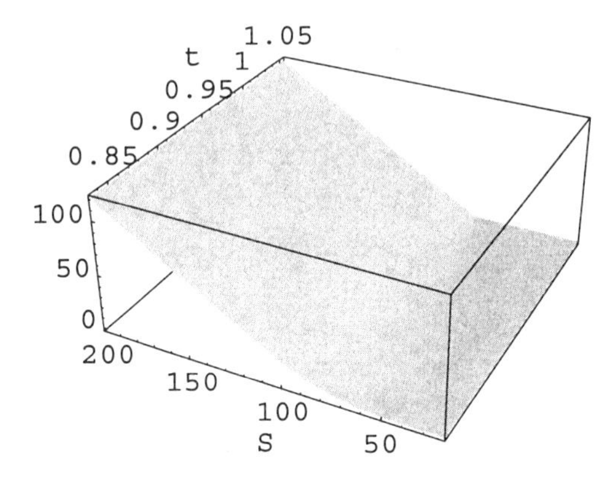

One should compare the last two plots to understand the "duality" between Black–Scholes and Dupire PDEs.

We end this chapter with the question: *Does stochastic volatility really make a difference?* For example, can the market-timing, as discussed before, be improved by considering strike-price i.e., stock-price dependencies? Or, does the price-dependent volatility bring something substantially new into the picture from the point of view of option hedging/trading? Instead of giving an answer to the second (sub)question we give a hint as to how to proceed in a search for one. The methodology presented here and in the next chapter, combined with the symbolic and numerical methodology to be described in Section 8.2 below (and in particular in Section 8.2.4.2 on Advanced Hedging of Options) can be used to give non-ambiguous, well-founded, and precise answers (depending on market scenarios) to such a question.

6 American Style Stock Options

Optimal Stopping, Obstacle Problems, Fast Numerical Solutions, and Implied Volatility via Optimal Control

6.1 Remarks

The previous Black–Scholes analysis was based on the premise that an early exercise of options is not allowed. These are so-called European style options. Their possible exercise date is fixed in advance. On the other hand, the fact is that options that are usually traded on the option market can be exercised at any time before the expiry, although most often they are not. Such options are called American options. As seen so far, the problem of pricing European options was solved by solving the associated Black–Scholes partial differential equations, while the problem of implied volatility in its most sophisticated form so far was solved by solving an optimal control problem for the associated Dupire partial differential equation. Both the equations Black–Scholes and Dupire were linear although the optimal control problem was non-linear.

In the case of American options the situation is more complicated. The possibility of early exercise changes the mathematical problem considerably. The Black–Scholes PDE becomes an *obstacle problem for the* Black–Scholes *PDE*, while the Dupire PDE becomes an *obstacle problem for the* Dupire *PDE*. Both problems are *non-linear*. Furthermore, determination of the implied volatility is more complicated as well —the problem now is formulated as an *optimal control problem for the obstacle problem for the* Dupire *partial differential equation*. Except for some simplified problems that carry basically only pedagogical value, and which we shall rely on quite a bit, there are no exact solutions for any of the practically relevant problems such as these. Therefore we shall study their numerical solutions.

The numerical solutions of obstacle problems are usually based on the Gauss–Seidel iterative method (see [26]). Here we introduce another method, building furthermore on our solution of parabolic equations developed in Chapter 5, and assisted again by the tridiagonal *Mathematica*® solver. Since the tridiagonal solver is exact and fast, our method appears to be more precise (error is due only to the finite difference discretization), and more efficient than the usual Gauss–Seidel method. As opposed to the Gauss–Seidel method, our method is not for the general obstacle problem, but rather only for solving specific obstacle problems, since, for example, it depends on the geometry of the free boundary. The disadvantage in reduced generality is justified by added efficiency and accuracy.

Optimal control of obstacle problems, employed here for solving the problem of finding an implied volatility of American put options is, theoretically, more difficult than the problem of optimal control of PDEs employed in the previous chapter, since the corresponding minimization functional is not differentiable at every control, but in general only Lipschitz continuous (see [1,56]). From the practical computational point of view (such as employing the steepest descent method), which is the view adopted here, this is really not an issue, and the computational method is an appropriate modification of the minimization method developed in the previous chapter for the case of PDEs. Free boundary can be even viewed here as a tool for reducing the considered region to only a part that is relevant—option trades are not to be made outside of the region bounded by the free boundary (there is empirical confirmation of this theoretical conclusion), and therefore the outside region is not fully considered (see the optimization algorithm below for details).

6.2 American Options and Obstacle Problems

6.2.1 American Options, Optimal Stopping, and Obstacle Problem

In Chapter 3 we derived the probabilistic representation of the fair price of the European style option at time t for the underlying stock that has current price S, according to which the fair price is equal to

$$V(t, S) = e^{-r(T-t)} E_{t,S}\, \psi(Z(T)) \tag{6.2.1}$$

if the interest rate r is constant, or, as it is not very difficult to see,

$$V(t, S) = E_{t,S}\big(e^{-\int_t^T r(s,Z(s))\,ds}\, \psi(Z(T))\big) \tag{6.2.2}$$

in general, where T is the option expiration time, and process Z has the stochastic differential

$$dZ(t) = (r(t, Z(t)) - D_0(t, Z(t)))\, Z(t)\, dt + \sigma(t, Z(t))\, Z(t)\, dB(t). \tag{6.2.3}$$

D_0 is the dividend rate, σ is the volatility, and B is the standard Brownian motion. Alternatively, we know from before that the same fair option price is the unique solution of the *linear* partial differential equation

$$\frac{\partial V(t,S)}{\partial t} + \frac{1}{2} S^2 \frac{\partial^2 V(t,S)}{\partial S^2} \sigma(t,S)^2$$
$$+ (r(t,S) - D_0(t,S))\, S\, \frac{\partial V(t,S)}{\partial S} - r(t,S)\, V(t,S) = 0$$

for $\{t, S\} \in Q_T = \{\{t, S\}, t < T, S > 0\}$, together with the terminal condition

$$V(T, S) = \psi(S)$$

where $\psi(S) = \text{Max}(0, S - k)$ in the case of calls, and $\psi(S) = \text{Max}(0, k - S)$ in the case of puts.

On the other hand, in the case of *American* options the holder does not have to wait for the expiration time T in order to exercise the option, but rather has the prerogative to possibly exercise the option early, at some time $\tau \le T$, collecting $\psi(S(\tau))$ instead of $\psi(S(T))$. What kind of *times* τ are considered here? The decision of the option holder to exercise has to be made based on the performance of the underlying stock. In particular, this implies that the time of exercise is in general going to be a random variable, since the performance of the stock is random. Furthermore, the decision to exercise at time τ cannot be based on stock performance in the future, but instead must be based on the performance up until the time τ. Such random times are very well understood in stochastic calculus (see, e.g., [18]), and they are called *stopping times*. It is important to have the right intuition about them right away. For example, for the above process Z, among random times

$$\tau_1 = \text{Min}(t, Z(t) \ge 1),\ \tau_2 = \tau_1 + 1,\ \tau_3 = \tau_1 - 1,$$

τ_1 and τ_2 are stopping times with respect to the information generated by the process Z, while τ_3 is not.

Considering the above probabilistic representation of the fair price of an European option, it should not be surprising, although we shall not venture here into proving it, that the fair price of an American option is equal to

$$V(t, S) = \underset{t \le \tau \le T}{\text{Max}}\, E_{t,S}\left(e^{-\int_t^\tau r(s,Z(s))\,ds}\, \psi(Z(\tau))\right) \tag{6.2.4}$$

for any $\{t, S\} \in Q_T$, where the maximum is taken over all *stopping times* τ, such that $t \le \tau \le T$; we shall prove below that indeed the maximum is taken, and moreover show how to compute this maximum, as well as to how to find the stopping time $\tau^\star$—the optimal stopping time—for which the maximum is obtained. The mathematical problems of maximization (or minimization) of expectations of various functions computed on top of the realizations of the stochastic processes are referred to as *stochastic control*. That maximization could be with respect to various parameters, but if it is, as above, with respect to the class of stopping times, then such a problem is called the problem of *optimal stopping*. In either case, the above *function* $\{t, S\} \mapsto \underset{t \le \tau \le T}{\text{Max}}\, E_{t,S}\left(e^{-\int_t^\tau r(s,Z(s))\,ds}\, \psi(Z(\tau))\right)$ is of crucial importance, and it is called the *value function*.

Theorem:

Suppose there exists a function $\varphi(t, S)$ such that

$$\begin{aligned}\operatorname{Max}\Big[&\frac{\partial\varphi(t,S)}{\partial t}+\frac{1}{2}S^2\frac{\partial^2\varphi(t,S)}{\partial S^2}\sigma(t,S)^2+(r(t,S)-D_0(t,S))\\ &S\frac{\partial\varphi(t,S)}{\partial S}-r(t,S)\,\varphi(t,S),\ \psi(S)-\varphi(t,S)\Big]=0\end{aligned}\tag{6.2.5}$$

a.e. in Q_T, and moreover the terminal condition

$$\varphi(T,S)=\psi(S)\tag{6.2.6}$$

is satisfied. Then φ is the value function, i.e.,

$$\varphi(t,S)=\operatorname*{Max}_{t\le\tau\le T}E_{t,S}\Big(e^{-\int_t^\tau r(s,Z(s))\,ds}\,\psi(Z(\tau))\Big)\tag{6.2.7}$$

and the optimal stopping time is equal to

$$\tau^\star=\operatorname{Min}(t\le T,\ \varphi(t,S(t))=\psi(S(t))).\tag{6.2.8}$$

Remarks: In (6.2.8), and similarly throughout the book, we make a small abuse of notation: if the set $\{t\le T,\ \varphi(t,S(t))=\psi(S(t))\}$ is empty, then, by definition, $\tau^\star=T$. Problem (6.2.5)–(6.2.6) is an example of the so-called *obstacle problems*. We shall study them in some detail below, at least on a very typical simple example. Equation (6.2.5) is an economical way of writing three more explicit conditions:

$$\begin{aligned}&\frac{\partial\varphi(t,S)}{\partial t}+\frac{1}{2}S^2\frac{\partial^2\varphi(t,S)}{\partial S^2}\sigma(t,S)^2+(r(t,S)-D_0(t,S))\,S\\ &\qquad\frac{\partial\varphi(t,S)}{\partial S}-r(t,S)\,\varphi(t,S)\le 0\end{aligned}\tag{6.2.9}$$

$$\varphi(t,S)\ge\psi(S)\tag{6.2.10}$$

$$\begin{aligned}\Big(&\frac{\partial\varphi(t,S)}{\partial t}+\frac{1}{2}S^2\frac{\partial^2\varphi(t,S)}{\partial S^2}\sigma(t,S)^2+(r(t,S)-D_0(t,S))\\ &S\frac{\partial\varphi(t,S)}{\partial S}-r(t,S)\,\varphi(t,S)\Big)(\psi(S)-\varphi(t,S))=0.\end{aligned}\tag{6.2.11}$$

The optimal stopping time $\tau^\star$ can also be written as

$$\tau^\star=\operatorname{Min}(t\le T,\ (t,S(t))\in\Lambda)\tag{6.2.12}$$

where $\Lambda\subset Q_T$ is called the *coincidence set* (coincidence of the obstacle and the solution of (6.2.5)–(6.2.6)), or *stopping region*, and it is equal to

$$\Lambda=\{\{t,S\}\in Q_T,\ \varphi(t,S)=\psi(S)\}.\tag{6.2.13}$$

The complement

$$N = Q_T \backslash \Lambda = \{\{t, S\} \in Q_T, \varphi(t, S) > \psi(S)\} \tag{6.2.14}$$

is called the *non-coincidence set.* The boundary separating them

$$\Gamma = \partial N \bigcap Q_T = \partial\{\{t, S\} \in Q_T, \varphi(t, S) > \psi(S)\} \bigcap Q_T \tag{6.2.15}$$

is called the *free boundary*.

Also, by writing $\frac{\partial\varphi(t,S)}{\partial t}$, $\frac{\partial\varphi(t,S)}{\partial S}$, $\frac{\partial^2\varphi(t,S)}{\partial S^2}$ what is silently implied is that the solution of the obstacle problem φ is differentiable enough for these derivatives to exist. It is interesting to mention that even though these derivatives exist, which is proved in the texts on obstacle problems (see, e.g., [23] for the *regularity theory* for obstacle problems), not all of them exist as continuous functions but rather only as bounded functions—second S-derivative $\frac{\partial^2\varphi(t,S)}{\partial S^2}$ and first t-derivative $\frac{\partial\varphi(t,S)}{\partial t}$ happen to be discontinuous across the free boundary Γ. On the other hand, the first S-derivative $\frac{\partial\varphi(t,S)}{\partial S}$ *is* continuous even across the free boundary implying, and this is very important, the *free boundary condition* on Γ

$$\varphi = \psi, \; \frac{\partial\varphi}{\partial S} = \frac{\partial\psi}{\partial S} = \pm 1 \tag{6.2.16}$$

depending on whether it is a call or a put.

To learn more about optimal stopping times the reader is referred to [5,13,22,54].

In the context of options, the optimal stopping time is the *optimal exercise time*. The above formula for the optimal stopping time is quite plausible: one exercises the option at the first time such that what can be collected is not less than what is the most that can be expected to be collected in the future (for the auxiliary process!)

Equation (6.2.11) implies that, the Black–Scholes partial differential equation

$$\begin{aligned} \mathcal{B}\,\varphi(t, S) = & \frac{\partial\varphi(t, S)}{\partial t} + \frac{1}{2}\, S^2\, \frac{\partial^2\varphi(t, S)}{\partial S^2}\, \sigma(t, S)^2 \\ & +(r(t, S) - D_0(t, S))\, S\, \frac{\partial\varphi(t, S)}{\partial S} - r(t, S)\,\varphi(t, S) = 0 \end{aligned} \tag{6.2.17}$$

holds in the non-coincidence set N. We shall use the notation $\mathcal{B}$ in the proof below.

Proof of the Theorem:

Let φ be a solution of (6.2.5)–(6.2.6). Then, differentiating with respect to t

$$\begin{aligned} & d\left(e^{\int_t^T r(s,Z(s))\,ds}\,\varphi(t, Z(t))\right) \\ & = -r(t, Z(t))\, e^{\int_t^T r(s,Z(s))\,ds}\,\varphi(t, Z(t))dt + e^{\int_t^T r(s,Z(s))\,ds}\, d\varphi(t, Z(t)) \\ & = -r(t, Z(t))\, e^{\int_t^T r(s,Z(s))\,ds}\,\varphi(t, Z(t))dt + e^{\int_t^T r(s,Z(s))\,ds}\Big(\varphi_t(t, Z(t))\, dt \\ & \qquad + \varphi_Z(t, Z(t))\, dZ(t) + \frac{1}{2}\,\varphi_{Z,Z}(t, Z(t))\,(dZ(t))^2\Big) \end{aligned}$$

$$= e^{\int_t^T r(s,Z(s))\,ds} (\mathcal{B}\,\varphi(t,\,Z(t))dt \\ + \varphi_Z(t,\,Z(t))\,Z(t)\,\sigma(t,\,Z(t))\,dB(t)). \tag{6.2.18}$$

Integrating in t, and dividing by $e^{\int_t^T r(s,Z(s))\,ds}$, we get

$$e^{-\int_t^T r(s,Z(s))\,ds}\,\varphi(\tau,\,Z(\tau)) - \varphi(t,\,Z(t)) \\ = \int_t^T e^{-\int_t^w r(s,Z(s))\,ds}\,\mathcal{B}\,\varphi(w,\,Z(w))\,dw \\ + \int_t^T e^{-\int_t^w r(s,Z(s))\,ds}\,\varphi_Z(w,\,Z(w))\,Z(w)\,\sigma(w,\,Z(w))\,dB(w). \tag{6.2.19}$$

Taking the conditional expectation

$$E_{t,S}\,e^{-\int_t^T r(s,Z(s))\,ds}\,\varphi(\tau,\,Z(\tau)) \\ = \varphi(t,\,S) + E_{t,S}\int_t^T e^{-\int_t^w r(s,Z(s))\,ds}\,\mathcal{B}\,\varphi(w,\,Z(w))\,dw. \tag{6.2.20}$$

Therefore, since by (6.2.9), $\mathcal{B}\,\varphi(w,\,Z(w)) \le 0$, and since by (6.2.10), $\psi(Z(\tau)) \le \varphi(\tau,\,Z(\tau))$ we get

$$E_{t,S}\,e^{-\int_t^T r(s,Z(s))\,ds}\,\psi(Z(\tau)) \le E_{t,S}\,e^{-\int_t^T r(s,Z(s))\,ds}\,\varphi(\tau,\,Z(\tau)) \le \varphi(t,\,S) \tag{6.2.21}$$

for *any* stopping time τ. Consequently,

$$\sup_{t\le\tau\le T}\,E_{t,S}\,e^{-\int_t^T r(s,Z(s))\,ds}\,\psi(Z(\tau)) \le \varphi(t,\,S). \tag{6.2.22}$$

On the other hand, for $\tau^\star$ given by (6.2.12), since for $w < \tau$, $\mathcal{B}\,\varphi(w,\,Z(w)) = 0$, we get from (6.2.20)

$$E_{t,S}\,e^{-\int_t^{T^\star} r(s,Z(s))\,ds}\,\varphi(\tau,\,Z(\tau)) = \varphi(t,\,S) \tag{6.2.23}$$

which together with (6.2.22) implies that

$$\operatorname*{Max}_{t\le\tau\le T} E_{t,S}\,e^{-\int_t^T r(s,Z(s))\,ds}\,\psi(Z(\tau)) \\ = E_{t,S}\,e^{-\int_t^{T^\star} r(s,Z(s))\,ds}\,\varphi(\tau,\,Z(\tau)) = \varphi(t,\,S) \tag{6.2.24}$$

which completes the proof of the theorem.

The literature on free boundary problems that can be formulated as obstacle problems is very extensive, and those problems were studied in great detail throughout the 70s and early 80s (see [23]). There are many equivalent mathematical formulations of those problems, some known widely, some not so well known, some very useful for proving theorems, some useful for computations. We shall discuss some of them below.

6.2.2 Equivalent Formulations of Obstacle Problems

To develop further intuition about obstacle problems we consider a very simple one: the one-dimensional problem. Everything that follows is true for general domains $\Omega \subset \mathbb{R}^n$ for $n \geq 1$, for general uniformly elliptic operators A instead of $A = -\frac{\partial^2}{\partial x^2}$, and for general obstacles ψ instead of $\psi = 0$ as below, although some of the formulations become more complicated in that case ((6.2.36) for example); also almost all can be reformulated for parabolic problems as well. So, let the domain be $\Omega = (-1, 1) \subset \mathbb{R}$, let the differential operator be $(A\,u)[x] = -\frac{\partial^2 u(x)}{\partial x^2}$, let the right-hand side be $f(x)$, let the obstacle be zero, and let the boundary condition for u be $u(-1) = 0,\ u(1) = 1$. All together one looks for the function $u \geq 0$ such that

$$(A\,u)[x] = -\frac{\partial^2 u(x)}{\partial x^2} = f(x)$$

in the set $\{x \in (-1, 1) \mid u(x) > 0\}$, together with the requirement that $u(x)$ *and* $u'(x)$ are continuous in the whole domain $\Omega = (-1, 1)$, which implies the free boundary condition. Here are several precise formulations, all equivalent to each other.

6.2.2.1 Variational Inequality Problem

Recall that $H^1(-1, 1) = \{u \in L^2(-1, 1),\ u' \in L^2(-1, 1)\}$. Then notice that the set

$$K = \{u \in H^1(-1, 1),\ u(-1) = 0,\ u(1) = 1,\ u \geq 0\} \tag{6.2.25}$$

is *convex and closed* in $H^1(-1, 1)$. The problem is, for given $f \in L^2(-1, 1)$, to find $u \in K$ such that

$$\int_{-1}^{1} \frac{\partial u(x)}{\partial x} \frac{\partial (v(x) - u(x))}{\partial x}\, dx \geq \int_{-1}^{1} f(x)\,(v(x) - u(x))\, dx \tag{6.2.26}$$

for every $v \in K$.

We show that there may be only one solution of the variational inequality. Indeed, suppose there are two solutions: u_1 and u_2. Then

$$\int_{-1}^{1} \frac{\partial u_1(x)}{\partial x} \frac{\partial (u_2(x) - u_1(x))}{\partial x}\, dx \geq \int_{-1}^{1} f(x)\,(u_2(x) - u_1(x))\, dx$$

and

$$\int_{-1}^{1} \frac{\partial u_2(x)}{\partial x} \frac{\partial (u_1(x) - u_2(x))}{\partial x}\, dx \geq \int_{-1}^{1} f(x)\,(u_1(x) - u_2(x))\, dx.$$

Adding

$$0 \leq \int_{-1}^{1} \left(\frac{\partial u_1(x)}{\partial x} - \frac{\partial u_2(x)}{\partial x} \right) \frac{\partial (u_2(x) - u_1(x))}{\partial x}\, dx$$

$$= -\int_{-1}^{1}\Big(\frac{\partial u_1(x)}{\partial x} - \frac{\partial u_2(x)}{\partial x}\Big)^2 dx$$

which implies $\frac{\partial u_1(x)}{\partial x} - \frac{\partial u_2(x)}{\partial x} = 0$, and since $u_1(-1) = u_2(-1)$, we conclude that

$$u_1(x) - u_2(x) = u_1(-1) - u_2(-1) + \int_{-1}^{x} \frac{\partial(u_1(\xi) - u_2(\xi))}{\partial \xi}\, d\xi = 0$$

for any $x \in [-1, 1]$.

6.2.2.2 Calculus of Variations Problem

Again, $K = \{v \in H^1(-1, 1),\ v(-1) = 0,\ v(1) = 1,\ v \geq 0\}$. Let $f \in L^2(-1, 1)$ be given, and let J be a variational functional defined by

$$J[v] = \int_{-1}^{1}\Big(\frac{1}{2}\Big(\frac{\partial v(x)}{\partial x}\Big)^2 - f(x)\, v(x)\Big) dx. \tag{6.2.27}$$

The problem is to find $u \in K$ such that

$$J[u] = \min\nolimits_{v \in K} J[v]. \tag{6.2.28}$$

The existence of a minimizer follows from the fact that the first term is positive quadratic and the second only linear. Indeed,

$$\Big|\int_{-1}^{1} f(x)\, v(x)\, dx\Big| \leq \Big(\int_{-1}^{1} f(x)^2\, dx\Big)^{1/2} \Big(\int_{-1}^{1} v(x)^2\, dx\Big)^{1/2}$$
$$= \Big(\int_{-1}^{1} f(x)^2\, dx\Big)^{1/2} \Big(\int_{-1}^{1}\Big(\int_{-1}^{x} \frac{\partial v(\xi)}{\partial \xi}\, d\xi\Big)^2 dx\Big)^{1/2}$$
$$\leq \Big(\int_{-1}^{1} f(x)^2\, dx\Big)^{1/2} \Big(\int_{-1}^{1}\Big((x+1)\int_{-1}^{x}\Big(\frac{\partial v(\xi)}{\partial \xi}\Big)^2 d\xi\Big) dx\Big)^{1/2}$$
$$\leq \Big(\int_{-1}^{1} f(x)^2\, dx\Big)^{1/2} \Big(\int_{-1}^{1}\Big(2\int_{-1}^{1}\Big(\frac{\partial v(\xi)}{\partial \xi}\Big)^2 d\xi\Big) dx\Big)^{1/2}$$
$$= 2\Big(\int_{-1}^{1} f(x)^2\, dx\Big)^{1/2} \Big(\int_{-1}^{1}\Big(\frac{\partial v(x)}{\partial x}\Big)^2 dx\Big)^{1/2}$$

and therefore, since $f \in L^2(-1, 1)$ is fixed

$$J[v] = \int_{-1}^{1}\Big(\frac{1}{2}\Big(\frac{\partial v(x)}{\partial x}\Big)^2 - f(x)\, v(x)\Big) dx$$
$$\geq \int_{-1}^{1} \frac{1}{2}\Big(\frac{\partial v(x)}{\partial x}\Big)^2 dx - 2\Big(\int_{-1}^{1} f(x)^2\, dx\Big)^{1/2} \Big(\int_{-1}^{1}\Big(\frac{\partial v(x)}{\partial x}\Big)^2 dx\Big)^{1/2}$$
$$= \frac{1}{2} A^2 - c\, A$$

for $A = \left(\int_{-1}^{1}\left(\frac{\partial v(x)}{\partial x}\right)^2 dx\right)^{1/2}$ and $c = 2\left(\int_{-1}^{1} f(x)^2\, dx\right)^{1/2}$. But $q(A) = \frac{1}{2}A^2 - c\,A$ is a simple quadratic function in A, and therefore it is simple to find its minimum:

```
In[1]:= A^2/2 - c A /. Solve[∂_A (A^2/2 - c A) == 0, A][[1]]

Out[1]= -c^2/2
```

This implies

$$
\begin{aligned}
J[v] &\geq \frac{1}{2}A^2 - c\,A \geq -\frac{c^2}{2} \\
&= -\frac{\left(2\left(\int_{-1}^{1} f(x)^2\, dx\right)^{1/2}\right)^2}{2} = -2\int_{-1}^{1} f(x)^2\, dx > -\infty
\end{aligned}
$$

no matter what the $v \in K$ is, and therefore

$$
\inf_{v\in K} J[v] \geq -2\int_{-1}^{1} f(x)^2\, dx > -\infty.
$$

Consequently there exists a minimizing sequence $\{v_n\} \subset K$ such that

$$
\infty > c_3 \geq J[v_k] \geq \lim_{n\to\infty} J[v_n] = \inf_{v\in K} J[v] > -\infty. \tag{6.2.29}
$$

The next claim is that the minimizing sequence is bounded in $H^1(-1, 1)$, i.e., that

$$
\left(\|v_n\|_{H^1(-1,1)}\right)^2 = \int_{-1}^{1}(v_n(x))^2\, dx + \int_{-1}^{1}\left(\frac{\partial v_n(x)}{\partial x}\right)^2 dx < c_2 < \infty \tag{6.2.30}
$$

for any n. Indeed, from above

$$
\begin{aligned}
&\frac{1}{2}\int_{-1}^{1}\left(\frac{\partial v_n(x)}{\partial x}\right)^2 dx - 2\left(\int_{-1}^{1} f(x)^2\, dx\right)^{1/2}\left(\int_{-1}^{1}\left(\frac{\partial v_n(x)}{\partial x}\right)^2 dx\right)^{1/2} \\
&\leq \frac{1}{2}\int_{-1}^{1}\left(\frac{\partial v_n(x)}{\partial x}\right)^2 dx + \int_{-1}^{1} f(x)\, v_n(x)\, dx = J\,(v_n) \leq c_3
\end{aligned}
$$

implies

$$
\int_{-1}^{1}\left(\frac{\partial v_n(x)}{\partial x}\right)^2 dx \leq 2\left(\int_{-1}^{1} f(x)^2\, dx\right)^{1/2} + \sqrt{2}\,\sqrt{2\int_{-1}^{1} f(x)^2\, dx + c_3}
$$

since

```
In[2]:= A /. Solve[A^2/2 - 2 f A == c, A][[2]]
```

Out[2]= $2f + \sqrt{2}\sqrt{2f^2 + c}$

Now, since also $\int_{-1}^{1}(v_n(x))^2\,dx \le c\int_{-1}^{1}\left(\frac{\partial v_n(x)}{\partial x}\right)^2 dx$, claim (6.2.30) follows. This in turn implies that there exists $u \in K$ such that at least a subsequence of v_n converges to u weakly in $H^1(-1, 1)$ and strongly in $L^2(-1, 1)$. This is non-trivial although not very difficult, and can be found in texts on real analysis. More explicitly, $u \in K$ and

$$\lim_{n\to\infty}\int_{-1}^{1}\frac{\partial v_n(x)}{\partial x}\,g(x)\,dx = \int_{-1}^{1}\frac{\partial u(x)}{\partial x}\,g(x)\,dx$$

for any $g \in L^2(-1, 1)$, and

$$\lim_{n\to\infty}\int_{-1}^{1}(v_n(x) - u(x))^2\,dx = 0.$$

Finally, again from real analysis, we know that the norm $|\ |_{H^1}$ is weakly lower semicontinuous, or explicitly

$$\liminf_{n\to\infty}\int_{-1}^{1}\left(\frac{\partial v_n(x)}{\partial x}\right)^2 dx \ge \int_{-1}^{1}\left(\frac{\partial u(x)}{\partial x}\right)^2 dx$$

Also, obviously,

$$\lim_{n\to\infty}\int_{-1}^{1}f(x)\,v_n(x)\,dx = \int_{-1}^{1}f(x)\,u(x)\,dx.$$

This implies

$$\inf_{v\in K} J[v] = \liminf_{n\to\infty} J[v_n] \ge J[u]$$

i.e., the existence of the minimizer $u \in K$ is proved.

Next, we show that the minimizer u solves the above variational inequality and in particular, therefore, it is unique. Indeed, in addition to the minimizer $u \in K$, consider any other $v \in K$. Then the convex combination

$$v_\epsilon = \epsilon\,v + (1-\epsilon)\,u = u + \epsilon\,(v - u) \in K$$

for any $\epsilon \in [0, 1]$ (since K is a convex set). Therefore, by the definition of the minimizer,

$$J[u] \le J[v_\epsilon] = \int_{-1}^{1}\left(\frac{1}{2}\left(\frac{\partial v_\epsilon(x)}{\partial x}\right)^2 - f(x)\,v_\epsilon(x)\right)dx$$

$$= \int_{-1}^{1}\left(\frac{1}{2}\left(\frac{\partial(u + \epsilon\,(v-u))\,(x)}{\partial x}\right)^2 - f(x)\,(u + \epsilon\,(v-u))\,(x)\right)dx = J[u]$$

$$+ \int_{-1}^{1}\left(\epsilon\,\frac{\partial u(x)}{\partial x}\,\frac{\partial(v-u)\,(x)}{\partial x} + \frac{\epsilon^2}{2}\left(\frac{\partial(v-u)\,(x)}{\partial x}\right)^2 - \epsilon\,f(x)\,(v-u)(x)\right)dx$$

and consequently

$$0 \leq \int_{-1}^{1}\left(\epsilon \frac{\partial u(x)}{\partial x} \frac{\partial (v-u)(x)}{\partial x} + \frac{\epsilon^2}{2}\left(\frac{\partial (v-u)(x)}{\partial x}\right)^2 - \epsilon f(x)(v-u)(x)\right)dx$$

for any $\epsilon \in [0, 1]$. Dividing by such $\epsilon \neq 0$ we get

$$0 \leq \int_{-1}^{1}\left(\frac{\partial u(x)}{\partial x} \frac{\partial (v-u)(x)}{\partial x} + \frac{\epsilon}{2}\left(\frac{\partial (v-u)(x)}{\partial x}\right)^2 - f(x)(v-u)(x)\right)dx.$$

Sending $\epsilon \to 0$ we conclude the variational inequality

$$\int_{-1}^{1} \frac{\partial u(x)}{\partial x} \frac{\partial (v-u)(x)}{\partial x}\, dx \geq \int_{-1}^{1} f(x)(v-u)(x)\, dx.$$

6.2.2.3 Complementarity Problem

Find $u \in H^2(-1, 1) = \{u \in L^2(-1, 1),\, u' \in L^2(-1, 1),\, u'' \in L^2(-1, 1)\}$, such that $u(-1) = 0$, $u(1) = 1$, and

$$-\frac{\partial^2 u(x)}{\partial x^2} \geq f(x), \tag{6.2.31}$$

$$u(x) \geq 0, \tag{6.2.32}$$

$$u(x)\left(-\frac{\partial^2 u(x)}{\partial x^2} - f(x)\right) = 0 \tag{6.2.33}$$

in $(-1, 1)$.

We shall show that if u is the solution of the variational inequality, and if additionally $u \in H^2(-1, 1)$, then u is a solution of the complementarity problem. By the way, if $f \in L^2(-1, 1)$ this is true always. We shall not prove this regularity result here—see, e.g., [23]. By choosing the test function v in the variational inequality to be $v = u + \zeta$ where $\zeta \in H_0^1(-1, 1)$, (i.e., in addition to be in $H^1(-1, 1)$, ζ is also equal to zero at the boundary) and also such that $\zeta \geq 0$ throughout $(-1, 1)$, we get

$$\int_{-1}^{1} \frac{\partial u(x)}{\partial x} \frac{\partial (u+\zeta-u)(x)}{\partial x}\, dx \geq \int_{-1}^{1} f(x)(u+\zeta-u)(x)\, dx$$

i.e.,

$$\int_{-1}^{1} \frac{\partial u(x)}{\partial x} \frac{\partial \zeta(x)}{\partial x}\, dx \geq \int_{-1}^{1} f(x)\zeta(x)\, dx$$

or

$$\int_{-1}^{1}\left(-\frac{\partial^2 u(x)}{\partial x^2} - f(x)\right)\zeta(x)\, dx \geq 0$$

for any $\zeta \in H_0^1(-1, 1)$ such that $\zeta \geq 0$, and therefore $-\frac{\partial^2 u(x)}{\partial x^2} \geq f(x)$ follows. On the other hand, consider functions $v_\epsilon = u + \epsilon\,\zeta$, where $\zeta \in C_0^\infty(\{u > 0\})$ extended as zero in Ω outside of the non-coincidence set $\{u > 0\}$. It is easy to see that for fixed ζ, there exists $\epsilon_0 > 0$, such that if $|\epsilon| \leq \epsilon_0$, then $v_\epsilon \in K$. Therefore

$$\epsilon \int_{-1}^{1} \frac{\partial u(x)}{\partial x} \frac{\partial \zeta(x)}{\partial x} \, dx \geq \epsilon \int_{-1}^{1} f(x)\, \zeta(x)\, dx$$

and

$$-\epsilon \int_{-1}^{1} \frac{\partial u(x)}{\partial x} \frac{\partial \zeta(x)}{\partial x} \, dx \geq -\epsilon \int_{-1}^{1} f(x)\, \zeta(x)\, dx$$

which is possible only if

$$\int_{-1}^{1} \frac{\partial u(x)}{\partial x} \frac{\partial \zeta(x)}{\partial x} \, dx = \int_{-1}^{1} f(x)\, \zeta(x)\, dx.$$

This equality can be concluded for any $\zeta \in C_0^\infty(\{u > 0\})$, which implies that $-\frac{\partial^2 u(x)}{\partial x^2}$ $-f(x) = 0$ in the set $\{u > 0\}$, and therefore also $u(x)\left(-\frac{\partial^2 u(x)}{\partial x^2} - f(x)\right) = 0$.

6.2.2.4 Fully Non-Linear PDE Problem

Find $u \in H^2(-1, 1) = \{u \in L^2(-1, 1), u' \in L^2(-1, 1), u'' \in L^2(-1, 1)\}$, such that $u(-1) = 0$, $u(1) = 1$, and

$$\text{Min}\left[u(x), -\frac{\partial^2 u(x)}{\partial x^2} - f(x)\right] = 0 \tag{6.2.34}$$

in $(-1, 1)$.

The alternative way of writing this equation, more consistent with optimal stopping problems that follow, is

$$\text{Max}\left[-u(x), \frac{\partial^2 u(x)}{\partial x^2} + f(x)\right] = 0. \tag{6.2.35}$$

Obviously, (6.2.34) is just the complementarity problem written more economically.

6.2.2.5 Semilinear PDE Problem 1

Notice that if u is a solution of the complementarity problem then, as shown above, $-\frac{\partial^2 u(x)}{\partial x^2} - f(x) = 0$ in the set $\{u > 0\}$, while obviously $-\frac{\partial^2 u(x)}{\partial x^2} = 0$ in the set $\{u = 0\}$. Since also, as it is well known $u \in H^2(-1, 1)$, and in particular u is twice weakly differentiable in $(-1, 1)$, the following semilinear partial differential *equation* is the consequence of the complementarity problem $-\frac{\partial^2 u(x)}{\partial x^2} - f(x)\, I_{\{u>0\}}(x) = 0$, where

$$I_U(x) = \begin{cases} 1, & x \in U \\ 0, & x \notin U \end{cases}$$

is the characteristic function of the set U. Let also $f^+ = \text{Max}[f, 0]$ be the positive part of f, and let $f^- = -\text{Min}[f, 0]$ be the negative part of f. Then, the other complementarity problem condition $-\frac{\partial^2 u(x)}{\partial x^2} - f(x) \geq 0$ will be violated unless $-f(x) \geq 0$ in the set $\{u = 0\}$, or which is the same, unless $f^+(x)\, I_{\{u=0\}} = 0$. Summarizing, consider the problem of finding $u \in H^2(-1, 1)$, such that $u(-1) = 0$, $u(1) = 1$, and

$$\begin{aligned} &-\frac{\partial^2 u(x)}{\partial x^2} - f(x)\, I_{\{u>0\}}(x) = 0 \\ &f^+(x)\, I_{\{u=0\}}(x) = 0 \end{aligned} \tag{6.2.36}$$

in $(-1, 1)$. Notice that no constraint $u \geq 0$ is imposed explicitly; nevertheless, the property is a consequence of the maximum principle. Also notice that if $f \leq 0$, then the second equation is always trivially fulfilled.

The second equation will not be needed in order to show that $u \geq 0$. Indeed, (notice u^- is equal to zero at both $x = \pm 1$)

$$\begin{aligned} 0 &= \int_{-1}^{1} \left(-\frac{\partial^2 u(x)}{\partial x^2} - f(x)\, I_{\{u>0\}}(x) \right) u^-(x)\, dx \\ &= \int_{-1}^{1} \left(\frac{\partial u(x)}{\partial x} \frac{\partial u^-(x)}{\partial x} - (f^+(x) - f^-(x))\, I_{\{u>0\}}(x)\, u^-(x) \right) dx \\ &= \int_{-1}^{1} \left(\left(\frac{\partial u^-(x)}{\partial x} \right)^2 + f^-(x)\, I_{\{u>0\}}(x)\, u^-(x) \right) dx \geq \int_{-1}^{1} \left(\frac{\partial u^-(x)}{\partial x} \right)^2 dx \end{aligned} \tag{6.2.37}$$

and therefore $\frac{\partial u^-(x)}{\partial x} = 0$; consequently as before $u^- = 0$, i.e., $u \geq 0$.

Moreover, uniqueness holds. Indeed, if u_1 and u_2 are two solution, then

$$0 = \int_{-1}^{1} \left(-\frac{\partial^2 u_1(x)}{\partial x^2} - f(x)\, I_{\{u_1>0\}}(x) \right) (u_1(x) - u_2(x))^-\, dx$$

and

$$0 = \int_{-1}^{1} \left(-\frac{\partial^2 u_2(x)}{\partial x^2} - f(x)\, I_{\{u_1>0\}}(x) \right) (u_2(x) - u_1(x))^-\, dx.$$

Adding

$$\begin{aligned} 0 &= \int_{-1}^{1} \left(-\frac{\partial^2 (u_1(x) - u_2(x))}{\partial x^2} - f(x)\, (I_{\{u_1>0\}}(x) - I_{\{u_2>0\}}(x)) \right) (u_2(x) - u_1(x))^- dx \\ &= \int_{-1}^{1} \left(\frac{\partial (u_1(x) - u_2(x))}{\partial x} \frac{\partial (u_2(x) - u_1(x))^-}{\partial x} \right. \end{aligned}$$

$$- (f^+(x) - f^-(x)) (I_{\{u_1>0\}}(x) - I_{\{u_2>0\}}(x)) (u_2(x) - u_1(x))^- \Big) dx$$

$$= \int_{-1}^{1} \left(\left(\frac{\partial (u_2(x) - u_1(x))^-}{\partial x} \right)^2 + f^-(x)(I_{\{u_1>0\}}(x) - I_{\{u_2>0\}}(x)) (u_2(x) - u_1(x))^- \right) dx$$

$$\geq \int_{-1}^{1} \left(\frac{\partial (u_2(x) - u_1(x))^-}{\partial x} \right)^2 dx$$

since the conditions $f^+ \, I_{\{u_i=0\}} = 0, \, i = 1, 2$ imply

$$f^+(I_{\{u_1>0\}} - I_{\{u_2>0\}}) = f^+ \, (I_{\{u_1>0, u_2=0\}} - I_{\{u_1=0, u_2>0\}}) = 0.$$

The conclusion is $(u_2 - u_1)^- = 0$ i.e. $u_2 \geq u_1$. But in the same fashion $u_1 \geq u_2$, and therefore $u_2 = u_1$ (cf. [55,57]).

6.2.2.6 Semilinear PDE Problem 2

Find the *maximal* $u \in H^2(-1, 1)$, such that $u(-1) = 0$, $u(1) = 1$, and

$$-\frac{\partial^2 u(x)}{\partial x^2} - f(x) \, I_{\{u>0\}}(x) = 0 \tag{6.2.38}$$

in $(-1, 1)$.

Again, no constraint $u \geq 0$ is imposed explicitly. If $f^+ = 0$, then as seen above only an equation is necessary to characterize the solution of the obstacle problem. In general, let u be the solution of the semilinear PDE problem 1, and let v be any other solution of (25). Then, similar to above,

$$0 = \int_{-1}^{1} \left(-\frac{\partial^2 (u(x) - v(x))}{\partial x^2} - f(x) \, (I_{\{u>0\}}(x) - I_{\{v>0\}}(x)) \right) (u(x) - v(x))^- \, dx$$

$$= \int_{-1}^{1} \Big(\frac{\partial (u(x) - v(x))}{\partial x} \; \frac{\partial (u(x) - v(x))^-}{\partial x}$$

$$- (f^+(x) - f^-(x)) \, (I_{\{u>0\}}(x) - I_{\{v>0\}}(x)) \, (u(x) - v(x))^- \Big) \, dx$$

$$= \int_{-1}^{1} \Big(\left(\frac{\partial (u(x) - v(x))^-}{\partial x} \right)^2 - f^+(x) \, I_{\{u>0, v=0\}}(x) \, (u(x) - v(x))^-$$

$$+ f^-(x) \, (I_{\{u>0\}}(x) - I_{\{v>0\}}(x)) \, (u(x) - v(x))^- \Big)$$

$$= \int_{-1}^{1} \left(\left(\frac{\partial (u(x) - v(x))^-}{\partial x} \right)^2 + f^-(x) \, (I_{\{u>0\}}(x) - I_{\{v>0\}}(x)) \, (u(x) - v(x))^- \right) dx$$

$$\geq \int_{-1}^{1} \left(\frac{\partial (u(x) - v(x))^-}{\partial x} \right)^2 dx$$

since

$$f^+(I_{\{u>0\}} - I_{\{v>0\}}) = f^+ \, (I_{\{u>0, v=0\}} - I_{\{u=0, v>0\}}) = f^+ \, I_{\{u>0, v=0\}}.$$

The conclusion is $(u-v)^- = 0$ i.e. $u \geq v$, and consequently the semilinear PDE solution 1 is the maximal semilinear PDE solution.

6.2.2.7 Free Boundary Value Problem

All of the above formulations are equivalent and all of them admit a unique solution of the obstacle problem. Furthermore, define a *free boundary* $\Gamma = \partial\{u > 0\} \cap (-1, 1)$. Notice carefully that by definition free boundary is in the *interior* of the considered domain $(-1, 1)$. Notice also that since $u \in H^2(-1, 1) \subset C^1(-1, 1)$, and since the obstacle is equal to zero, then $u(x) = \frac{\partial u(x)}{\partial x} = 0$, for any $x \in \Gamma$. This is the *free boundary condition*. We conclude that either one of the above formulations of the obstacle problem *implies* the following set of conditions: $u \in H^2(-1, 1)$, $u(0)$, $u(1) = 1$,

$$-\frac{\partial^2 u(x)}{\partial x^2} - f(x) = 0$$

in the non-coincidence set $N = \{x \in (-1, 1); u(x) > 0\}$, and

$$u(x) = \frac{\partial u(x)}{\partial x} = 0$$

on the free boundary $\Gamma = \partial N \cap (-1, 1)$.

Now we emphasize that the free boundary value problem formulation does not necessarily determine the solution uniquely, and therefore although a consequence of any of the above formulations of the obstacle problem, it is not, *in general*, equivalent to them; it is merely a necessary condition, but not sufficient. In particular cases the free boundary value problem produces the unique solution. Below we shall see even in this simple case, depending on the right-hand side $f(x)$, a free boundary value problem sometimes does and sometimes does not determine the unique solution. In either case, the free boundary value problem formulation is the practical way to compute the solution. We shall introduce an additional practical way of computing the solution below.

6.2.2.8 Computing Solution: Free Boundary Value Problem

```
In[3]:= Clear["Global`*"]
```

Let $f = -10$. Relying somewhat on the geometry of the problem we proceed as follows. First, use the differential equation to find a family of solutions,

```
In[4]:= v[x_] = u[x] /. DSolve[-∂{x,2} u[x] == -10, u[x], x]〚1〛
```

Out[4]= $5x^2 + c_2 x + c_1$

satisfying only one boundary condition, the one at $x = 1$,

```
In[5]:= w[x_] = v[x] /. Solve[v[1] == 1, C[2]]〚1〛
```

Out[5]= $5x^2 + (-c_1 - 4)x + c_1$

Among those solutions we seek the one that satisfies two free boundary conditions:

```
In[6]:= FBCond1 = w[Γ] == 0
```

Out[6]= $5\,\Gamma^2 + (-c_1 - 4)\,\Gamma + c_1 == 0$

and

```
In[7]:= FBCond2 = (∂x w[x] /. x → Γ) == 0
```

Out[7]= $10\,\Gamma - c_1 - 4 == 0$

There are two solutions:

```
In[8]:= Solve[{FBCond1, FBCond2}, {C[1], Γ}]
```

Out[8]= $\left\{\left\{c_1 \to 2\left(3-\sqrt{5}\right), \Gamma \to \frac{1}{5}\left(5-\sqrt{5}\right)\right\}, \left\{c_1 \to 2\left(3+\sqrt{5}\right), \Gamma \to \frac{1}{5}\left(5+\sqrt{5}\right)\right\}\right\}$

but only one whose free boundary is inside (0, 1)

```
In[9]:= Sol = {Γ, w[x]} /. %[[1]]
```

Out[9]= $\left\{\frac{1}{5}\left(5-\sqrt{5}\right), 5\,x^2 + \left(-4-2\left(3-\sqrt{5}\right)\right)x + 2\left(3-\sqrt{5}\right)\right\}$

We use it to define the solution of the problem in the non-coincidence set

```
In[10]:= u[x_ /; x ≥ Sol[[1]]] := Sol[[2]]
```

extending it to be equal to the obstacle in the coincidence set

```
In[11]:= u[x_ /; x < Sol[[1]]] := 0
```

Finally, here is the plot of the solution (together with the free boundary point):

```
In[12]:= g = Graphics[{AbsolutePointSize[5],
             RGBColor[1, 0, 0], Point[{Sol[[1]], 0}]}];
         p = Plot[u[x], {x, -1, 1}, DisplayFunction → Identity];
         Show[p, g,
           DisplayFunction → $DisplayFunction, PlotRange → All];
```

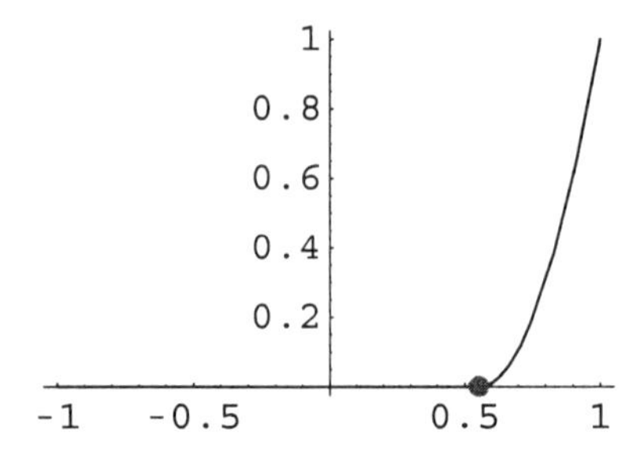

6.2.2.9 Computing Solution: Maximal Boundary Value Problem

```
In[15]:= Clear["Global`*"]
```

Now solve the same differential equation

```
In[16]:= v[x_] = u[x] /. DSolve[-∂{x,2} u[x] == -10, u[x], x][[1]]
```

Out[16]= $5x^2 + c_2 x + c_1$

with *two* boundary conditions: one as before at $x = 1$ and another on the variable point a inside $(-1, 1)$:

```
In[17]:= w[a_, x_] = v[x] /.
          Solve[{v[1] == 1, v[a] == 0}, {C[1], C[2]}][[1]]
```

Out[17]= $5x^2 - \frac{(5a^2 - 4)x}{a - 1} - \frac{4a - 5a^2}{a - 1}$

Also, extend these functions as zero elsewhere in (0, 1):

```
In[18]:= AuxSol[a_, x_] := If[x < a, 0, w[a, x]]
```

Here are the plots of some of these auxiliary functions:

```
In[19]:= l = Plot[Evaluate[
            (AuxSol[#1, x] &) /@Range[-1, .99, .3]],
           {x, -1, 1}, PlotRange → All];
```

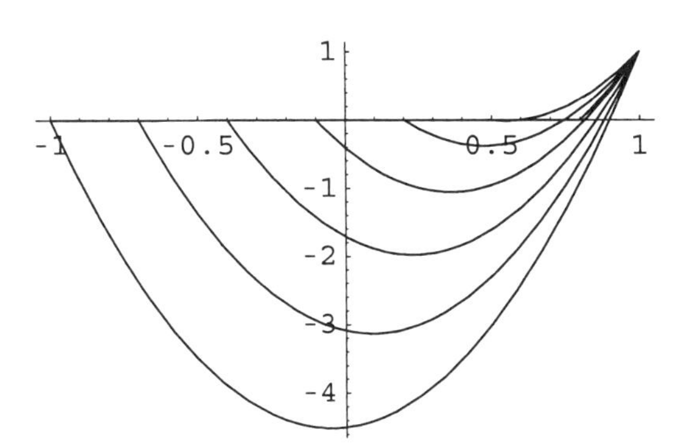

Neither of these functions is a solution of the obstacle problem since the second free boundary condition $\frac{\partial u(a)}{\partial x} = 0$ is not satisfied, or, which amounts to the same, neither of these functions belongs to $H^2(0, 1)$. Now, *maximize* these boundary value problem solutions with respect to the parameter a. To this end, in the present example it suffices to just find critical points:

```
In[20]:= z = Solve[∂a w[a, x] == 0, a]
```

Out[20]= $\{\{a \to \frac{1}{5}(5-\sqrt{5})\}, \{a \to \frac{1}{5}(5+\sqrt{5})\}\}$

Only the first critical point is inside (0, 1) and indeed yields the maximum among all $a \in (0, 1)$, and therefore we find the solution of the maximization problem to be

```
In[21]:= v[x_] = AuxSol[a, x] /. z[[1]]
```

Out[21]= $\text{If}[x < \frac{1}{5}(5-\sqrt{5}), 0, w(\frac{1}{5}(5-\sqrt{5}), x)]$

which is the same as the solution derived solving the free boundary value problem formulation, and which looks like, compared with the above auxiliary functions:

```
In[22]:= g = Graphics[{AbsolutePointSize[6],
             RGBColor[1, 0, 0], Point[{a /. z[[1]], 0}]}];
         p = Plot[v[x], {x, -1, 1}, PlotStyle →
             {RGBColor[1, 0, 0], AbsoluteThickness[2]},
            DisplayFunction → Identity];
         Show[p, g, l, DisplayFunction → $DisplayFunction,
           PlotRange → All];
```

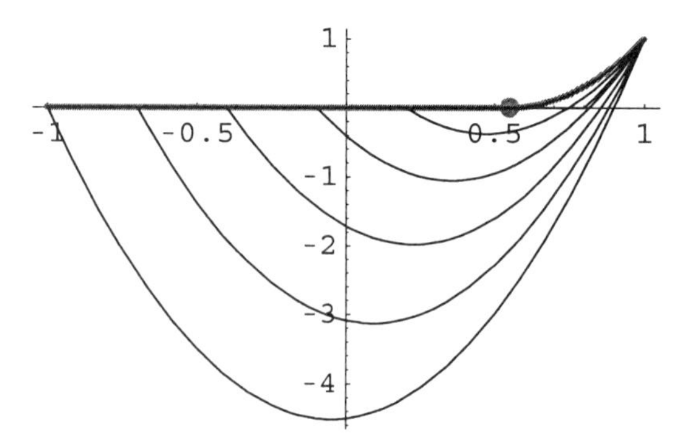

This, in addition to giving more insight, promises to be useful, since each auxiliary problem is linear, and therefore easy and efficient to solve either exactly or numerically, while maximality is numerically a better defined property than the second free boundary condition; $\frac{\partial u(a)}{\partial x} = 0$ will never hold exactly in numerical calculations. We shall use the adaptation of this method for our numerical solution of obstacle problems for Black–Scholes and Dupire PDEs arising in American put options.

6.2.2.10 Free Boundary Value Problem: Non-Uniqueness

Now we show that the free boundary value problem formulation is not enough in general to determine the unique solution of the obstacle problem. To this end, let the right-hand side this time be equal to

```
In[25]:= Clear["Global`*"]
         g = 10; f[x_] := 2 g UnitStep[-x] - g
```

whose plot looks like

```
In[27]:= Plot[f[x], {x, -1, 1}];
```

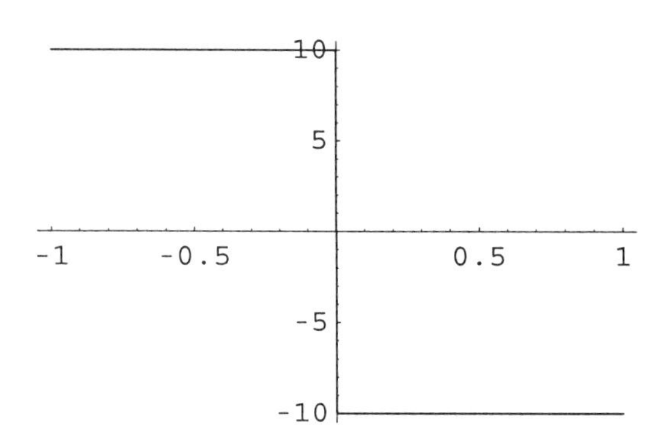

Solving the corresponding ODE for the data in $\{x < 0\}$, we get

```
In[28]:= v[x_] = u[x] /. DSolve[-∂{x,2} u[x] == g, u[x], x][[1]]
```

Out[28]= $-5x^2 + c_2 x + c_1$

while together with the boundary condition on $x = -1$, we get a solution in terms of one constant yet to be determined:

```
In[29]:= w[x_] = v[x] /. Solve[v[-1] == 0, C[2]][[1]] /. C[1] → A
```

Out[29]= $-5x^2 + (A-5)x + A$

Continuing the solution of the problem from the left to the right, now in $\{x > 0\}$, we first solve the ODE

```
In[30]:= v[x_] = u[x] /. DSolve[-∂{x,2} u[x] == -g, u[x], x][[1]]
```

Out[30]= $5x^2 + c_2 x + c_1$

and find constants A, c_1, c_2 and the first free boundary F such that the two pieces of the solution match smoothly at $x = 0$, and such that the free boundary condition holds at F:

```
In[31]:= Sol1 = {w[x], v[x], F} /.
            Solve[{v[0] == w[0], v'[0] == w'[0], v[F] == 0,
               v'[F] == 0}, {A, C[1], C[2], F}][[1]]
```

Out[31]= $\{-5x^2 + (-5 + 5(3 - 2\sqrt{2}))x + 5(3 - 2\sqrt{2}), 5x^2 + 10(1 - \sqrt{2})x + 5(3 - 2\sqrt{2}), -1 + \sqrt{2}\}$

Continuing the solution of the problem from the left to the right, now in $\{x > F\}$, we use the same general solution $v(x)$ of the ODE, but imposing the boundary condition at $x = 1$ and a free boundary condition at another free boundary point:

```
In[32]:= Sol2 =
          {v[x], F} /. Solve[{v[1] == 1, v[F] == 0, v'[F] == 0},
             {C[1], C[2], F}][[1]]
```

Out[32]= $\left\{5x^2 + 2\left(-5 + \sqrt{5}\right)x + 2\left(3 - \sqrt{5}\right), \frac{1}{5}\left(5 - \sqrt{5}\right)\right\}$

and now define the solution of the obstacle problem via

```
In[33]:= u[x_ /; -1 ≤ x < 0] = Sol1[[1]];
         u[x_ /; 0 ≤ x < Sol1[[3]]] = Sol1[[2]];
         u[x_ /; Sol1[[3]] ≤ x < Sol2[[2]]] = 0;
         u[x_ /; Sol2[[2]] ≤ x] = Sol2[[1]];
```

The plot of the solution looks like

```
In[34]:= FBs = (Graphics[{AbsolutePointSize[5],
               RGBColor[1, 0, 0], Point[{#1, 0}]}] &) /@
           {Sol1[[3]], Sol2[[2]]}; up = Plot[u[x],
           {x, -1, 1}, DisplayFunction → Identity,
           PlotLabel → "u[x]"];
         ups = Show[up, FBs,
            DisplayFunction → $DisplayFunction];
```

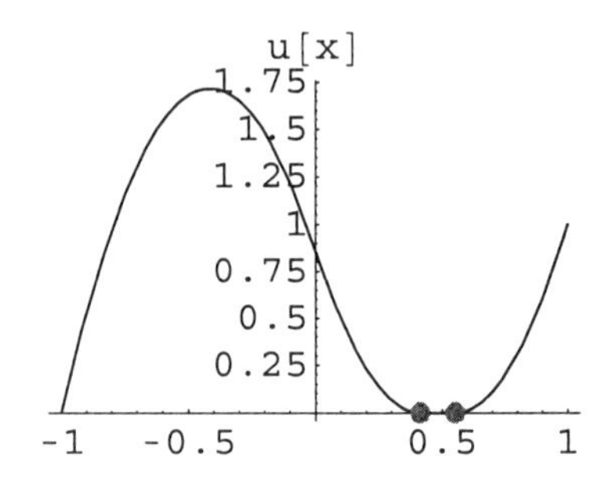

Now define

```
In[36]:= v[x_ /; x ≥ Sol2[[2]]] = u[x]; v[x_ /; x < Sol2[[2]]] = 0;
         up2 = Plot[v[x], {x, -1, 1}, DisplayFunction →
             Identity, PlotLabel → "v[x]"];
         Show[up2, FBs[[2]], DisplayFunction →
            $DisplayFunction, PlotRange → {0, 1.75}];
```

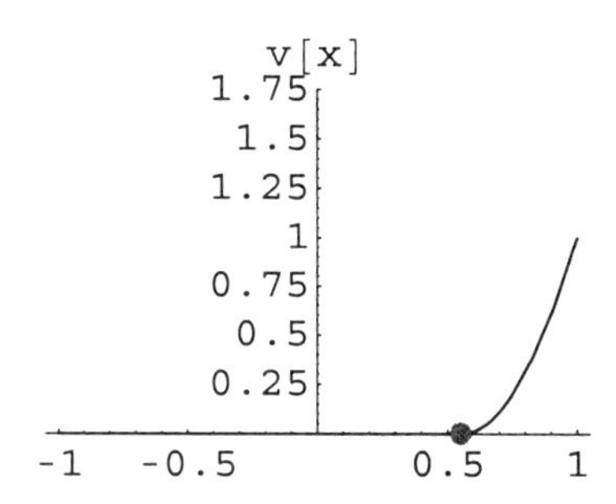

Both u and v fulfill all of the conditions in the free boundary value problem formulation. On the other hand, for example, in the semilinear PDE formulation 1

$$f^+(x)\, I_{\{u=0\}}(x) = 0$$

but

$$f^+(x)\, I_{\{v=0\}}(x) \neq 0.$$

Also, in semilinear PDE formulation 2, u is maximal, while v is not. Therefore only u is the solution of the obstacle problem, while both u and v are fulfilling the conditions of the free boundary value problem formulation.

6.2.2.11 Maximal Boundary Value Problem: Uniqueness

```
In[39]:= Clear[u, v]
```

Now we show that the above deficiency of the free boundary value formulation does not hold for the maximal boundary value formulation. The mathematical proof can be found in [55]. To this end solve the differential equation

```
In[40]:= v[x_] =
           u[x] /. DSolve[-∂{x,2} u[x] == f[x], u[x], x][[1]]
```

Out[40]= $10\,\theta(x)\,x^2 - 5\,x^2 + c_2\,x + c_1$

with *two* boundary conditions: one as before at $x = 1$ and another on the variable point a inside $(-1, 1)$:

```
In[41]:= w[a_, x_] = v[x] /.
            Solve[{v[1] == 1, v[a] == 0}, {C[1], C[2]}][[1]]
```

Out[41]= $10\,\theta(x)\,x^2 - 5\,x^2 - \dfrac{(10\,\theta(a)\,a^2 - 5\,a^2 - 4)\,x}{a-1} - \dfrac{-10\,\theta(a)\,a^2 + 5\,a^2 + 4\,a}{a-1}$

Also, extend these functions as zero (obstacle) elsewhere in $(0, 1)$:

```
In[42]:= AuxSol[a_, x_] := If[x < a, 0, w[a, x]]
```

Here are some functions

```
In[43]:= l = Plot[Evaluate[
           (AuxSol[#1, x] &) /@ Range[-1, .99, .01]],
          {x, -1, 1}, PlotRange → All,
          PlotStyle → RGBColor[.6, .6, .6]];
```

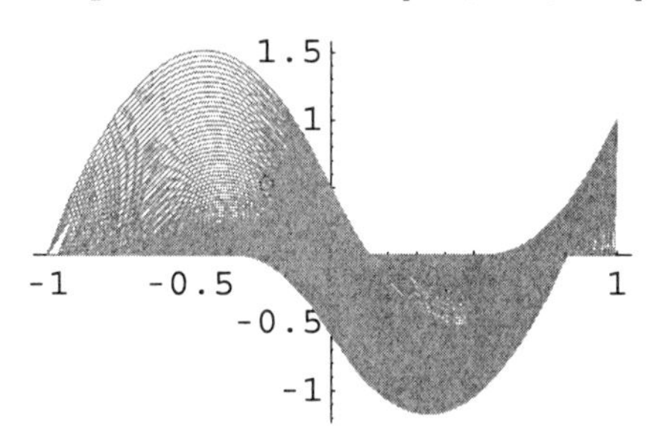

We can see that if the only maximization of these functions with respect to a was performed, one of the free boundary conditions would not be satisfied. The reason is that not all possible boundary value problems were considered. Therefore consider also the boundary conditions at $x = -1$ and another on the variable point a inside $(-1, 1)$:

```
In[44]:= w2[a_, x_] = v[x] /.
          Solve[{v[-1] == 0, v[a] == 0}, {C[1], C[2]}]〚1〛
```

Out[44]= $10\,\theta(x)\,x^2 - 5\,x^2 - \frac{5\,(2\,\theta(a)\,a^2 - a^2 + 1)\,x}{a+1} - \frac{5\,(2\,\theta(a)\,a^2 - a^2 - a)}{a+1}$

Again, extend these functions as zero elsewhere in $(0, 1)$:

```
In[45]:= AuxSol2[a_, x_] := If[x < a, w2[a, x], 0]
```

Here are some functions

```
In[46]:= l2 = Plot[Evaluate[
           (AuxSol2[#1, x] &) /@ Range[-1, .99, .01]],
          {x, -1, 1}, PlotRange → All,
          PlotStyle → RGBColor[.6, .6, .6]];
```

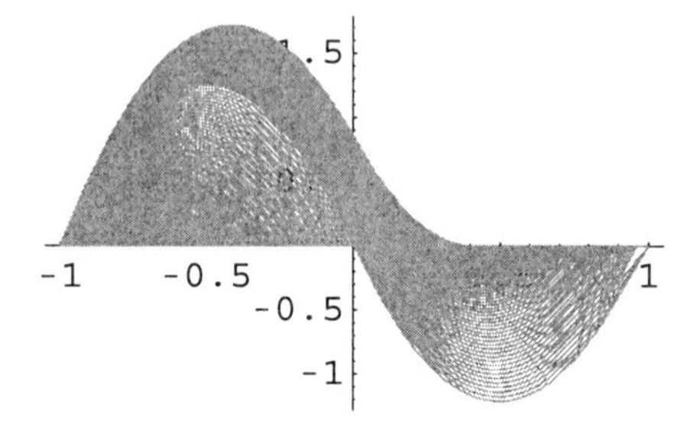

Putting them all together, i.e., taking the maximum of all possible solutions of boundary value problems, and superimposing the previously computed solution of the obstacle problem, we get

```
In[47]:= Show[l, l2, ups];
```

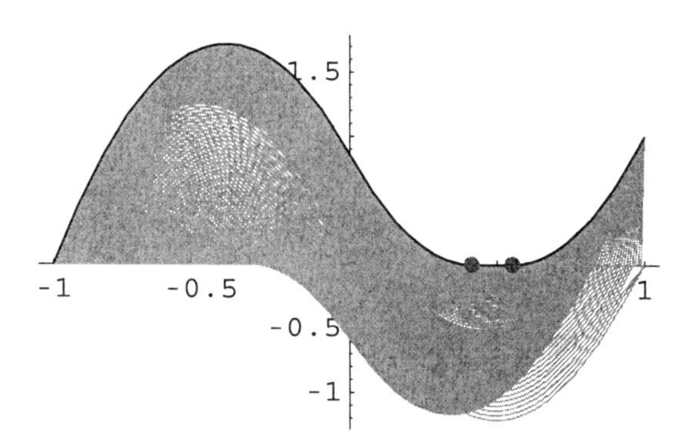

We conclude that, as opposed to the free boundary value problem formulation, the maximal boundary value formulation does identify the unique solution of the obstacle problem. As already stated we shall use the maximal boundary value formulation in our algorithm for solving the obstacle problem for the Black–Scholes and Dupire PDEs. The simple geometry of these problems will allow for very efficient calculation of the above maximal boundary value solutions.

6.2.3 "Perpetual American Options"

6.2.3.1 Steady State Obstacle Problem for Black–Scholes PDE

```
In[48]:= Clear["Global`*"]
```

Assume r, the interest rate, is constant, and D_0, the dividend rate, is equal to zero. Considering our discussion on various equivalent formulations of obstacle problems, it should not be surprising now that the obstacle problem (6.2.5) characterizing the value function of the optimal stopping problem for the Black–Scholes PDE is equivalent to the free boundary problem:

$$\frac{\partial\psi(t,S)}{\partial t} + \frac{1}{2} S^2 \frac{\partial^2\psi(t,S)}{\partial S^2}\sigma^2 + r S \frac{\partial\psi(t,S)}{\partial S} - r\psi(t,S) = 0$$

$$\text{in } N = \{\{t, S\}, \psi(t,S) > \phi(S)\}$$

$$\psi(t,S) \geq \phi(S)$$

$$\psi(t,S) = \phi(S) \quad \text{on } \Gamma = \partial N \cap \{\{t,S\}, t < T, S > 0\}$$

$$\frac{\partial\psi(t,S)}{\partial S} = \frac{\partial\phi(S)}{\partial S} \quad \text{on } \Gamma$$

$$\psi(T,S) = \phi(S).$$

Also, it is not difficult to see that if the solution of the Black–Scholes PDE happens to be above the obstacle ϕ, then it solves the above free boundary problem, with the free boundary being an empty set (or any of its more precise formulations, discussed in the previous section). In option terms, if the Black–Scholes option value is above the exercise profile at all times before the expiration T, then it is not reasonable to exercise the option early. This is exactly what happens in the case of calls (if $D_0 = 0$). Calls are never exercised early (under the condition that no dividends are paid). Looking back in Chapter 3, one can see that in the case of puts, on the other hand, it happens that the fair option price crosses over ϕ, i.e., the above observation about calls does not hold. So the above free boundary problem is not trivial in the case of puts. Therefore, we shall concentrate on puts, although everything that will be done could be extended easily to the case of calls when dividends are paid.

It is not possible to solve the above problem exactly, but rather only numerically. We shall consider such a problem later in the present chapter. But now we first consider a much simpler problem: we shall be looking for the steady state solution $\psi(t, S) = \psi(S)$. Intuitively, this would be the case when the expiration time is very large, and even infinite (perpetual options do not exist in practice; this is just a theoretical exercise).

So, the steady state problem is to find $\psi(S)$ such that:

$$\begin{aligned}
&\frac{1}{2} S^2 \frac{\partial^2 \psi(S)}{\partial S^2} \sigma^2 + r S \frac{\partial \psi(S)}{\partial S} - r \psi(S) = 0 \\
&\qquad \text{in } N = \{S, \psi(S) > \text{Max}(\text{Strike} - S, 0)\} \\
&\psi(S) \geq \text{Max}(\text{Strike} - S, 0) \\
&\psi(S) = \text{Max}(\text{Strike} - S, 0) \quad \text{on } \Gamma = \partial N \cap \{S, S > 0\} \\
&\frac{\partial \psi(S)}{\partial S} = -1 \text{ on } \Gamma.
\end{aligned} \tag{6.2.39}$$

The simplification is dramatic: there exists an easy explicit solution. Indeed, finding the general solution of the ODE above

```
In[49]:= ψ[S] /.
          DSolve[1/2 S^2 ∂{S,2} ψ[S] σ^2 + r S ∂S ψ[S] - r ψ[S] == 0,
            ψ[S], S][[1]]
```

Out[49]= $c_1 S^{-\frac{2r}{\sigma^2}} + c_2 S$

and knowing that the value of the put option must vanish as the price of the underlying S goes to infinity, we eliminate one of the constants of integration

```
In[50]:= solveODE[S_] = % /. C[2] → 0
```

Out[50]= $S^{-\frac{2r}{\sigma^2}} c_1$

The next issue is to find the free boundary and the unknown constant in such a manner so that the two free boundary conditions

```
In[51]:= FreeBoundaryCond = {solveODE[Γ] == Strike - Γ,
            (∂S solveODE[S] /. S → Γ) == -1}
```

Out[51]= $\left\{\Gamma^{-\frac{2r}{\sigma^2}} c_1 == \text{Strike} - \Gamma,\ -\frac{2\,r\,\Gamma^{-\frac{2r}{\sigma^2}-1}\, c_1}{\sigma^2} == -1\right\}$

are satisfied. So we shall need to fix data. Let

```
In[52]:= r = .05; σ = .4; Strike = 120;
```

Then

```
In[53]:= Sol = {Γ, solveODE[S]} /.
            Solve[FreeBoundaryCond, {C[1], Γ}][[1]]
```

Out[53]= $\left\{46.1538,\ \frac{809.949}{S^{0.625}}\right\}$

and we define the solution of the obstacle problem in the non-coincidence set

```
In[54]:= u[S_ /; S ≥ Sol[[1]]] := Sol[[2]]
```

extending it to be equal to the obstacle in the coincidence set

```
In[55]:= u[S_ /; S < Sol[[1]]] := Strike - S
```

The solution looks like

```
In[56]:= a = (Graphics[{RGBColor[0, 0, 1], AbsolutePointSize[5],
                Point[{Sol[[1]], #1}]}] &) /@ {Strike - Sol[[1]], 0};
         b = Plot[{Max[Strike - S, 0], u[S]}, {S, 0, 200},
            PlotStyle → {RGBColor[0, 0, 0], RGBColor[0, 0, 1]},
            DisplayFunction → Identity];
         Show[b, a, DisplayFunction → $DisplayFunction];
```

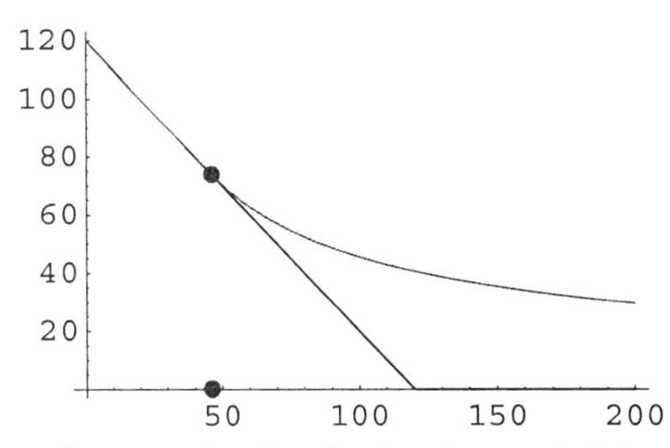

In particular the results are saying that for the given underlying stock volatility and interest rate

```
In[59]:= {σ, r}
Out[59]= {0.4, 0.05}
```

for the put option with

```
In[60]:= Strike
Out[60]= 120
```

owning the option does not make sense if the price of the underlying stock falls below

```
In[61]:= Sol⟦1⟧
Out[61]= 46.1538
```

no matter how far the expiration date is. This threshold is always lower than the strike price, of course, but gets higher and higher as expiration approaches. We shall see how to compute the threshold (free boundary) precisely, depending on the expiration date.

Now let us compare the computed steady state solution with the Black–Scholes fair European put option price with finite expiration time

```
In[62]:= << "CFMLab`BlackScholes`"
```

For example, let expiration times be:

```
In[63]:= Step = 10; Initial = 1;
         ExpTimes = Table[T, {T, Initial, 50, Step}]
Out[63]= {1, 11, 21, 31, 41}
```

Then

```
In[64]:= c =
           (Plot[Evaluate[VP[0, S, #1, Strike, r, σ]], {S, 0,
                200}, PlotRange → All, DisplayFunction →
                Identity, PlotStyle → RGBColor[
                 #1/100 + 1/2, #1/100 + 1/2, 1]] &) /@ ExpTimes;
         Show[b, a, c, DisplayFunction → $DisplayFunction,
           PlotRange → All];
```

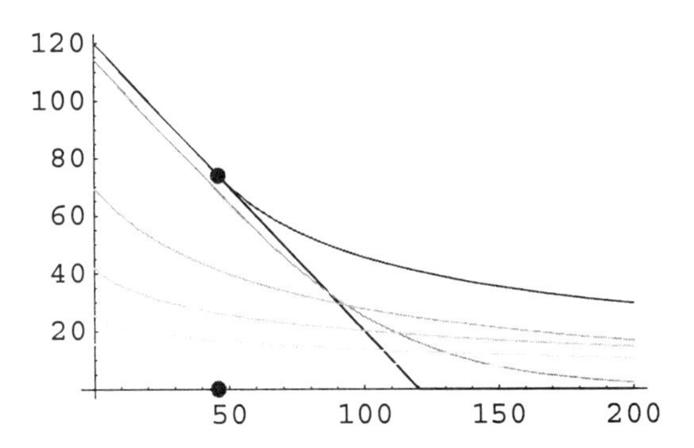

The reader is invited to make this plot with many more expiration times, to see the gap between taking a supremum over fixed-nonrandom (stopping) times, and the supremum over all (random) stopping times (the solution of the obstacle problem).

6.2.3.2 Steady State Obstacle Problem for Dupire PDE

Assume no dividends are paid. As discussed in Chapter 5, the fair put option price *as a function of expiration time and of the strike price* satisfies

$$
\begin{aligned}
&-\frac{\partial V(T,k)}{\partial T} + \frac{1}{2}k^2\frac{\partial^2 V(T,k)}{\partial k^2}\sigma(T,k)^2 - r\,k\,\frac{\partial V(T,k)}{\partial k} = 0\\
&\qquad \text{in } N = \{\{T,k\},\ V(T,k) > \mathrm{Max}(0, k-S)\}\\
&V(T,k) \geq \mathrm{Max}(0, k-S)\\
&V(T,k) = \mathrm{Max}(0, k-S)\\
&\qquad \text{on } \Gamma = \partial N \cap \{\{T,k\},\ t_0 < T,\ k > 0\}\\
&\frac{\partial V(T,k)}{\partial k} = 1 \ \text{ on } \Gamma
\end{aligned}
\tag{6.2.40}
$$

together with the *initial* condition

$$V(t_0, k) = \mathrm{Max}(0, k-S).$$

Again we find the steady state solution first (constant volatility), i.e., the solution of the free boundary value problem

$$
\begin{aligned}
&\frac{1}{2}k^2\frac{\partial^2 V(k)}{\partial k^2}\sigma^2 - r\,k\,\frac{\partial V(k)}{\partial k} = 0\\
&\qquad \text{in } N = \{S,\ V(k) > \mathrm{Max}(0, k-S)\}\\
&V(k) \geq \mathrm{Max}(0, k-S)\\
&V(k) = \mathrm{Max}(0, k-S) \ \text{ on } \Gamma = \partial N \cap \{k,\ k > 0\}\\
&\frac{\partial V(k)}{\partial k} = 1 \ \text{ on } \Gamma.
\end{aligned}
\tag{6.2.41}
$$

First, solve the ODE

```
In[66]:= ψ[k] /. DSolve[
             1/2 k^2 ∂{k,2} ψ[k] σ^2 - r k ∂k ψ[k] == 0, ψ[k], k][[1]]
```

Out[66]= $0.615385\, c_1\, k^{1.625} + c_2$

and knowing that the value of the put option must vanish as the strike price goes to zero, we eliminate one of the constants of integration:

```
In[67]:= solveODE[k_] = % /. C[2] → 0
```

Out[67]= $0.615385\, k^{1.625}\, c_1$

The next issue is to find the free boundary and the unknown constant in such a way so that the two free boundary conditions

```
In[68]:= FreeBoundaryCond =
           {solveODE[Γ] == Γ - S, (∂k solveODE[k] /. k → Γ) == 1}
```

Out[68]= $\{0.615385\, \Gamma^{1.625}\, c_1 == \Gamma - S,\ 1.\, \Gamma^{0.625}\, c_1 == 1\}$

For the same interest rate, volatility as before, and the current stock price,

```
In[69]:= S = 120;
```

we compute

```
In[70]:= Sol2 = {Γ, solveODE[k]} /.
            Solve[FreeBoundaryCond, {C[1], Γ}][[1]]
```

Out[70]= $\{312.,\ 0.0169942\, k^{1.625}\}$

and define the solution of the obstacle problem in the non-coincidence set

```
In[71]:= v[k_ /; k ≤ Sol2[[1]]] := Sol2[[2]]
```

extending it to be equal to the obstacle in the coincidence set

```
In[72]:= v[k_ /; k > Sol2[[1]]] := k - S
```

The solution looks like

```
In[73]:= f = (Graphics[{RGBColor[1, 0, 1],
                 AbsolutePointSize[5], Point[
                  {Sol2[[1]], #1}]}] &) /@ {Sol2[[1]] - S, 0};
         g = Plot[{Max[k - S, 0], v[k]}, {k, 0, 1.5 Sol2[[1]]},
            PlotStyle →
             {RGBColor[0, 0, 0], RGBColor[0, 0, 1]},
            DisplayFunction → Identity];
         Show[g, f, DisplayFunction → $DisplayFunction];
```

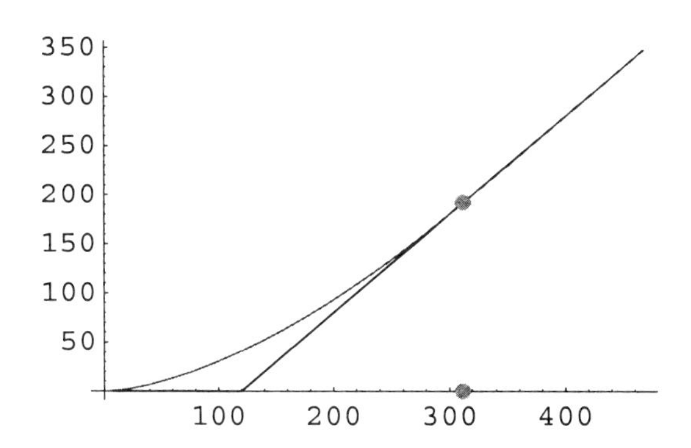

In particular, this results is saying that for the given underlying stock volatility and interest rate

```
In[76]:= {σ, r}
Out[76]= {0.4, 0.05}
```

for the underlying stock that has a current price

```
In[77]:= S
Out[77]= 120
```

owning the put option with the strike price above

```
In[78]:= Sol2[[1]]
Out[78]= 312.
```

does not make sense, no matter how far away the expiration date is. That threshold is always higher than the underlying stock price, of course, but gets lower and lower as expiration approaches, and again we shall see how to compute the threshold (free boundary), which depends on the expiration date when we solve the full time-dependent problem.

Again comparing the computed steady state solution of the obstacle problem with the Black–Scholes fair European put option price with finite expiration time, for expiration times:

```
In[79]:= Step = .1; Initial = .01;
         ExpTimes = Table[T, {T, Initial, 50, Step}];
```

we get

```
In[80]:= h = (Plot[Evaluate[VP[0, S, #1, k, r, σ]],
             {k, 0, 1.5 Sol2[[1]]}, PlotRange → All,
             DisplayFunction → Identity, PlotStyle →
```

```
    RGBColor[#1/50, #1/50, 1]] &) /@ExpTimes;
Show[g, h, f, DisplayFunction → $DisplayFunction,
  PlotRange → All];
```

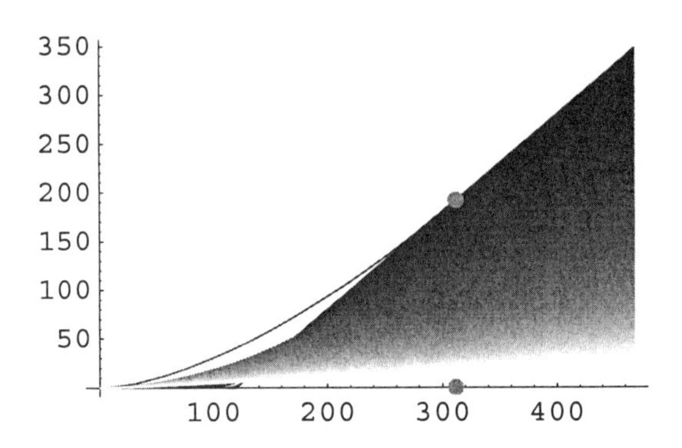

showing the gap between taking a supremum over fixed-nonrandom (stopping) times, and the supremum over all (random) stopping times (the solution of the obstacle problem).

6.2.4 Fast Numerical Solution of Obstacle Problems for Black–Scholes PDE

```
In[82]:= Clear["Global`*"]
```

```
In[83]:= << "LinearAlgebra`Tridiagonal`"
```

Again, the dynamics of the underlying stock is assumed to be

$$dS(t) = S(t)\,((a(t, S(t)) - D_0(t, S(t)))\,dt + \sigma(t, S(t))\,dB(t))$$
$$S(0) = s_0$$

where $a(t, S)$ is the appreciation rate, $D_0(t, S)$ is the dividend payment rate (dividend payment reduces the value of the stock), and $\sigma(t, S)$ is the volatility. It was shown that the fair call and put American option prices $V(t, S)$ obey, and moreover they are *the unique solutions of* the (backward parabolic) obstacle problem

$$\begin{aligned} \text{Max}\Big[&\frac{\partial V(t, S)}{\partial t} + \frac{1}{2} S^2 \frac{\partial^2 V(t, S)}{\partial S^2} \sigma(t, S)^2 \\ &+ (r(t, S) - D_0(t, S))\, S \frac{\partial V(t, S)}{\partial S} - r(t, S)\, V(t, S),\, \psi(S) \\ &- V(t, S)\Big] = \text{Max}[\mathcal{B}\, V(t, S),\, \psi(S) - V(t, S)] = 0 \end{aligned} \tag{6.2.42}$$

for $\{t, S\} \in Q_T = \{\{t, S\}, t < T, S > 0\}$, where T is the expiration time, together with the terminal condition

$$V(T, S) = \psi(S)$$

where $\psi(S) = \text{Max}(0, S - k)$ in the case of calls, and $\psi(S) = \text{Max}(0, k - S)$ in the case of puts. No boundary condition is needed due to degeneracy. Also, we use the notation

$$\mathcal{B} = \frac{\partial}{\partial t} + \frac{1}{2} S^2 \sigma(t, S)^2 \frac{\partial^2}{\partial S^2} + (r(t, S) - D_0(t, S))\, S \frac{\partial}{\partial S} - r(t, S) \tag{6.2.43}$$

—the Black-Scholes operator. The first issue that arises when attempting to solve such a problem numerically is the issue of truncation of the domain Q_T. So let $x_0 > 0$, $x_1 < \infty$ and $t_0 < T$ be fixed, and instead of the above equation considered in Q_T, we shall solve the same obstacle problem in $Q = \{(t, S), t_0 < t < T, x_0 < S < x_1\}$. This implies the necessity of imposing (truncated) boundary conditions on $S = x_0$ and $S = x_1$, which together with the terminal condition can be stated as

$$V(t, S) = \psi(S)$$

on $\partial_{\mathcal{B}} Q$, the backward parabolic boundary of Q, which is equal to $\partial_{\mathcal{B}} Q = \{T\} \times (x_0, x_1) \bigcup (t_0, T) \times \{x_0, x_1\}$. The above parabolic obstacle problem is discretized in the time direction first. So, let $dt = \frac{T - t_0}{N_t}$, for some integer N_t, and consider a sequence of N_t one-dimensional obstacle problems

$$\begin{aligned}
0 = \text{Max}\Big[& \frac{V(t + dt, S) - V(t, S)}{dt} + \frac{1}{2} S^2 \frac{\partial^2 V(t, S)}{\partial S^2} \sigma(t, S)^2 \\
& + (r(t, S) - D_0(t, S))\, S \frac{\partial V(t, S)}{\partial S} - r(t, S)\, V(t, S), \psi(S) - V(t, S)\Big] \\
= \text{Max}\Big[& \frac{V(t + dt, S) - V(t, S)}{dt} + a_2(t, S) \frac{\partial^2 V(t, S)}{\partial S^2} \\
& + a_1(t, S) \frac{\partial V(t, S)}{\partial S} + a_0(t, S)\, V(t, S), \psi(S) - V(t, S)\Big]
\end{aligned}$$

for

$$\begin{aligned}
a_2(t, S) &= \frac{1}{2} S^2 \sigma(t, S)^2 \\
a_1(t, S) &= (r(t, S) - D_0(t, S))\, S \\
a_0(t, S) &= -r(t, S)
\end{aligned}$$

or, which is more readable,

$$\begin{aligned}
\text{Max}\Big[& a_2(t, S) \frac{\partial^2 V(t, S)}{\partial S^2} + a_1(t, S) \frac{\partial V(t, S)}{\partial S} \\
& + \left(a_0(t, S) - \frac{1}{dt}\right) V(t, S) + \frac{V(t + dt, S)}{dt}, \psi(S) - V(t, S)\Big] = 0
\end{aligned} \tag{6.2.44}$$

together with the boundary conditions

$$V(t, x_0) = \psi(x_0)$$
$$V(t, x_1) = \psi(x_1)$$

for $t = t_1 - dt, t_1 - 2\,dt, \ldots, t_0 + dt, t_0$. Of course, this sequence of problems is initialized by the terminal condition

$$V(t_1, S) = \psi(S)$$

for $x_0 < S < x_1$. In solving each of these one-dimensional obstacle problems we shall not use the usual Gauss–Seidel method (see [26]), but instead we shall use in a substantial manner the geometry of the free boundary, and more precisely, its monotonicity as a function of time (heuristically, the option's exercise region shrinks as time approaches the expiration), to design an original, more efficient, and precise algorithm, based on the above introduced the maximal boundary value formulation of the obstacle problem.

To fix ideas, consider American calls. Introduce the data

```
In[84]:= σ[t_, S_] := .5; r[t_, S_] := .05;
         D0[t_, S_] := .04; s0 = 20; s1 = 200;
         Ns = 200; t0 = 0; t1 = 3/12; Nt = 200; k = 100;
```

the corresponding coefficients

```
In[85]:= a2[t_, S_] := 1/2 σ[t, S]^2 S^2
```

```
In[86]:= a1[t_, S_] := (r[t, S] - D0[t, S]) S
```

```
In[87]:= a0[t_, S_] := -r[t, S]
```

and the obstacle, as well as the (backward) parabolic boundary value

```
In[88]:= p[t_, S_] := Max[S - k, 0]
```

Compute the grid

```
In[89]:= ds = (s1 - s0)/Ns; ss = N[Range[s0, s1, ds]];
         dt = (t1 - t0)/Nt; tt = N[Range[t1, t0, -dt]];
         Arguments = Outer[{#1, #2} &, tt, ss];
```

as well as the above data on the grid:

```
In[90]:= PPList = ((p @@ #1 &) /@ #1 &) /@ Arguments;
```

```
In[91]:= A2List = ((a2 @@ #1 &) /@ #1 &) /@ Drop[Arguments, 1];
```

```
In[92]:= A1List = ((a1 @@ #1 &) /@ #1 &) /@ Drop[Arguments, 1];
```

```
In[93]:= A0List = ((a0 @@ #1 &) /@ #1 &) /@ Drop[Arguments, 1];

In[94]:= A1PlusList = ((Max[#1, 0] &) /@ #1 &) /@ A1List;

In[95]:= A1MinusList = ((-Min[#1, 0] &) /@ #1 &) /@ A1List;
```

Compute the terminal condition, which is the initial condition for our sequence of one-dimensional obstacle problems:

```
In[96]:= Prev = (p @@ #1 &) /@ Arguments[[1]];
```

as well as the (approximation of the) "initial condition" for the free boundary

```
In[97]:= FB0 = {Select[ss, #1 ≥ k &][[1]],
           Position[ss, Select[ss, #1 ≥ k &][[1]]][[1, 1]]}

Out[97]= {100.1, 90}
```

where the first number is the approximate value of the free boundary point, and the second number is its position in the S-grid. It is time to compute the entries for the tri-diagonal solver we are going to use (they will have to be modified in each time-step, depending where the free boundary point is). What is happening, of course, is that the above one-dimensional obstacle problem is discretized into a *system of non-linear algebraic problems*

$$\begin{aligned}\mathrm{Max}\Big[& a_2(t,S)\,\frac{V(t,S-dS)-2\,V(t,S)+V(t,dS+S)}{dS^2} \\ &+(a_1(t,S))^+\,\frac{V(t,S)-V(t,S-dS)}{dS}-(a_1(t,S))^-\,\frac{V(t,S+dS)-V(t,S)}{dS} \\ &+\left(a_0(t,S)-\frac{1}{dt}\right)V(t,S)+\frac{V(t+dt,S)}{dt},\ \psi(S)-V(t,S)\Big]=0\end{aligned}$$

$$V(t,x_0)=p(t,x_0)$$

$$V(t,x_1)=p(t,x_1)$$

where $a_1(t,S)=(a_1(t,S))^+-(a_1(t,S))^-$, $(a_1(t,S))^{\pm}\geq 0$. We shall use the maximal boundary value formulation to solve this problem, utilizing the special geometry of the free boundary; for each time-iterate there is a single free boundary point only, and also, due to the monotonicity of the free boundary with respect to time, it is known where it is in comparison to the previously computed one. Also for solving each boundary value problem (not necessarily the maximal one) we shall use the tri-diagonal solver, as in the case of equations in Chapter 5. To this end we prepare the entries for the tri-diagonal solver; they will have to be adjusted in each boundary value calculation to take into the account the location of the free-boundary-point candidate, as well as the previously (timewise) computed solution.

```
In[98]:= LowerDiagonal = (ReplacePart[#1, 0, -1] &) /@
           (Drop[#1, 1] &) /@ (A2List/ds^2 - A1PlusList/ds);

In[99]:= Diagonal = (ReplacePart[#1, 1, 1] &) /@
           (ReplacePart[#1, 1, -1] &) /@ (-(2 A2List)/ds^2 +
             A1PlusList/ds + A1MinusList/ds + A0List - 1/dt);

In[100]:= UpperDiagonal = (ReplacePart[#1, 0, 1] &) /@
            (Drop[#1, -1] &) /@ (A2List/ds^2 - A1MinusList/ds);

In[101]:= FoldingList = Transpose[{LowerDiagonal,
              Diagonal, UpperDiagonal, Drop[PPList, 1]}];
```

Next, we define a function that is searching for the maximal among computed solutions of the boundary value problems corresponding to different candidates for the free boundary; this search is easy due to the geometry of the problem:

```
In[102]:= MaximalBoundaryValueSolver[FB0_, LowerDiagonal_,
             Diagonal_, UpperDiagonal_, Prev_] :=
            Module[{y, x}, y = {}; Do[x = {FB0 + {i ds, i},
                BoundaryValueSolver[FB0 + {i ds, i}, LowerDiagonal,
                 Diagonal, UpperDiagonal, Prev]}; y = Append[y, x];
              If[i ≥ 1 && y[[-2, 2]][[FB0[[2]]]] > y[[-1, 2]][[FB0[[2]]]],
               Break[]], {i, 0, 1000}]; y[[-2]]]
```

On the other hand, each boundary value problem needs to be solved, which is done by the following function: For a candidate (single) free boundary point, the corresponding coincidence set is easy to identify; in the coincidence set the solution is equal to the obstacle, which, due to the geometry of the free boundary, is equal to the previously (timewise) computed solution; the equality to the obstacle is handled in the same way as the equality to the boundary condition, and therefore the system matrix and the right-hand side are adjusted accordingly; the rest is the same as in Chapter 5—just a tri-diagonal solve):

```
In[103]:= BoundaryValueSolver[FB0_, LowerDiagonal_,
             Diagonal_, UpperDiagonal_, Prev_] :=
            Module[{ModifiedLowerDiagonal, ModifiedDiagonal,
              ModifiedUpperDiagonal, RHS, ModifiedRHS, k},
             k = FB0[[2]]; RHS = -Prev/dt; ModifiedLowerDiagonal =
              ReplaceAtPlaces[LowerDiagonal,
               Table[0, {Length[LowerDiagonal] + 1 - k}],
               Range[k - 1, Length[LowerDiagonal] - 1]];
             ModifiedDiagonal = ReplaceAtPlaces[Diagonal,
               Table[1, {Length[Diagonal] - k}], Range[k,
```

```
      Length[Diagonal] - 1]]; ModifiedUpperDiagonal =
    ReplaceAtPlaces[UpperDiagonal,
     Table[0, {Length[UpperDiagonal] + 1 - k}],
     Range[k, Length[UpperDiagonal]]];
   ModifiedRHS = ReplaceAtPlaces[RHS, (Prev⟦#1⟧ &) /@
      Range[k, Length[RHS]], Range[k, Length[RHS]]];
   TridiagonalSolve[ModifiedLowerDiagonal,
    ModifiedDiagonal,
    ModifiedUpperDiagonal, ModifiedRHS]]
 ReplaceAtPlaces[expr_, new_, places_] :=
  Fold[ReplacePart[#1, #2⟦1⟧, #2⟦2⟧] &,
   expr, Transpose[{new, places}]]
```

It is time to do the actual computation:

```
In[105]:= Timing[ObstacleProblemSolution =
             FoldList[MaximalBoundaryValueSolver[
                #1⟦1⟧, #2⟦1⟧, #2⟦2⟧, #2⟦3⟧, #1⟦2⟧] &,
              {FB0, PPList⟦1⟧}, FoldingList];]
```

```
Out[105]= {5.27 Second, Null}
```

```
In[106]:= AmericanCall =
            Interpolation[({#1⟦1, 1⟧, #1⟦1, 2⟧, #1⟦2⟧} &) /@
              Transpose[{(Flatten[#1, 1] &)[Arguments],
                Flatten[(#1⟦2⟧ &) /@ ObstacleProblemSolution]}],
             InterpolationOrder → {1, 2}];
```

The solution looks like this

```
In[107]:= ACPlot = Plot3D[AmericanCall[t, S], {t, t0, t1},
             {S, s0, s1}, Mesh → False, PlotPoints → 50];
```

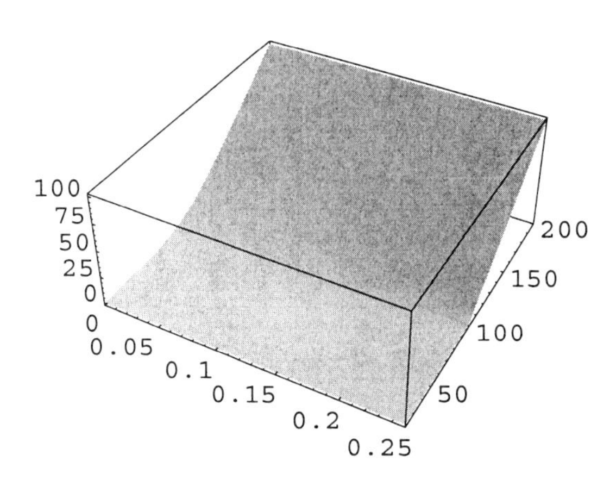

and it has a free boundary

```
In[108]:= FreeBoundary = Interpolation[Transpose[
              {tt, (#1[[1, 1]] &) /@ObstacleProblemSolution}],
             InterpolationOrder → 1];
```

that looks like this

```
In[109]:= FBPlot = Plot[FreeBoundary[t], {t, t0, t1},
             PlotRange → {0, s1}, PlotPoints → Nt];
```

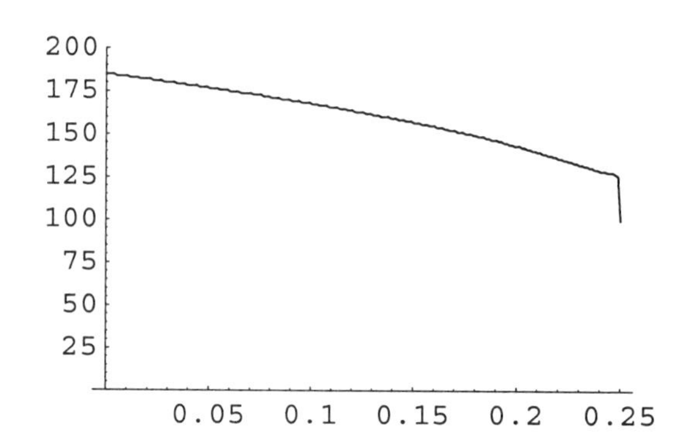

Notice how singular the problem is (and the free boundary) at the expiration time —knowing that the computational error is only due the discretization is quite comforting. One can also look at the solution from various angles:

```
In[110]:= PP3D = ParametricPlot3D[{t, FreeBoundary[t], 0},
              {t, t0, t1}, DisplayFunction → Identity];
          << "Graphics`Animation`";
          qw = Show[ACPlot, PP3D, Boxed → False, Axes → False,
            DisplayFunction → Identity]; SpinShow[qw];
```

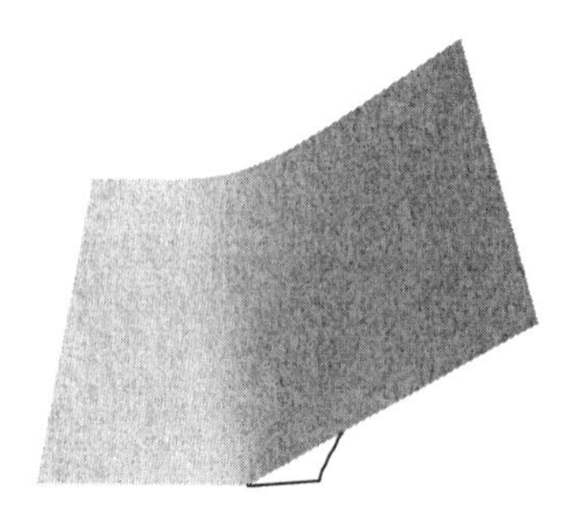

The above procedure solved backward obstacle problem for the Black–Scholes operator and for the call terminal and boundary data. It might be useful to wrap the whole procedure into a function called, for example, BlackScholesObstacleCall and to place it into the package CFMLab`NumericalBlackScholes`. We shall not do this here: function BlackScholesObstacleCall and the analogous function for the case of puts BlackScholesObstaclePut should be available later at http://CFMLab.com.

6.2.5 Fast Numerical Solution of Obstacle Problems for the Dupire PDE

```
In[113]:= Clear["Global`*"]
```

```
In[114]:= << "LinearAlgebra`Tridiagonal`"
```

Again, the dynamics of the underlying stock is assumed to be

$$dS(t) = S(t)\ ((a(t, S(t)) - D_0\,(t, S(t)))\ dt + \sigma(t, S(t))\, dB\,(t))$$
$$S(0) = s_0$$

where $a(t, S)$ is the appreciation rate, $D_0(t, S)$ is the dividend payment rate (dividend payment reduces the value of the stock), and $\sigma(t, S)$ is the volatility. Similarly, as in the case of European options, fair call and put American option prices, when considered as functions of expiration time T, and the strike price k, i.e., $V(T, k)$ obey, and moreover they are *the unique solutions of* the (forward parabolic) obstacle problem

$$\begin{aligned}\text{Max}\Big[-\frac{\partial V(T,k)}{\partial T} + \frac{1}{2}k^2\,\frac{\partial^2 V(T,k)}{\partial k^2}\,\sigma(T,k)^2 - (r(T,k) - D_0\\ (T,k))\,k\,\frac{\partial V(T,k)}{\partial k} - D_0\,(T,k)\ V(T,k),\ \psi(k) - V(T,k)\Big]\\ = \text{Max}[\mathcal{D}\,V(T,k),\ \psi(k) - V(T,k)] = 0\end{aligned} \tag{6.2.45}$$

for $\{T, k\} \in Q_{t_0} = \{\{T, k\}, T > t_0, k > 0\}$, where t_0 is the current time, together with the initial condition

$$V(t_0, k) = \psi(k)$$

where $\psi(k) = \text{Max}(0, S - k)$ in the case of calls, and $\psi(k) = \text{Max}(0, k - S)$ in the case of puts, and where S is the current underlying price. No boundary condition is needed due to the degeneracy. Also, we use notation

$$\begin{aligned}\mathcal{D} = -\frac{\partial}{\partial T} + \frac{1}{2}\,k^2\,\sigma(T,k)^2\,\frac{\partial^2}{\partial k^2}\\ -(r(T,k) - D_0\,(T,k))\,k\,\frac{\partial}{\partial k} - D_0\,(T,k)\end{aligned} \tag{6.2.46}$$

for the Dupire operator. Again we truncate the domain Q_{t_0} into $Q = \{\{T, k\}, t_0 < T < t_1, x_0 < k < x_1\}$. This implies the necessity of imposing (truncated) boundary conditions on $k = x_0$ and $k = x_1$, which together with the initial condition can be stated as

$$V(T, k) = \psi(k)$$

on $\partial_{\mathcal{D}} Q$, the (forward) parabolic boundary of Q, which is equal to $\partial_{\mathcal{D}} Q = \{t_0\} \times (x_0, x_1) \bigcup (t_0, t_1) \times \{x_0, x_1\}$. The above parabolic obstacle problem is discretized first in the time direction. So, let $dT = \frac{t_1 - t_0}{N_T}$, for some integer N_T, and consider a sequence of N_T one-dimensional obstacle problems

$$\begin{aligned}
0 = \mathrm{Max}\Big[& -\frac{V(T, k) - V(T - dT, k)}{dT} + \frac{1}{2}\, k^2\, \frac{\partial^2 V(T, k)}{\partial k^2}\, \sigma(T, k)^2 \\
& - (r(T, k) - D_0(T, k))\, k\, \frac{\partial V(T, k)}{\partial k} - D_0(T, k)\, V(T, k), \\
& \psi(k) - V(T, k)\Big] = \mathrm{Max}\Big[\frac{V(T - dT, k) - V(T, k)}{dT} + a_2(T, k) \\
& \frac{\partial^2 V(T, k)}{\partial k^2} + a_1(T, k)\, \frac{\partial V(T, k)}{\partial k} + a_0(T, k)\, V(T, k), \\
& \psi(k) - V(T, k)\Big]
\end{aligned} \tag{6.2.47}$$

for

$$\begin{aligned}
a_2(T, k) &= \frac{1}{2}\, k^2\, \sigma(T, k)^2 \\
a_1(T, k) &= -(r(T, k) - D_0(T, k))\, k \\
a_0(T, k) &= -D_0(T, k)
\end{aligned}$$

or, which is more readable,

$$\begin{aligned}
\mathrm{Max}\Big[& a_2(T, k)\, \frac{\partial^2 V(T, k)}{\partial k^2} + a_1(T, k)\, \frac{\partial V(T, k)}{\partial k} + \left(a_0(T, k) - \frac{1}{dT}\right) V(T, k) \\
& + \frac{V(T - dT, k)}{dt}, \psi(k) - V(T, k)\Big] = 0
\end{aligned}$$

together with the boundary conditions

$$\begin{aligned}
V(T, x_0) &= \psi(x_0) \\
V(T, x_1) &= \psi(x_1)
\end{aligned}$$

for $t = t_0 + dt,\ t_0 + 2\, dt,\ \ldots,\ t_1 - dt,\ t_1$. Of course, this sequence of problems is initialized by the initial condition

$$V(t_0, k) = \psi(k)$$

for $x_0 < k < x_1$. We use the same method as in the case of the Black–Scholes obstacle problem.

To fix ideas, consider American puts. Introduce the data

```
In[115]:= σ[t_, S_] := .5; r[t_, S_] := .05;
          D0[t_, S_] := .0; x0 = 20; x1 = 200;
          Nx = 200; t0 = 0; t1 = 3/12; Nt = 200; S = 100;
```

and the corresponding coefficients

```
In[116]:= a2[T_, k_] := 1/2 σ[T, k]^2 k^2

In[117]:= a1[T_, k_] := -(r[T, k] - D0[T, k]) k

In[118]:= a0[T_, k_] := -D0[T, k]
```

the obstacle, as well as the (forward) parabolic boundary value

```
In[119]:= p[T_, k_] := Max[k - S, 0]
```

Compute the grid

```
In[120]:= ds = (x1 - x0)/Nx; ss = N[Range[x0, x1, ds]];
          dt = (t1 - t0)/Nt; tt = N[Range[t0, t1, dt]];
          Arguments = Outer[{#1, #2} &, tt, ss];
```

as well as the above data on the grid:

```
In[121]:= PPList = ((p @@ #1 &) /@ #1 &) /@ Arguments;

In[122]:= A2List = ((a2 @@ #1 &) /@ #1 &) /@ Drop[Arguments, 1];

In[123]:= A1List = ((a1 @@ #1 &) /@ #1 &) /@ Drop[Arguments, 1];

In[124]:= A0List = ((a0 @@ #1 &) /@ #1 &) /@ Drop[Arguments, 1];

In[125]:= A1PlusList = ((Max[#1, 0] &) /@ #1 &) /@ A1List;

In[126]:= A1MinusList = ((-Min[#1, 0] &) /@ #1 &) /@ A1List;
```

Compute the initial condition, which is the initial condition for our sequence of one-dimensional obstacle problems:

```
In[127]:= Prev = (p @@ #1 &) /@ Arguments[[1]];
```

as well as the (approximation of the) "initial condition" for the free boundary

```
In[128]:= FB0 = {Select[ss, #1 ≥ S &][[1]],
            Position[ss, Select[ss, #1 ≥ S &][[1]]][[1, 1]]}

Out[128]= {100.1, 90}
```

Proceeding, we define as before

```
In[129]:= LowerDiagonal = (ReplacePart[#1, 0, -1] &) /@
            (Drop[#1, 1] &) /@ (A2List/ds^2 - A1PlusList/ds);
```

```
In[130]:= Diagonal = (ReplacePart[#1, 1, 1] &) /@
            (ReplacePart[#1, 1, -1] &) /@ (-(2 A2List)/ds^2 +
              A1PlusList/ds + A1MinusList/ds + A0List - 1/dt);
```

```
In[131]:= UpperDiagonal = (ReplacePart[#1, 0, 1] &) /@
            (Drop[#1, -1] &) /@ (A2List/ds^2 - A1MinusList/ds);
```

```
In[132]:= FoldingList = Transpose[{LowerDiagonal,
            Diagonal, UpperDiagonal, Drop[PPList, 1]}];
```

and

```
In[133]:= MaximalBoundaryValueSolver[FB0_,
           LowerDiagonal_, Diagonal_, UpperDiagonal_,
           Prev_] := Module[{y, x}, y = {};
           Do[x = {FB0 + {i ds, i}, BoundaryValueSolver[
               FB0 + {i ds, i}, LowerDiagonal, Diagonal,
               UpperDiagonal, Prev]}; y = Append[y, x];
            If[i ≥ 1 && y[[-2, 2]][[FB0[[2]]]] > y[[-1, 2]][[FB0[[2]]]],
             Break[]], {i, 0, 1000}]; y[[-2]]]
```

as well as

```
In[134]:= BoundaryValueSolver[FB0_, LowerDiagonal_,
            Diagonal_, UpperDiagonal_, Prev_] :=
           Module[{ModifiedLowerDiagonal, ModifiedDiagonal,
             ModifiedUpperDiagonal, RHS, ModifiedRHS, k},
            k = FB0[[2]]; RHS = -Prev/dt; ModifiedLowerDiagonal =
             ReplaceAtPlaces[LowerDiagonal,
              Table[0, {Length[LowerDiagonal] + 1 - k}],
              Range[k - 1, Length[LowerDiagonal] - 1]];
            ModifiedDiagonal = ReplaceAtPlaces[Diagonal,
              Table[1, {Length[Diagonal] - k}], Range[k,
               Length[Diagonal] - 1]]; ModifiedUpperDiagonal =
             ReplaceAtPlaces[UpperDiagonal,
              Table[0, {Length[UpperDiagonal] + 1 - k}],
              Range[k, Length[UpperDiagonal]]];
```

```
        ModifiedRHS = ReplaceAtPlaces[RHS, (Prev[[#1]] &) /@
          Range[k, Length[RHS]], Range[k, Length[RHS]]];
        TridiagonalSolve[ModifiedLowerDiagonal,
         ModifiedDiagonal,
         ModifiedUpperDiagonal, ModifiedRHS]];
     ReplaceAtPlaces[expr_, new_, places_] :=
       Fold[ReplacePart[#1, #2[[1]], #2[[2]]] &,
        expr, Transpose[{new, places}]];
```

It is time to do the actual computation:

```
In[136]:= Timing[DupireObstacleProblemSolution =
            FoldList[MaximalBoundaryValueSolver[
               #1[[1]], #2[[1]], #2[[2]], #2[[3]], #1[[2]]] &,
             {FB0, PPList[[1]]}, FoldingList];]
```

Out[136]= {6.54 Second, Null}

```
In[137]:= DupirePut =
           Interpolation[({#1[[1, 1]], #1[[1, 2]], #1[[2]]} &) /@
             Transpose[{(Flatten[#1, 1] &)[Arguments], Flatten[
                (#1[[2]] &) /@ DupireObstacleProblemSolution]}],
            InterpolationOrder → {1, 2}];
```

The solution looks like this:

```
In[138]:= ACPlot = Plot3D[DupirePut[T, k], {T, t0, t1},
            {k, x0, x1}, Mesh → False, PlotPoints → 50];
```

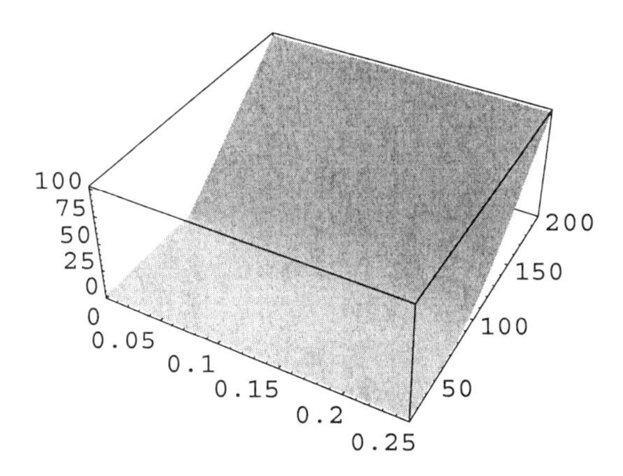

and it has a free boundary

```
In[139]:= FreeBoundary = Interpolation[Transpose[
             {tt, (#1[[1, 1]] &) /@ DupireObstacleProblemSolution}],
           InterpolationOrder → 1];
```

that looks like this:

```
In[140]:= FBPlot =
            Plot[FreeBoundary[t], {t, t0, t1}, PlotRange → {0, x1},
             PlotPoints → Nt, AxesOrigin → {t1, 0}];
```

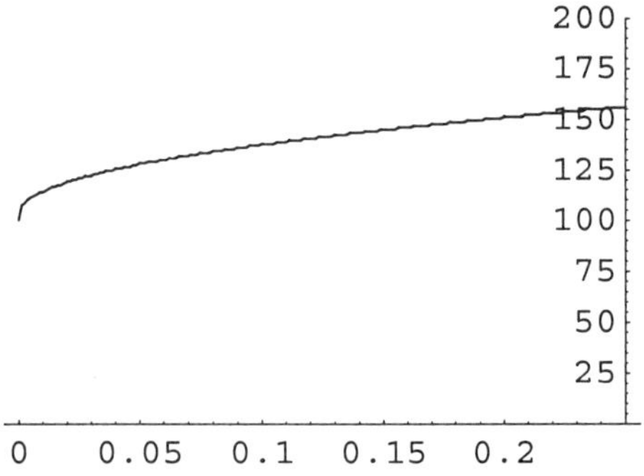

Finally, we indulge again to see a movie

```
In[141]:= PP3D = ParametricPlot3D[{t, FreeBoundary[t], 0},
              {t, t0, t1}, DisplayFunction → Identity];
          << "Graphics`Animation`";
          qw = Show[ACPlot, PP3D, Boxed → False, Axes → False,
            DisplayFunction → Identity]; SpinShow[qw];
```

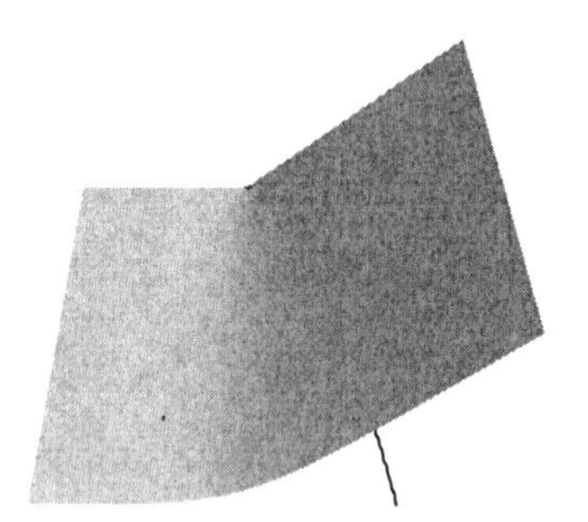

All of the essentials in the above were wrapped into a function DupireObstaclePut which is placed in the package CFMLab`NumericalBlackScholes` for future reference. On the other hand, function DupireObstacleCall might be available later at http://CFMLab.com.

6.3 General Implied Volatility for American Options

6.3.1 Implied Volatility via Optimal Control of Obstacle Problems

Again, the dynamics of the underlying stock is assumed to be

$$\begin{aligned} dS(t) &= S(t)\,(a(t, S(t))\,dt + \sigma(t, S(t))\,dB(t)) \\ S(0) &= s_0 \end{aligned}$$

i.e., for simplicity, we assume no dividends are paid. The state equations of the soon to be defined optimal control problem is the corresponding "Dupire obstacle equation":

$$\begin{aligned} \text{Max}\Big[&-\frac{\partial V(T, k)}{\partial T} + \frac{1}{2}\,k^2\,\frac{\partial^2 V(T, k)}{\partial k^2}\,\sigma(T, k)^2 - r\,k\,\frac{\partial V(T, k)}{\partial k}, \\ &\psi(k) - V(T, k)\Big] = 0 \end{aligned} \tag{6.3.1}$$

for $\{T, k\} \in Q_{t_0} = \{\{T, k\},\ T > t_0,\ k > 0\}$, where t_0 is the current time, together with the initial condition

$$V(t_0, k) = \psi(k) = \text{Max}(0, k - S). \tag{6.3.2}$$

As before, we truncate the domain into, say $Q = \{\{T, k\};\ t_0 < T < t_1,\ k_0 < k < k_1\}$, for some $0 < k_0 < k_1$, and introduce boundary conditions

$$V(T, k_0) = 0,\ V(T, k_1) = k_1 - S. \tag{6.3.3}$$

Let the class of admissible volatilities be the same as before:

$$\begin{aligned} K = \Big\{ &\sigma \in L^2((t_0, t_1) \times (k_0, k_1)),\ \sigma(T, k) \geq \sigma_{\min}, \\ &\int_{k_0}^{k_1} \int_{t_0}^{t_1} \left(\left(\frac{\partial \sigma(T, k)}{\partial k} \right)^2 + \sigma(T, k)^2 \right) dT\,dk < \infty \Big\}. \end{aligned} \tag{6.3.4}$$

For each $\sigma \in K$, let $V(\sigma)$ denote the unique solution of the Dupire obstacle problem (6.3.1); this is formal and motivated/justified by the numerical scheme. For the sake of theoretical results it might be necessary to impose more restrictions on the set of admissible volatilities, which in turn would make the numerical scheme less natural; we do not venture into those considerations here. So, from the discrete option price observations, as before, we construct the *target* functions $v_i(k)$, $i = 1, \ldots, n$, which the solution of the Dupire obstacle problem is supposed to match. We adopt the same cost functional as in the case of calls, i.e., as in the case of the Dupire *equation*,

$$J(\sigma) = \frac{1}{2} \sum_{i=1}^{n} \int_{k_0}^{k_1} (V(\sigma)(T_i, k) - v_i(k))^2\, q_i(k)\, dk$$

$$+\frac{1}{2}\int_{k_0}^{k_1}\int_{t_0}^{t_1}\left(\epsilon_1\left(\frac{\partial\sigma(T,k)}{\partial k}\right)^2+\epsilon_2\left(\sigma(T,k)-\sigma_{\text{stat}}\right)^2\right)dT\,dk \tag{6.3.5}$$

where ϵ_1, ϵ_2, $q_i(k) > 0$ are given and σ_{stat} is the statistical estimate of the constant volatility.

The optimal control problem is again to find $\sigma_{\text{opt}} \in K$ such that

$$J(\sigma_{\text{opt}}) = \operatorname*{Min}_{\sigma\in K} J(\sigma). \tag{6.3.6}$$

The state equation being what it is, the functional $J(\sigma)$ is *not* differentiable. This is a very well-known phenomenon in the theory of optimal control of variational inequalities (see [1,56] and references given there). Nevertheless from the practical point of view, i.e., from the point of view of designing a working numerical optimization algorithm, this is not a problem. The reason is that non-differentiability happens *rarely*, and therefore as one goes down the steepest descent path, the chances of actually encountering a non-differentiability point are null. The derivation given for the case of the Dupire equation in Chapter 5 needs only a modification. Indeed, for a fixed $\sigma \in K$, such that J is differentiable at σ (see [56]) and $\overline{\sigma}$ such that $\sigma + \delta\overline{\sigma} \in K$, compute the directional derivative of J in the direction $\overline{\sigma}$:

$$J'(\sigma;\overline{\sigma}) = \lim_{\delta\to 0}\frac{J(\sigma+\delta\overline{\sigma})-J(\sigma)}{\delta}$$

$$=\sum_{i=1}^{n}\int_{k_0}^{k_1}(V(T_i,k)-v_i(s))\,w(T_i,k)\,q_i(k)\,dk$$

$$+\int_{t_0}^{t_1}\int_{k_0}^{k_1}\left(\epsilon_1\frac{\partial\sigma(T,k)}{\partial k}\frac{\partial\overline{\sigma}(T,k)}{\partial k}+\epsilon_2\left(\sigma(T,k)-\sigma_{\text{stat}}\right)\overline{\sigma}(T,k)\right)dk\,dT$$

where

$$w(T,k) = \lim_{\delta\to 0}\frac{V(\sigma+\delta\overline{\sigma})(T,k)-V(\sigma)(T,k)}{\delta} \tag{6.3.7}$$

is the unique solution of the forward equation

$$\begin{aligned}&-\frac{\partial w(T,k)}{\partial T}+\frac{1}{2}k^2\frac{\partial^2 w(T,k)}{\partial k^2}\sigma(T,k)^2-rk\frac{\partial w(T,k)}{\partial k}\\&=-k^2\frac{\partial^2 V(\sigma)(T,k)}{\partial k^2}\sigma(T,k)\,\overline{\sigma}(T,k)\end{aligned} \tag{6.3.8}$$

in the non-coincidence set

$$N = \{\{T,k\},\ V(T,k) > \operatorname{Max}(0,k-S)\}\bigcap Q \tag{6.3.9}$$

and

$$w(T, k) = 0$$

in the coincidence set

$$\Lambda = \{\{T, k\}, V(T, k) = \text{Max}(0, k - S)\}. \tag{6.3.10}$$

For any $1 \le i \le n$, let $p_i(T, k)$ be the unique solution of the *backward equations* (not obstacle problems) in the non-coincidence region (so-called adjoint equations):

$$\begin{aligned} \mathcal{A}\, p = {} & \frac{\partial p_i(T, k)}{\partial T} + \frac{1}{2} k^2 \frac{\partial^2 p_i(T, k)}{\partial k^2} \sigma(T, k)^2 \\ & + (2\,\sigma(T, k)^2 + 2\,k\,\sigma^{(0,1)}(T, k)\,\sigma(T, k) + r)\,k\, \frac{\partial p_i(T, k)}{\partial k} \\ & + \big(\sigma(T, k)^2 + k\,(4\,\sigma^{(0,1)}(T, k) + k\,\sigma^{(0,2)}(T, k))\,\sigma(T, k) \\ & \qquad + k^2\,\sigma^{(0,1)}(T, k)^2 + r\big)\,p_i(T, k) = 0 \end{aligned} \tag{6.3.11}$$

in $N_i = N \bigcap \{\{T, k\}, t_0 < T < T_i\}$, with the *terminal* conditions

$$p_i(T_i, k) = -(V(\sigma)(T_i, k) - v_i(k))\, q_i(k) \tag{6.3.12}$$

and (non-cylindrical) boundary condition

$$p_i = 0 \tag{6.3.13}$$

on the lateral boundary of N_i. Each p_i is furthermore extended to zero outside of the non-coincidence region. By means of integration by parts, this terminal boundary value problem is equivalent to its weak formulation:

$$\iint_{N_i} \left(-\frac{\partial \phi(T, k)}{\partial T} + \frac{1}{2} k^2 \frac{\partial^2 \phi(T, k)}{\partial k^2} \sigma(T, k)^2 - r\,k\, \frac{\partial \phi(T, k)}{\partial k} \right) p_i(T, k)\,dk\,dT$$

$$= -\int_{k_0}^{k_1} \phi(T_i, k)\, p_i(T, k)\, dk = \int_{k_0}^{k_1} (V(\sigma)(T_i, k) - v_i(k))\, q_i(k)\, \phi(T_i, k)\, dk$$

for any smooth test function ϕ such that $\phi = 0$ on the backward parabolic boundary of the non-coincidence set.

In particular, if $\phi = w$,

$$J'(\sigma; \overline{\sigma}) = \sum_{i=1}^{n} \int_{k_0}^{k_1} (V(\sigma)(T_i, k) - v_i(k))\, w(T_i, k)\, q_i(k)\, dk$$

$$+ \int_{t_0}^{t_1} \int_{k_0}^{k_1} \left(\epsilon_1 \frac{\partial \sigma(T, k)}{\partial k} \frac{\partial \overline{\sigma}(T, k)}{\partial k} + \epsilon_0\, (\sigma(T, k) - \sigma_{\text{stat}})\, \overline{\sigma}(T, k) \right) dk\, dT$$

$$= \sum_{i=1}^{n} \iint_{N_i} \left(-\frac{\partial w(T,k)}{\partial T} + \frac{1}{2} k^2 \frac{\partial^2 w(T,k)}{\partial k^2} \sigma(T,k)^2 - r k \frac{\partial w(T,k)}{\partial k} \right)$$

$$p_i(T,k)\, dk\, dT + \int_{t_0}^{t_1} \int_{k_0}^{k_1} \left(\epsilon_1 \frac{\partial \sigma(T,k)}{\partial k} \frac{\partial \overline{\sigma}(T,k)}{\partial k} + \epsilon_0 (\sigma(T,k) - \sigma_{\text{stat}}) \overline{\sigma}(T,k) \right)$$

$$dk\, dT = \sum_{i=1}^{n} \iint_{N_i} \left(-k^2 \frac{\partial^2 V(\sigma)(T,k)}{\partial k^2} \sigma(T,k)\, p_i(T,k) \right) \overline{\sigma}(T,k)\, dk\, dT$$

$$+ \int_{t_0}^{t_1} \int_{k_0}^{k_1} \left(-\epsilon_1 \frac{\partial^2 \sigma(T,k)}{\partial k^2} + \epsilon_0 (\sigma(T,k) - \sigma_{\text{stat}}) \right) \overline{\sigma}(T,k)\, dk\, dT$$

$$+ \epsilon_1 \left(\int_{t_0}^{t_1} \frac{\partial \sigma(T,k_1)}{\partial k} \overline{\sigma}(T,k_1)\, dT - \int_{t_0}^{t_1} \frac{\partial \sigma(T,k_0)}{\partial k} \overline{\sigma}(T,k_0)\, dT \right)$$

$$= \int_{t_0}^{t_1} \int_{k_0}^{k_1} \left(-k^2 \frac{\partial^2 V(\sigma)(T,k)}{\partial k^2} \sigma(T,k) \sum_{i=1}^{n} (p_i(T,k)\, \chi_{(t_0,T_i)}) \right.$$

$$\left. + \left(-\epsilon_1 \frac{\partial^2 \sigma(T,k)}{\partial k^2} + \epsilon_0 (\sigma(T,k) - \sigma_{\text{stat}}) \right) \right) \overline{\sigma}(T,k)\, dk\, dT$$

$$+ \epsilon_1 \left(\int_{t_0}^{t_1} \frac{\partial \sigma(T,k_1)}{\partial k} \overline{\sigma}(T,k_1)\, dT - \int_{t_0}^{t_1} \frac{\partial \sigma(T,k_0)}{\partial k} \overline{\sigma}(T,k_0)\, dT \right).$$

On the other hand, exactly as before, there exists z such that

$$\nabla J(\sigma) \cdot \overline{\sigma} = \int_{t_0}^{t_1} \int_{k_0}^{k_1} \left(\epsilon_1 \frac{\partial z(T,k)}{\partial k} \frac{\partial \overline{\sigma}(T,k)}{\partial k} + \epsilon_0\, z(T,k) \overline{\sigma}(T,k) \right) dk\, dT$$

$$= \int_{t_0}^{t_1} \int_{k_0}^{k_1} \left(-\epsilon_1 \frac{\partial^2 z(T,k)}{\partial k^2} + \epsilon_0\, z(T,k) \right) \overline{\sigma}(T,k)\, dk\, dT$$

$$+ \epsilon_1 \left(\int_{t_0}^{t_1} \frac{\partial z(T,k_1)}{\partial k} \overline{\sigma}(T,k_1)\, dT - \int_{t_0}^{t_1} \frac{\partial z(T,k_0)}{\partial k} \overline{\sigma}(T,k_0)\, dT \right).$$

Since $J'(\sigma; \overline{\sigma}) = \nabla J_\epsilon(\sigma) \cdot \overline{\sigma}$ for any admissible $\overline{\sigma}$, we conclude that, for any T, $z(T, \cdot)$ can be computed as a unique solution of the boundary value problem for an ordinary differential equation (regularization equation):

$$\begin{aligned} -\epsilon_1 \frac{\partial^2 z(T,k)}{\partial k^2} + \epsilon_0\, z(T,k) = -k^2 \frac{\partial^2 V(\sigma)(T,k)}{\partial k^2} \sigma(T,k) \\ \sum_{i=1}^{n} (p_i(T,k)\, \chi_{(t_0,T_i)}) + \left(-\epsilon_1 \frac{\partial^2 \sigma(T,k)}{\partial k^2} + \epsilon_0 (\sigma(T,k) - \sigma_{\text{stat}}) \right) \end{aligned} \tag{6.3.14}$$

with boundary conditions

$$\frac{\partial z(T, k_0)}{\partial k} = \frac{\partial \sigma(T, k_0)}{\partial k}, \; \frac{\partial z(T, k_1)}{\partial k} = \frac{\partial \sigma(T, k_1)}{\partial k}. \tag{6.3.15}$$

We again summarize the algorithm. The single iterate in the steepest descent method is done in 4 steps:

1) For given $\sigma(T, k)$, $V(\sigma)(T, k)$ is computed as the unique solution of the Dupire obstacle problem.

2) Next, the $p_i(T, k)$'s are computed using $V(\sigma)(T, k)$ as solutions of the adjoint equations in the non-coincidence region.

3) $V(\sigma)(T, k)$ and all of the $p_i(T, k)$'s are used in the regularization ODEs computing $z(T, k)$.

4) $\sigma_{\text{next}}(T, k) = \sigma(T, k) - \rho\, z(T, k)$, for some $\rho > 0$.

6.3.2 Tridiagonal Solver for Parabolic PDEs in Non-Cylindrical Domains

```
In[144]:= Clear["Global`*"]
```

```
In[145]:= << CFMLab`NumericalBlackScholes`
```

Equations (6.3.11) need to be solved in the subsets of the non-coincidence set N: $N_i = N \cap \{\{T, k\}, t_0 < T < T_i\}$. Each N_i is a non-cylindrical domain, and therefore the solution we have developed in Chapter 5, for solving parabolic PDEs has to be modified. Looking forward at the type of non-cylindrical domains we shall encounter, we consider time/space domains Q of the form

$$Q = \{\{t, x\}, t_0 < t < t_1, x_0 < x < f(t)\}. \tag{6.3.16}$$

For example, if $x_0 = 0$, $t_0 = 0$, $t_1 = 1$, $f(t) = \frac{1}{4}\sin(2\pi t) + \frac{3}{4}$, the domain Q looks like this:

```
In[146]:= With[{t0 = 0, t1 = 1, f = 1/4 Sin[2 π #1] + 3/4 &},
    Plot[f[t], {t, t0, t1}, PlotRange → {0, 1},
     FrameLabel → {"t", "x"},
     RotateLabel → False, Frame → True,
     Epilog → Text["Q", {.4, Sin[.4]}]]];
```

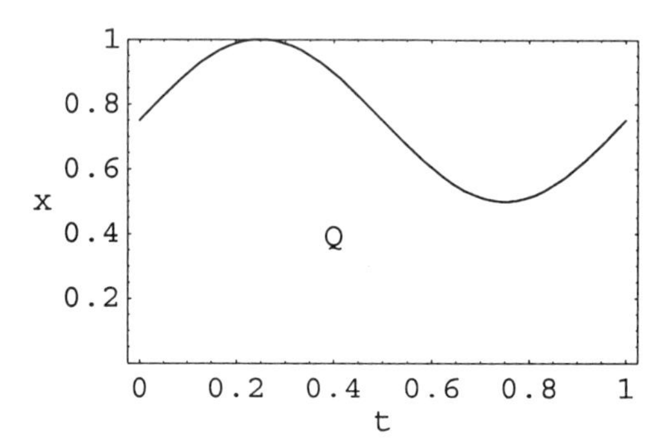

and we are after the numerical solution of the forward parabolic PDE:

$$\mathcal{P}u = b(t,x)\frac{\partial u(t,x)}{\partial t} - \left(a_2(t,x)\frac{\partial^2 u(t,x)}{\partial x^2} + a_1(t,x)\frac{\partial u(t,x)}{\partial x} + a_0(t,x)\,u(t,x)\right) = f(t,x) \tag{6.3.17}$$

in Q, together with the condition

$$u = p \tag{6.3.18}$$

on $\partial_{\mathcal{P}} Q$, the forward parabolic boundary of Q. Above, $b(t,x) > 0$ and $a_2(t,x) > 0$. The solution is computed similarly as in the cylindrical domain case. The only difference is that one has to impose the boundary condition throughout the region outside Q.

```
In[147]:= << "LinearAlgebra`Tridiagonal`"

In[148]:= ForwardNonCylindricalParabolicSolver[
            g_, {b_, a2_, a1_, a0_}, {f_, p_},
            {s0_, s1_, Ns_}, {t0_, t1_, Nt_}, IOrder_ : 1] :=
           Module[{ds, ss, dt, tt, Arguments, PPList,
             A2List, A1List, A0List, RightHandSide,
             A1PlusList, A1MinusList, LowerDiagonal,
             Diagonal, UpperDiagonal, AAA,
             EllipticSolver, ParabolicSolution},
            ds = (s1 - s0)/Ns; ss = N[Range[s0, s1, ds]];
            dt = (t1 - t0)/Nt; tt = N[Range[t0, t1, dt]];
            Arguments = Outer[{#1, #2} &, tt, ss];
            PPList = ((p @@ #1 &) /@ #1 &) /@ Arguments; zxc =
             Function[X, Select[X, #1[[2]] ≥ g[#1[[1]]] &]] /@
              Arguments; DirichletCondAt =
```

```
 Drop[(#1 ⋃ {1} ⋃ {Ns + 1} &) /@ Table[
    Flatten[(Position[Arguments⟦i⟧, #1] &) /@
      zxc⟦i⟧, 2], {i, 1, Length[zxc]}], 1];
DirichletCondAt2 = Flatten[Table[
   ({i, #1} &) /@ DirichletCondAt⟦i⟧, {i, 1,
    Length[DirichletCondAt]}], 1]; BCoefList =
 ((b @@ #1 &) /@ #1 &) /@ Drop[Arguments, 1];
A2List = ((a2 @@ #1 &) /@ #1 &) /@
  Drop[Arguments, 1]; A1List =
 ((a1 @@ #1 &) /@ #1 &) /@ Drop[Arguments, 1];
A0List = ((a0 @@ #1 &) /@ #1 &) /@
  Drop[Arguments, 1]; RightHandSide =
 ((f @@ #1 &) /@ #1 &) /@ Drop[Arguments, 1];
A1PlusList = ((Max[#1, 0] &) /@ #1 &) /@ A1List;
A1MinusList = ((-Min[#1, 0] &) /@ #1 &) /@ A1List;
LowerDiagonal =
 (Drop[#1, 1] &) /@ Fold[ReplacePart[#1, 0, #2] &,
   -A1PlusList/ds + A2List/ds^2, DirichletCondAt2];
Diagonal = Fold[ReplacePart[#1, 1, #2] &,
  A0List + A1MinusList/ds + A1PlusList/ds -
   (2 A2List)/ds^2 - BCoefList/dt, DirichletCondAt2];
UpperDiagonal = (Drop[#1, -1] &) /@
  Fold[ReplacePart[#1, 0, #2] &,
   -A1PlusList/ds + A2List/ds^2, DirichletCondAt2];
AAA = Transpose[{LowerDiagonal, Diagonal,
   UpperDiagonal, RightHandSide, Drop[
    PPList, 1], BCoefList, DirichletCondAt}];
EllipticSolver := Function[{X, Y},
  TridiagonalSolve[Y⟦1⟧, Y⟦2⟧, Y⟦3⟧,
   Fold[ReplacePart[#1, Y⟦5, #2⟧, #2] &,
    -Y⟦4⟧ - (Y⟦6⟧ X)/dt, Y⟦7⟧]]]; ParabolicSolution =
 FoldList[EllipticSolver, PPList⟦1⟧, AAA];
Interpolation[(Append[#1⟦1⟧, #1⟦2⟧] &) /@
  Transpose[{Flatten[Arguments, 1],
    Flatten[ParabolicSolution]}],
 InterpolationOrder → IOrder]]
```

For example, for the above domain Q, i.e., for

```
In[149]:= g[t_] := 1/4 Sin[2 π t] + 3/4
```

and for $b = 1$, $a_2 = 1$, $a_1 = 0$, $a_0 = 0$, $f = 50$, $p = 0$, we compute the solution of (6.3.17) and (6.3.18):

```
In[150]:= Off[General::"spell1"]

In[151]:= Timing[
            sol = ForwardNonCylindricalParabolicSolver[
              g, {1 &, 1 &, 0 &, 0 &}, {50 &, 0 &},
              {0, 1, 200}, {0, 1., 200}]]

Out[151]= {28.23 Second, InterpolatingFunction[(0. 1.
                                                0. 1.), <>]}
```

The solution looks like this:

```
In[152]:= Plot3D[sol[t, x], {x, 0, 1},
             {t, 0, 1}, PlotRange → All, Mesh → False,
             ViewPoint → {2.591, 1.174, 1.833},
             PlotPoints → 200, AxesLabel → {x, t, ""}];
```

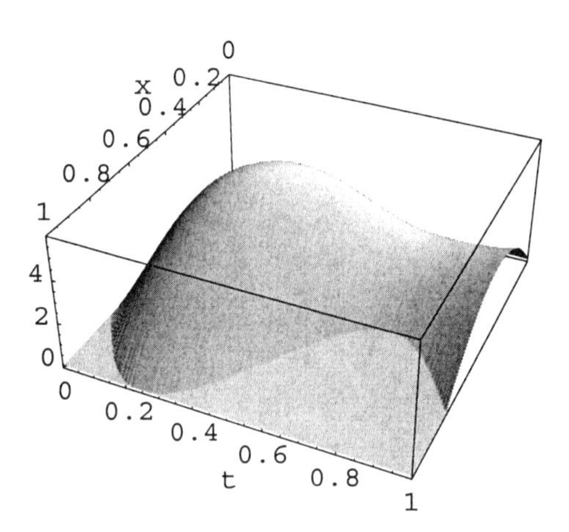

Functions ForwardNonCylindricalParabolicSolver and the solver for the backward problem named BackwardNonCylindricalParabolicSolver are stored in the package CFMLab`NumericalBlackScholes`.

Actually, let us be more specific about the backward problem. The problem is to solve the PDE

$$\mathcal{B}u = b(t, x)\,\frac{\partial u(t, x)}{\partial t} + a_2(t, x)\,\frac{\partial^2 u(t, x)}{\partial x^2}$$
$$+a_1(t, x)\,\frac{\partial u(t, x)}{\partial x} + a_0(t, x)\,u(t, x) = f(t, x) \qquad (6.3.19)$$

in (the above) Q, together with the condition $u = p$ on $\partial_{\mathcal{B}}Q$, the *backward* parabolic boundary of Q. The solution is easily obtained from the forward solver:

```
In[153]:= BackwardNonCylindricalParabolicSolver[g_,
    {b_, a2_, a1_, a0_}, {f_, p_}, {s0_, s1_, Ns_},
    {t0_, t1_, Nt_}] := Module[{ForwardSol},
    ForwardSol = ForwardNonCylindricalParabolicSolver[
      g[t1 - #1] &, {b[t1 - #1, #2] &, a2[t1 - #1, #2] &,
       a1[t1 - #1, #2] &, a0[t1 - #1, #2] &},
      {-f[t1 - #1, #2] &, p[t1 - #1, #2] &}, {s0, s1, Ns},
      {0, t1 - t0, Nt}]; ForwardSol[t1 - #1, #2] &]
```

and checking:

```
In[154]:= Timing[
   sol2 = BackwardNonCylindricalParabolicSolver[g, {1 &,
      1 &, 0 &, 0 &}, {50 &, 0 &}, {0, 1, 200}, {0, 1., 200}]]
```

Out[154]= {31.2 Second, ForwardSol$3865(1. – #1, #2) &}

the solution looks like this:

```
In[155]:= Plot3D[sol2[t, x], {x, 0, 1}, {t, 0, 1}, PlotRange → All,
    Mesh → False, ViewPoint → {2.591, 1.174, 1.833},
    PlotPoints → 200, AxesLabel → {x, t, ""}];
```

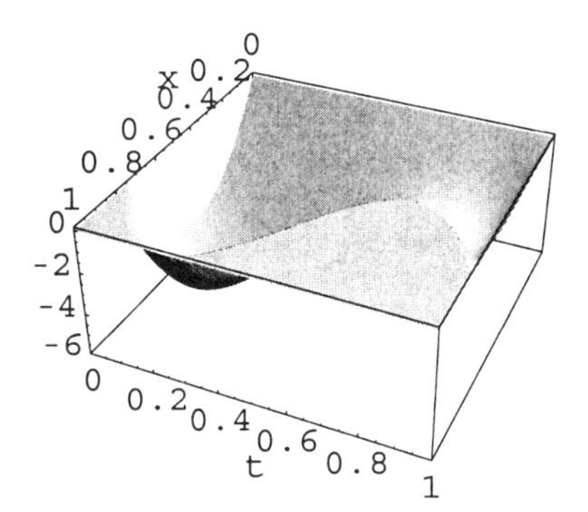

which is what was expected (notice the same data was used; $f > 0$ implies now the negativity of the solution; also comparing the two pictures above, notice the difference between $\partial_{\mathcal{P}}Q$ and $\partial_{\mathcal{B}}Q$; to this end recall that $p = 0$; more light on this will be shed in the context of hypoelliptic equations, in Section 8.3, below).

6.3.3 Computational Example: Put Implied Volatility for QQQ

6.3.3.1 Data Import

We need functions developed before

```
In[156]:= Clear["Global`*"];
          << CFMLab`NumericalBlackScholes`;
          << CFMLab`DataImport`;
```

Next, we import the market data (recall that the liquidity of options is defined in terms of the Volume)

```
In[159]:= PutData = GetAllData["QQQ", Put, 10]
```

Out[159]= {{QQQ, Put, {0.883437, 0.960099, 1.05593}, 0.831417, 76.7656},

{

0.883437	64	0.6875	67
0.883437	65	0.78125	47
0.883437	66	0.9375	100
0.883437	67	1.09375	72
0.883437	68	1.28125	66
0.883437	69	1.40625	450
0.883437	70	1.6875	1178
0.883437	71	1.96875	221
0.883437	72	2.1875	475
0.883437	73	2.5	382
0.883437	74	2.875	5552
0.883437	75	3.3125	4288
0.883437	76	3.8125	789
0.883437	77	4.1875	2306
0.883437	78	4.6875	1560
0.883437	79	5.25	4661
0.883437	80	5.875	2184
0.883437	81	6.375	667
0.883437	82	7.1875	147
0.883437	83	7.8125	321
0.883437	84	8.5625	301
0.883437	85	9.3125	669
0.883437	86	10.0625	234
0.883437	87	10.9375	324
0.883437	88	11.8125	212
0.883437	89	12.6875	201
0.883437	90	13.5625	367
0.883437	91	14.5625	42

,

0.960099	64	1.6875	108
0.960099	65	1.8125	20
0.960099	68	2.5625	35
0.960099	69	2.8125	601
0.960099	70	3.09375	91
0.960099	71	3.5	33
0.960099	73	4.125	20
0.960099	74	4.5	52
0.960099	75	4.875	367
0.960099	76	5.375	53
0.960099	77	5.875	14
0.960099	78	6.3125	236
0.960099	79	6.875	47
0.960099	80	7.3125	773
0.960099	81	7.9375	10
0.960099	85	10.5625	201
0.960099	86	11.3125	44
0.960099	87	11.9375	44
0.960099	88	12.6875	162
0.960099	89	13.4375	20
0.960099	90	14.3125	81
0.960099	91	15.1875	24
0.960099	95	18.5625	33
0.960099	96	19.5625	10
0.960099	100	23.25	10

,

1.05593	50	0.625	10
1.05593	53	0.84375	10
1.05593	60	1.75	30
1.05593	66	3.0625	50
1.05593	68	3.625	24
1.05593	70	4.25	134
1.05593	71	4.625	250
1.05593	73	5.375	20
1.05593	75	6.125	60
1.05593	76	6.625	66
1.05593	77	7.0625	40
1.05593	78	7.5625	35
1.05593	80	8.5625	553
1.05593	85	11.5625	115
1.05593	90	15.1875	35
1.05593	100	23.5	13

}}

where again by default we consider only the next 3 expiration dates. Let

```
In[160]:= TradingData = PutData[[2]];
```

The first step is to construct the *target* functions $v_i(k)$, $i = 1, ..., n$, which the solution of the Dupire obstacle problem is supposed to match:

```
In[161]:= Observations = {#[[1, 1]],
              Interpolation[{#[[2]], #[[3]]} & /@ #,
               InterpolationOrder → 1]} & /@ TradingData;
```

They look like

```
In[162]:= Plot[#[[2]][k],
             {k, #[[2, 1, 1, 1]], #[[2, 1, 1, 2]]},
             AxesLabel → {"Strike k", "$"},
             PlotLabel → StringJoin["T = ",
               ToString[#[[1]]]]] & /@ Observations;
```

$ T = 0.883437

14 12 10 8 6 4 2

Strike k

70 75 80 85 90

$ T = 0.960099

20 15 10 5

Strike k

70 75 80 85 90 95 100

$ T = 1.05593

20 15 10 5

Strike k

60 70 80 90 100

The steepest descent method is initialized by selecting the volatility function $\sigma(T, k) = \sigma_{\text{stat}}$:

```
In[163]:= σstat = Σ[Select[GetPriceData[qqq],
              #[[1]] ≤ YearClock[{2000, 10, 30, 16, 0, 0}] &]]

Out[163]= 0.494515
```

and the computation now starts.

6.3.3.2 Single Iterative Steepest Descent Step

Compute the solution of the Dupire obstacle problem, corresponding to the above σ:

```
In[164]:= x0 = 20; x1 = 150; r = .05; Nx = 200; Nt = 200;
          t1 = Transpose[Observations][[1, -1]];
          S = PutData[[1, -1]];
          t0 = PutData[[1, -2]];

In[168]:= DupirePut =
            DupireObstaclePut[{σstat &, r &, 0 &, S},
             {x0, x1, Nx}, {t0, t1, Nt}];
```

The free boundary looks like

```
In[169]:= jk = Graphics[{AbsolutePointSize[5],
              RGBColor[0, 0, 1], Point[{t0, S}]}];
          Show[Plot[DupirePut[[1]][t], {t, t0, t1},
              DisplayFunction → Identity,
              AxesLabel → {"T", "k"}, AxesOrigin → {t0, 0},
              PlotLabel → "Free Boundary", PlotPoints → Nt,
              PlotRange → {0, DupirePut[[1]][t1]}],
            jk, DisplayFunction → $DisplayFunction];
```

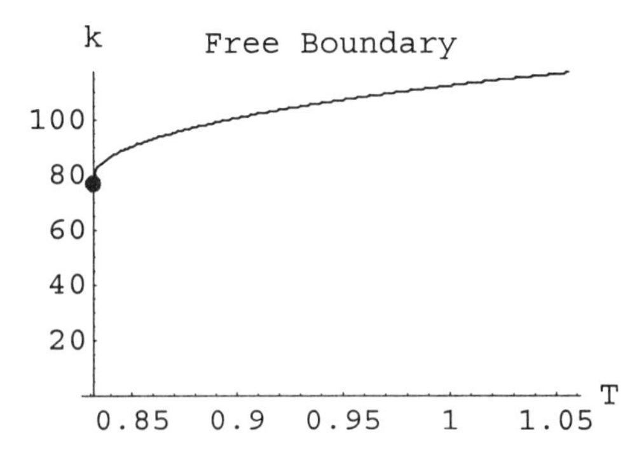

while the solution of the obstacle problem looks like

```
In[171]:= jk3D = Graphics3D[{AbsolutePointSize[5],
              RGBColor[0, 0, 1], Point[{t0, S, 0}]}];
```

```
o = ParametricPlot3D[{t, DupirePut[[1]][t],
    .03 + DupirePut[[2]][t, DupirePut[[1]][t]]},
   {t, t0, t1}, PlotPoints → Nt,
   DisplayFunction → Identity];
o2 = ParametricPlot3D[{t, DupirePut[[1]][t], 0},
    {t, t0, t1}, PlotPoints → Nt,
    DisplayFunction → Identity];
DupirePlot = Plot3D[DupirePut[[2]][t, x],
    {t, t0, t1}, {x, x0, x1}, PlotRange → All,
    Mesh → False, PlotPoints → 80,
    DisplayFunction → Identity,
    ViewPoint -> {-2.622, -1.350, 1.659},
    AxesLabel → {"T", "k", ""}];
Show[DupirePlot, o, o2, jk3D,
   DisplayFunction → $DisplayFunction];
```

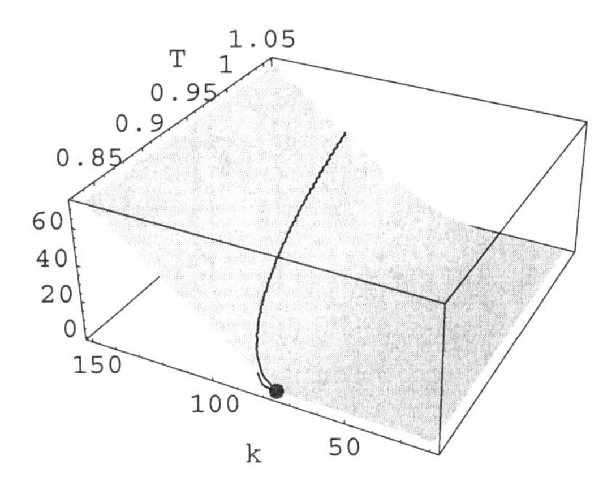

It is interesting to notice the monotonicity of the free boundary, which was announced before and used in the algorithm. Also, it is very interesting to see that theory is confirmed by practice: the realized trades are all below free boundary as they should be—all the options with strikes above the free boundary should be exercised, and therefore should not be traded. Looking closely, one can see the discrepancy between the theoretical and observed prices:

```
In[175]:= Plot[{#[[2]][k], DupirePut[[2]][#[[1]], k]},
            {k, #[[2, 1, 1, 1]], #[[2, 1, 1, 2]]},
            AxesLabel → {"Strike k", "$"}, PlotStyle →
             {RGBColor[0, 0, 0], RGBColor[1, 0, 0]},
            PlotLabel → StringJoin["T = ",
              ToString[#[[1]]]]] & /@ Observations;
```

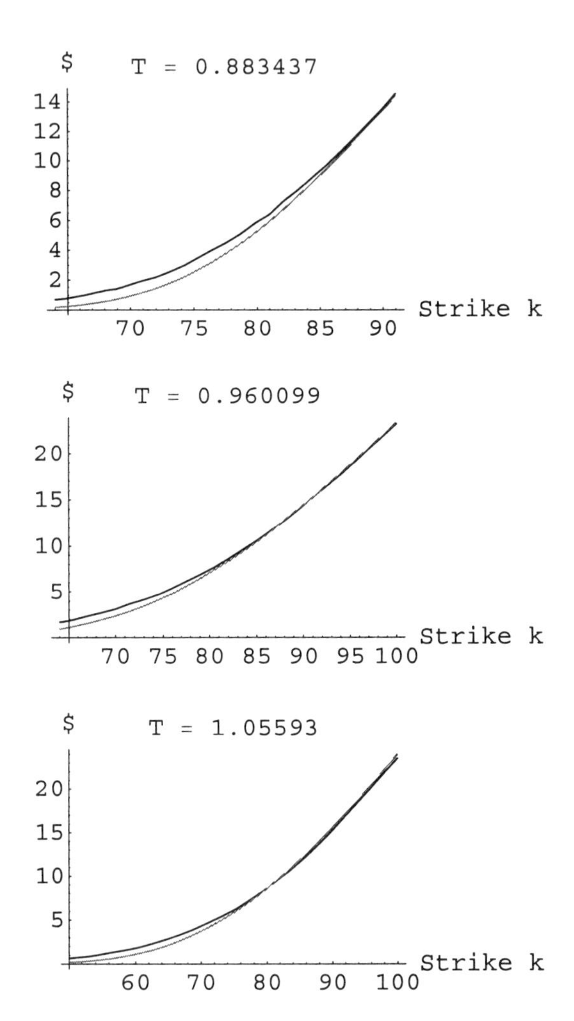

The goal is the make the error as small as possible, and to identify the corresponding implied volatility.

Next we compute the coefficients of the adjoint equations (6.3.11), and also data on the backward parabolic boundary, i.e., the terminal conditions of the adjoint equations (in the non-coincidence region):

```
In[176]:= σ = σstat &;
ExpiryTimes = Transpose[Observations][[1]];
AA = FunctionInterpolation[
  (1/2 #2^2 σ[#1, #2]^2 &)[t, s], {t, t0, t1},
  {s, x0, x1}, InterpolationOrder → {1, 2}];
```

```
In[177]:= BB = FunctionInterpolation[
            ((2 σ[#1, #2]^2 + 2 #2 σ^(0,1)[#1, #2] σ[#1, #2] + r)
               #2 &)[t, s], {t, t0, t1},
            {s, x0, x1}, InterpolationOrder → 1];

In[178]:= CC = FunctionInterpolation[(σ[#1, #2]^2 +
               #2 (4 σ^(0,1)[#1, #2] + #2 σ^(0,2)[#1, #2])
                σ[#1, #2] + #2^2 σ^(0,1)[#1, #2]^2 + r &)[
            t, s], {t, t0, t1}, {s, x0, x1},
           InterpolationOrder → 1];

In[179]:= Clear[t]; CostFunctions =
            If[#[[2, 1, 1, 1]] < s < #[[2, 1, 1, 2]] &&
               t == #[[1]],
              - (DupirePut[[2]][#[[1]], s] - #[[2]][s]),
              0] & /@ Observations;

In[180]:= Plot[CostFunctions[[#]] /. t → ExpiryTimes[[#]],
              {s, 0, 200}] & /@ {1, 2, 3};
```

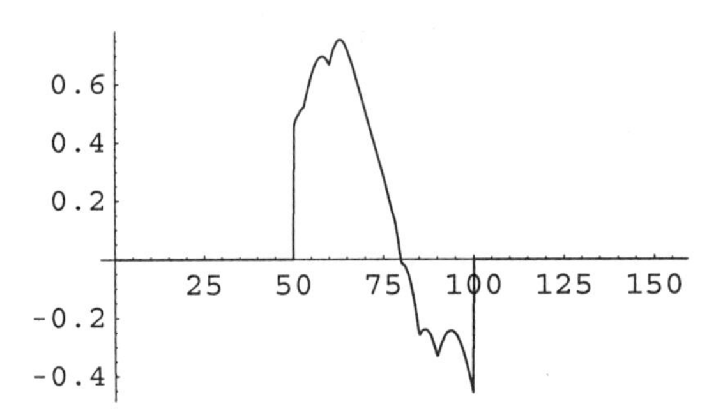

We define the solver of the adjoint equation (notice that now free boundary enters the calculation)

```
In[181]:= AdjointSolver[FB_, Target_, t0_, t1_] :=
    BackwardNonCylindricalParabolicSolver[
     FB, {1 &, AA, BB, CC}, {0 &, Target},
     {x0, x1, 200}, {t0, t1, 200}];
```

and solve them all:

```
In[182]:= AdjointList =
    Table[AdjointSolver[DupirePut[[1]], Function[
       {t, s}, Evaluate[CostFunctions⟦i⟧]], t0,
      ExpiryTimes⟦i⟧], {i, 1, Length[ExpiryTimes]}];
```

Since these solutions are defined only until expiration times, we need to extend them to be zero otherwise (they are already zero outside the noncoincidence set for the times before expiration):

```
In[183]:= AdjointList2 = Function[{T, s},
       If[T ≤ #[[2]], #[[1]][T, s], 0]] &/@
     Transpose[{AdjointList, ExpiryTimes}];
```

and sum them up in a single adjoint function

```
In[184]:= AdjointP =
    Function[{T, s}, Sum[AdjointList2⟦i⟧[T, s],
      {i, 1, Length[ExpiryTimes]}]];
```

which, in the first iterate, looks like

```
In[185]:= Plot3D[AdjointP[T, s], {T, t0, t1}, {s, x0, x1},
    PlotRange → All, PlotPoints → 60, ViewPoint ->
     {-2.608, 1.509, 1.540}, Mesh → False];
```

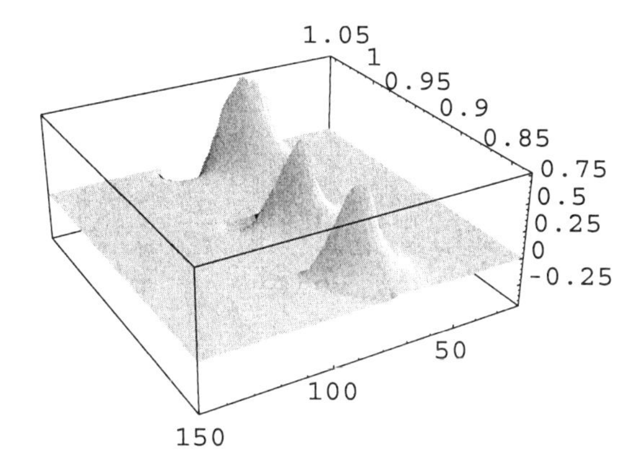

Notice the free boundary from the state equation is quite transparent here; now it is not a free boundary but a fixed, non-cylindrical boundary for the adjoint equations.

After adjoint function p is computed, we need to compute the gradient. To this end we choose the regularization parameters

```
In[186]:= ∈1 = 10000; ∈2 = 1;
```

and compute the right-hand side of the regularization ODE:

```
In[187]:= RHS1[T_, s_] = -s^2 ∂{s,2} DupirePut[[2]][T, s]
             σ[T, s] AdjointP[T, s];

In[188]:= RHS2[T_, s_] = -∈1 ∂{s,2} σ[T, s] + ∈2 (σ[T, s] - σstat);

In[189]:= RH = FunctionInterpolation[
             RHS1[T, s] + RHS2[T, s], {T, t0, t1},
             {s, x0, x1}, InterpolationOrder → 1];
```

Next we compute the grid for the ODE solvers

```
In[190]:= dx = (x1 - x0)/Nx; dt = (t1 - t0)/Nt; tt = N[Range[t0, t1, dt]];
          ss = N[Range[x0, x1, dx]];
```

For the ODE solver we employ the Tridiagonal solver again:

```
In[191]:= A2List = (-∈1 &) /@ ss; A0List = (∈2 &) /@ ss;

In[192]:= LowerDiagonal = (ReplacePart[#1, -1, -1] &)[
             (Drop[#1, 1] &)[A2List/dx^2]];
          Diagonal = (ReplacePart[#1, 1, 1] &)[
```

```
        (ReplacePart[#1, 1, -1] &)[A0List - 2 A2List/dx^2]];
    UpperDiagonal = (ReplacePart[#1, -1, 1] &)[
        (Drop[#1, -1] &)[A2List/dx^2]];
```

```
In[195]:= Regularization[t_] :=
            TridiagonalSolve[LowerDiagonal, Diagonal,
             UpperDiagonal, ReplacePart[ReplacePart[
               Chop[(RH[t, #1] &) /@ ss], 0, 1], 0, -1]]
```

```
In[196]:= Reg = Regularization /@ tt;
```

Once the gradient is computed on the grid points, it can be interpolated into a function:

```
In[197]:= Arguments = Outer[{#1, #2} &, tt, ss];
          z = Interpolation[
             (Append[#1[[1]], #1[[2]]] &) /@ Transpose[
               {Flatten[Arguments, 1], Flatten[Reg]}],
             InterpolationOrder → 1];
```

and viewed

```
In[199]:= Plot3D[z[T, s], {T, t0, t1}, {s, x0, x1},
            PlotRange → All, PlotPoints → 60, ViewPoint ->
             {-2.622, -1.350, 1.659}, Mesh → False];
```

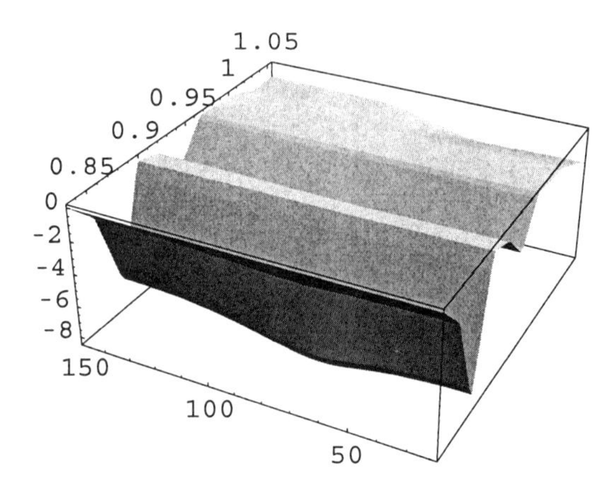

The success of the steepest descent iteration can again be measured by the smallness of the gradient $z(T, k)$. Of course the first one is very large. Finally, the new volatility $\sigma(T, k)$ is computed by

$$\sigma_{\text{new}}(T, k) = \sigma_{\text{old}}(T, k) - \rho\, z(T, k)$$

for some $\rho > 0$. Making sure that $\sigma_{\text{new}} > 0$, let

```
In[200]:= ρ = .005; σmin = .1;
          newσ[T_, s_] = Max[σ[T, s] - ρ z[T, s], σmin];
```

which looks like

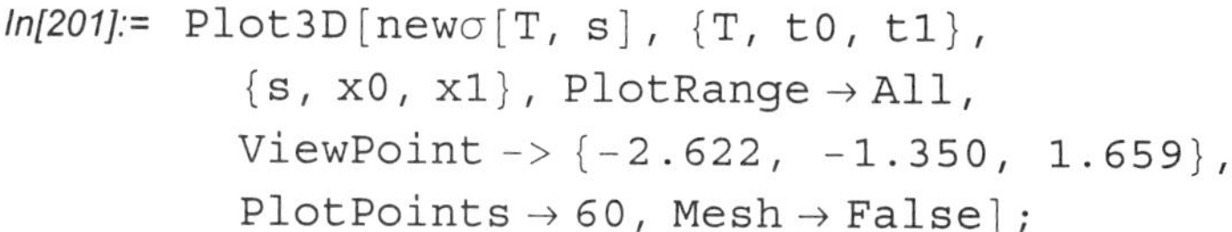

```
In[201]:= Plot3D[newσ[T, s], {T, t0, t1},
            {s, x0, x1}, PlotRange → All,
            ViewPoint -> {-2.622, -1.350, 1.659},
            PlotPoints → 60, Mesh → False];
```

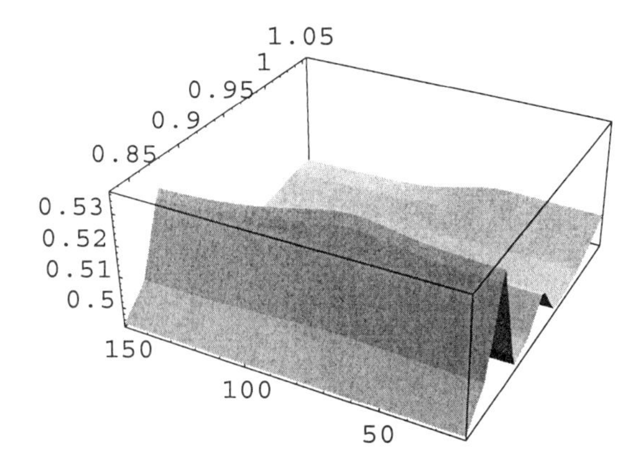

and we need to go all over again.

6.3.3.3 Iteration

Define all of the essential steps in the single iteration as

```
In[202]:= F[σ_, Nx_, Nt_] :=
           (x0 = 20; x1 = 150; r = .05; S = PutData[[1, -1]];
          t0 = PutData[[1, -2]];
          DupirePut = DupireObstaclePut[
              {σ, r &, 0 &, S}, {x0, x1, Nx}, {t0, t1, Nt}];
          ExpiryTimes = Transpose[Observations][[1]];
            AA = FunctionInterpolation[
              (1/2 #2^2 σ[#1, #2]^2 &) [t, s], {t, t0, t1},
              {s, x0, x1}, InterpolationOrder → {1, 2}];
          BB = FunctionInterpolation[
              ((2 σ[#1, #2]^2 + 2 #2 σ^(0,1) [#1, #2] σ[#1, #2] + r) #2 &)[
               t, s], {t, t0, t1},
              {s, x0, x1}, InterpolationOrder → 1];
```

```
CC = FunctionInterpolation[
    (σ[#1, #2]^2 + #2 (4 σ^(0,1)[#1, #2] + #2 σ^(0,2)[#1, #2])
        σ[#1, #2] + #2^2 σ^(0,1)[#1, #2]^2 + r &)[t, s],
    {t, t0, t1}, {s, x0, x1}, InterpolationOrder → 1];
Clear[t]; CostFunctions =
   If[#[[2, 1, 1, 1]] < s < #[[2, 1, 1, 2]] && t == #[[1]],
     - (DupirePut[[2]][#[[1]], s] - #[[2]][s]),
     0] & /@ Observations;
AdjointSolver[FB_, Target_, t0_, t1_] :=
   BackwardNonCylindricalParabolicSolver[FB, {1 &, AA,
     BB, CC}, {0 &, Target}, {x0, x1, Nx}, {t0, t1, Nt}];
AdjointList = Table[AdjointSolver[DupirePut[[1]],
     Function[{t, s}, Evaluate[CostFunctions⟦i⟧]], t0,
     ExpiryTimes⟦i⟧], {i, 1, Length[ExpiryTimes]}];
AdjointList2 = Function[{T, s},
      If[T ≤ #[[2]], #[[1]][T, s], 0]] & /@
    Transpose[{AdjointList, ExpiryTimes}];
AdjointP = Function[{T, s}, Sum[AdjointList2⟦i⟧[T, s],
     {i, 1, Length[ExpiryTimes]}]];
ϵ1 = 10000; ϵ2 = 1;
RHS1[T_, s_] = -s^2
    ∂{s,2} DupirePut[[2]][T, s] σ[T, s] AdjointP[T, s];
RHS2[T_, s_] = -ϵ1 ∂{s,2} σ[T, s] + ϵ2 (σ[T, s] - σstat);
RH = FunctionInterpolation[RHS1[T, s] + RHS2[T, s],
    {T, t0, t1}, {s, x0, x1}, InterpolationOrder → 1];
dx = (x1 - x0)/Nx; dt = (t1 - t0)/Nt; tt = N[Range[t0, t1, dt]];
  ss = N[Range[x0, x1, dx]];
A2List = (-ϵ1 &) /@ ss; A0List = (ϵ2 &) /@ ss;
LowerDiagonal = (ReplacePart[#1, -1, -1] &)[
    (Drop[#1, 1] &)[A2List/dx^2]];
Diagonal = (ReplacePart[#1, 1, 1] &)[
    (ReplacePart[#1, 1, -1] &)[A0List - (2 A2List)/dx^2]];
UpperDiagonal = (ReplacePart[#1, -1, 1] &)[
    (Drop[#1, -1] &)[A2List/dx^2]];
Regularization[t_] := TridiagonalSolve[LowerDiagonal,
    Diagonal, UpperDiagonal, ReplacePart[ReplacePart[
      Chop[(RH[t, #1] &) /@ ss], 0, 1], 0, -1]];
Reg = Regularization /@ tt;
Arguments = Outer[{#1, #2} &, tt, ss];
z = Interpolation[(Append[#1⟦1⟧, #1⟦2⟧] &) /@
```

```
      Transpose[{Flatten[Arguments, 1], Flatten[Reg]}],
     InterpolationOrder → 1];
 ρ = .005; σmin = .1; newσ[T_, s_] =
    Max[σ[T, s] - ρ z[T, s], σmin];
 FunctionInterpolation[newσ[T, s], {T, t0, t1},
    {s, x0, x1}, InterpolationOrder → 1])
```

Many iterations are needed. The *multigrid* approach sometimes aids efficiency: one starts with the coarse grid first to get quickly in the neighborhood of the minimizer, and then by increasing the number of grid points, one recovers the finer features of the optimal solution as well. This is subtle though: if for example the initial grid is too coarse, instead of getting into a neighborhood of the solution, one might end up further away (try the following calculation on a 10×10 grid). So the first computation will be on a 20×20 grid

```
In[203]:= (σ1 = Nest[F[#1, 20, 20] &, σstat &, 50]) // Timing
```

```
Out[203]= {71.96 Second, InterpolatingFunction[( 0.831417  1.05593 ),
                                                  ( 20.       150.    )
           <>]}
```

```
In[204]:= Plot3D[σ1[T, s], {T, t0, t1}, {s, x0, x1},
            PlotRange → All, ViewPoint → {-2.622, -1.350, 1.659},
            PlotPoints → 60, Mesh → False];
```

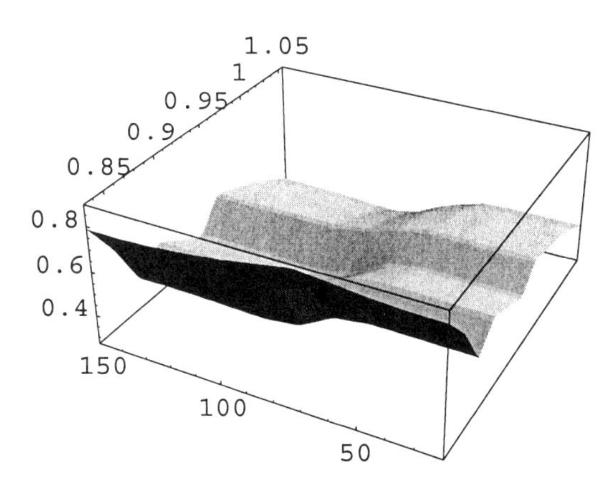

The second calculation was done on a 50×50 grid:

```
In[205]:= (σ2 = Nest[F[#1, 50, 50] &, σ1, 250];) // Timing
```

```
Out[205]= {1776.18 Second, Null}
```

The computed solution looks like this:

```
In[206]:= Plot3D[σ2[T, s], {T, t0, t1}, {s, x0, x1},
            PlotRange → All, ViewPoint -> {-2.622, -1.350, 1.659},
            PlotPoints → 60, Mesh → False];
```

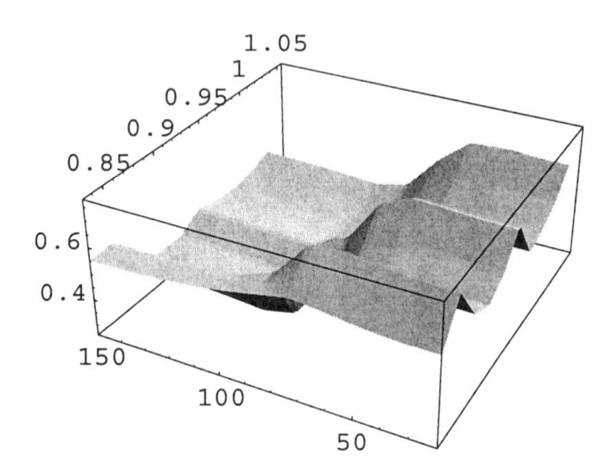

Comparing the times of executions just performed with the ones at the end of Chapter 5, we can see that the optimizer for the Dupire obstacle problem is not inferior to the optimizer for the Dupire (*linear*) equation. This is surprising. It is a testimony to the efficiency of the obstacle solver we have introduced here, and also related to the fact that the adjoint equations are *simpler* in the case of the obstacle problem. Indeed, the unknown values are only in the non-coincidence region.

6.3.3.4 Final Result

The final computed QQQ PUT implied volatility, based on the data as of end-of-day on 10/30/00, can be seen above, while the corresponding solution of the Dupire obstacle problem is now recomputed on a finer grid

```
In[207]:= FinalDupirePut = DupireObstaclePut[
              {σ2, r &, 0 &, S}, {x0, x1, Nx}, {t0, t1, Nt}];
```

The free boundary looks like

```
In[208]:= jk = Graphics[{AbsolutePointSize[5],
               RGBColor[0, 0, 1], Point[{t0, S}]}];
          Show[Plot[FinalDupirePut[[1]][t], {t, t0, t1},
              DisplayFunction → Identity,
              AxesLabel → {"T", "k"}, AxesOrigin → {t0, 0},
              PlotLabel → "Free Boundary", PlotPoints → Nt,
              PlotRange → {0, FinalDupirePut[[1]][t1]}],
            jk, DisplayFunction → $DisplayFunction];
```

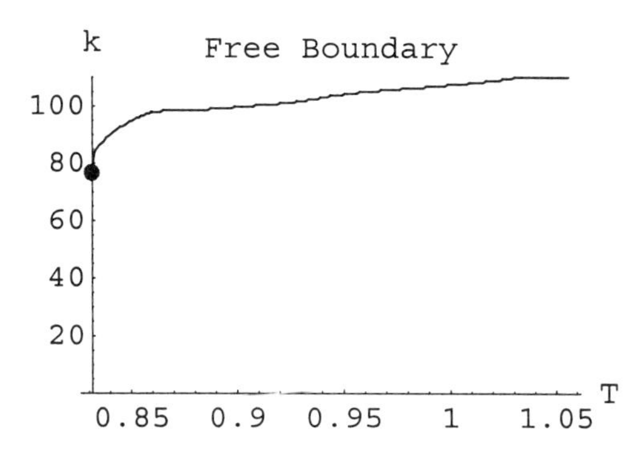

while the solution of the obstacle problem looks like

```
In[210]:= jk3D = Graphics3D[{AbsolutePointSize[5],
             RGBColor[0, 0, 1], Point[{t0, S, 0}]}];
          o = ParametricPlot3D[{t, FinalDupirePut⟦1⟧[t],
             .03 + FinalDupirePut⟦2⟧[t, FinalDupirePut⟦1⟧[t]]},
            {t, t0, t1}, PlotPoints → Nt,
            DisplayFunction → Identity];
          o2 = ParametricPlot3D[{t, FinalDupirePut⟦1⟧[t], 0},
             {t, t0, t1}, PlotPoints → Nt,
             DisplayFunction → Identity];
          DupirePlot = Plot3D[FinalDupirePut[[2]][t, x],
             {t, t0, t1}, {x, x0, x1}, PlotRange → All,
             Mesh → False, PlotPoints → 80, DisplayFunction →
              Identity, ViewPoint -> {-2.622, -1.350, 1.659},
             AxesLabel → {"T", "k", ""}];
          Show[DupirePlot, o, o2, jk3D,
            DisplayFunction → $DisplayFunction];
```

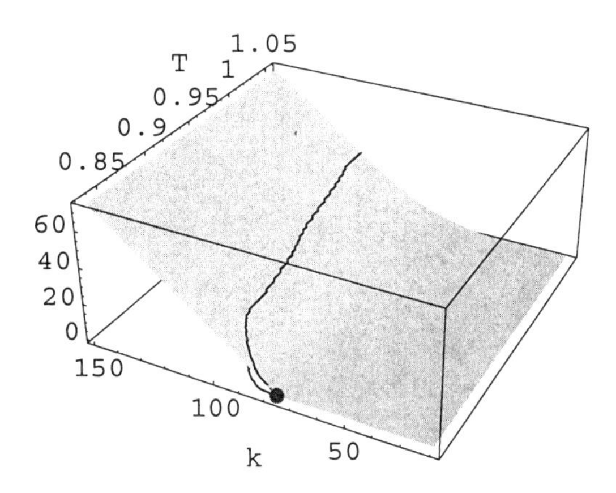

Choosing parameters of the minimization procedure such as ρ, the size of the initial grid, the number of iterative steps, may effect quite a bit the efficiency of the procedure. Therefore it is important that the objectives of the minimization procedure were accomplished:

```
In[214]:= Plot[{#[[2]][k], FinalDupirePut[[2]][#[[1]], k]},
            {k, #[[2, 1, 1, 1]], #[[2, 1, 1, 2]]},
            AxesLabel → {"Strike k", "$"}, PlotStyle →
             {RGBColor[0, 0, 0], RGBColor[1, 0, 0]},
            PlotLabel → StringJoin["T = ",
              ToString[#[[1]]]]] & /@ Observations;
```

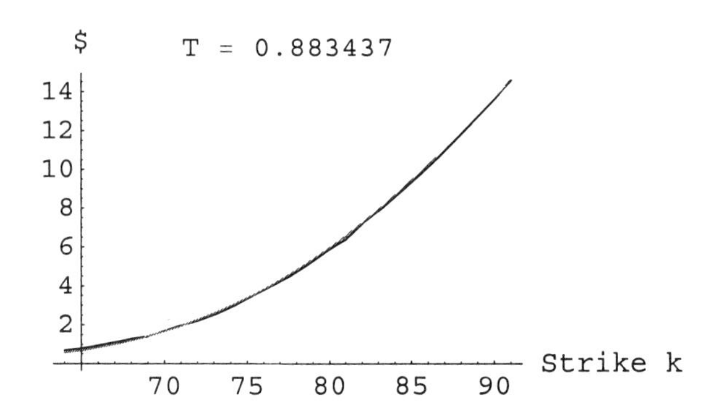

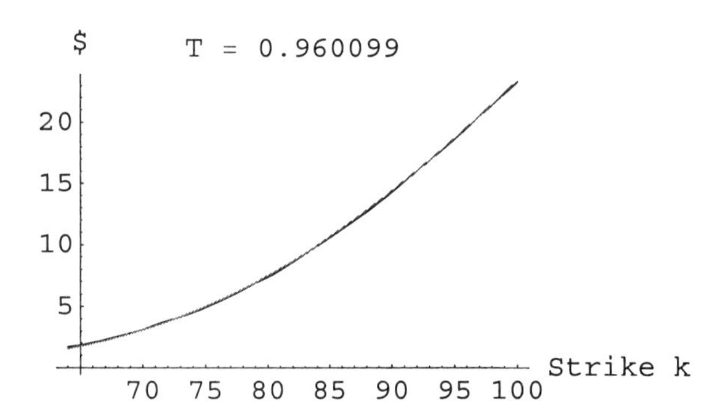

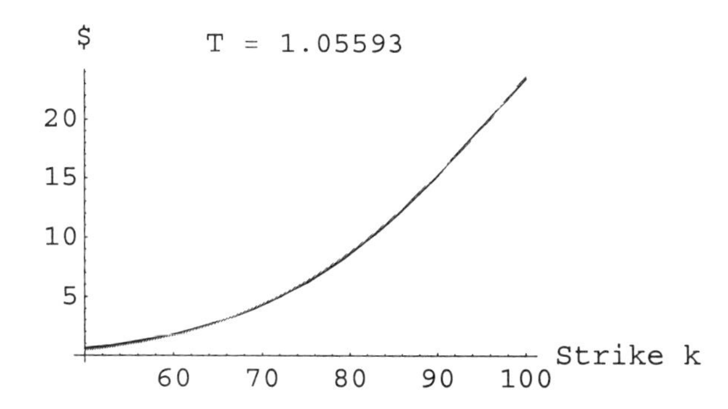

which appears quite satisfactory, validating the proposed methodology for finding price-dependent implied volatility for American options.

7 Optimal Portfolio Rules

Symbolic Solutions of Stochastic Control Problems in Portfolio Management

7.1 Remarks

The classical portfolio theory goes back to Markowitz and his mean-variance portfolio theory. Portfolio theory based on stochastic control goes back to Merton's classical paper in the early 70s [see, e.g., Ch. 5 of 46].

This chapter is an overview, *Mathematica*® implementation, and an extension of the Merton theory. All the results are based on the Log-Normal asset-price dynamics. This assumption yields beautiful explicit/symbolic solutions that are extremely efficient computationally. On the other hand, the same assumption is a source of criticism, since the asset-price dynamics Log-Normality is often not sufficiently descriptive. We shall remove this assumption completely in the next Chapter 8 (Section 8.2), where we introduce a new theory for optimal portfolio hedging for assets with general SDE price dynamics based on numerical solutions of *reduced* Monge–Ampère PDEs.

In our search for explicit/symbolic solutions to portfolio optimization problems we shall employ two methods. One is by finding explicit solutions of Monge–Ampère PDEs, while the other way is by direct maximization of the expected gain with respect to the portfolio rules. The first method is based on partial differential equations, while the second is probability theory based, although it is facilitated by some ordinary differential equations, and even by calculus of variations. One method will be checking the other, and/or continuing when the other cannot be applied.

Explicit/symbolic solutions of portfolio optimization problems, in addition to being very fast computationally and therefore potentially very convenient for day to day trading computations, have high educational value. By understanding and solving them in detail, and experimenting, one can grasp what the right stochastic control problems are to pose also in more complicated situations when each experiment is expensive computationally. One can also better determine reasonable parameter values, the kind of constraints needed and what to expect from them. Finally, one comes to understand the interplay among different kinds of constraints, the connection between utility functions and various constraints, and so on.

7.2 Utility of Wealth

At first it may appear quite surprising how much of what follows depends on the notion of (and on the selection of) the *utility function*. The utility function measures the utility of wealth. We shall always denote the total wealth or the total balance in the investor's brokerage account at time t by $X = X(t)$. Now one may notice right away that what $X(t)$ is exactly makes a whole lot of difference. It is quite different from the point of view of how much risk it is prudent to take: whether $X(t)$ is a gambling allowance, the balance in one of say several brokerage accounts that an investor holds, or if the risk involves the total family fortune. Alternatively, as we shall see in detail, explicit mathematical constraints on the investment portfolio may change the nature of $X(t)$ and the amount of risk one can afford to take.

So what is the proper way of measuring the utility of wealth, i.e., risk-taking? In a way, the entire chapter that follows is, in addition to many other issues, also about answering this question posteriori, meaning that after calculating the strategy implied by a particular utility function, we can judge whether it seems reasonable (see for example final thoughts in Section 7.4.1, and even Section 8.2), and whether the experimental wealth evolutions it generates seem acceptable, thereby judging the utility function used.

On the other hand, a priori, from the psychological point of view, or alternatively, from the trading experience, it is known that the pain experienced by the loss of $-dX$ is greater than the pleasure experienced from the gain of $dX > 0$. This is particularly true if X is the total wealth and no additional constraints on risk-taking are imposed. This fact translates into the mathematical property of (strict) *concavity* for any reasonable utility function under such circumstances. If, on the other hand, for example, X is the balance in an account among several accounts and is designated for risk-taking or experimenting, then a different much more aggressive utility function may be appropriate.

There exists a body of literature that considers, in addition to accumulation of the wealth also its consumption, both in an optimal manner balancing the two. Although we could, using the same methods that are presented in this chapter and in Section 8.2, we do not consider the issue of wealth consumption, but only the issue of wealth accumulation. Consequently, the most important class of utility functions to be considered here is the class of functions:

In[1]:= $\psi_{\gamma\_}[X\_] := \frac{X^{1-\gamma}}{1-\gamma}$

for $0 < \gamma \neq 1$, while for $\gamma = 1$

In[2]:= $\psi_1[X\_] := \text{Log}[X]$

Notice right away that

In[3]:= $\partial_{\{X,2\}}\psi_\gamma[X]$

Out[3]= $-X^{-\gamma-1}\gamma$

and therefore, since $X > 0$, the strict concavity, the desired property of the utility functions as it was justified above, or explicitly the mathematical property that $\frac{\partial^2 \psi_\gamma(X)}{\partial X^2} < 0$ holds provided, as it was assumed, $\gamma > 0$. Here are some examples:

In[4]:= `UtilityFunctions = Table[`ψ_γ`[X], {`γ`, `$\frac{1}{2}$`, 10, `$\frac{1}{2}$`}]`

Out[4]= $\left\{2\sqrt{X}, \log(X), -\frac{2}{\sqrt{X}}, -\frac{1}{X}, -\frac{2}{3X^{3/2}}, -\frac{1}{2X^2}, -\frac{2}{5X^{5/2}}, -\frac{1}{3X^3}, -\frac{2}{7X^{7/2}}, -\frac{1}{4X^4}, -\frac{2}{9X^{9/2}}, -\frac{1}{5X^5}, -\frac{2}{11X^{11/2}}, -\frac{1}{6X^6}, -\frac{2}{13X^{13/2}}, -\frac{1}{7X^7}, -\frac{2}{15X^{15/2}}, -\frac{1}{8X^8}, -\frac{2}{17X^{17/2}}, -\frac{1}{9X^9}\right\}$

They look like:

In[5]:= `Plot[Evaluate[UtilityFunctions], {X, .0001, 10},`
`PlotRange → {-20, 10}, AxesLabel → {X,` ψ_γ`}];`

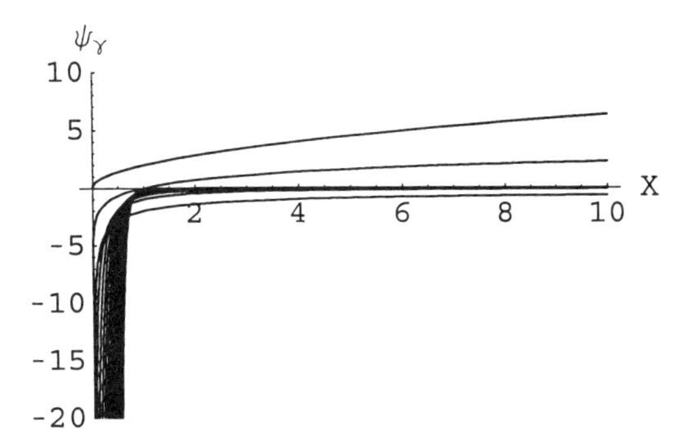

A larger γ implies more conservative investing, while the smaller the γ the more aggressive the trading decisions that will follow. This can be inferred from the shape of the above utility functions, and in particular from three different regions for γ: $0 < \gamma < 1$, $\gamma = 1$, and $1 < \gamma < \infty$. Indeed, in the first case, when $0 < \gamma < 1$, $\lim_{X\to 0}\psi_\gamma(X) = 0$ while $\lim_{X\to\infty}\psi_\gamma(X) = \infty$, implying that, heuristically, the "pain" from bankruptcy is finite; the fact that it is equal to zero does not matter since only the relative differences of the values of the utility function matter and not its absolute value; but "pleasure" from the infinite gain is infinite. In the second case, when $\gamma = 1$, $\lim_{X\to 0}\psi_\gamma(X) = -\infty$ and $\lim_{X\to\infty}\psi_\gamma(X) = \infty$ says that the "pain" from bankruptcy is now infinite, but the "pleasure" from infinite gain is still infinite. Finally, in the third case, when $1 < \gamma < \infty$, $\lim_{X\to 0}\psi_\gamma(X) = -\infty$ and

$\lim_{X\to\infty} \psi_\gamma(X) = 0$, meaning that the "pleasure" from infinite wealth is finite, which indeed seems to be reasonable. It is quite interesting how these somewhat ad hoc observations are justified later on the basis of the trading strategies they imply, which in turn cause quite different fortunes when trading. This can be seen without exposure to the market risks using Monte–Carlo trading simulations, to be performed below.

We shall need a name for γ. The *safety exponent* seems appropriate.

The most conservative investing, i.e., risk minimizing investing, is done when the safety exponent $\gamma \to \infty$, or as we shall denote when $\gamma = \infty$. As a matter of fact, in such a case the utility function neither exists nor is it needed. By adopting trading strategies inferred by safety exponents $0 < \gamma < \infty$, an investor increases the risk as well as the potential for a gain. However, depending on the problem (whether we are trading with total wealth or only with part of it), and on the lack of adopted constraints, risk can become enormous and unbearable as $\gamma \to 0$. Case $\gamma = 0$ (i.e., the innocent looking utility function $\psi_0(X) = X$), or any other $\gamma < 0$, as opposed to the case $\gamma = \infty$ will not be allowed, unless there are other ways to stop or at least inhibit trading. Although we shall not be concerned with such problems, conceptually this would work as a utility function (see also Section 8.3):

```
In[6]:= With[{b = 10}, Plot[Evaluate[
          Table[Min[b, b^γ X^(1-γ)], {γ, 0, -10, -1/2}]],
        {X, 0, 2 b}, PlotRange → {0, 1.2 b},
        AxesLabel → {X, ψ}]];
```

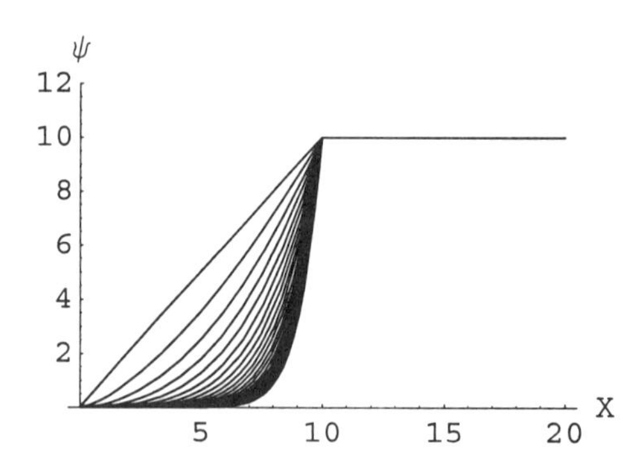

7.3 Merton's Optimal Portfolio Rule Derived and Implemented

7.3.1 Derivation of Equations: Cash Transactions, Wealth Evolution

Stock market prices are assumed to be governed by the Log-Normal model:

$$dS(t) = S(t)\,a\,dt + S(t)\,\sigma.dB(t) \tag{7.3.1}$$

where $S(t) = \{S_1(t), \ldots, S_m(t)\}$ are the stock prices, and the vector $a = \{a_1, \ldots, a_m\} \in \mathbb{R}^m$ is the vector (one-dimensional array) of appreciation rates of the corresponding stocks, $\sigma \in \mathbb{R}^m \times \mathbb{R}^n$ is the (volatility) matrix (two-dimensional array), and $B(t) = \{B_1(t), \ldots, B_n(t)\}$ is the vector of n independent Brownian motion. We assume that the $m \times m$ matrix $\sigma.\sigma^T$ is invertible (this would be impossible if $m > n$); in practical terms this means that no assets with perfectly correlated prices are allowed (cf. Section 8.2.4.2)

An investor has a cash account whose balance at time t is denoted by $C(t)$. Describing precisely the dynamics for $C(t)$ is going to be very important, but for now we can state that

$$dC(t) = r\,C(t)\,dt + dc(t) \tag{7.3.2}$$

where r is the interest rate, and where much more importantly the $dc(t)$ are the cash transactions in ($dc(t) > 0$), and out ($dc(t) < 0$) of the cash account as a consequence of selling and buying stocks. We notice that $C(t)$ may take negative values, meaning borrowing (at the same interest rate r).

The basic quantity to be considered in investing is the total, or available, wealth that shall be denoted by $X(t)$. It represents the cash value of the whole stock portfolio, plus the balance in the cash account.

The problem is to design a trading strategy. What kind of trading strategies are considered? The trading or hedging strategy is going to be any vector-valued function $\Pi(t, X) = \{\Pi_1(t, X), \Pi_2(t, X), \ldots, \Pi_m(t, X)\}$, where $\Pi_j(t, X)$ is the cash value, positive or negative, depending on whether it is a long or short position of the investment in the jth stock considered. On the other hand, in the present chapter strategies are not allowed to depend on particular stock prices. Consequently, simple strategies such as the prescribed number of stocks of a particular company are not considered here. Instead, an example of allowed strategies would be to prescribe percentages of the total wealth to be invested in particular companies. Maintaining such a strategy will require perpetual trading to account for changes in the asset prices, reminiscent of what professional traders are doing on the floor and not of what individual investors can afford to do (buy-and-hold kind of strategies are not considered here; see Section 8.3 for different kinds of strategies). This justifies indeed calling them *portfolio hedging strategies*.

In such a framework the total wealth can be written as

$$X(t) = C(t) + \sum_{i=1}^{m} \Pi_i(t, X(t)) = C(t) + \Pi(t, X(t)).\mathrm{I}_m \tag{7.3.3}$$

where, for the sake of calculations below, I_m is introduced as an m-vector with all components equal to 1: $\mathrm{I}_m = \{1, 1, \ldots, 1\}$. Obviously then, the cash account balance is equal to

$$C(t) = X(t) - \Pi(t, X(t)).\mathrm{I}_m.$$

We shall make an effort to completely understand what so-called self-financing strategies really mean and imply, by studying carefully the cash transactions $dc(t)$ from (7.3.2). Usually in the literature this is ignored and instead one jumps from (7.3.3) to (7.3.6) declaring it a consequence of *self-financing*.

To this end, let $v(t, X(t))$ be the vector of numbers (fractions being allowed) of stocks of each particular company in the portfolio

$$v(t, X(t)) = \frac{\Pi(t, X(t))}{S(t)}.$$

For example, if $S = \{50, 60\}$, i.e., the current price of two stocks considered for trading are \$50 and \$60, if $\Pi = \{5000, 12000\}$, i.e., the current investment in the first stock is \$5000 and in the second stock \$12000, then $v = \{100, 200\}$, i.e., the investor owns 100 stocks of the first company and 200 stocks of the second. During the time interval dt, the vector of all cash transactions out or in the particular company in the portfolio is equal to (recall the *Mathematica*® convention: vectors are multiplied component-wise)

$$\begin{aligned}&(v(t, X(t)) - v(t + dt, X(t + dt)))\, S(t + dt)\\ &= \Big(\frac{\Pi(t, X(t))}{S(t)} - \frac{\Pi(t + dt, X(t + dt))}{S(t + dt)}\Big) S(t + dt).\end{aligned} \tag{7.3.4}$$

Notice carefully that the trading decision $v(t, X(t)) - v(t + dt, X(t + dt))$ could be made only at the time $t + dt$, and not before, and therefore the stock prices $S(t + dt)$ are used rather than $S(t)$; the latter choice, by the way, would lead to a wrong conclusion since in the stochastic calculus they are not equivalent even when $dt \to 0$. All those particular transactions sum up to the influx $dc(t) > 0$, or out-flux $dc(t) < 0$, from the cash account during the time interval dt, which is equal to

$$\begin{aligned}dc(t) &= (v(t, X(t)) - v(t + dt, X(t + dt))).S(t + dt)\\ &= \Big(\frac{\Pi(t, X(t))}{S(t)} - \frac{\Pi(t + dt, X(t + dt))}{S(t + dt)}\Big).S(t + dt)\\ &= \Pi(t, X(t)).\frac{S(t + dt)}{S(t)} - \Pi(t + dt, X(t + dt)).\mathrm{I}_m\\ &= \Pi(t, X(t)).\frac{S(t) + dS(t)}{S(t)} - \Pi(t + dt, X(t + dt)).\mathrm{I}_m\\ &= \Pi(t, X(t)).\Big(\mathrm{I}_m + \frac{dS(t)}{S(t)}\Big) - \Pi(t + dt, X(t + dt)).\mathrm{I}_m\\ &= (\Pi(t, X(t)) - \Pi(t + dt, X(t + dt))).\mathrm{I}_m + \Pi(t, X(t)).\frac{dS(t)}{S(t)}.\end{aligned} \tag{7.3.5}$$

Having calculated $dc(t)$, we are now in a position to carefully derive the wealth-evolution SDE.

$$\begin{aligned} dX(t) &= dC(t) + (\Pi(t+dt, X(t+dt)) - \Pi(t, X(t))).1_m \\ &= r\,C(t)\,dt + dc(t) + (\Pi(t+dt, X(t+dt)) - \Pi(t, X(t))).1_m \\ &= r\,C(t)\,dt + (\Pi(t, X(t)) - \Pi(t+dt, X(t+dt))).1_m + \Pi(t, X(t)).\frac{dS(t)}{S(t)} \end{aligned}$$

$$+(\Pi(t+dt, X(t+dt)) - \Pi(t, X(t))).1_m = r\,C(t)\,dt + \Pi(t, X(t)).\frac{dS(t)}{S(t)}$$

and thereafter, as it is usually done,

$$dX(t) = r\,C(t)\,dt + \Pi(t, X(t)).\frac{dS(t)}{S(t)} = r\,(X(t) - \Pi(t, X(t)).1_m)\,dt$$
$$+\Pi(t, X(t)).(a\,dt + \sigma.dB(t))$$

concluding the wealth evolution SDE

$$dX(t) = (\Pi(t, X(t)).(a-r) + r\,X(t))\,dt + \Pi(t, X(t)).\sigma.dB(t). \tag{7.3.6}$$

The final SDE for $X(t)$ does not depend on the cash transactions, but the consequence of the above careful derivation is not just the full understanding of the argument, but also results in explicit formulas for the cash transactions, which are the most important quantities in any trading; also they will allow us to do Monte–Carlo simulations below to give us a complete understanding of what is going on. One should realize the magnitude in simplifying the problem: considering possibly thousands of different stocks, and thereby as many SDEs in the stock market SDE system (7.3.1), has been reduced to considering a *single* (scalar-valued) SDE (7.3.6).

The stochastic process $\{X(t)\}$, corresponding to the fixed hedging strategy $\Pi = \Pi(t, X)$, is to be denoted as $\{X^{\Pi}(t)\}$ whenever we wish to emphasize its dependence on the strategy. Also, $X^{\Pi}(t) \neq X(t)^{\Pi}$ where the latter would denote the Πth power of $X(t)$.

7.3.2 A Non-Optimal Hedging Strategy Implemented

7.3.2.1 Data

To confirm all the details in the above (unorthodox) derivation of the wealth evolution, and to have the right intuition about the whole model and its details, we present an immediate example.

```
In[7]:= Off[General::"spell1"];
```

We need the standard *Mathematica*® package

```
In[8]:= << "Statistics`NormalDistribution`"
```

Let m be the number of stocks in the market considered for trading, n the number of sources of randomness, T_1 the time horizon, k the number of time subintervals in the discretization, X_0 the initial wealth, and finally, let r be the interest rate. For example,

```
In[9]:= m = 3; n = 5; T1 = 3; k = 2000;
        X0 = 1000000; r = 0.05; dt = T1/k;
```

Let the volatility matrix be

```
In[10]:= σ = 1/4 Array[Random[Real, {-1, 1}] &, {m, n}]
```

$$Out[10]= \begin{pmatrix} 0.138937 & 0.186114 & -0.0761569 & 0.146026 & 0.157908 \\ -0.107189 & 0.190008 & 0.159677 & -0.223419 & -0.135206 \\ 0.210771 & -0.153473 & -0.0868801 & -0.203035 & -0.127446 \end{pmatrix}$$

To check whether it is reasonable, compute the individual volatilities

```
In[11]:= Volatilities = (√Plus @@ #1^2 &) /@ σ
Out[11]= {0.325576, 0.37588, 0.364681}
```

and the correlation matrix

```
In[12]:= σ.Transpose[σ] / Outer[Times, Volatilities, Volatilities]
```

$$Out[12]= \begin{pmatrix} 1. & -0.373146 & -0.357412 \\ -0.373146 & 1. & -0.022126 \\ -0.357412 & -0.022126 & 1. \end{pmatrix}$$

Let the appreciation rates be (in particular, we assume that they are known by the investor)

```
In[13]:= a = Array[Random[Real, {-0.1, .5}] &, {m}] - .2
Out[13]= {-0.00319635, 0.0595035, -0.256589}
```

and let the initial stock prices be

```
In[14]:= s = Array[Random[Real, {15, 110}] &, {m}]
Out[14]= {101.677, 19.2194, 23.3271}
```

7.3.2.2 Market Evolution

Instead of calling on the SDESolver, we solve the stock market SDE explicitly (it is not difficult and we shall need to compute many other quantities on the same randomization, so it is easier to do everything explicitly):

```
In[15]:= dB = Table[Random[NormalDistribution[0, 1]],
              {k}, {n}] √dt ;
         G[{t_, S_}, dB_] := {dt + t, a dt S + σ.dB S + S};
         MarketEvolution = FoldList[G, {0, s}, dB];
```

The stock price evolution turns out to be

```
In[18]:= ithStock[i_] :=
           ({#1⟦1⟧, #1⟦2, i⟧} &) /@ MarketEvolution;
         PS = Table[RGBColor[Random[], 0, Random[]], {m}];
         Show[(ListPlot[ithStock[#1], PlotRange →
                 {0, Max[(#1⟦2⟧ &) /@ MarketEvolution]},
               PlotJoined → True, PlotStyle → PS⟦#1⟧,
               DisplayFunction → Identity] &) /@ Range[m],
           DisplayFunction → $DisplayFunction];
```

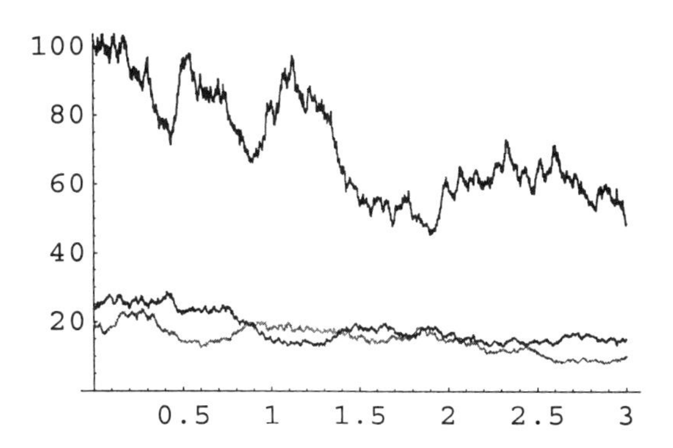

7.3.2.3 Monte–Carlo Simulation of Trading

We choose, for no particular reason, a simple trading (hedging) strategy: the fixed fractions of the total wealth will be kept in each stock and the rest in the cash account. For example, let the stock wealth fractions be

```
In[21]:= P = {1/4, 1/5, 1/6};
```

the fraction of the wealth to be always kept in the cash account is equal to

```
In[22]:= CashFraction = 1 - Plus @@ P
```

Out[22]= $\frac{23}{60}$

The trading strategy is therefore equal to

```
In[23]:= Π[t_, X_] = P X
```

Out[23]= $\{\frac{X}{4}, \frac{X}{5}, \frac{X}{6}\}$

Recalling the wealth evolution SDE (7.3.6), we see what the folding function needs to be:

```
In[24]:= F[{t_, X_}, dB_] := {t + dt, Max[X, 0] +
          (Π[t, Max[X, 0]].(a - r) + r Max[X, 0]) dt +
          Π[t, Max[X, 0]].σ.dB}
```

generating the trajectory of the wealth process $X(t)$:

```
In[25]:= WealthEvolution = FoldList[F, {0, X0}, dB];
```

which looks like

```
In[26]:= lpwe = ListPlot[WealthEvolution,
          PlotJoined → True, PlotRange →
           {0, Max[Transpose[WealthEvolution]〚2〛]}];
```

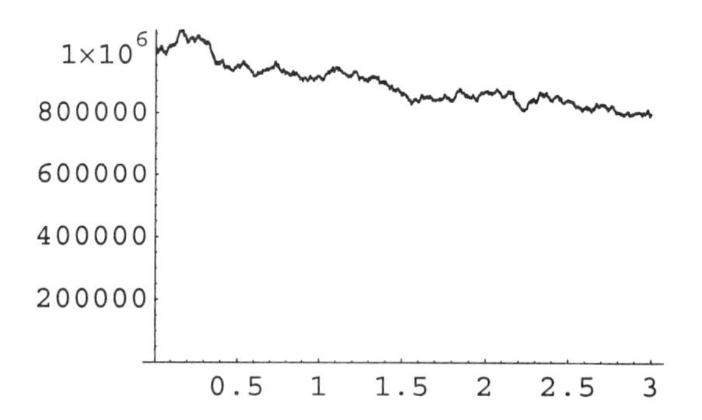

Let

```
In[27]:= T = (#1〚1〛 &) /@ MarketEvolution;
         S = (#1〚2〛 &) /@ MarketEvolution; St = Drop[S, -1];
         Stdt = Drop[S, 1]; X = (#1〚2〛 &) /@ WealthEvolution;
         ΠtXt = (P #1 &) /@ Drop[X, -1];
         ΠtdtXtdt = (P #1 &) /@ Drop[X, 1];
```

The evolution of cash value of the investments in different stocks looks like

```
In[28]:= StockBalancePlot = Show[
          Module[{x}, x = Table[({#1〚1〛, #1〚2, i〛} &) /@
              Transpose[{Drop[T, -1], ΠtXt}], {i, 1, m}];
           Table[ListPlot[x〚i〛, PlotJoined → True,
```

```
        DisplayFunction → Identity,
        PlotStyle → PS⟦i⟧], {i, 1, m}]],
    DisplayFunction → $DisplayFunction];
```

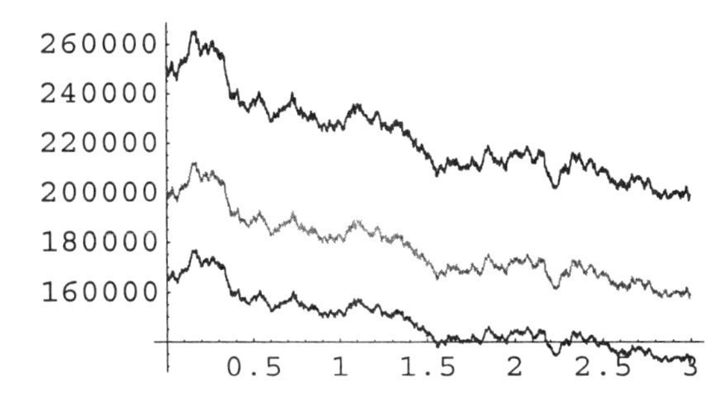

On the other hand, referring to the cash transactions equations (7.3.3) and (7.3.4), we compute for example

```
In[29]:= NumberOfStocksTraded = (#1⟦3⟧/#1⟦1⟧ - #1⟦4⟧/#1⟦2⟧ &) /@
           Transpose[{St, Stdt, ΠtXt, ΠtdtXtdt}];
```

The size of typical trades may indicate whether, for example, the wealth size is appropriate for such a hedging strategy (by experimenting with k one can also see that the size of typical trades will depend on $dt = T/k$: if the hedging is done more often, an average size of trades is smaller; in the present calculation, one may think as if the hedging is done approximately

```
In[30]:= IntegerPart[1/(365 dt)]

Out[30]= 1
```

times each day). So, here are the realized trades:

```
In[31]:= Do[ListPlot[Transpose[{Drop[T, 1],
             Transpose[NumberOfStocksTraded]⟦i⟧}],
           PlotRange → All, PlotStyle → PS⟦i⟧,
           PlotLabel → "Stocks Sold/Bought Company " <>
             ToString[i]], {i, 1, m}];
```

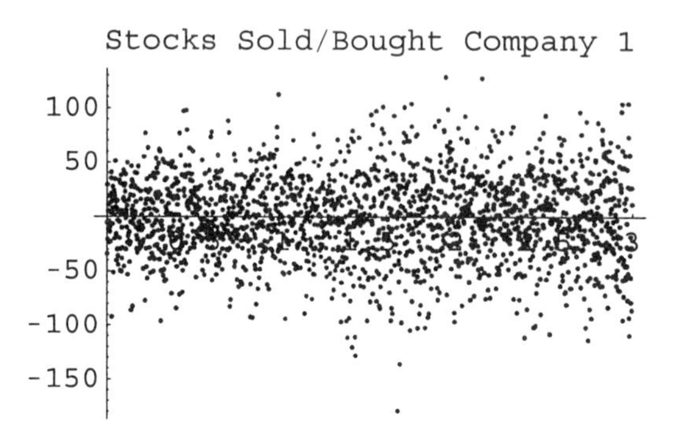

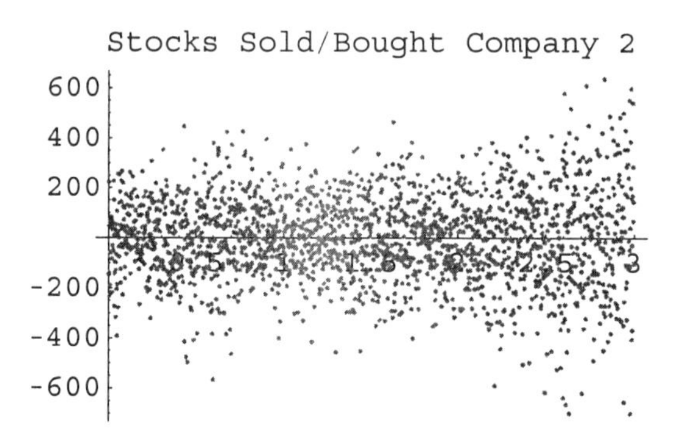

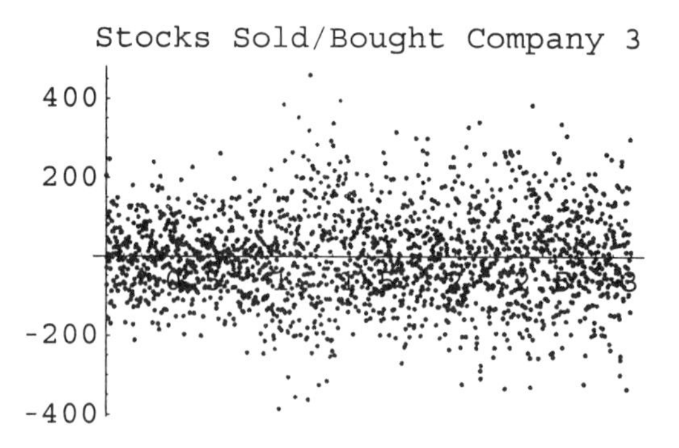

The corresponding cash transactions were (positive transaction means a sale)

```
In[32]:= DC = ((#1⟦3⟧/#1⟦1⟧ - #1⟦4⟧/#1⟦2⟧) #1⟦2⟧ &) /@
            Transpose[{St, Stdt, ΠtXt, ΠtdtXtdt}];
         Do[ListPlot[Transpose[{Drop[T, 1],
```

```
    Transpose[DC]⟦i⟧}], PlotStyle → PS⟦i⟧,
  PlotRange → All, PlotLabel → ToString[i] <>
    "th Stock Caused Cash Flow"], {i, 1, m}];
```

1th Stock Caused Cash Flow

2th Stock Caused Cash Flow

3th Stock Caused Cash Flow

The cash transactions out and in particular stocks add up into the cash influx into the cash account $dc(t)$:

```
In[34]:= dc = ((#1⟦3⟧/#1⟦1⟧ - #1⟦4⟧/#1⟦2⟧).#1⟦2⟧ &) /@
      Transpose[{St, Stdt, ΠtXt, ΠtdtXtdt}];
    ListPlot[Transpose[{Drop[T, 1], dc}],
      PlotLabel → "Cash Account Influx/Outflux"];
```

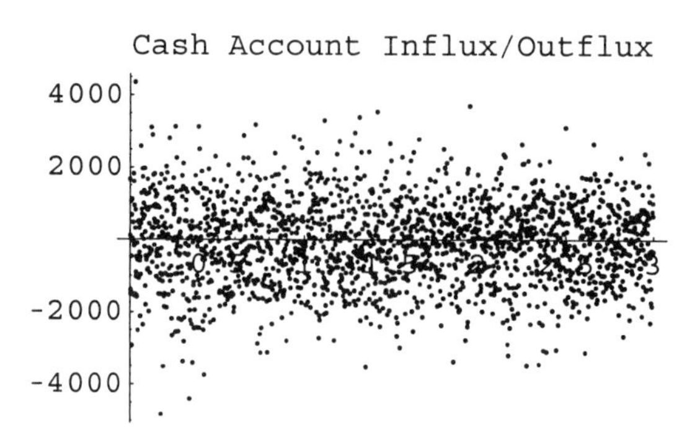

Finally, notice that the initial cash holding is equal to

```
In[36]:= lm_ = Table[1, {m}]; C0 = N[X0 - X0 P.lm]
Out[36]= 383333.
```

and integrating $dC(t) = r\,C(t)\,dt + dc(t)$, we get the cash account balance trajectory, which checks out to be the same, i.e., $X(t) * \text{CashFraction} = C(t)$, as it should, and thereby confirming the above derivation and calculations:

```
In[37]:= CashBalance = FoldList[dt r #1 + #1 + #2 &, C0, dc];
         lpcb = ListPlot[Transpose[{T, CashBalance}],
            DisplayFunction → Identity,
            PlotRange → {Min[CashBalance],
              Max[CashBalance]}, PlotJoined → True,
            PlotLabel → "Cash Acount Balance"];
         ListPlot[({#1[[1]], CashFraction #1[[2]]} &) /@
            WealthEvolution,
           DisplayFunction → Identity, PlotJoined → True];
         Show[%, %%, DisplayFunction → $DisplayFunction];
```

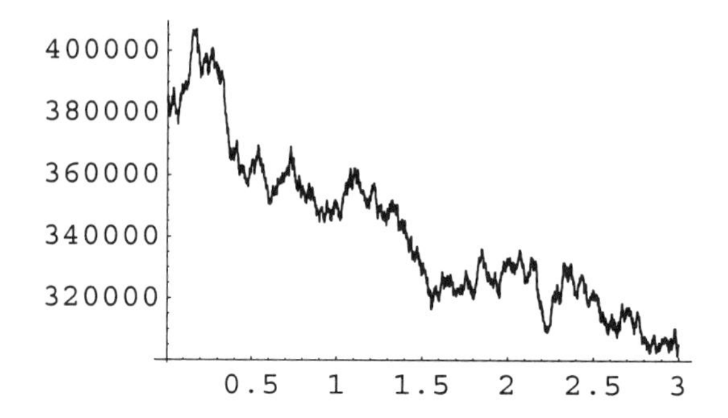

So, having understood the model, the kind of trading strategies considered, and all the trading action that is going on behind the evolution of $X(t)$, it is now time to ask what a better strategy would be than the one just applied, or what the best possible strategy would be for an adopted investor's attitude towards risk-taking (to be quantified by the safety exponent γ).

We store the experiment outcomes for later comparison with the optimal strategy:

```
In[41]:= {dB, MarketEvolution, X0, dt,
    r, a, σ, PS, m, lpwe} >> "Experiment"
```

7.3.3 Stochastic Control Problem

7.3.3.1 Derivation of the Hamilton–Jacobi–Bellman PDE

Similarly, as in the case of optimal stopping, since we cannot maximize the final outcome $X^{\Pi}(T)$, stochastic control theory confines itself to maximizing expressions such as the *expected value* of the utility of the final outcome. So, for a given safety exponent $\gamma > 0$, the problem is to find a strategy $\Pi^{\star}(t, X)$ such as

$$\sup_{\Pi} E_{t,X}\, \psi_{\gamma}\,(X^{\Pi}(T)) = \underset{\Pi}{\mathrm{Max}}\, E_{t,X}\, \psi_{\gamma}\,(X^{\Pi}(T)) = E_{t,X}\, \psi_{\gamma}\left(X^{\Pi^{\star}}(T)\right)$$

where ψ_{γ} is the utility of wealth function. As in the case of optimal stopping, the crucial construct to study in solving stochastic control problems such as this one is to study the *value function*

$$\varphi(t, X) = \sup_{\Pi} E_{t,X}\, \psi_{\gamma}\,(X^{\Pi}(T)).$$

We shall characterize the value function. To this end, recall that for any fixed strategy Π and any differentiable function φ (not necessarily the value function), since $(d X(t))^2 = \Pi(t, X(t)).\sigma.\sigma^T.\Pi(t, X(t))\, dt$, we have

$$\begin{aligned} E_{s,X}\, \varphi(T, X(T)) - \varphi(s, x) = E_{s,X} \int_s^T \Big(&\varphi_t(t, X(t)) \\ &+ \varphi_x(t, X(t))\,(\Pi(t, X(t)).(a - r) + r\, X(t)) \\ &+ \frac{1}{2}\, \varphi_{x,x}(t, X(t))\, \Pi(t, X(t)).\sigma.\sigma^T.\Pi(t, X(t))\Big)\, dt \end{aligned}$$

so that if $\varphi(T, X) = \psi_{\gamma}(X)$, then

$$\begin{aligned} &E_{s,X}\, \psi_{\gamma}(X^{\Pi}(T)) \\ &= \varphi(s, x) + E_{s,x} \int_s^T \Big(\varphi_t(t, X(t)) + \varphi_x(t, X(t))\,(\Pi(t, X(t)).(a - r) + r\, X(t)) \\ &\qquad + \frac{1}{2}\, \varphi_{x,x}(t, X(t))\, \Pi(t, X(t)).\sigma.\sigma^T.\Pi(t, X(t))\Big)\, dt. \end{aligned}$$

Also, if additionally, the linear PDE holds:

$$\varphi_t(t, X) + \varphi_x(t, X)\,(\Pi(t, X).(a - r) + r\,X) + \frac{1}{2}\,\varphi_{x,x}(t, X)\,\Pi(t, X).\sigma.\sigma^T.\Pi(t, X) = 0$$

and then

$$E_{s,X}\,\psi_\gamma(X^\Pi(T)) = \varphi(s, x). \tag{7.3.7}$$

Now suppose we are able to solve the very difficult PDE (Hamilton–Jacobi–Bellman PDE; fully non-linear PDE), i.e., suppose a differentiable function φ can be found such that

$$\begin{aligned}\operatorname*{Max}_{\Pi(t,X)}\Big[&\varphi_t(t, X) + \varphi_x(t, X)\,(\Pi(t, X).(a - r) + r\,X)\\ &+ \frac{1}{2}\,\varphi_{x,x}(t, X)\,\Pi(t, X).\sigma.\sigma^T.\Pi(t, X)\Big] = 0\end{aligned} \tag{7.3.8}$$

together with the terminal condition

$$\varphi(T, X) = \psi_\gamma(X). \tag{7.3.9}$$

Let $\Pi^\star(X, t)$ be the strategy for which the maximum in (7.3.8) is obtained. This means that

$$\begin{aligned}&\varphi_t(t, X) + \varphi_x(t, X)\,(\Pi^\star(X, t).(a - r) + r\,X) + \frac{1}{2}\,\varphi_{x,x}(t, X)\\ &\quad\Pi^\star(X, t).\sigma.\sigma^T.\Pi^\star(X, t) = 0\end{aligned} \tag{7.3.10}$$

while

$$\begin{aligned}&\varphi_t(t, X) + \varphi_x(t, X)\,(\Pi(X, t).(a - r) + r\,X) + \frac{1}{2}\,\varphi_{x,x}(t, X)\\ &\quad\Pi(X, t).\sigma.\sigma^T.\Pi(X, t) \le 0\end{aligned} \tag{7.3.11}$$

for any (other) Π. Therefore, by (7.3.7), (7.3.10) and (7.3.11),

$$\begin{aligned}E_{s,X}\,\psi\big(X^{\Pi^\star}(T)\big) &= \varphi(s, x) \ge \varphi(s, x)\\ &+ E_{s,X}\int_s^T \Big(\varphi_t(t, X(t)) + \varphi_x(t, X(t))\,(\Pi(X(t), t).(a - r) + r\,X(t))\\ &\qquad + \frac{1}{2}\,\varphi_{x,x}(t, X(t))\,\Pi(X(t), t).\sigma.\sigma^T.\Pi(X(t), t)\Big)\,dt\\ &= E_{s,X}\,\psi(X^\Pi(T));\end{aligned}$$

consequently

$$E_{s,X}\,\psi\big(X^{\Pi^\star}(T)\big) = \operatorname*{Max}_{\Pi} E_{s,X}\,\psi(X^\Pi(T)) = \varphi(t, X)$$

and $\Pi^\star$ is the optimal trading strategy. Equation (7.3.8) is called the Hamilton–Jacobi–Bellman PDE, or shortly HJB PDE. It is quite difficult in general. It is fully non-linear and possibly degenerate. Usually, it is quite hopeless to seek exact solutions, and numerical ones are not easy. It is remarkable that in the present case it is going to be possible to solve quite easily the equation in an explicit way.

7.3.3.2 Monge–Ampère PDEs

The HJB PDE is not solved directly. Instead, it is first transformed into another kind of fully-nonlinear and even possibly degenerate equation, but yet a quite a bit easier one. Such equations can be referred to as Monge–Ampère PDEs, and much of what follows will be based on such equations. The terminology is due to the classical, or the so-called *simplest* Monge–Ampère equation (see [39,29])

$$\mathrm{Det}(D^2\varphi) = h \tag{7.3.12}$$

in some convex domain $\Omega \subset \mathbb{R}^n$, for a given function $h > 0$. Det is of course the determinant. We pause here to make few remarks about this equation. For example in $\Omega \subset \mathbb{R}^2$ the equation becomes

$$u^{(0,2)}(x, y)\, u^{(2,0)}(x, y) - u^{(1,1)}(x, y)^2 = h \tag{7.3.13}$$

together with the boundary condition $u = u_b$ on $\partial\Omega$. It is easy to see that if u is a solution of (7.3.12), with $u_b = 0$, then $-u$ is a solution as well. Therefore additional conditions are needed if uniqueness is desired, i.e., if the *right* solution is sought. The additional condition (in $\mathbb{R}^n$) is that the Hessian matrix $D^2\varphi$ is, for example, negative—requesting the positivity of $D^2\varphi$ identifies an alternative solution.

Indeed, the equation (7.3.13) with $h = 10$, in domain $\Omega = (0, 1)\times(0, 1)$, with boundary condition $u_b = 0$ on $\partial\Omega$, together with the concavity condition $D^2 u \le 0$, has a *numerical* solution

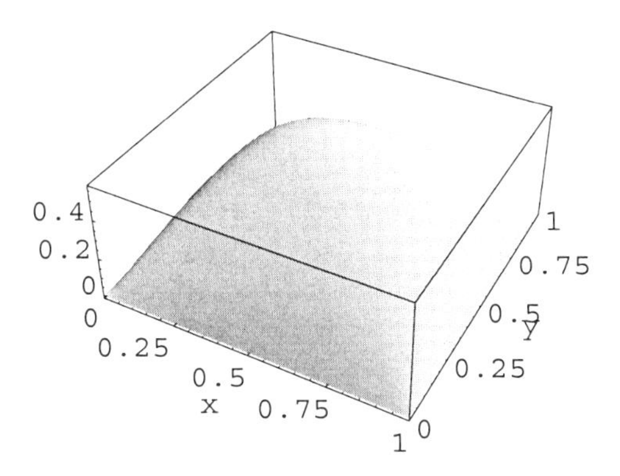

while, if instead the convexity $D^2\varphi \ge 0$ is requested, the unique solution is equal to

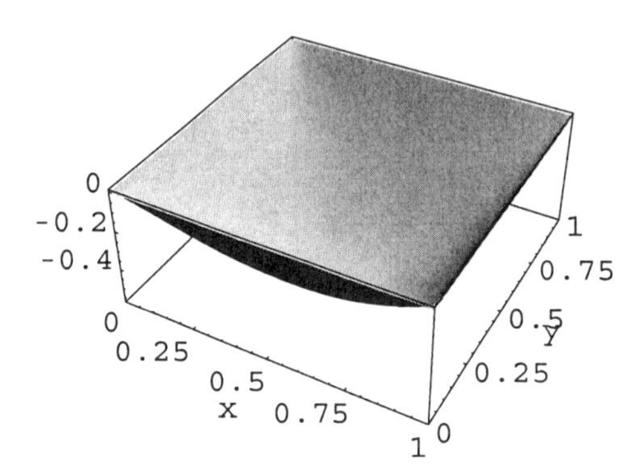

If only $h \geq 0$ is assumed, then the equation becomes *degenerate* and smooth, i.e., a differentiable solution may not exist. Indeed, for example let us look for concave solutions, i.e., for the solutions such that

$$D^2 u \leq 0 \tag{7.3.14}$$

in Ω, and with a non-zero boundary condition; for example, let the boundary condition be $u(x, y) = x\,y$ on $\partial\Omega$. This shows solutions for different right-hand sides:

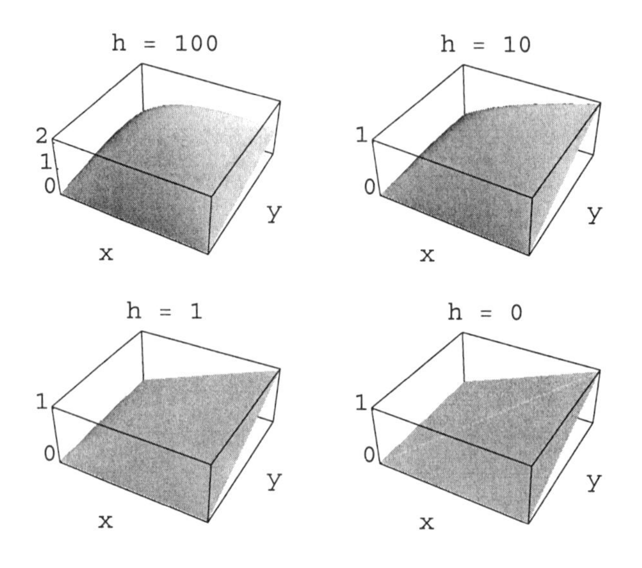

where, by definition,

$$u_0(x, y) := \lim_{h \to 0} u_h(x, y) = \text{Min}[x, y].$$

Since $u_0 \notin C^1(\Omega)$, i.e., since u_0 is not differentiable, u_0 cannot solve (7.3.13) for $h = 0$ in a classical sense; nevertheless u_0 is a solution of (7.3.13) in other ways; also (7.3.14) holds in a generalized sense: u_0 is concave. We notice that if, instead of (7.3.14), $D^2 u \geq 0$, i.e., convexity is assumed, then instead of Min[x, y] we would have $u_0(x, y) = \text{Max}[0, x + y - 1]$, and again $u_0 \notin C^1(\Omega)$.

7.3.3.3 Derivation of Merton's Monge–Ampère PDE

The equation that is equivalent to the above HJB equation (7.3.8) is going to have a similar *multiplicative structure* as (7.3.13). Indeed, let $\Pi^\star$ be the strategy for which the maximum in the HJB equation is realized. Then the gradient with respect to Π of the expression under the Max in the HJB equation, or equivalently, the gradient $\nabla_\Pi f$ of

$$f(\Pi) = \varphi_x(t, X)\,\Pi.(a - r) + \frac{1}{2}\,\varphi_{x,x}(t, X)\,\Pi.\sigma.\sigma^T.\Pi \tag{7.3.15}$$

(for any fixed (t, X)) has to be equal to zero:

$$\nabla_\Pi f(\Pi^\star) = \varphi_x(t, X)\,(a - r) + \varphi_{x,x}(t, X)\,\sigma.\sigma^T.\Pi^\star = 0.$$

There is an interesting detail here, with consequences later on: function f is maximized, and since $\sigma.\sigma^T$ is positive definite, in order for quadratic function f to have the maximum, the condition

$$\varphi_{x,x}(t, X) < 0 \tag{7.3.16}$$

for any (t, X) is necessary. Consequently, since we assume $\sigma.\sigma^T$ to be invertible, we conclude that

$$\Pi^\star(X, t) = -\frac{\varphi_x(t, X)}{\varphi_{x,x}(t, X)}\,(\sigma.\sigma^T)^{-1}.(a - r). \tag{7.3.17}$$

This is not the end of the derivation since the derived strategy depends on the still unknown function φ, the value function. So, plugging (7.3.17) back into the HJB equation, or, which is equivalent, into (7.3.10), we get

$$\varphi_t\,(t, X) + \varphi_x(t, X)\left(\left(-\frac{\varphi_x(t, X)}{\varphi_{x,x}(t, X)}\,(a - r).(\sigma.\sigma^T)^{-1}.(a - r)\right) + r\,X\right)$$
$$+\frac{1}{2}\left(\frac{\varphi_x(t, X)}{\varphi_{x,x}(t, X)}\,(a - r).(\sigma.\sigma^T)^{-1}.(a - r)\right)\varphi_x(t, X) = 0$$

and consequently

$$\varphi_t(t, X)\,\varphi_{x,x}(t, X) + \varphi_x(t, X)\,\varphi_{x,x}(t, X)\,r\,X - \frac{1}{2}\,(\varphi_x(t, X))^2 \\ (a - r).(\sigma.\sigma^T)^{-1}.(a - r) = 0. \tag{7.3.18}$$

We shall refer to (7.3.18), together with the terminal condition

$$\varphi(T, X) = \psi_\gamma(X) \tag{7.3.19}$$

and the additional condition (7.3.16), as Merton's Monge–Ampère PDE. No boundary condition is needed nor is it possible to impose one at $X = 0$, due to degeneracy.

The condition (7.3.16) suggests the following interesting possibility: if $\psi_\gamma(X)$ is strictly convex, instead of strictly concave as assumed in the present chapter ($\gamma > 0$), the terminal condition (7.3.19) cannot be taken continuously since (7.3.16) would be violated. This actually happens in some problems. Condition (7.3.16) is analogous to the concavity condition in the case of the simplest Monge–Ampère equation.

7.3.3.4 Symbolic Solution of Merton's Monge–Ampère PDE

```
In[42]:= Clear["Global`*"];
```

If $\gamma \neq 1$ and $\gamma > 0$, one looks for an explicit solution in the form

$$\text{In[43]:= } \phi_{\gamma\_}[t\_, X\_] := \frac{f[t]\,X^{1-\gamma}}{1-\gamma}$$

i.e., as a product of the terminal condition and some function of t only. The terminal condition (7.3.19) for $\varphi = \phi_\gamma$, $\gamma \neq 1$, implies that $f(T) = 1$. If $\gamma = 1$, we shall search for the explicit solution of the above PDE in the form

```
In[44]:= φ1[t_, X_] := f[t] + Log[X]
```

i.e., as a sum of the terminal condition and some function of t only. The terminal condition (7.3.19) for $\varphi = \phi_1$ implies that $f(T) = 0$. By substitution, we get

$$\text{In[45]:= LHS} = -\frac{1}{2}\,(a - r).\text{Inverse}[\sigma.\text{Transpose}[\sigma]].(a - r) \\ (\partial_X v[t, X])^2 + r\,X\,\partial_{\{X,2\}} v[t, X]\,\partial_X v[t, X] + \\ \partial_t v[t, X]\,\partial_{\{X,2\}} v[t, X]\ /.\ v \to \phi_\gamma$$

$$\text{Out[45]= } -r\,\gamma\,f(t)^2\,X^{-2\gamma} - \frac{1}{2}\,(a - r).(\sigma.\sigma^T)^{-1}.(a - r)\,f(t)^2\,X^{-2\gamma} \\ - \frac{\gamma\,f(t)\,f'(t)\,X^{-2\gamma}}{1-\gamma}$$

and furthermore

```
In[46]:= ODE1 = Simplify[Numerator[Together[LHS]]] == 0
```

Out[46]= $-f(t)\left((\gamma-1)\left(2r\gamma+(a-r).(\sigma.\sigma^T)^{-1}.(a-r)\right)f(t)-2\gamma f'(t)\right)$
$== 0$

in the case $\gamma \neq 1$, and

```
In[47]:= ODE2 = Simplify[Numerator[Together[
             - 1/2 (a - r).Inverse[σ.Transpose[σ]].(a - r)
               (∂X v[t, X])^2 + r X ∂{X,2} v[t, X] ∂X v[t, X] +
              ∂t v[t, X] ∂{X,2} v[t, X] /. v → ϕ1]]] == 0
```

Out[47]= $-(a-r).(\sigma.\sigma^T)^{-1}.(a-r)-2(r+f'(t)) == 0$

in the case $\gamma = 1$. ODE1 can be written more conveniently as

```
In[48]:= ODE1 /.
          {(γ - 1) (2 r γ + (a - r).Inverse[σ.Transpose[σ]].
                (a - r)) → A, -2 γ → B}
```

Out[48]= $-f(t)(A\,f(t)+B\,f'(t)) == 0$

and then it can be solved, together with the terminal condition:

```
In[49]:= Off[DSolve::"bvnul"];
         DSolve[{f[T] == 1, %}, f[t], t][[1]]
```

Out[49]= $\left\{f(t)\to e^{\frac{AT}{B}-\frac{At}{B}}\right\}$

Therefore the explicit solution of the above Monge–Ampère and also of the above HJB equation is then equal to

```
In[50]:= φγ_[t_, X_] = ϕγ[t, X] /. %
```

Out[50]= $\dfrac{e^{\frac{AT}{B}-\frac{At}{B}}X^{1-\gamma}}{1-\gamma}$

if $\gamma \neq 1$, while in the case $\gamma = 1$

```
In[51]:= φ1[t_, X_] =
          FullSimplify[ϕ1[t, X] /. DSolve[{ODE2 /.
                (a - r).Inverse[σ.Transpose[σ]].(a - r) → A,
              f[T] == 0}, f[t], t][[1]]]
```

Out[51]= $\log(X)-\dfrac{1}{2}(A+2r)(t-T)$

We can compute the optimal trading rules

```
In[52]:= Πγ_[t_, X_] =
    - ∂X φγ[t, X] Inverse[σ.Transpose[σ]].(a - r) / ∂{X,2} φγ[t, X]
```

Out[52]= $\frac{X\,(\sigma.\sigma^T)^{-1}.(a-r)}{\gamma}$

in the case $\gamma \neq 1$, and

```
In[53]:= Π1[t_, X_] =
    - ∂X φ1[t, X] Inverse[σ.Transpose[σ]].(a - r) / ∂{X,2} φ1[t, X]
```

Out[53]= $X\,(\sigma.\sigma^T)^{-1}.(a-r)$

if $\gamma = 1$. They coincide for $\gamma = 1$, confirming that the Log utility function is justly named ψ_1. We can also compute the value functions φ_γ in terms of the original data by means of back-substitution

```
In[54]:= Simplify[
    φγ[t, X] /. {A → (γ - 1) (2 r γ + (a - r).Inverse[
          σ.Transpose[σ]].(a - r)), B → -2 γ}]
```

Out[54]= $-\frac{e^{\frac{(t-T)(\gamma-1)\left(2r\gamma+(a-r).(\sigma.\sigma^T)^{-1}.(a-r)\right)}{2\gamma}}\,X^{1-\gamma}}{\gamma-1}$

in the case $\gamma \neq 1$, and

```
In[55]:= φ1[t, X] /.
    {A → (a - r).Inverse[σ.Transpose[σ]].(a - r)}
```

Out[55]= $\log(X) - \frac{1}{2}\,(t-T)\left(2\,r + (a-r).(\sigma.\sigma^T)^{-1}.(a-r)\right)$

in the case $\gamma = 1$. These explicit solutions of the fully non-linear PDE of Monge–Ampère type are very interesting. Indeed, if $\gamma \geq 1$ they blow up at $X = 0$; recall no boundary condition is imposed there. This is possible since the equation is degenerate.

The extraordinary success of the above derivation, the explicit solution of the HJB and/or Monge–Ampère type equation, and consequently the explicit solution of the optimal portfolio rule problem is due to the special structure of the equation and of the terminal condition. If either one of these is perturbed, the above analysis would not be applicable and one would have to resort to numerical solutions.

The reader is invited to find where the condition $\gamma > 0$ was used in the above derivation. That indeed the condition was used can be seen from the derived result: if $\gamma < 0$ the investment decisions would be the *opposite* of what is recommended in the case $\gamma > 0$.

7.3.4 Optimal Portfolio Hedging Strategy Implemented

Now we implement the Merton's optimal trading rule. We use the same data, and the same randomization as in the case of an ad-hoc strategy implemented earlier, hoping to see an improvement.

```
In[56]:= Clear["Global`*"];
         {dB, MarketEvolution, X0, dt, r, a, σ, PS, m, lpwe} =
          << "Experiment";
```

So now, instead of applying an ad hoc strategy, we define Merton's optimal strategy:

```
In[57]:= Πγ_[t_, X_] := Pγ X;
         Pγ_ = Inverse[σ.Transpose[σ]].(a - r) / γ
```

$$\text{Out[58]= } \left\{-\frac{1.94907}{\gamma}, -\frac{0.625835}{\gamma}, -\frac{2.94151}{\gamma}\right\}$$

and prepare for the generation of the corresponding wealth trajectory:

```
In[59]:= Fγ_[{t_, X_}, dB_] := {t + dt, Max[X, 0] +
           (Πγ[t, Max[X, 0]].(a - r) + r Max[X, 0]) dt +
           Πγ[t, Max[X, 0]].σ.dB}
```

So, this time, if we choose the safety exponent to be

```
In[60]:= γ = 7;
```

the corresponding wealth evolution is equal to

```
In[61]:= WealthEvolution =
           ({#1[[1]], #1[[2]]} &) /@ FoldList[Fγ, {0, X0}, dB];
         lpwe2 = ListPlot[WealthEvolution, PlotJoined →
             True, PlotStyle → RGBColor[0, 0, 1], PlotRange →
             {0, Max[Transpose[WealthEvolution][[2]]]}];
```

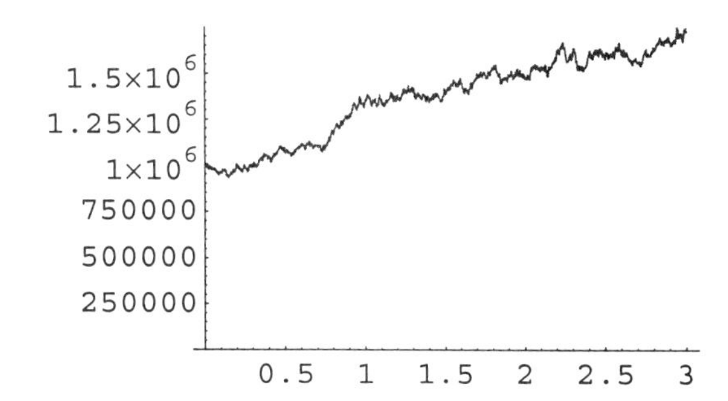

Or, comparing with the outcome of trading according to the previous ad hoc strategy, discussed in Section 7.3.2.

```
In[63]:= Show[lpwe, lpwe2, PlotRange → All];
```

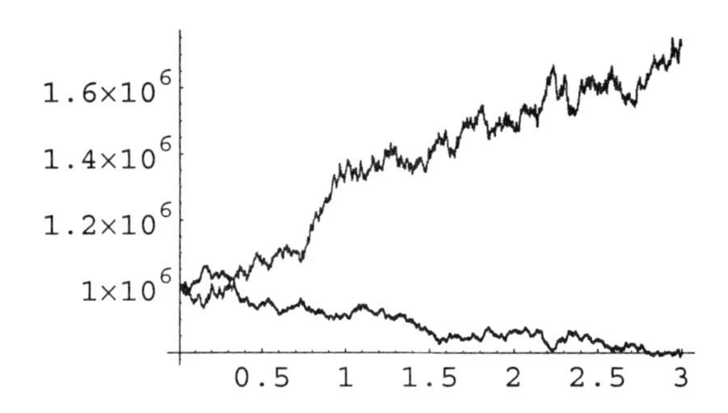

Similarly, as before, define

```
In[64]:= T = (#1[[1]] &) /@ MarketEvolution;
         S = (#1[[2]] &) /@ MarketEvolution; St = Drop[S, -1];
         Stdt = Drop[S, 1]; X = (#1[[2]] &) /@ WealthEvolution;
         ΠtXt = (Pγ #1 &) /@ Drop[X, -1];
         ΠtdtXtdt = (Pγ #1 &) /@ Drop[X, 1];
```

yielding, for example, the evolution of the components of $\Pi_\gamma(t, X(t))$, where negative investments represent selling short:

```
In[65]:= StockBalancePlot = Show[
           Module[{x}, x = Table[({#1[[1]], #1[[2, i]]} &) /@
               Transpose[{Drop[T, -1], ΠtXt}], {i, 1, m}];
            Table[ListPlot[x[[i]], PlotJoined → True,
              DisplayFunction → Identity,
              PlotStyle → PS[[i]]], {i, 1, m}]],
           DisplayFunction → $DisplayFunction];
```

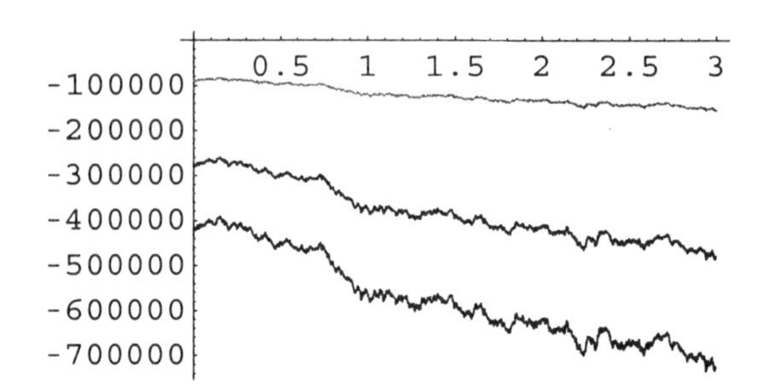

We compute also

```
In[66]:= NumberOfStocksTraded = (#1[[3]]/#1[[1]] - #1[[4]]/#1[[2]] &) /@
           Transpose[{St, Stdt, ΠtXt, ΠtdtXtdt}];
```

which look like

```
In[67]:= Do[ListPlot[Transpose[{Drop[T, 1],
             Transpose[NumberOfStocksTraded][[i]]}],
           PlotRange → All, PlotStyle → PS[[i]],
           PlotLabel → "Stocks Sold/Bought Company " <>
             ToString[i]], {i, 1, m}];
```

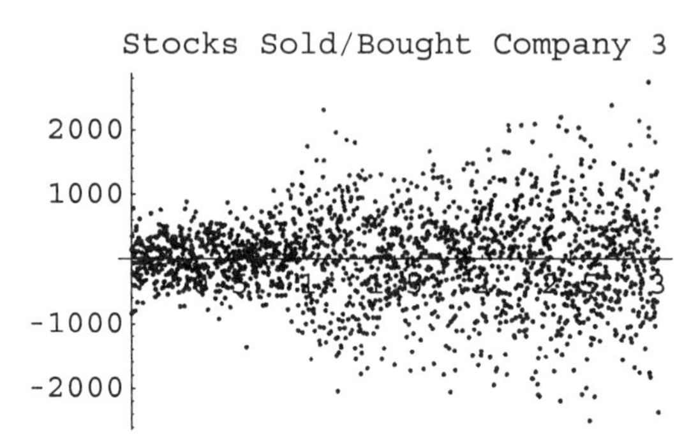

and the corresponding cash transactions were (positive transaction means a sale)

```
In[68]:= DC = ((#1[[3]]/#1[[1]] - #1[[4]]/#1[[2]]) #1[[2]] &) /@
      Transpose[{St, Stdt, ΠtXt, ΠtdtXtdt}];
    Do[ListPlot[Transpose[{Drop[T, 1], Transpose[DC][[i]]}],
      PlotStyle → PS[[i]], PlotRange → All,
      PlotLabel → ToString[i] <>
        "th Stock Caused Cash Flow"], {i, 1, m}];
```

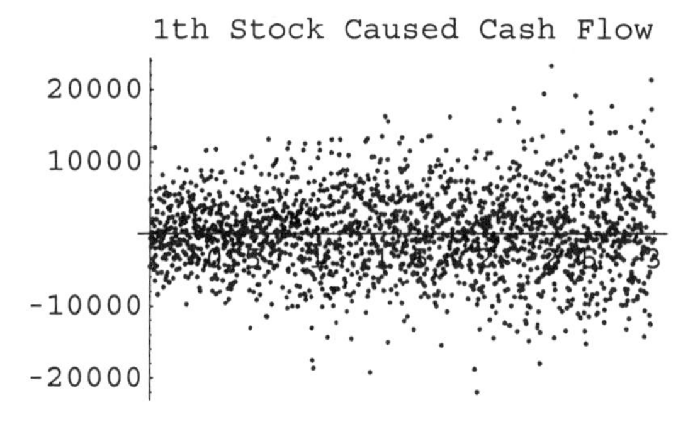

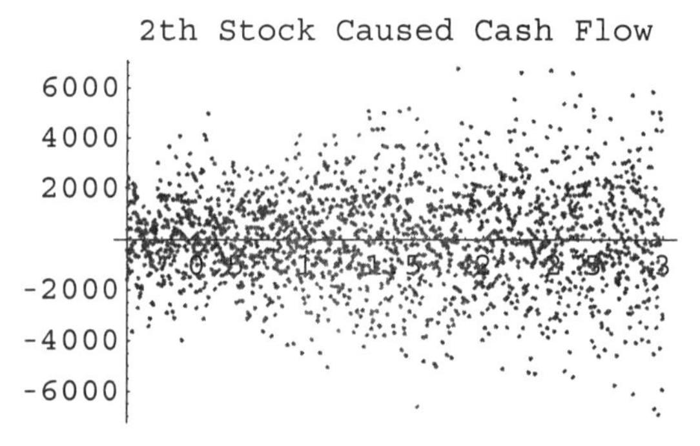

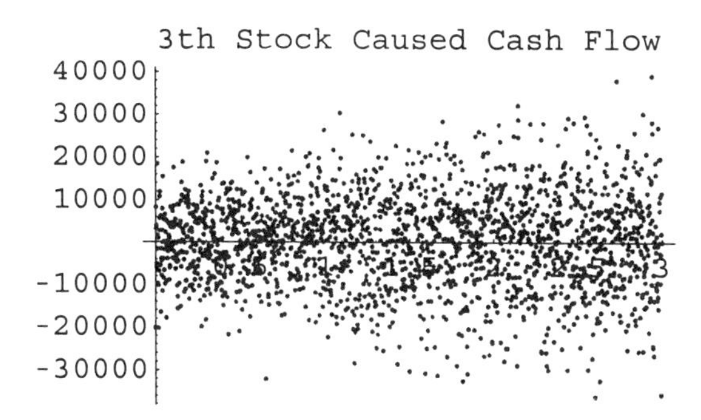

as well as their cumulative cash influx into the cash account $dc(t)$ (one can see that many of the cash transactions are offset between sales and purchases of the stocks, and only the remaining balance is deposited or withdrawn from the cash account):

```
In[70]:= dc = ((#1[[3]]/#1[[1]] - #1[[4]]/#1[[2]]) .#1[[2]] &) /@
      Transpose[{St, Stdt, ΠtXt, ΠtdtXtdt}];
    ListPlot[Transpose[{Drop[T, 1], dc}],
      PlotLabel → "Cash Account Influx/Outflux"];
```

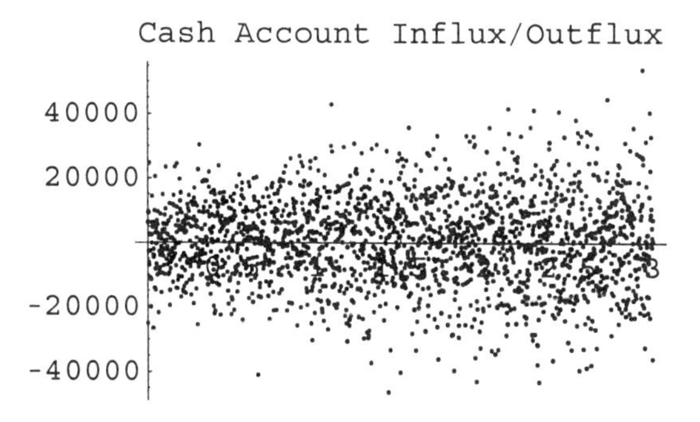

Since the value function $\varphi_\gamma(t, X)$:

```
In[72]:= φγ[t_, x_] =
   - e^((t-T[[-1]]) (γ-1) (2 r γ+ (a-r).Inverse[σ.Transpose[σ]].(a-r)) / (2 γ)) x^(1-γ) / (γ - 1)
```

$$Out[72]= -\frac{e^{0.728387\,(t-3)}}{6\,x^6}$$

is the central object in stochastic control theory, we make one plot of φ_γ (in the restricted domain) for an overall understanding:

```
In[73]:= Plot3D[Evaluate[φγ[t, x]], {t, 0, T[[-1]]},
    {x, 1000, 2000}, PlotRange → All,
    PlotPoints → 80, Mesh → False,
    AxesLabel → {"t", "X", ""}, Ticks → {Automatic,
      Range[1000., 2000, 1000 / 3], Automatic}];
```

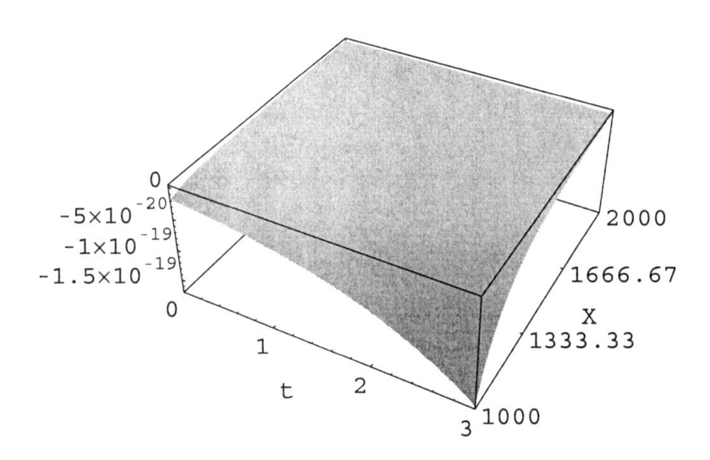

One may argue that for a *central object* this value function looks quite uneventful. Furthermore, as a numerical analyst we would start to worry when looking at the values it takes and how flat it is if numerical calculations for related problems are planned, especially since the optimal portfolio rule is computed using first and even second derivatives of the value function. We shall *settle* these issues in Chapter 8.

Regardless of these concerns, putting all that was developed so far together, we gain a fairly complete description of the nature of Merton's optimal trading strategy. The strategy appears to be quite successful provided the assumed stock price Log-Normal model is correct, and providing that the investor's inference about the appreciation rate vector a is precise (which is unrealistic, as we have seen earlier), and provided one can afford its implementation.

7.3.5 Fringe Issues

7.3.5.1 An Alternative Approach

Here is an another method for finding the optimal trading strategy which does not utilize PDEs at all. Assume $\Pi(t, X) = P(t)\, X$.

Indeed, from the wealth evolution equation

$$\begin{aligned} dX(t) &= (\Pi(t, X(t)).(a - r) + r\, X(t))\, dt + \Pi(t, X(t)).\sigma.dB(t) \\ &= (X(t)\, P(t).(a - r) + r\, X(t))\, dt + X(t)\, P(t).\sigma.dB(t), \end{aligned}$$

we see, applying the Itô chain rule, that

$$E_{s,X}\, f(X(T)) = f(X) + E_{s,X} \int_s^T \left(\frac{\partial f(X(t))}{\partial X} (r\, X(t) + P(t).(a-r) X(t)) + \frac{1}{2} \frac{\partial^2 f(X(t))}{\partial X^2} X(t)^2\, P(t).\sigma.\sigma^T.P(t) \right) dt$$

for any function f regular enough. As an introduction, we shall present the method here in the case of the Log utility function. We shall come back to this method later many times to solve many different problems and for different utility functions. So, if $f(X) = \log(X)$, then

$$E_{s,X} \log(X(T)) = \log(X) + \int_s^T \left(r + P(t).(a-r) - \frac{1}{2} P(t).\sigma.\sigma^T.P(t)\right) dt.$$

This expectation is maximized if the integrand is maximized (with respect to $P(t)$) for any time t. Differentiating the integrand with respect to $P(t)$, and setting it equal to zero, we get

$$(a-r) - \sigma.\sigma^T.P(t) = 0$$

i.e.,

$$P(t) = P = (\sigma.\sigma^T)^{-1}.(a-r).$$

So, even though we have allowed time dependency of P, in the end it turned out that is not needed—the optimal P is time independent. The value function, which is not really necessary now, in the case of the Log utility can be re-computed in the following way:

$$\begin{aligned} E_{s,X} \log(X(T)) &= \log(X) + \left(r + P.(a-r) - \frac{1}{2} P.\sigma.\sigma^T.P\right) \int_s^T dt \\ &= \log(X) + \Big(r + \left((\sigma.\sigma^T)^{-1}.(a-r)\right).(a-r) \\ &\quad - \frac{1}{2} \left((\sigma.\sigma^T)^{-1}.(a-r)\right).\sigma.\sigma^T.\left((\sigma.\sigma^T)^{-1}.(a-r)\right)\Big)(T-s) \\ &= \log(X) + \Big(r + (a-r).(\sigma.\sigma^T)^{-1}.(a-r) \\ &\quad - \frac{1}{2} (a-r).(\sigma.\sigma^T)^{-1}.(a-r)\Big)(T-s) \\ &= \log(X) + \left(r + \frac{1}{2} (a-r).(\sigma.\sigma^T)^{-1}.(a-r)\right)(T-s) \end{aligned}$$

which is consistent with the above computation. In the case $\gamma \neq 1$ the calculation is more complicated and we postpone it until one of the subsequent sections when it is actually necessary.

7.3.5.2 Optimal Portfolio Balance Evolution

Applying the optimal strategy

$$\Pi(t, X) = \frac{(\sigma.\sigma^T)^{-1}.(a-r)}{\gamma} X = \frac{(a-r).(\sigma.\sigma^T)^{-1}}{\gamma} X \tag{7.3.20}$$

in the wealth evolution SDE

$$dX(t) = (\Pi(t, X(t)).(a-r) + r X(t))\, dt + \Pi(t, X(t)).\sigma.dB(t),$$

we arrive at

$$\begin{aligned} dX(t) &= \left(X(t) \frac{(a-r).(\sigma.\sigma^T)^{-1}}{\gamma}.(a-r) + r X(t) \right) dt \\ &+ X(t) \frac{(a-r).(\sigma.\sigma^T)^{-1}}{\gamma}.\sigma.dB(t) \\ &= X(t) \left(r + \frac{(a-r).(\sigma.\sigma^T)^{-1}.(a-r)}{\gamma} \right) dt \\ &+ X(t) \frac{\sqrt{(a-r).(\sigma.\sigma^T)^{-1}.(a-r)}}{\gamma} dB_0(t) \end{aligned} \tag{7.3.21}$$

where

$$B_0(t) = \frac{\frac{(a-r).(\sigma.\sigma^T)^{-1}}{\gamma}.\sigma.B(t)}{\frac{\sqrt{(a-r).(\sigma.\sigma^T)^{-1}.(a-r)}}{\gamma}} = \frac{(a-r).(\sigma.\sigma^T)^{-1}.\sigma.B(t)}{\sqrt{(a-r).(\sigma.\sigma^T)^{-1}.(a-r)}}$$

is one-dimensional Brownian motion. Indeed,

$$(dB_0(t))^2 = \frac{(a-r).(\sigma.\sigma^T)^{-1}.\sigma}{\sqrt{(a-r).(\sigma.\sigma^T)^{-1}.(a-r)}} \cdot \frac{\sigma^T.(\sigma.\sigma^T)^{-1}.(a-r)}{\sqrt{(a-r).(\sigma.\sigma^T)^{-1}.(a-r)}} dt$$

$$= \frac{(a-r).(\sigma.\sigma^T)^{-1}.(a-r)}{(a-r).(\sigma.\sigma^T)^{-1}.(a-r)} dt = dt.$$

From (7.3.18), in particular, we conclude that the wealth volatility is equal to (computed on the data from Section 7.3.4)

```
In[74]:= √(a - r).Inverse[σ.Transpose[σ]].(a - r)
         ----------------------------------------
                            γ

Out[74]= 0.142826
```

while the wealth appreciation rate is equal to (on the same data)

```
In[75]:= r + (a - r).Inverse[σ.Transpose[σ]].(a - r) / γ
```

```
Out[75]= 0.192796
```

Those two numbers are good measures for evaluating the deduced strategy Π.

7.3.5.3 Time-Dependent Market Dynamics

```
In[76]:= Clear["Global`*"]
```

It is easy to generalize the above solution to the optimal portfolio problem under time-dependent market dynamics:

$$d C(t) = r(t)\, C(t)\, d t \tag{7.3.22}$$

(in the absence of trading, i.e., when $d c(t) = 0$), while the stock prices $S(t) = \{S_1(t), \ldots, S_m(t)\}$ obey the SDE

$$d S(t) = S(t)\, a(t)\, d t + S(t)\, \sigma(t).d B(t) \tag{7.3.23}$$

where the vector-valued function $a(t) = \{a_1(t), \ldots, a_m(t)\}$ is the vector (one-dimensional array) of appreciation rates of the corresponding stocks, $\sigma(t) \in \mathbb{R}^m \times \mathbb{R}^n$ is the (volatility) matrix (two-dimensional array), and $B(t) = \{B_1(t), \ldots, B_n(t)\}$ is the vector of n independent Brownian motions.

As above, let the utility function for the terminal wealth be

$$\psi_\gamma(X) = \frac{X^{1-\gamma}}{1-\gamma}$$

suggesting again that the value function is in the form

```
In[77]:= ϕγ_[t_, X_] := f[t] X^(1-γ) / (1 - γ)
```

with $f(T) = 1$. By substitution, in the Monge–Ampère equation with time-dependent coefficients, we get

```
In[78]:= LHS = FullSimplify[
           Numerator[Together[-1/2 (a[t] - r[t]).Inverse[
               σ[t].Transpose[σ[t]]].(a[t] - r[t])
              (∂X φ[t, X])^2 + X ∂{X,2} φ[t, X] r[t] ∂X φ[t, X] +
             ∂t φ[t, X] ∂{X,2} φ[t, X] /. φ → ϕγ]]]
```

$$\text{Out[78]= } f(t)\left(2\,\gamma\, f'(t) - (\gamma-1)\, f(t)\left((a(t)-r(t)).(\sigma(t).\sigma(t)^T)^{-1} .(a(t)-r(t)) + 2\,\gamma\, r(t)\right)\right)$$

It is easy to notice in the above equation an ODE that can be written more conveniently after the substitution as

```
In[79]:= LHS2 = (LHS /.
           {(a[t] - r[t]).Inverse[σ[t].Transpose[σ[t]]].
              (a[t] - r[t]) + 2 γ r[t] →
            A[t], -2 γ → B}) / f[t]
```

Out[79]= $2\gamma f'(t) - (\gamma - 1) A(t) f(t)$

and then solved, together with the terminal condition:

```
In[80]:= Simplify[
          f[t] /. DSolve[{f[T] == 1, LHS2 == 0}, f[t], t][[1]]]
```

Out[80]= $e^{\frac{(\gamma-1)\left(\int_{K\$345}^{t} A(K\$344)\, dK\$344 - \int_{K\$345}^{T} A(K\$344)\, dK\$344\right)}{2\gamma}}$

Plugging the just-computed $f(t)$ into the expression for $\phi_\gamma(t, X)$ above,

```
In[81]:= φγ_[t_, X_] = % X^(1-γ) / (1 - γ)
```

Out[81]= $\frac{e^{\frac{(\gamma-1)\left(\int_{K\$411}^{T} A(K\$410)\, dK\$410 - \int_{K\$411}^{t} A(K\$410)\, dK\$410\right)}{B}} X^{1-\gamma}}{1-\gamma}$

we get the solution of the Monge–Ampère PDE. We can compute the optimal portfolio rule (even without using the substitution)

```
In[82]:= Πγ_[t_, X_] =
          -(∂X φγ[t, X] Inverse[σ[t].Transpose[σ[t]]].
              (a[t] - r[t])) / ∂{X,2} φγ[t, X]
```

Out[82]= $\frac{X\, (\sigma(t).\sigma(t)^T)^{-1}.(a(t) - r(t))}{\gamma}$

for any $\gamma > 0$, which is the same formula as before in the case of the time-independent market dynamics. This seems to be important conceptually. It implies that even though the market conditions (the coefficients in the above equation) change over time, if at each time we apply "current understanding" of the market in an optimal fashion, this is the best we can do also in the long run. We emphasize that variability of the market conditions here is only with respect to time. If for example market conditions depend also on an market index, e.g., Dow Jones or Nasdaq, the conclusion is not quite as simple (see Section 8.2).

7.4 Portfolio Rules under Appreciation Rate Uncertainty

7.4.1 Statement of the Problem and Solution of the Easy Case

As discussed before, statistical estimates of volatility are quite accurate even over very short periods of time provided enough data is gathered, while the statistical estimates of appreciation rates on the contrary are unfortunately inadequate. Although estimates can be improved if some prior knowledge/assumption is available (and correct), they still can provide only a probability distribution for the appreciation rate, with hopefully a reasonably small standard deviation. How can we incorporate this reality into portfolio optimization techniques? What can be said if no exact information about stock appreciation rates is available, as certainly is the case in practice? So, we assume that the stock prices obey SDE:

$$\frac{dS(t)}{S(t)} = (a + b)\,dt + \sigma.dB(t) \tag{7.4.1}$$

where $a = \{a_1, ..., a_m\} \in \mathbb{R}^m$ is the constant vector of *expected* appreciation rates, and now $b = \{b_1, ..., b_m\} \in \mathbb{R}^m$ is a random vector normally distributed, $b \sim N(0, S.S^T)$ where $S.S^T \geq 0$ is the covariance matrix, measuring our lack of confidence in the appreciation rates estimates a. Also, as before, the $m \times n$ matrix $\sigma = (\sigma_{i,j})$ is used for quantifying the volatility and correlation between the stocks. Recall that the standard assumption in Merton's model was that $\sigma.\sigma^T$ is invertible. In the present case there is an analogous but somewhat different assumption to be made when it becomes necessary, at the end of the derivation below. Also $B(t) = \{B_1(t), ..., B_n(t)\}$ is the vector of n independent Brownian motions. It is assumed that the random vector b is independent of the Brownian motion $B(t)$. The cash account balance, as before, obeys

$$dC(t) = r\,C(t)\,dt \tag{7.4.2}$$

(in the absence of trading). The motivation for this section comes from the following natural practical question: since, as seen before, the volatility, i.e., the price instability, or in a way the price uncertainty, inhibits investments, should the uncertainty of the knowledge do the same? Or maybe not, depending on the investors attitude towards risk? Furthermore, one way or the other, one would like to quantify precisely the trading rule, taking into account the measure of the uncertainty of the knowledge (about appreciation rates).

Indeed, proceeding as before we can see that now that the wealth, i.e., the brokerage account balance evolution obeys

$$\begin{aligned} dX(t) &= r\,(X(t) - \Pi(t, X(t)).\mathrm{I}_m)\,dt + \Pi(t, X(t)).((a + b)\,dt + \sigma.dB(t)) \\ &= (\Pi(t, X(t)).(a + b - r) + r\,X(t))\,dt + \Pi(t, X(t)).\sigma.dB(t). \end{aligned} \tag{7.4.3}$$

Justified by the outcome of the analysis in the case of Merton's portfolio problem, we assume that an optimal trading rule can be found among those that are linear in wealth, i.e., we assume that $\Pi(X, t) = P(t)\, X$, for some, possibly time-dependent, vector $P(t)$. The wealth evolution equation becomes

$$dX(t) = (X(t)\, P(t).(a + b - r) + r\, X(t))\, dt + X(t)\, P(t).\sigma.dB(t). \tag{7.4.4}$$

Sometimes, we shall use the notation $X^P(t)$ to denote the wealth process generated by a particular trading strategy $\Pi(X, t) = P(t)\, X$, as opposed to $X(t)^p$ which would mean the pth power of $X(t)$. The problem we are going to solve is the following: For any $\gamma > 0$, find $P^\star(t)$ such that

$$E_{s,X}\, \psi_\gamma\big(X^{P^\star}(T)\big) = \operatorname*{Max}_P E_{s,X}\, \psi_\gamma(X^P(T))$$

where $\psi_\gamma(X)$ is the usual wealth utility function. Applying the Itô chain rule, we get

$$E_{s,X}\, f(X(T)) = f(X) + E_{s,X} \int_s^T \left(\frac{\partial f\,(X(t))}{\partial X}\, (r\, X(t) + P(t).(a + b - r)\, X(t)) \right.$$
$$\left. + \frac{1}{2}\, \frac{\partial^2 f\,(X(t))}{\partial X^2}\, X(t)^2\, P(t).\sigma.\sigma^T.P(t) \right) dt$$

for any regular function f. Consider first the Log utility function $\psi_1(X) = \log(X)$. This case is rather simple. Setting $f(X) = \psi_1(X) = \log(X)$, we get

$$\begin{aligned}
&E_{s,X} \log(X^P(T)) = E_{s,X} \log(X(T)) \\
&= \log(X) + E_{s,X} \int_s^T \left(r + P(t).(a + b - r) - \frac{1}{2}\, P(t).\sigma.\sigma^T.P(t) \right) dt \\
&= \log(X) + \int_s^T \left(r + P(t).(a + E_{s,X}\, b - r) - \frac{1}{2}\, P(t).\sigma.\sigma^T.P(t) \right) dt \\
&= \log(X) + \int_s^T \left(r + P(t).(a - r) - \frac{1}{2}\, P(t).\sigma.\sigma^T.P(t) \right) dt
\end{aligned}$$

since $E_{s,X}\, b = E\, b = 0$. This yields that the optimal trading strategy, i.e., the one maximizing $E_{s,X} \log(X^P(T))$, is the one maximizing the integrand

$$r + P(t).(a - r) - \frac{1}{2}\, P(t).\sigma.\sigma^T.P(t)$$

for every $t \in (s, T)$. The maximizer is obviously a constant vector

$$P(t) = P = (\sigma.\sigma^T)^{-1}.(a - r). \tag{7.4.5}$$

So, in the case of the Log utility function nothing changes as the doubts about the appreciation rate are introduced into consideration. This suggests that the Log utility function is not conservative enough. Indeed, the lack of knowledge (about the exact appreciation rate) should cause concern, and therefore should reduce the risk exposure.

7.4.2 A New Portfolio Rule via Calculus of Variations

In the case $\gamma \neq 1$ the calculation is quite a bit more complicated than the one above. We have, by means of the Itô chain rule,

$$
\begin{aligned}
d X(t)^{1-\gamma} &= (1-\gamma) X(t)^{-\gamma} ((X(t) P(t).(a+b-r) + r X(t)) dt \\
&\quad + X(t) P(t).\sigma.dB(t)) - \frac{1}{2}(1-\gamma)\gamma X(t)^{-\gamma-1} X(t)^2 P(t).\sigma.\sigma^T.P(t) dt \\
&= (1-\gamma) X(t)^{1-\gamma}\left(r + P(t).(a+b-r) - \frac{1}{2}\gamma P(t).\sigma.\sigma^T.P(t)\right) dt \\
&\quad + (1-\gamma) X(t)^{1-\gamma} P(t).\sigma.dB(t).
\end{aligned}
$$

Taking the conditional expectation of both sides, we get

$$
\begin{aligned}
& d E_{s,X}[X(t)^{1-\gamma} \mid b=\beta] = (1-\gamma) E_{s,X}[X(t)^{1-\gamma} \mid b=\beta] \\
& \left(r + P(t).(a+\beta-r) - \frac{1}{2}\gamma P(t).\sigma.\sigma^T.P(t)\right) dt.
\end{aligned}
$$

So, the function $g(t) = E_{s,X}[X(t)^{1-\gamma} \mid b=\beta]$ satisfies an ODE (initial value problem)

$$
\begin{aligned}
& \frac{\partial g(t)}{\partial t} = (1-\gamma) g(t)\left(r + P(t).(a+\beta-r) - \frac{1}{2}\gamma P(t).\sigma.\sigma^T.P(t)\right) \\
& g(s) = X^{1-\gamma}.
\end{aligned}
$$

This ODE has the unique solution

$$
\begin{aligned}
& E_{s,X}[X(t)^{1-\gamma} \mid b=\beta] = g(t) = X^{1-\gamma} e^{\int_s^t (1-\gamma)\left(r+P(\tau).(a+\beta-r) - \frac{1}{2}\gamma P(\tau).\sigma.\sigma^T.P(\tau)\right) d\tau} \\
& = X^{1-\gamma} e^{\int_s^t (1-\gamma)\left(r+P(\tau).(a-r) - \frac{1}{2}\gamma P(\tau).\sigma.\sigma^T.P(\tau)\right) d\tau} e^{\int_s^t (1-\gamma) P(\tau)\, d\tau.\beta}.
\end{aligned}
$$

Therefore,

$$
\begin{aligned}
& E_{s,X}(X^P(t))^{1-\gamma} = E_{s,X}(X(t))^{1-\gamma} = E_{s,X}(E_{s,X}[(X(t))^{1-\gamma} \mid b=\beta]) \\
& = X^{1-\gamma} e^{\int_s^t (1-\gamma)\left(r+P(\tau).(a-r) - \frac{1}{2}\gamma P(\tau).\sigma.\sigma^T.P(\tau)\right) d\tau} E_{s,X}\, e^{\int_s^t (1-\gamma) P(\tau)\, d\tau.b} \qquad (7.4.6) \\
& = X^{1-\gamma} e^{\int_s^t (1-\gamma)\left(r+P(\tau).(a-r) - \frac{1}{2}\gamma P(\tau).\sigma.\sigma^T.P(\tau)\right) d\tau} E\, e^{\int_s^t (1-\gamma) P(\tau)\, d\tau.b}
\end{aligned}
$$

for any strategy $P(t)$. We claim that for any (deterministic) vector $\pi = \{\pi_1, \ldots, \pi_m\}$, since b is $N(0, S.S^T)$ distributed, the following beautiful formula holds:

$$
E\, e^{\pi.b} = e^{\frac{1}{2}|\pi.S|^2}. \qquad (7.4.7)
$$

Notice also $|\pi.S|^2 = \pi.S.S^T.\pi$. To prove (7.4.7), we do the (symbolic) calculation in the case $m = 2$ only. To this end we shall need the package

Define

```
In[83]:= S = Array[s#1,#2 &, {2, 2}]
```

$$\text{Out[83]=}\quad \begin{pmatrix} s_{1,1} & s_{1,2} \\ s_{2,1} & s_{2,2} \end{pmatrix}$$

We need the probability density function for the 2-dimensional normal distribution:

```
In[84]:= << "Statistics`MultinormalDistribution`"
         f[b1, b2] = PDF[MultinormalDistribution[
            {0, 0}, S.Transpose[S]], {b1, b2}]
```

$$\text{Out[85]=}\quad e^{\frac{1}{2}\left(-b_2\left(\frac{b_2\,(s_{1,1}^2+s_{1,2}^2)}{s_{1,2}^2 s_{2,1}^2-2 s_{1,1} s_{1,2} s_{2,2} s_{2,1}+s_{1,1}^2 s_{2,2}^2}+\frac{b_1\,(-s_{1,1} s_{2,1}-s_{1,2} s_{2,2})}{s_{1,2}^2 s_{2,1}^2-2 s_{1,1} s_{1,2} s_{2,2} s_{2,1}+s_{1,1}^2 s_{2,2}^2}\right)-b_1\left(\frac{b_2\,(-s_{1,1} s_{2,1}-s_{1,2} s_{2,2})}{s_{1,2}^2 s_{2,1}^2-2 s_{1,1} s_{1,2} s_{2,2} s_{2,1}+s_{1,1}^2 s_{2,2}^2}+\frac{b_1\,(s_{2,1}^2+s_{2,2}^2)}{s_{1,2}^2 s_{2,1}^2-2 s_{1,1} s_{1,2} s_{2,2} s_{2,1}+s_{1,1}^2 s_{2,2}^2}\right)\right)}$$

$$\Big/\left(2\pi\sqrt{s_{1,2}^2 s_{2,1}^2-2 s_{1,1} s_{1,2} s_{2,2} s_{2,1}+s_{1,1}^2 s_{2,2}^2}\right)$$

We compute $E\,e^{\pi.b}-e^{\frac{1}{2}\pi.S.S^T.\pi}$, i.e.,

$$\text{In[86]:=}\quad \int_{-\infty}^{\infty}\int_{-\infty}^{\infty} e^{\{\pi_1,\pi_2\}.\{b_1,b_2\}}\, f[b_1,\,b_2]\, db_2\, db_1 - e^{\frac{1}{2}\{\pi_1,\pi_2\}.S.\text{Transpose}[S].\{\pi_1,\pi_2\}}\ //\ \text{Simplify}[\#1,\, s_{\_,\_}\in \text{Reals}]\ \&$$

```
Out[86]= 0
```

which proves (7.4.7) in the case $n=2$. Setting $\pi=\int_s^t(1-\gamma)\,P(\tau)\,d\tau$ in (7.4.7), we get

$$E\,e^{\left(\int_s^t(1-\gamma)\,P(\tau)\,d\tau\right).b}=e^{\frac{1}{2}\left|\left(\int_s^t(1-\gamma)\,P(\tau)\,d\tau\right).S\right|^2}=e^{\frac{1}{2}(1-\gamma)^2\left|\left(\int_s^t P(\tau)\,d\tau\right).S\right|^2}$$

Consequently, continuing (7.4.6),

$$\begin{aligned} E_{s,X}(X^P(t))^{1-\gamma} &= X^{1-\gamma}\,e^{\int_s^t(1-\gamma)\left(r+P(\tau).(a-r)-\frac{1}{2}\gamma\,P(\tau).\sigma.\sigma^T.P(\tau)\right)d\tau} \\ E\,e^{\int_s^t(1-\gamma)\,P(\tau)\,d\tau.b} &= X^{1-\gamma}\,e^{\int_s^t(1-\gamma)\left(r+P(\tau).(a-r)-\frac{1}{2}\gamma\,|P(\tau).\sigma|^2\right)d\tau} \\ & e^{\frac{1}{2}(1-\gamma)^2\left|\left(\int_s^t P(\tau)\,d\tau\right).S\right|^2}. \end{aligned} \tag{7.4.8}$$

Therefore

$$E_{s,X} \frac{(X^P(t))^{1-\gamma}}{1-\gamma}$$
$$= \frac{X^{1-\gamma}}{1-\gamma} e^{\int_s^t (1-\gamma)\left(r+P(\tau).(a-r)-\frac{1}{2}\gamma|P(\tau).\sigma|^2\right)d\tau+\frac{1}{2}(1-\gamma)^2 |(\int_s^t P(\tau)\,d\tau).S|^2}$$

for any strategy $P(\tau)$. So, we are looking for $P^\star(\tau)$, such that

$$\frac{X^{1-\gamma}}{1-\gamma} e^{\int_s^t (1-\gamma)\left(r+P^\star(\tau).(a-r)-\frac{1}{2}\gamma|P^\star(\tau).\sigma|^2\right)d\tau+\frac{1}{2}(1-\gamma)^2 |(\int_s^t P^\star(\tau)\,d\tau).S|^2}$$
$$= \operatorname*{Max}_P \frac{X^{1-\gamma}}{1-\gamma} e^{\int_s^t (1-\gamma)\left(r+P(\tau).(a-r)-\frac{1}{2}\gamma|P(\tau).\sigma|^2\right)d\tau+\frac{1}{2}(1-\gamma)^2 |(\int_s^t P(\tau)\,d\tau).S|^2}.$$

Consider two cases: $\gamma < 1$, and $\gamma > 1$. If $\gamma < 1$, then $1-\gamma > 0$, and therefore we need to maximize $e^{\int_s^t (1-\gamma)\left(r+P(\tau).(a-r)-\frac{1}{2}\gamma|P(\tau).\sigma|^2\right)d\tau+\frac{1}{2}(1-\gamma)^2 |(\int_s^t P(\tau)\,d\tau).S|^2}$, and consequently, we need to maximize

$$(1-\gamma)\Big(\int_s^t \Big(r+P(\tau).(a-r)-\frac{1}{2}\gamma\,\big|P(\tau).\sigma\big|^2\Big)d\tau + \frac{1}{2}(1-\gamma)\,\Big|\Big(\int_s^t P(\tau)\,d\tau\Big).S\Big|^2\Big) \tag{7.4.9}$$

and consequently, again using $1-\gamma > 0$, we need to maximize

$$\int_s^t \Big(r+P(\tau).(a-r)-\frac{1}{2}\gamma\,\big|P(\tau).\sigma\big|^2\Big)d\tau + \frac{1}{2}(1-\gamma)\,\Big|\Big(\int_s^t P(\tau)d\tau\Big).S\Big|^2. \tag{7.4.10}$$

On the other hand, if $\gamma > 1$, then $1-\gamma < 0$, and therefore we need to minimize $e^{\int_s^t (1-\gamma)\left(r+P(\tau).(a-r)-\frac{1}{2}\gamma|P(\tau).\sigma|^2\right)d\tau+\frac{1}{2}(1-\gamma)^2 |(\int_s^t P(\tau)\,d\tau).S|^2}$, and consequently, we need to minimize (7.4.9), and consequently, using $1-\gamma < 0$, we need to maximize (7.4.10). So, in either case, we need to maximize (7.4.10), or which is equivalent, we need to solve the problem: Find $P^\star(\tau)$ such that

$$\begin{aligned} J(P^\star) &= \int_s^t \Big(P^\star(\tau).(a-r)-\frac{1}{2}\gamma\,\big|P^\star(\tau).\sigma\big|^2\Big)d\tau \\ &+\frac{1}{2}(1-\gamma)\,\Big|\Big(\int_s^t P^\star(\tau)\,d\tau\Big).S\Big|^2 \\ &= \operatorname*{Max}_P \int_s^t \Big(P(\tau).(a-r)-\frac{1}{2}\gamma\,\big|P(\tau).\sigma\big|^2\Big)d\tau \\ &+\frac{1}{2}(1-\gamma)\,\Big|\Big(\int_s^t P(\tau)\,d\tau\Big).S\Big|^2 = \operatorname*{Max}_P J(P). \end{aligned} \tag{7.4.11}$$

This is an interesting *calculus of variations* problem, which can be solved by looking for a critical point, i.e., for the function $P^\star(\tau)$ such that $\nabla_P J(P^\star) = 0$, or which is the same $J'(P^\star, Q) = \nabla_P J(P^\star).Q = 0$ for any Q. More explicitly, we need to solve the problem: Find an m-vector valued function $P(\tau)$, such that

$$
\begin{aligned}
0 = J'(P, Q) &= \lim_{\delta\to 0} \frac{J(P + Q\,\delta) - J(P)}{\delta} \\
&= \int_s^t (Q(\tau).(a-r) - \gamma\, P(\tau).\sigma.\sigma^T.Q(\tau))\, d\tau + (1-\gamma) \int_s^t P(\tau)\, d\tau.S.S^T. \int_s^t Q(\tau) d\tau \\
&= \int_s^t (Q(\tau).(a-r) - \gamma\, Q(\tau).\sigma.\sigma^T.P(\tau))\, d\tau + (1-\gamma) \int_s^t Q(\tau)\, d\tau.S.S^T. \int_s^t P(\tau) d\tau \\
&= \int_s^t (Q(\tau).((a-r) - \gamma\, \sigma.\sigma^T.P(\tau)))\, d\tau + (1-\gamma) \int_s^t Q(\tau).S.S^T. \int_s^t P(u) du\, d\tau \\
&= \int_s^t Q(\tau).\Big(a - r - \gamma\, \sigma.\sigma^T.P(\tau) + (1-\gamma)\, S.S^T. \int_s^t P(u)\, du\Big)\, d\tau
\end{aligned}
$$

for any m-vector-valued function $Q(\tau)$. Therefore,

$$
0 = a - r - \gamma\, \sigma.\sigma^T.P(\tau) + (1-\gamma)\, S.S^T. \int_s^t P(u)\, du \tag{7.4.12}
$$

for any $\tau \in (s, t)$. Since $\int_s^t P(u)\, du$ is a constant, the only non-constant term in (7.4.12) is $P(\tau)$, and therefore $P(\tau)$ has to be a constant as well; say $P(\tau) = P$. Consequently,

$$
\begin{aligned}
0 &= a - r - \gamma\, \sigma.\sigma^T.P + (1-\gamma)\,(t-s)\, S.S^T.P \\
&= a - r - (\gamma\, \sigma.\sigma^T + (\gamma - 1)\,(t-s)\, S.S^T).P
\end{aligned}
$$

yielding the optimal trading strategy, i.e., the one maximizing $E_{s,X} \frac{X^{P}(t)^{1-\gamma}}{1-\gamma}$, to be equal to

$$
P^\star = P_\gamma = ((\gamma - 1)\,(t-s)\, S.S^T + \gamma\, \sigma.\sigma^T)^{-1}.(a-r) \tag{7.4.13}
$$

provided, as assumed here, that the $m \times m$ matrix $(\gamma - 1)\,(t-s)\, S.S^T + \gamma\, \sigma.\sigma^T$ is invertible.

Some additional comments could be useful here. First, comparing the way matrices $\sigma.\sigma^T$ and $S.S^T$ appear in the above formula, notice that the lack of knowledge (about the appreciation rates) affects an investment decision in a similar fashion as does volatility, but not in exactly the same way. Also, from the practical point of view, they manifest themselves differently on the market, and consequently in general they are estimated differently, as seen before.

Next, analyzing the term $(\gamma - 1)\,(t-s)\, S.S^T$ we see that the utility function plays a fundamental role, and moreover as anticipated above, $\gamma = 1$ is a special case since it separates *aggressive* investing $\gamma < 1$ from *reasonable* investing $\gamma > 1$. Proposing such a terminology is motivated by the above mathematical result. If $\gamma < 1$, *not knowing* appreciation rates exactly increases investments into risky assets (long or short positions), while if $\gamma > 1$, not knowing appreciation rates exactly decreases such investments. It was seen before, and the above formula confirms it, that if $\gamma = 1$, not knowing the precise appreciation rates does not change anything —investing is done on the basis of *expected* appreciation rates. Speaking of γ, notice finally that even under the standard assumption on σ, i.e., under the assumption that $\sigma.\sigma^T$ is invertible, the invertibility of $(\gamma - 1)(t - s) S.S^T + \gamma\, \sigma.\sigma^T$ may fail if $\gamma < 1$.

The final observation is that the remaining trading time $t - s$ appears in the formula as well. This is quite new compared with Merton's formula: the true appreciation rate $a + b$ will affect trading much more during longer investment periods than during the short ones. Over short periods volatility dominates. So over the short periods of time, it is as if trading with the expected appreciation rate a, while over the longer periods of time, the true appreciation rate will affect the outcome more profoundly, and therefore needs to be taken into account proportional to the total length of trading/investing.

7.5 Portfolio Optimization under Equality Constraints

7.5.1 Portfolio Optimization under General Affine Constraints

7.5.1.1 Affine Constraints: Introduction

```
In[87]:= Remove["Global`*"]
```

We start with the same setup as in Merton's problem. More precisely, consider the time-independent market dynamics:

$$d C(t) = r\, C(t)\, d t + d c(t) \tag{7.5.1}$$

is the cash account dynamics (in the presence of trading), while the stock prices $S(t) = \{S_1(t), \ldots, S_m(t)\}$ obey the SDE

$$d S(t) = S(t)\, a\, d t + S(t)\, \sigma.d B(t) \tag{7.5.2}$$

where the vector $a = \{a_1, \ldots, a_m\} \in \mathbb{R}^m$ is the vector (one-dimensional array) of appreciation rates of the corresponding stocks, $\sigma \in \mathbb{R}^m \times \mathbb{R}^n$ is the (volatility) matrix (two-dimensional array), such that $\sigma.\sigma^T$ is invertible, and $B(t) = \{B_1(t), \ldots, B_n(t)\}$ is a vector of n independent Brownian motions.

Here, as everywhere, m, the number of stocks considered, can be quite large. The admissible portfolio strategies

$$\Pi(t, X) = \{\Pi_1(t, X), \ldots, \Pi_m(t, X)\}$$

are now constrained by an underdetermined system of *linear* equations, written in the matrix form as

$$\mu.\Pi(t, X) = \xi\, X \tag{7.5.3}$$

or, which is the same and also useful

$$\Pi(t, X).\mu^T = \xi X \tag{7.5.4}$$

where μ is $k \times m$ matrix, and ξ is k-vector (like Π, ξ is a one-dimensional array, and therefore they are not affected by the transposition) for some $k \le m$. In addition to assuming as before that $\sigma.\sigma^T$ is invertible, it will be necessary to assume that the $k \times k$ matrix

$$\mu.(\sigma.\sigma^T)^{-1}.\mu^T$$

is invertible as well. Under the above assumption that $\sigma.\sigma^T$ is invertible, this is equivalent to the condition that the rows of μ be linearly independent, or equivalently, that $\mu.\mu^T$ be invertible. Such a condition is quite natural. Indeed, it guarantees that there are no contradicting constraints, i.e., that the above constraint equation has at least one solution.

A few examples will help appreciate the usefulness of such a class of constraints.

An example:

Let the number of stock investments considered and the above μ matrix be

In[88]:= `n = 10; 𝟙n_ := Table[1, {n}]; μ = {𝟙n}`

Out[88]= $(1 \ 1 \ 1 \ 1 \ 1 \ 1 \ 1 \ 1 \ 1 \ 1)$

Let ξ be the 1-vector

In[89]:= `ξ = {.5};`

This means that if the portfolio rule is equal to $\Pi(t, X) = \Pi_n$

In[90]:= `Πn_ := Array[π#1 &, {n}]`

then the affine constraint $\mu.\Pi = \xi X$ means

In[91]:= `μ.Πn == ξ X`

Out[91]= $\{\pi_1 + \pi_2 + \pi_3 + \pi_4 + \pi_5 + \pi_6 + \pi_7 + \pi_8 + \pi_9 + \pi_{10}\} == \{0.5\, X\}$

i.e., the sum of all investments into the risky assets is equal to 50% of the total wealth. Notice that indeed the transposition of the same condition is exactly the same condition

In[92]:= `Πn.Transpose[μ] == ξ X`

Out[92]= $\{\pi_1 + \pi_2 + \pi_3 + \pi_4 + \pi_5 + \pi_6 + \pi_7 + \pi_8 + \pi_9 + \pi_{10}\} == \{0.5\, X\}$

Another example:

Define the unit n-vector in direction i:

In[93]:= `en_,i_ := ReplacePart[Table[0, {n}], 1, i]`

For example

```
In[94]:= e10,2
```

Out[94]= {0, 1, 0, 0, 0, 0, 0, 0, 0, 0}

Define μ as, for example,

```
In[95]:= μ = {17, e7,3, e7,5}
```

$$\text{Out[95]=} \begin{pmatrix} 1 & 1 & 1 & 1 & 1 & 1 & 1 \\ 0 & 0 & 1 & 0 & 0 & 0 & 0 \\ 0 & 0 & 0 & 0 & 1 & 0 & 0 \end{pmatrix}$$

and corresponding ξ:

```
In[96]:= ξ = {.7, .2, .2};
```

Then the affine constraint $\mu.\Pi = \xi\, X$ can be read as

```
In[97]:= TableForm[Thread[μ.Π7 == ξ X]]
```

Out[97]//TableForm=

$$\pi_1 + \pi_2 + \pi_3 + \pi_4 + \pi_5 + \pi_6 + \pi_7 == 0.7\, X$$
$$\pi_3 == 0.2\, X$$
$$\pi_5 == 0.2\, X$$

meaning: the sum of all risky investments is equal to 70% of the total wealth, while 30% is in the cash account, and moreover 20% of the wealth is in each of the third and fifth stock considered.

A very special example (*Mean-Variance Analysis*):

$$\mu = \{a - r\};\ \xi = \{\alpha - r\}$$

in which case

$$\mu.\Pi(t,\, X) = \{a - r\}.\Pi(t,\, X) = \{(a - r).\Pi(t,\, X)\} = \xi\, X = \{\alpha - r\}\, X$$

i.e.,

$$r\, X + (a - r).\Pi(t,\, X) = \alpha\, X$$

meaning: the wealth appreciation rate is prescribed to be equal to α. We shall see later that in this example the utility function ceases to be a contributing factor to the solution of the problem, and the problem is then equivalent to minimizing the volatility of the corresponding portfolio balance evolution. The same is true if only one of the rows of μ is equal to $a - r$.

7.5.1.2 Wealth Volatility Minimization

```
In[98]:= Remove["Global`*"]
```

Before going into stochastic control considerations, consider first the problem of *volatility of wealth minimization*. For example, by writing the wealth equation as

$$dX(t) = (\Pi(t, X(t)).(a - r) + r\,X(t))\,dt + \Pi(t, X(t)).\sigma.dB(t)$$
$$= a_0\,X(t)\,dt + \sigma_0\,X(t)\,dB_0(t)$$

and computing

$$(dX(t))^2 = \Pi(t, X(t)).\sigma.\sigma^T.\Pi(t, X(t))\,dt = {\sigma_0}^2\,X(t)^2\,dt,$$

we see that any portfolio rule Π yields wealth volatility

$$\sigma_0 = \frac{\sqrt{\Pi(t, X(t)).\sigma.\sigma^T.\Pi(t, X(t))}}{X(t)}.$$

The problem then is to find $\Pi(t, X)$ which minimizes the volatility, i.e.,

$$\frac{\sqrt{\Pi(t, X).\sigma.\sigma^T.\Pi(t, X)}}{X} \to \text{Min}$$

under the constraint

$$\mu.\Pi(t, X) = \Pi(t, X).\mu^T = \xi\,X.$$

More conveniently, we have a family of minimization problems parametrized by (t, X): Find the minimum of the function

$$f(\Pi) = \Pi.\sigma.\sigma^T.\Pi$$

under the constraint

$$g(\Pi) = \Pi.\mu^T - \xi\,X = \mu.\Pi - \xi\,X = 0.$$

Notice that if we assume, as we do, that $\sigma.\sigma^T$ is positive definite, then finding a critical point is equivalent to finding the unique minimizer of f. Also notice that if there are no constraints on the portfolio, then the problem is trivial since it would yield the solution $\Pi(t, X) = 0$. This is why the analogous problem was not considered when portfolios without constraints were studied. Now, since the above minimization is constrained, we apply the method of Lagrange multipliers. To this end notice first that there is a qualitative difference between f and g: f is scalar valued, while $g = \{g_1, \ldots, g_k\}$ is k-vector valued. Consequently, the gradient of f is an m-vector $\nabla f = D f = \{\frac{\partial f}{\partial \Pi_1}, \ldots, \frac{\partial f}{\partial \Pi_m}\}$, while the gradient of g is defined to be an $m \times k$ matrix

$$\nabla g = D g = \{\frac{\partial g}{\partial \Pi_1}, \ldots, \frac{\partial g}{\partial \Pi_m}\} = \begin{pmatrix} \frac{\partial g_1}{\partial \Pi_1} & \cdots & \frac{\partial g_k}{\partial \Pi_1} \\ \cdots & \cdots & \cdots \\ \frac{\partial g_1}{\partial \Pi_m} & \cdots & \frac{\partial g_k}{\partial \Pi_m} \end{pmatrix}.$$

This, by the way, is consistent with our definition of gradient made in Chapter 2:

```
In[99]:= D_V_[u_] := (∂_#1 u &) /@ V
```

when applied on a vector-valued function; for example

```
In[100]:= D_{Π1,Π2,Π3}[{g1[Π1, Π2, Π3], g2[Π1, Π2, Π3]}]
```

Out[100]=
$$\begin{pmatrix} g_1^{(1,0,0)}(\Pi_1, \Pi_2, \Pi_3) & g_2^{(1,0,0)}(\Pi_1, \Pi_2, \Pi_3) \\ g_1^{(0,1,0)}(\Pi_1, \Pi_2, \Pi_3) & g_2^{(0,1,0)}(\Pi_1, \Pi_2, \Pi_3) \\ g_1^{(0,0,1)}(\Pi_1, \Pi_2, \Pi_3) & g_2^{(0,0,1)}(\Pi_1, \Pi_2, \Pi_3) \end{pmatrix}$$

Set $\mu^T = \{(\mu^T)_1, \ldots, (\mu^T)_m\}$, where $(\mu^T)_i$ are the rows of the matrix μ^T. Notice that then we can write $g(\Pi) = \sum_{i=1}^m (\mu^T)_i \Pi_i - \xi X$, and consequently $\frac{\partial g}{\partial \Pi_j} = (\mu^T)_j$, and furthermore

$$\nabla g = \{\frac{\partial g}{\partial \Pi_1}, \ldots, \frac{\partial g}{\partial \Pi_m}\} = \{(\mu^T)_1, \ldots, (\mu^T)_m\} = \mu^T.$$

Now we are ready to apply the method of Lagrange multipliers, according to which there exists a k-vector λ such that $\sigma.\sigma^T.\Pi = \nabla f(\Pi) = \nabla g(\Pi).\lambda = \mu^T.\lambda$, and consequently

$$\sigma.\sigma^T.\Pi = \mu^T.\lambda.$$

Solving this equation for Π, we see that the optimal trading strategy has the form

$$\Pi = (\sigma.\sigma^T)^{-1}.\mu^T.\lambda$$

with λ still unknown. Using the affine constraint $\mu.\Pi = \xi X$ we get

$$\mu.(\sigma.\sigma^T)^{-1}.\mu^T.\lambda = \xi X$$

and therefore, since it is assumed that the $k \times k$ matrix $\mu.(\sigma.\sigma^T)^{-1}.\mu^T$ is invertible,

$$\lambda = \lambda(X) = \left(\mu.(\sigma.\sigma^T)^{-1}.\mu^T\right)^{-1}.\xi X.$$

We conclude that the volatility-of-wealth-minimizing trading rule under the general affine constraint on a portfolio is equal to

$$\Pi(t, X) = (\sigma.\sigma^T)^{-1}.\mu^T.\lambda = (\sigma.\sigma^T)^{-1}.\mu^T.\left(\mu.(\sigma.\sigma^T)^{-1}.\mu^T\right)^{-1}.\xi X$$

or

$$\begin{aligned} &\Pi(t, X) = P X \\ &P = (\sigma.\sigma^T)^{-1}.\mu^T.\left(\mu.(\sigma.\sigma^T)^{-1}.\mu^T\right)^{-1}.\xi = \xi.\left(\mu.(\sigma.\sigma^T)^{-1}.\mu^T\right)^{-1}.\mu.(\sigma.\sigma^T)^{-1}. \end{aligned} \tag{7.5.5}$$

Consequently, notice finally that if this wealth volatility minimization strategy is applied, then the actual wealth volatility is equal to

$$\frac{\sqrt{\Pi(t, X).\sigma.\sigma^T.\Pi(t, X)}}{X} = \frac{\sqrt{(P\,X).\sigma.\sigma^T.P\,X}}{X} = \sqrt{P.\sigma.\sigma^T.P}$$
$$= \sqrt{\left(\xi.\left(\mu.(\sigma.\sigma^T)^{-1}.\mu^T\right)^{-1}.\mu.(\sigma.\sigma^T)^{-1}.\sigma.\sigma^T.(\sigma.\sigma^T)^{-1}.\mu^T.\left(\mu.(\sigma.\sigma^T)^{-1}.\mu^T\right)^{-1}.\xi\right)}$$
$$= \sqrt{\left(\xi.\left(\mu.(\sigma.\sigma^T)^{-1}.\mu^T\right)^{-1}.\mu.(\sigma.\sigma^T)^{-1}.\mu^T.\left(\mu.(\sigma.\sigma^T)^{-1}.\mu^T\right)^{-1}.\xi\right)}$$
$$= \sqrt{\xi.(\mu.(\sigma.\sigma^T)^{-1}.\mu^T)^{-1}.\xi}$$

while the wealth appreciation rate is equal to

$$\frac{\Pi(t, X).(a-r)}{X} + r = \frac{(P\,X).(a-r)}{X} + r = P.(a-r) + r$$
$$= \xi.\left(\mu.(\sigma.\sigma^T)^{-1}.\mu^T\right)^{-1}.\mu.(\sigma.\sigma^T)^{-1}.(a-r) + r.$$

7.5.1.3 Stochastic Control under Affine Constraints on the Portfolio

In[101]:= `Clear["Global`*"]`

Back to the stochastic control approach. The problem now is to find an optimal trading rule $\Pi^\star(t, X)$ such that $\mu.\Pi^\star(t, X) = \xi\, X$, and furthermore such that

$$\sup_{\Pi:\mu.\Pi\,(t,X)=\xi\,X} E_{t,X}\,\psi_\gamma\,(X^\Pi(T))$$
$$= \underset{\Pi:\mu.\Pi(t,X)=\xi\,X}{\text{Max}} E_{t,X}\,\psi_\gamma\,(X^\Pi(T)) = E_{t,X}\,\psi_\gamma\left(X^{\Pi^\star}(T)\right)$$

where ψ_γ is the utility of wealth function. As before, we study the value function

$$\varphi(t, X) = \sup_{\Pi:\mu.\Pi(t,X)=\xi\,X} E_{t,X}\,\psi_\gamma\,(X^\Pi(T)).$$

Obviously, the Hamilton–Jacobi–Bellman PDE, characterizing the value function, now reads as

$$\underset{\{\Pi(t,X),\mu.\Pi(t,X)=\xi\,X\}}{\text{Max}}\Big[\varphi_t(t, X) + \varphi_x(t, X)\,(\Pi(t, X).(a-r) + r\,X)$$
$$+ \frac{1}{2}\,\varphi_{x,x}(t, X)\,\Pi(t, X).\sigma.\sigma^T.\Pi(t, X)\Big] = 0$$

together with the terminal condition

$$\varphi(T, X) = \psi_\gamma(X).$$

So, for every (t, X), we need to maximize

$$f(\Pi) = \varphi_x(t, X)\,\Pi.(a-r) + \frac{1}{2}\,\varphi_{x,x}(t, X)\,\Pi.\sigma.\sigma^T.\Pi$$

(again, it is necessary therefore that $\varphi_{x,x}(t, X) < 0$) under the constraint

$$g(\Pi) = \Pi.\mu^T - \xi X = 0.$$

Again, according to the Lagrange multiplier method, there exists a k-vector $\lambda = \lambda(t, X)$, such that $\nabla_\Pi f(\Pi) = \nabla_\Pi g(\Pi).\lambda = \nabla_\Pi g(\Pi).\lambda(t, X)$, i.e.,

$$\nabla f(\Pi) = \varphi_x(t, X)(a - r) + \varphi_{x,x}(t, X)\,\sigma.\sigma^T.\Pi = \nabla g(\Pi).\lambda(t, X) = \mu^T.\lambda(t, X),$$

and consequently

$$\Pi(t, X) = \frac{1}{\varphi_{x,x}(t, X)}\,(\sigma.\sigma^T)^{-1}.\mu^T.\lambda(t, X) - \frac{\varphi_x(t, X)}{\varphi_{x,x}(t, X)}\,(\sigma.\sigma^T)^{-1}.(a - r)$$

with λ, as well as the value function φ, still unknown. Using the affine constraint $\mu.\Pi(t, X) = \xi X$ we get

$$\mu.\left(\frac{1}{\varphi_{x,x}(t, X)}\,(\sigma.\sigma^T)^{-1}.\mu^T.\lambda(t, X) - \frac{\varphi_x(t, X)}{\varphi_{x,x}(t, X)}\,(\sigma.\sigma^T)^{-1}.(a - r)\right) = \xi X,$$

and therefore

$$\mu.(\sigma.\sigma^T)^{-1}.\mu^T.\lambda(t, X) = \varphi_x(t, X)\,\mu.(\sigma.\sigma^T)^{-1}.(a - r) + \xi X\,\varphi_{x,x}(t, X).$$

It is assumed that the $k \times k$ matrix $\mu.(\sigma.\sigma^T)^{-1}.\mu^T$ is invertible. Therefore

$$\begin{aligned}\lambda(t, X) = {} & \varphi_x(t, X)\left(\mu.(\sigma.\sigma^T)^{-1}.\mu^T\right)^{-1}.\mu.(\sigma.\sigma^T)^{-1}.(a - r)\\ & + \varphi_{x,x}(t, X)\,X\left(\mu.(\sigma.\sigma^T)^{-1}.\mu^T\right)^{-1}.\xi,\end{aligned}$$

where one should notice that, since μ is a not a square matrix, it cannot be invertible in general, and therefore there is no simplification above. Going back into the expression for Π:

$$\begin{aligned}\Pi^\star(t, X) = {} & \frac{1}{\varphi_{x,x}(t, X)}\,(\sigma.\sigma^T)^{-1}.\mu^T.\Big(\varphi_x(t, X)\left(\mu.(\sigma.\sigma^T)^{-1}.\mu^T\right)^{-1}.\mu.(\sigma.\sigma^T)^{-1}.(a - r)\\ & + \varphi_{x,x}(t, X)\,X\left(\mu.(\sigma.\sigma^T)^{-1}.\mu^T\right)^{-1}.\xi\Big) - \frac{\varphi_x(t, X)}{\varphi_{x,x}(t, X)}\,(\sigma.\sigma^T)^{-1}.(a - r)\\ = {} & (\sigma.\sigma^T)^{-1}.\Big(X\,\mu^T.\left(\mu.(\sigma.\sigma^T)^{-1}.\mu^T\right)^{-1}.\xi\\ & + \frac{\varphi_x(t, X)}{\varphi_{x,x}(t, X)}\left(\mu^T.\left(\mu.(\sigma.\sigma^T)^{-1}.\mu^T\right)^{-1}.\mu.(\sigma.\sigma^T)^{-1} - \mathbb{I}_m\right).(a - r)\Big)\end{aligned}$$

where $\mathbb{I}_m$ is the m-dimensional identity matrix, or which is the same (we shall use them both)

$$= \Bigg(X\,\xi.\big(\mu.(\sigma.\sigma^T)^{-1}.\mu^T\big)^{-1}.\mu$$

$$+ \frac{\varphi_x(t, X)}{\varphi_{x,x}(t, X)}\,(a - r).\big((\sigma.\sigma^T)^{-1}.\mu^T.\big(\mu.(\sigma.\sigma^T)^{-1}.\mu^T\big)^{-1}.\mu - \mathbb{I}_m\big)\Bigg).(\sigma.\sigma^T)^{-1}.$$

So the optimal portfolio rule is determined modulo computation of the value function φ. Knowing that in Merton's unconstrained case there was a way to compute the value function explicitly, we proceed in an analogous way here.

Going back into the HJB PDE, the Max is realized for $\Pi^\star(t, X)$, and therefore, after multiplying the equation with $\varphi_{x,x}(t, X)$, we get

$$\begin{aligned}&\varphi_t(t, X)\,\varphi_{x,x}(t, X) + \varphi_x(t, X)\,\varphi_{x,x}(t, X)\,(\Pi^\star(t, X).(a - r) + r\,X)\\&\quad + \frac{1}{2}\,\varphi_{x,x}(t, X)^2\,\Pi^\star(t, X).\sigma.\sigma^T.\Pi^\star(t, X) = 0.\end{aligned} \tag{7.5.6}$$

Some significant simplifications are in order now. To this end, compute first

$$\begin{aligned}&\varphi_{x,x}(t, X)\,\varphi_x(t, X)\,\Pi^\star(t, X).(a - r)\\&= \varphi_{x,x}(t, X)\,\varphi_x(t, X)\Bigg(X\,\xi.\big(\mu.(\sigma.\sigma^T)^{-1}.\mu^T\big)^{-1}.\mu + \frac{\varphi_x(t, X)}{\varphi_{x,x}(t, X)}\,(a - r)\\&\qquad .\big((\sigma.\sigma^T)^{-1}.\mu^T.\big(\mu.(\sigma.\sigma^T)^{-1}.\mu^T\big)^{-1}.\mu - \mathbb{I}_m\big)\Bigg).(\sigma.\sigma^T)^{-1}.(a - r)\\&= \varphi_{x,x}(t, X)\,\varphi_x(t, X)\,X\,\xi.\big(\mu.(\sigma.\sigma^T)^{-1}.\mu^T\big)^{-1}.\mu.(\sigma.\sigma^T)^{-1}.(a - r)\\&+\varphi_x(t, X)^2\,(a - r).(\sigma.\sigma^T)^{-1}.\mu^T.\big(\mu.(\sigma.\sigma^T)^{-1}.\mu^T\big)^{-1}.\mu.(\sigma.\sigma^T)^{-1}.(a - r)\\&-\varphi_x(t, X)^2\,(a - r).(\sigma.\sigma^T)^{-1}.(a - r)\end{aligned}$$

and also, after some work

$$\begin{aligned}&\varphi_{x,x}(t, X)^2\,\Pi^\star(t, X).\sigma.\sigma^T.\Pi^\star(t, X) = \big(\varphi_{x,x}(t, X)\,X\,\xi.\big(\mu.(\sigma.\sigma^T)^{-1}.\mu^T\big)^{-1}.\mu\\&\qquad + \varphi_x(t, X)\,(a - r).\big((\sigma.\sigma^T)^{-1}.\mu^T.\big(\mu.(\sigma.\sigma^T)^{-1}.\mu^T\big)^{-1}.\mu - \mathbb{I}_m\big)\big)\\&.(\sigma.\sigma^T)^{-1}.\big(\varphi_{x,x}(t, X)\,X\,\mu^T.\big(\mu.(\sigma.\sigma^T)^{-1}.\mu^T\big)^{-1}.\xi\\&\qquad + \varphi_x(t, X)\,\big(\mu^T.\big(\mu.(\sigma.\sigma^T)^{-1}.\mu^T\big)^{-1}.\mu.(\sigma.\sigma^T)^{-1} - \mathbb{I}_m\big).(a - r)\big)\\&= \varphi_{x,x}(t, X)^2\,X^2\,\xi.\big(\mu.(\sigma.\sigma^T)^{-1}.\mu^T\big)^{-1}.\xi - \varphi_x(t, X)^2\,(a - r).(\sigma.\sigma^T)^{-1}.\mu^T\\&.\big(\mu.(\sigma.\sigma^T)^{-1}.\mu^T\big)^{-1}.\mu.(\sigma.\sigma^T)^{-1}.(a - r) + \varphi_x(t, X)^2\,(a - r).(\sigma.\sigma^T)^{-1}.(a - r).\end{aligned}$$

Now we can go back into (7.5.6) , and after some further simplifications, we derive

$$\begin{aligned}&0 = \varphi_t(t, X)\,\varphi_{x,x}(t, X) + \varphi_x(t, X)\,\varphi_{x,x}(t, X)\,(\Pi^\star(t, X).(a - r) + r\,X)\\&+\frac{1}{2}\,\varphi_{x,x}(t, X)^2\,\Pi^\star(t, X).\sigma.\sigma^T.\Pi^\star(t, X) = \varphi_t(t, X)\,\varphi_{x,x}(t, X)\end{aligned}$$

$$+\varphi_{x,x}(t, X)\,\varphi_x(t, X)\,X\,\xi.\left(\mu.(\sigma.\sigma^T)^{-1}.\mu^T\right)^{-1}.\mu.(\sigma.\sigma^T)^{-1}.(a-r)$$
$$+\varphi_x(t, X)^2\,(a-r).(\sigma.\sigma^T)^{-1}.\mu^T.\left(\mu.(\sigma.\sigma^T)^{-1}.\mu^T\right)^{-1}.\mu.(\sigma.\sigma^T)^{-1}.(a-r)$$
$$-\varphi_x(t, X)^2\,(a-r).(\sigma.\sigma^T)^{-1}.(a-r) + r\,X\,\varphi_x(t, X)\,\varphi_{x,x}(t, X)$$
$$+\frac{1}{2}\Big(\varphi_{x,x}(t, X)^2\,X^2\,\xi.\left(\mu.(\sigma.\sigma^T)^{-1}.\mu^T\right)^{-1}.\xi$$
$$-\varphi_x(t, X)^2\,(a-r).(\sigma.\sigma^T)^{-1}.\mu^T.\left(\mu.(\sigma.\sigma^T)^{-1}.\mu^T\right)^{-1}.\mu.(\sigma.\sigma^T)^{-1}.(a-r)$$
$$+\varphi_x(t, X)^2\,(a-r).(\sigma.\sigma^T)^{-1}.(a-r)\Big) = \varphi_t(t, X)\;\varphi_{x,x}(t, X)$$
$$+\varphi_{x,x}(t, X)\,\varphi_x(t, X)\,X\,\left(\xi.\left(\mu.(\sigma.\sigma^T)^{-1}.\mu^T\right)^{-1}.\mu.(\sigma.\sigma^T)^{-1}.(a-r)+r\right)$$
$$-\frac{1}{2}\,\varphi_x(t, X)^2\,(a-r).(\sigma.\sigma^T)^{-1}.\left(\mathbb{I}_m-\mu^T.\left(\mu.(\sigma.\sigma^T)^{-1}.\mu^T\right)^{-1}.\mu.(\sigma.\sigma^T)^{-1}\right).(a-r)$$
$$+\frac{1}{2}\,\varphi_{x,x}(t, X)^2\,X^2\,\xi.\left(\mu.(\sigma.\sigma^T)^{-1}.\mu^T\right)^{-1}.\xi$$

and we finally arrive at the Monge–Ampère PDE

$$\begin{aligned}&\varphi_t(t, X)\;\varphi_{x,x}(t, X)+\varphi_{x,x}(t, X)\,\varphi_x(t, X)\,X\,\left(\xi.\left(\mu.(\sigma.\sigma^T)^{-1}.\mu^T\right)^{-1}\right.\\&\qquad\left..\mu.(\sigma.\sigma^T)^{-1}.(a-r)+r\right)-\frac{1}{2}\,\varphi_x(t, X)^2\,(a-r).(\sigma.\sigma^T)^{-1}\\&.\left(\mathbb{I}_m-\mu^T.\left(\mu.(\sigma.\sigma^T)^{-1}.\mu^T\right)^{-1}.\mu.(\sigma.\sigma^T)^{-1}\right).(a-r)\\&+\frac{1}{2}\,\varphi_{x,x}(t, X)^2\,X^2\,\xi.\left(\mu.(\sigma.\sigma^T)^{-1}.\mu^T\right)^{-1}.\xi=0\end{aligned}\tag{7.5.7}$$

It is easier to see the structure of this equation in the form

```
In[102]:= MongeAmpereEqn =
            ∂t φ[t, X] ∂{X,2} φ[t, X] + ∂X φ[t, X] ∂{X,2} φ[t, X] X A +
              (∂X φ[t, X])^2 B + (∂{X,2} φ[t, X])^2 X^2 C == 0
```

Out[102]= $B\,\varphi^{(0,1)}(t, X)^2 + A\,X\,\varphi^{(0,2)}(t, X)\,\varphi^{(0,1)}(t, X) + C\,X^2\,\varphi^{(0,2)}(t, X)^2 + \varphi^{(0,2)}(t, X)\,\varphi^{(1,0)}(t, X) == 0$

obtained from (7.5.7) by means of the substitution

```
In[103]:= Sub = {A → r + ξ.Inverse[
                 μ.Inverse[σ.Transpose[σ]].Transpose[μ]].
               μ.Inverse[σ.Transpose[σ]].(a - r),
            B → -1/2 (a - r).Inverse[σ.Transpose[σ]].
               (𝕀m - Transpose[μ].Inverse[μ.
                    Inverse[σ.Transpose[σ]].Transpose[μ]].
```

```
        μ.Inverse[σ.Transpose[σ]]).(a - r),
    C → 1/2 ξ.Inverse[μ.Inverse[σ.Transpose[σ]].
        Transpose[μ]].ξ}
```

Out[103]= $\{A \to r + \xi.(\mu.(\sigma.\sigma^T)^{-1}.\mu^T)^{-1}.\mu.(\sigma.\sigma^T)^{-1}.(a-r),\ B \to -\frac{1}{2}(a-r).(\sigma.\sigma^T)^{-1}.(\mathbb{1}_m - \mu^T.(\mu.(\sigma.\sigma^T)^{-1}.\mu^T)^{-1}.\mu.(\sigma.\sigma^T)^{-1}).(a-r),\ C \to \frac{1}{2}\xi.(\mu.(\sigma.\sigma^T)^{-1}.\mu^T)^{-1}.\xi\}$

Compared with Merton's Monge–Ampère PDE (7.3.18), which as derived before, has the form

$$\varphi_t(t, X)\,\varphi_{x,x}(t, X) + \varphi_x(t, X)\,\varphi_{x,x}(t, X)\,X\,A + \varphi_x(t, X)^2\,B = 0,$$

we can see an extra term $\varphi_{x,x}(t, X)^2\,X^2\,C$. Nevertheless, that term has just the right form so that everything works out exactly in the same fashion as in the unconstrained case. Indeed, if moreover, ψ_γ the utility function i.e., the terminal condition, is equal to

$$\varphi(T, X) = \psi_\gamma(X) = \frac{X^{1-\gamma}}{1-\gamma} \tag{7.5.8}$$

for some $0 < \gamma \neq 1$, (the Log utility, corresponding to $\gamma = 1$, can be considered as well) we look for the solution in the form

```
In[104]:= ϕ[t_, X_] := f[t] X^(1-γ) / (1 - γ)
```

So, by substitution, we get

```
In[105]:= LHS = MongeAmpereEqn[[1]] /. φ → ϕ
```

Out[105]= $C\gamma^2 f(t)^2 X^{-2\gamma} + B f(t)^2 X^{-2\gamma} - A\gamma f(t)^2 X^{-2\gamma} - \frac{\gamma f(t) f'(t) X^{-2\gamma}}{1-\gamma}$

or

```
In[106]:= ODE = Simplify[Numerator[Together[LHS]]] / f[t] == 0
```

Out[106]= $(\gamma - 1)(B + \gamma(C\gamma - A)) f(t) + \gamma f'(t) == 0$

Solving the ODE (together with the terminal condition $f(T) = 1$) in the above equation we get

```
In[107]:= Simplify[
            f[t] /. DSolve[{f[T] == 1, ODE}, f[t], t][[1]]]
```

Out[107]= $e^{-\frac{(t-T)(\gamma-1)(B+\gamma(C\gamma-A))}{\gamma}}$

This yields the solution of the Monge–Ampère equation, i.e., the value function

```
In[108]:= φ[t_, X_] = % X^(1-γ)/(1 - γ)
```

Out[108]= $\frac{e^{-\frac{(t-T)(\gamma-1)(B+\gamma(C\gamma-A))}{\gamma}} X^{1-\gamma}}{1-\gamma}$

The value function in turn yields the optimal trading (hedging) rule

```
In[109]:= Πγ_[t_, X_] = Simplify[Inverse[σ.Transpose[σ]].
            ((∂X φ[t, X] (Transpose[μ].Inverse[μ.Inverse[σ.
                    Transpose[σ]].Transpose[μ]].μ.
                  Inverse[σ.Transpose[σ]] - Im).(a - r))/
              ∂{X,2} φ[t, X] + X Transpose[μ].
               Inverse[μ.Inverse[σ.Transpose[σ]].
                 Transpose[μ]].ξ)]
```

Out[109]= $(\sigma.\sigma^T)^{-1}.\left(X\left(\mu^T.(\mu.(\sigma.\sigma^T)^{-1}.\mu^T)^{-1}.\xi - \frac{(\mu^T.(\mu.(\sigma.\sigma^T)^{-1}.\mu^T)^{-1}.\mu.(\sigma.\sigma^T)^{-1} - \mathbb{I}_m).(a-r)}{\gamma}\right)\right)$

It is interesting to notice that we did not even have to use back-substitution (for A, B, C) to compute the optimal trading rule. On the other hand, the substitution can be used to express the value function in terms of the original data, which we skip.

It will be convenient to introduce the notation

$$\Pi_\gamma(t, X) = \frac{\Pi_*(t, X)}{\gamma} + \Pi_\infty(t, X)$$

where

```
In[110]:= Π*[t_, X_] =
          X Inverse[σ.Transpose[σ]].(Transpose[μ].Inverse[
                μ.Inverse[σ.Transpose[σ]].Transpose[μ]].
               μ.Inverse[σ.Transpose[σ]] - Im).(a - r)
```

Out[110]= $X\,(\sigma.\sigma^T)^{-1}.(\mu^T.(\mu.(\sigma.\sigma^T)^{-1}.\mu^T)^{-1}.\mu.(\sigma.\sigma^T)^{-1} - \mathbb{I}_m).(a-r)$

and

```
In[111]:= Π∞[t_, X_] =
            X Inverse[σ.Transpose[σ]].Transpose[μ].Inverse[
              μ.Inverse[σ.Transpose[σ]].Transpose[μ]].ξ
```

Out[111]= $X\,(\sigma.\sigma^T)^{-1}.\mu^T.\left(\mu.(\sigma.\sigma^T)^{-1}.\mu^T\right)^{-1}.\xi$

Finally, notice that

$$\lim_{\gamma\to\infty}\Pi_\gamma(t, X) = \lim_{\gamma\to\infty}\left(\frac{\Pi_*(t, X)}{\gamma} + \Pi_\infty(t, X)\right)$$
$$= \Pi_\infty(t, X) = X\,(\sigma.\sigma^T)^{-1}.\mu^T.\left(\mu.(\sigma.\sigma^T)^{-1}.\mu^T\right)^{-1}.\xi$$

and comparing with the previous section we see that the "infinitely conservative" strategy $\Pi_\infty(t, X)$ is equal to the wealth-volatility minimization strategy, i.e., the wealth-volatility minimization strategy is the most conservative. Stochastic control theory therefore provides a precise framework for incorporating varying degrees of an investor's willingness to risk. This is another justification for referring to $\gamma : 0 < \gamma \le \infty$ as the safety exponent (the greater the safety exponent the less risky is the position).

7.5.1.4 Examples

```
In[112]:= Clear["Global`*"]
```

We shall implement and experiment with the above strategies for $0 < \gamma \le \infty$. Consider a stock market:

```
In[113]:= m = 20; a = Array[Random[Real, {-.2, .5}] &, {m}];
          σ = 1/10 Array[Random[Real, {-1, 1}] &, {m, m}];
          1_k_ := Array[1 &, {m}]; T = 1/12;
          e_k_,i_ := ReplacePart[Table[0, {k}], 1, i];
          I_k_ := IdentityMatrix[k];
          r = .05;
```

Check whether the volatilities are reasonable:

```
In[115]:= Max[(√(Plus @@ #1^2) &) /@ σ]
```

Out[115]= 0.290617

Define an affine constraint $\mu.\Pi = \xi\, X$ with

```
In[116]:= ξ = {1, 0, .2};
          μ = {1_m, e_m,3, e_m,5}
```

Out[117]= $\begin{pmatrix} 1&1&1&1&1&1&1&1&1&1&1&1&1&1&1&1&1&1&1&1 \\ 0&0&1&0&0&0&0&0&0&0&0&0&0&0&0&0&0&0&0&0 \\ 0&0&0&0&1&0&0&0&0&0&0&0&0&0&0&0&0&0&0&0 \end{pmatrix}$

meaning that the sum of all risky investments, long and short positions, add up to 100% of the wealth, no investment is made in the third stock, while 20% of the wealth is in the fifth stock.

As derived above, the optimal portfolio rule is equal to $\Pi_\gamma(t, X) = P_\gamma X$, with

```
In[118]:= Pγ_ := Chop[
    Inverse[σ.Transpose[σ]].(Transpose[μ].Inverse[
        μ.Inverse[σ.Transpose[σ]].Transpose[μ]].
      ξ - 1/γ ((Transpose[μ].Inverse[μ.Inverse[
            σ.Transpose[σ]].Transpose[μ]].μ.
          Inverse[σ.Transpose[σ]] - 𝕀m).(a - r)))]
```

For example

```
In[119]:= P∞
```

```
Out[119]= {0.0420255, -0.0198418, 0, 0.072418, 0.2, 0.0696879,
   0.0775671, -0.0938596, -0.032231, 0.00290067, 0.0931764,
   0.211053, 0.0905896, -0.0691444, 0.0609671, 0.20781,
   0.123707, -0.103722, -0.00686445, 0.0737607}
```

while

```
In[120]:= P1
```

```
Out[120]= {9.59311, -1.62244, 0, -3.87924, 0.2, -8.13301, 5.89067,
   -25.4524, -2.76416, -22.275, 8.29475, 9.56228, 16.4853,
   -13.3997, 26.7029, 9.15168, 21.5321, -22.026, -22.9221,
   16.0612}
```

One can see how very different those two strategies are. Checking one of the constraints

```
In[121]:= Plus @@ P∞
```

```
Out[121]= 1.
```

Now generate the market history

```
In[122]:= << "Statistics`NormalDistribution`";
    s = Array[Random[Real, {15, 120}] &, {m}];
```

```
k = 1000; T = 1; dt = T/k;
dB = Table[
    Random[NormalDistribution[0, 1]], {k}, {m}] √dt ;
```

We do not evaluate the particular price trajectories but rather only the corresponding wealth trajectory. To this end the folding function, with the safety parameter γ is

```
In[126]:= Fγ_[{t_, X_}, dB_] :=
          {t + dt, X + (Max[X, 0] Pγ.(a - r) + r Max[X, 0]) dt +
            Max[X, 0] Pγ.σ.dB}
```

yielding, for example, the following two trajectories

```
In[127]:= Show[(ListPlot[FoldList[F#1, {0, 10000}, dB],
               PlotJoined → True, DisplayFunction → Identity,
               PlotStyle → RGBColor[Random[], 0, Random[]]] &) /@
             {100, ∞}, DisplayFunction → $DisplayFunction];
```

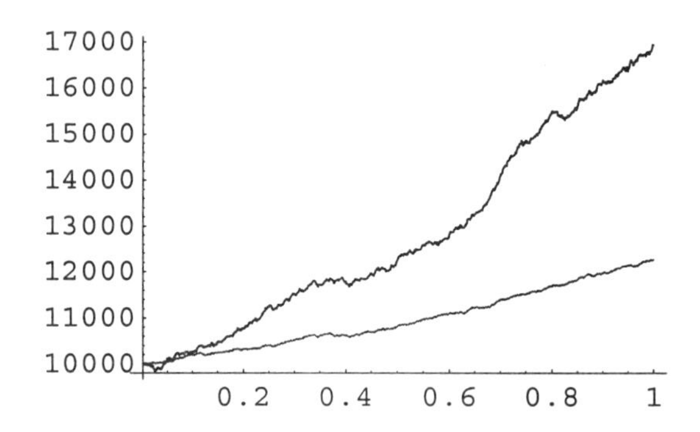

7.5.1.5 Time-Dependent Market Dynamics

Not much will be different here, so we shall be quite brief. As always, consider the cash account

$$dC(t) = r(t)\, C(t)\, dt \tag{7.5.9}$$

(in the absence of trading), and (many) stocks are considered for trading, whose dynamics obeys SDEs

$$dS(t) = S(t)\, a(t)\, dt + S(t)\, \sigma(t).dB(t). \tag{7.5.10}$$

The portfolio rule

$$\Pi(t, X) = \{\Pi_1(t, X), \;\ldots, \Pi_m(t, X)\}$$

is now constrained by

$$\mu(t).\Pi(t, X) = \xi(t)\, X \tag{7.5.11}$$

or, which is the same, $\Pi(t, X).\mu(t)^T = \xi(t) X$. Similarly as before, the wealth evolves according to

$$dX(t) = (\Pi(t, X(t)).(a(t) - r(t)) + r(t) X(t)) dt + \Pi(t, X(t)).\sigma(t).dB(t).$$

We just state the conclusion: under the above circumstances, the optimal trading rule is equal to

$$\begin{aligned}\Pi_\gamma(t, X) = \frac{1}{\gamma} \Big(X \big((\sigma(t).\sigma(t)^T)^{-1} - (\sigma(t).\sigma(t)^T)^{-1}.\mu(t)^T.\big(\mu(t).(\sigma(t).\sigma(t)^T)^{-1} \\ .\mu(t)^T\big)^{-1}.\mu(t).(\sigma(t).\sigma(t)^T)^{-1}\big).(b(t) - r(t))\Big) \\ + X (\sigma(t).\sigma(t)^T)^{-1}.\mu(t)^T.\big(\mu(t).(\sigma(t).\sigma(t)^T)^{-1}.\mu(t)^T\big)^{-1}.\xi(t)\end{aligned} \tag{7.5.12}$$

and we have the same conclusion as in the case without constraints: if at each time we apply the "current understanding" of the market in an optimal fashion, this is the best we can also do in the long run. Same warning appears as before: this is true if the market dynamics depends only on time.

7.5.2 Portfolios Under Affine Constraints and Appreciation Uncertainty

Back to the time-independent market dynamics, but with uncertainty about stock appreciation rates. So, as in (7.4.3), the wealth evolution equation is

$$dX(t) = (\Pi(t, X(t)).(a + b - r) + r X(t)) dt + \Pi(t, X(t)).\sigma.dB(t) \tag{7.5.13}$$

where a is a given constant vector (appreciation rate point estimate), while $b \sim N(0, S.S^T)$ is a random variable, modeling our lack of confidence in a. The portfolio rules are constrained by the affine constraint $\mu.\Pi(t, X) = \xi X$, for given matrix μ and vector ξ, just as before.

Motivated by previous developments, wc shall assume that

$$\Pi(t, X) = P X. \tag{7.5.14}$$

This implies that the affine constraint reduces to

$$\mu.\Pi(t, X) = \mu.P X = \xi X$$

or, which is the same,

$$\mu.P = P.\mu^T = \xi. \tag{7.5.15}$$

Then, as shown before,

$$\begin{aligned}E_{s,X} \log(X^\Pi(T)) &= E_{s,X} \log(X^P(T)) \\ &= \log(X) + (T - s)\left(r + P.(a - r) - \frac{1}{2} P.\sigma.\sigma^T.P\right)\end{aligned}$$

since $E_{s,X}\, b = E\, b = 0$. This yields that the optimal trading strategy under the affine constraint is obtained provided that

$$\begin{aligned}
&\underset{P:\mu.P=\xi}{\text{Max}}\ E_{s,X} \log(X^P(T)) \\
&= \underset{P:\mu.P=\xi}{\text{Max}} \left(\log(X) + (T-s)\left(r + P.(a-r) - \frac{1}{2} P.\sigma.\sigma^T.P\right)\right) \\
&= \log(X) + (T-s)\left(r + \underset{P:\mu.P=\xi}{\text{Max}} \left(P.(a-r) - \frac{1}{2} P.\sigma.\sigma^T.P\right)\right)
\end{aligned}$$

or

$$\begin{aligned}
&\underset{P:\mu.P=\xi}{\text{Max}} \left(P.(a-r) - \frac{1}{2} P.\sigma.\sigma^T.P\right) \\
&= \underset{P:g(\Pi)=\mu.P-\xi=0}{\text{Max}} f(P) = \underset{P:g(\Pi)=P.\mu^T-\xi=0}{\text{Max}} f(P).
\end{aligned}$$

To perform this constrained maximization, the Lagrange multiplier method can be applied, according to which $\nabla f(P) = \nabla g(P).\lambda$, and consequently

$$a - r - \sigma.\sigma^T.P = \mu^T.\lambda$$

yielding

$$P = (\sigma.\sigma^T)^{-1}.(a-r) - (\sigma.\sigma^T)^{-1}.\mu^T.\lambda$$

with λ still unknown. Using the affine constraint above in the form $\mu.P = \xi$ we get

$$\mu.\left((\sigma.\sigma^T)^{-1}.(a-r) - (\sigma.\sigma^T)^{-1}.\mu^T.\lambda\right) = \xi$$

and therefore

$$\lambda = \left(\mu.(\sigma.\sigma^T)^{-1}.\mu^T\right)^{-1}.\mu.(\sigma.\sigma^T)^{-1}.(a-r) - \left(\mu.(\sigma.\sigma^T)^{-1}.\mu^T\right)^{-1}.\xi$$

since it is assumed that the $k \times k$ matrix $\mu.(\sigma.\sigma^T)^{-1}.\mu^T$ is invertible. Therefore

$$\begin{aligned}
P = {} & (\sigma.\sigma^T)^{-1}.(a-r) - (\sigma.\sigma^T)^{-1}.\mu^T.\left(\left(\mu.(\sigma.\sigma^T)^{-1}.\mu^T\right)^{-1}.\mu.(\sigma.\sigma^T)^{-1}\right. \\
& \left. .(a-r) - \left(\mu.(\sigma.\sigma^T)^{-1}.\mu^T\right)^{-1}.\xi\right) = (\sigma.\sigma^T)^{-1}.(a-r) \\
& -(\sigma.\sigma^T)^{-1}.\mu^T.\left(\mu.(\sigma.\sigma^T)^{-1}.\mu^T\right)^{-1}.\mu.(\sigma.\sigma^T)^{-1}.(a-r) \\
& +(\sigma.\sigma^T)^{-1}.\mu^T.\left(\mu.(\sigma.\sigma^T)^{-1}.\mu^T\right)^{-1}.\xi = (\sigma.\sigma^T)^{-1}.\mu^T.\left(\mu.(\sigma.\sigma^T)^{-1}.\mu^T\right)^{-1} \\
& .\xi + (\sigma.\sigma^T)^{-1}.\left(I - \mu^T.\left(\mu.(\sigma.\sigma^T)^{-1}.\mu^T\right)^{-1}.\mu.(\sigma.\sigma^T)^{-1}\right).(a-r)
\end{aligned}$$

which is the optimal strategy under general affine constraint, and under either no uncertainty about the appreciation rate, or under the Log utility function. If, on the other hand, the power function is used as a utility, i.e., if $\psi_\gamma(X) = \frac{X^{1-\gamma}}{1-\gamma}$ for $\gamma \neq 1$, we have from before:

$$\begin{aligned}
E_{s,X}\, X^P(T)^{1-\gamma} &= E\big(E_{s,X}\big[X^P(T)^{1-\gamma} \,\big|\, b = \beta\big]\big) \\
&= X^{1-\gamma}\, e^{\int_s^T (1-\gamma)\left(r+P.(a-r)-\frac{1}{2}\gamma P.\sigma.\sigma^T.P\right) d\tau + \frac{1}{2}(1-\gamma)^2 \left|\int_s^T P\, d\tau.S\right|^2} \\
&= X^{1-\gamma}\, e^{(1-\gamma)(T-s)\left(r+P.(a-r)-\frac{1}{2}\gamma P.\sigma.\sigma^T.P\right) + \frac{1}{2}(1-\gamma)^2 (T-s)^2 |P.S|^2} \\
&= X^{1-\gamma}\, e^{(1-\gamma)(T-s)\left(r+P.(a-r)-\frac{1}{2}\gamma |P.\sigma|^2\right) + \frac{1}{2}(1-\gamma)^2 (T-s)^2 |P.S|^2}.
\end{aligned}$$

Therefore

$$E_{s,X}\, \frac{X^P(T)^{1-\gamma}}{1-\gamma} = X^{1-\gamma}\, \frac{1}{1-\gamma}\, e^{(1-\gamma)(T-s)\left(r+P.(a-r)-\frac{1}{2}\gamma |P.\sigma|^2\right) + \frac{1}{2}(1-\gamma)^2 (T-s)^2 |P.S|^2}.$$

Therefore, similarly as before, we end up with the optimization problem

$$\begin{aligned}
&\underset{P:\mu.P=\xi}{\mathrm{Max}} \left(P.(a-r) - \frac{1}{2}\gamma \left|P.\sigma\right|^2 + \frac{1}{2}(1-\gamma)(T-s)\left|P.S\right|^2\right) \\
&= \underset{P:g(\Pi)=\mu.P-\xi=0}{\mathrm{Max}} f(P) = \underset{P:g(\Pi)=P.\mu^T-\xi=0}{\mathrm{Max}} f(P).
\end{aligned} \tag{7.5.16}$$

Again, we apply the Lagrange multiplier method, according to which there exists a vector λ such that

$$\nabla f(P) = a - r - \gamma\, \sigma.\sigma^T.P + (1-\gamma)(T-s)\, S.S^T.P = \nabla g(P).\lambda = \mu^T.\lambda$$

implying

$$(\gamma\, \sigma.\sigma^T + (\gamma-1)(T-s)\, S.S^T).P = -\mu^T.\lambda + (a-r)$$

or

$$\begin{aligned}
P &= ((\gamma-1)(T-s)\, S.S^T + \gamma\, \sigma.\sigma^T)^{-1}.(a-r) \\
&\quad -((\gamma-1)(T-s)\, S.S^T + \gamma\, \sigma.\sigma^T)^{-1}.\mu^T.\lambda
\end{aligned}$$

with λ still unknown. Using the affine constraint above in the form $\mu.P = \xi$ we get

$$\begin{aligned}
\mu.\big(((\gamma-1)(T-s)\, S.S^T + \gamma\, \sigma.\sigma^T)^{-1}.(a-r) - ((\gamma-1)(T-s)\, S.S^T \\
+ \gamma\, \sigma.\sigma^T)^{-1}.\mu^T.\lambda\big) = \xi
\end{aligned}$$

and consequently

$$\begin{aligned}
\lambda = \big(\mu.((\gamma-1)(T-s)\, S.S^T + \gamma\, \sigma.\sigma^T)^{-1}.\mu^T\big)^{-1}.\mu.((\gamma-1)(T-s)\, S.S^T \\
+ \gamma\, \sigma.\sigma^T)^{-1}.(a-r) - \big(\mu.((\gamma-1)(T-s)\, S.S^T + \gamma\, \sigma.\sigma^T)^{-1}.\mu^T\big)^{-1}.\xi
\end{aligned}$$

under the assumption that the $k \times k$ matrix $\mu.((1-\gamma)(T-s)\, S.S^T - \gamma\, \sigma.\sigma^T)^{-1}.\mu^T$ is invertible. Therefore

$$\begin{aligned} P_\gamma = P = {} & ((\gamma-1)(T-s)\, S.S^T + \gamma\, \sigma.\sigma^T)^{-1}.(a-r) - ((\gamma-1)(T-s)\, S.S^T \\ & + \gamma\, \sigma.\sigma^T)^{-1}.\mu^T.\big((\mu.((\gamma-1)(T-s)\, S.S^T + \gamma\, \sigma.\sigma^T)^{-1}.\mu^T\big)^{-1} \\ & .\mu.((\gamma-1)(T-s)\, S.S^T + \gamma\, \sigma.\sigma^T)^{-1}.(a-r) \\ & - \big(\mu.((\gamma-1)(T-s)\, S.S^T + \gamma\, \sigma.\sigma^T)^{-1}.\mu^T\big)^{-1}.\xi\big) \end{aligned}$$

and we get the optimal trading strategy under general affine constraint and under uncertainty (normal prior) of the appreciation rates. Of course, if $\gamma = 1$ this is the same as the above P computed in the case of the Log utility when the uncertainty of the appreciation rate estimates does not affect the strategy.

7.5.3 A Quadratic Constraint: Constraint on Wealth Volatility

7.5.3.1 Solution under the Constraint on Wealth Volatility

Recall that $\varphi(t, X) = E_{t,X}\, \psi(X^\Pi(T))$ solves

$$\frac{\partial \varphi(t, X)}{\partial t} + \frac{\partial \varphi(t, X)}{\partial X}\, (r\, X + \Pi(t, X).(a-r)) + \frac{1}{2}\, \frac{\partial^2 \varphi(t, X)}{\partial X^2}\, \Pi(t, X).\sigma.\sigma^T.\Pi(t, X) = 0$$

for any regular function ψ, and any $T > t$. If constraint on wealth-volatility

$$\frac{\sqrt{\Pi(t, X).\sigma.\sigma^T.\Pi(t, X)}}{X} = \eta$$

i.e.,

$$\Pi(t, X).\sigma.\sigma^T.\Pi(t, X) = |\, \Pi(t, X).\sigma\, |^2 = X^2\, \eta^2 \tag{7.5.17}$$

is imposed, the Hamilton–Jacobi–Bellman PDE reads as

$$\underset{\{\Pi:\, |\Pi(t,X).\sigma|^2 = X^2\, \eta^2\}}{\text{Max}} \left(\frac{\partial \varphi(t, X)}{\partial t} + \frac{\partial \varphi(t, X)}{\partial X}\, (r\, X + \Pi.(a-r)) + \frac{1}{2}\, \frac{\partial^2 \varphi(t, X)}{\partial X^2}\, X^2\, \eta^2 \right) = 0.$$

Notice that the constraint can be written as $g(\Pi) = |\, \Pi(t, X).\sigma\, |^2 - X^2\, \eta^2 = 0$, and that g is a scalar-valued function. We employ yet again the Lagrange multiplier method, according to which there exists a *number* λ such that

$$\frac{\partial \varphi(t, X)}{\partial X}(a-r) = 2\,\lambda\,\sigma.\sigma^T.\Pi(t, X)$$

and furthermore

$$\Pi(t, X) = \frac{\frac{\partial \varphi(t,X)}{\partial X}\,(\sigma.\sigma^T)^{-1}.(a-r)}{2\,\lambda} = \frac{\frac{\partial \varphi(t,X)}{\partial X}\,(a-r).(\sigma.\sigma^T)^{-1}}{2\,\lambda}.$$

Therefore, from the constraint we get

$$\left(\frac{\frac{\partial \varphi(t,X)}{\partial X}\,(a-r).(\sigma.\sigma^T)^{-1}}{2\,\lambda}\right).\sigma.\sigma^T.\frac{\frac{\partial \varphi(t,X)}{\partial X}\,(\sigma.\sigma^T)^{-1}.(a-r)}{2\,\lambda} = X^2\,\eta^2$$

i.e.,

$$\frac{\left(\frac{\partial \varphi(t,X)}{\partial X}\right)^2 (a-r).(\sigma.\sigma^T)^{-1}.(a-r)}{(2\,\lambda)^2} = X^2\,\eta^2$$

yielding, since we assume $\frac{\partial \varphi(t,X)}{\partial X} > 0$, that

$$2\,\lambda = \frac{\sqrt{(a-r).(\sigma.\sigma^T)^{-1}.(a-r)}\;\frac{\partial \varphi(t,X)}{\partial X}}{X\,\eta}.$$

Therefore

$$\Pi\,(t, X) = \frac{\frac{\partial \varphi(t,X)}{\partial X}\,(\sigma.\sigma^T)^{-1}.(a-r)}{2\,\lambda} = \frac{\frac{\partial \varphi(t,X)}{\partial X}\,(\sigma.\sigma^T)^{-1}.(a-r)}{\frac{\sqrt{(a-r).(\sigma.\sigma^T)^{-1}.(a-r)}\;\frac{\partial \varphi(t,X)}{\partial X}}{X\,\eta}} \tag{7.5.18}$$

$$= \frac{X\,\eta\,(\sigma.\sigma^T)^{-1}.(a-r)}{\sqrt{(a-r).(\sigma.\sigma^T)^{-1}.(a-r)}}.$$

So, since the final formula does not depend on the value function φ, the problem is solved suddenly, without even going back to the Monge–Ampère PDE to compute the value function, and consequently regardless of what the utility function is (utility function does, on the other hand affects the value function, but this is irrelevant).

Alternatively, if we maximize the wealth appreciation rate

$$f(\Pi) = r\,X + \Pi.(a-r)$$

under the constraint (7.5.16) we get

$$(a-r) = 2\,\lambda\,\sigma.\sigma^T.\Pi$$

and therefore

$$\Pi = \frac{(\sigma.\sigma^T)^{-1}.(a-r)}{2\lambda} = \frac{(a-r).(\sigma.\sigma^T)^{-1}}{2\lambda}.$$

Again utilizing the constraint, we get

$$\begin{aligned} X^2\eta^2 = \Pi(t, X).\sigma.\sigma^T.\Pi(t, X) = \frac{(a-r).(\sigma.\sigma^T)^{-1}}{2\lambda}.\sigma.\sigma^T. \\ \frac{(\sigma.\sigma^T)^{-1}.(a-r)}{2\lambda} = \frac{(a-r).(\sigma.\sigma^T)^{-1}.(a-r)}{4\lambda^2} \end{aligned} \tag{7.5.19}$$

implying

$$\lambda = \frac{\sqrt{(a-r).(\sigma.\sigma^T)^{-1}.(a-r)}}{2X\eta},$$

and arriving again at the optimal portfolio rule

$$\Pi(t, X) = \frac{(\sigma.\sigma^T)^{-1}.(a-r)}{2\lambda} = \frac{X\eta\,(\sigma.\sigma^T)^{-1}.(a-r)}{\sqrt{(a-r).(\sigma.\sigma^T)^{-1}.(a-r)}}.$$

So, under the wealth-volatility constraint, the stochastic control portfolio theory is equivalent to the appreciation rate maximization portfolio, regardless of the utility of the wealth function.

7.5.3.2 Affine Constraint together with Constraint on Wealth Volatility

Consider now the general affine constraint

$$\mu.\Pi(t, X) = \xi X \tag{7.5.20}$$

where μ is the $k \times m$ matrix, and ξ is the k-vector, together with the special form of the quadratic constraint

$$\Pi(t, X).\sigma.\sigma^T.\Pi(t, X) = X^2\eta^2 \tag{7.5.21}$$

where η is a given (positive) number. It will be necessary to assume that

$$\eta^2 - \xi^T.\left(\mu.(\sigma.\sigma^T)^{-1}.\mu^T\right)^{-1}.\xi > 0. \tag{7.5.22}$$

The condition (7.5.22) can be thought of geometrically: if the affine condition is thought of as a plane (of dimension $m-k$), and if the quadratic (volatility) constraint is thought of as an ellipsoid (of dimension $m-1$) centered at the origin, then the first condition above can be thought of as the geometric condition that the plane intersects the ellipse, and moreover, not just tangentially.

The Hamilton–Jacobi–Bellman PDE reads as

$$\underset{\{\Pi:\mu.\Pi(t,X)=\xi X;\,\Pi(t,X).\sigma.\sigma^T.\Pi(t,X)=X^2\eta^2\}}{\text{Max}} \left(\frac{\partial \varphi(t,X)}{\partial t} + \frac{\partial \varphi(t,X)}{\partial X} (r\,X + \Pi.(a-r)) + \frac{1}{2} \frac{\partial^2 \varphi(t,X)}{\partial X^2} \Pi.\sigma.\sigma^T.\Pi \right) = 0$$

which is possible to simplify from the start, due to the special form of the above quadratic constraint, to the following Hamilton–Jacobi–Bellman PDE, which is no longer even fully non-linear (the Max is taken only for the first order term in the PDE operator, and not for the second order term):

$$\underset{\{\Pi:\mu.\Pi(t,X)=\xi X;\,\Pi(t,X).\sigma.\sigma^T.\Pi(t,X)=X^2\eta^2\}}{\text{Max}} \left(\frac{\partial \varphi(t,X)}{\partial t} + \frac{\partial \varphi(t,X)}{\partial X} (r\,X + \Pi.(a-r)) + \frac{1}{2} \frac{\partial^2 \varphi(t,X)}{\partial X^2} X^2 \eta^2 \right) = 0. \tag{7.5.23}$$

Formally seeking a maximum, i.e., an extremum, of the expression under the Max in the above Hamilton–Jacobi–Bellman PDE, and using the Lagrange multiplier method to that end, one knows that there exist a k-vector λ_1 and a number λ_2 such that

$$\frac{\partial \varphi(t,X)}{\partial X} (a-r) = \mu^T.\lambda_1 \frac{\partial \varphi(t,X)}{\partial X} + 2 \frac{\frac{\partial \varphi(t,X)}{\partial X} \lambda_2}{X} \sigma.\sigma^T.\Pi(t,X)$$

i.e.,

$$a - r = \mu^T.\lambda_1 + 2 \frac{\lambda_2}{X} \sigma.\sigma^T.\Pi(t,X).$$

(Recall that $a - r := a - r\,\mathrm{I}_m$.) In the above the Lagrange multipliers were deliberately chosen in a form so that simplification occurs. Therefore

$$\Pi(t,X) = \frac{X\,(\sigma.\sigma^T)^{-1}.((a-r) - \mu^T.\lambda_1)}{2\,\lambda_2} = \frac{X\,((a-r) - \lambda_1.\mu).(\sigma.\sigma^T)^{-1}}{2\,\lambda_2} \tag{7.5.24}$$

Plugging this back into (7.5.20) and (7.5.21) we get

$$\frac{X\,((a-r) - \lambda_1.\mu).(\sigma.\sigma^T)^{-1}}{2\,\lambda_2}.\sigma.\sigma^T.\frac{X\,(\sigma.\sigma^T)^{-1}.((a-r) - \mu^T.\lambda_1)}{2\,\lambda_2} = X^2\,\eta^2$$

as well as

$$\mu.\frac{X\,(\sigma.\sigma^T)^{-1}.((a-r) - \mu^T.\lambda_1)}{2\,\lambda_2} = \xi\,X,$$

and therefore

$$(a-r).(\sigma.\sigma^T)^{-1}.(a-r) - 2\,(a-r).(\sigma.\sigma^T)^{-1}.\mu^T.\lambda_1 + \lambda_1.\mu \\ .(\sigma.\sigma^T)^{-1}.\mu^T.\lambda_1 = 4\,\lambda_2{}^2\,\eta^2 \tag{7.5.25}$$

together with

$$\mu.(\sigma.\sigma^T)^{-1}.(a-r) - \mu.(\sigma.\sigma^T)^{-1}.\mu^T.\lambda_1 = 2\,\lambda_2\,\xi. \tag{7.5.26}$$

Solving (7.5.26) for λ_1, we get

$$\begin{aligned}\lambda_1 &= -2\,\lambda_2\left(\mu.(\sigma.\sigma^T)^{-1}.\mu^T\right)^{-1}.\xi + \left(\mu.(\sigma.\sigma^T)^{-1}.\mu^T\right)^{-1}.\mu.(\sigma.\sigma^T)^{-1}.(a-r) \\ &= \left(\mu.(\sigma.\sigma^T)^{-1}.\mu^T\right)^{-1}.\left(-2\,\lambda_2\,\xi + \mu.(\sigma.\sigma^T)^{-1}.(a-r)\right) \\ &= \left(-2\,\lambda_2\,\xi + (a-r).(\sigma.\sigma^T)^{-1}.\mu^T\right).\left(\mu.(\sigma.\sigma^T)^{-1}.\mu^T\right)^{-1}.\end{aligned} \tag{7.5.27}$$

Plugging this back into (7.5.25) we get

$$\begin{aligned}&(a-r).(\sigma.\sigma^T)^{-1}.(a-r) - 2\,(a-r).(\sigma.\sigma^T)^{-1}.\mu^T.\left(\mu.(\sigma.\sigma^T)^{-1}.\mu^T\right)^{-1} \\ &.\left(-2\,\lambda_2\,\xi + \mu.(\sigma.\sigma^T)^{-1}.(a-r)\right) + \left(-2\,\lambda_2\,\xi + (a-r).(\sigma.\sigma^T)^{-1}.\mu^T\right) \\ &.\left(\mu.(\sigma.\sigma^T)^{-1}.\mu^T\right)^{-1}.\mu.(\sigma.\sigma^T)^{-1}.\mu^T.\left(\mu.(\sigma.\sigma^T)^{-1}.\mu^T\right)^{-1} \\ &.\left(-2\,\lambda_2\,\xi + \mu.(\sigma.\sigma^T)^{-1}.(a-r)\right) = 4\,\lambda_2{}^2\,\eta^2\end{aligned}$$

and consequently

$$\begin{aligned}&(a-r).(\sigma.\sigma^T)^{-1}.(a-r) + 4\,\lambda_2\,(a-r).(\sigma.\sigma^T)^{-1}.\mu^T.\left(\mu.(\sigma.\sigma^T)^{-1}.\mu^T\right)^{-1}.\xi \\ &-2\,(a-r).(\sigma.\sigma^T)^{-1}.\mu^T.\left(\mu.(\sigma.\sigma^T)^{-1}.\mu^T\right)^{-1}.\mu.(\sigma.\sigma^T)^{-1}.(a-r) + 4\,\lambda_2{}^2\,\xi \\ &.\left(\mu.(\sigma.\sigma^T)^{-1}.\mu^T\right)^{-1}.\xi - 4\,\lambda_2\,\xi.\left(\mu.(\sigma.\sigma^T)^{-1}.\mu^T\right)^{-1}.\mu.(\sigma.\sigma^T)^{-1}.(a-r) \\ &+(a-r).(\sigma.\sigma^T)^{-1}.\mu^T.\left(\mu.(\sigma.\sigma^T)^{-1}.\mu^T\right)^{-1}.\mu.(\sigma.\sigma^T)^{-1}.(a-r) = 4\,\lambda_2{}^2\,\eta^2\end{aligned}$$

which simplifies to

$$\begin{aligned}&4\,\lambda_2{}^2\left(\eta^2 - \xi.\left(\mu.(\sigma.\sigma^T)^{-1}.\mu^T\right)^{-1}.\xi\right) = (a-r).(\sigma.\sigma^T)^{-1}.(a-r) \\ &-(a-r).(\sigma.\sigma^T)^{-1}.\mu^T.\left(\mu.(\sigma.\sigma^T)^{-1}.\mu^T\right)^{-1}.\mu.(\sigma.\sigma^T)^{-1}.(a-r)\end{aligned}$$

yielding (we choose $\lambda_2 > 0$ by studying the sign of quantities in (7.5.24)

$$\begin{aligned}\lambda_2 = &\Big(\sqrt{}\Big((a-r).(\sigma.\sigma^T)^{-1}.(a-r) - (a-r).(\sigma.\sigma^T)^{-1}.\mu^T \\ &.\left(\mu.(\sigma.\sigma^T)^{-1}.\mu^T\right)^{-1}.\mu.(\sigma.\sigma^T)^{-1}.(a-r)\Big)\Big) \\ &\Big/\left(2\sqrt{\eta^2 - \xi.(\mu.(\sigma.\sigma^T)^{-1}.\mu^T)^{-1}.\xi}\right).\end{aligned} \tag{7.5.28}$$

Putting (7.5.24), (7.5.27) and (7.5.28) together we get the optimal portfolio rule (without ever finding the value function, which is not surprising since that was already seen in the volatility-only-constraint case):

$$\begin{aligned}\Pi(t, X) = \Big(X\,(\sigma.\sigma^T)^{-1}.\Big((a-r) - \mu^T.\Big(-2\,\big(\surd\big((a-r).(\sigma.\sigma^T)^{-1}.(a-r) \\ - (a-r).(\sigma.\sigma^T)^{-1}.\mu^T.\big(\mu.(\sigma.\sigma^T)^{-1}.\mu^T\big)^{-1}.\mu \\ .(\sigma.\sigma^T)^{-1}.(a-r)\big)\big) \\ \Big/\Big(2\sqrt{\eta^2 - \xi.(\mu.(\sigma.\sigma^T)^{-1}.\mu^T)^{-1}.\xi}\Big)\xi \\ + (a-r).(\sigma.\sigma^T)^{-1}.\mu^T\Big).\big(\mu.(\sigma.\sigma^T)^{-1}.\mu^T\big)^{-1}\Big)\Big) \\ \Big/\Big(2\,\big(\surd\big((a-r).(\sigma.\sigma^T)^{-1}.(a-r) - (a-r).(\sigma.\sigma^T)^{-1} \\ .\mu^T.\big(\mu.(\sigma.\sigma^T)^{-1}.\mu^T\big)^{-1}.\mu.(\sigma.\sigma^T)^{-1}.(a-r)\big)\big) \\ \Big/\Big(2\sqrt{\eta^2 - \xi.(\mu.(\sigma.\sigma^T)^{-1}.\mu^T)^{-1}.\xi}\Big)\Big).\end{aligned} \tag{7.5.29}$$

Since volatility of the wealth is prescribed, it is only left to compute the resulting wealth appreciation rate:

$$r + \frac{\Pi(t, X).(a-r)}{X} = r + \frac{\frac{X\,((a-r)-\lambda_1.\mu).(\sigma.\sigma^T)^{-1}}{2\,\lambda_2}.(a-r)}{X}$$

$$= r + \frac{((a-r) - \lambda_1.\mu).(\sigma.\sigma^T)^{-1}.(a-r)}{2\,\lambda_2}$$

with λ_1 and λ_2 as given in (7.5.27) and (7.5.28). There is no dependence on the value function at all. The reason is that volatility of wealth is constrained, and therefore the problem reduces to maximization of the appreciation rate, independently of the particular value function.

7.6 Portfolio Optimization under Inequality Constraints

More natural than the above solved problems with equality constraints on the portfolio are the problems with a mix of equalities and *inequality* constraints. Looking into the list of problems solved above, there are many different possibilities to consider in the case of inequality constraints as well. Of all of these possibilities, we choose to modify the very last problem considered: wealth-volatility (equality quadratic) constraint, together with the budget (equality affine) constraint, under no-short-selling (*inequality* affine) constraints. We shall explain the solution in an example.

```
In[128]:= Clear["Global`*"];
          Needs["LinearAlgebra`MatrixManipulation`"];
          e_i_ := {ReplacePart[Table[0, {Dim}], 1, i]};
          1_m := Array[1 &, {Dim}];
```

Consider

```
In[130]:= Dim = 100;
```

stocks, with appreciation rates

```
In[131]:= a = Array[Random[Real, {-.2, .6}] &, {Dim}];
```

and with

```
In[132]:= σ = 1/12 Array[Random[Real, {-1, 1}] &, {Dim, Dim}];
```

yielding the individual volatilities:

```
In[133]:= Volatilities = (√(Plus @@ #1^2) &) /@ σ

Out[133]= {0.476448, 0.466564, 0.438571, 0.453066, 0.472377, 0.487561,
  0.498874, 0.476499, 0.478431, 0.494706, 0.479469, 0.471277,
  0.495111, 0.454631, 0.51686, 0.510847, 0.479275, 0.481703,
  0.476499, 0.48236, 0.50309, 0.475562, 0.48761, 0.464864,
  0.468933, 0.474954, 0.478678, 0.508842, 0.526233, 0.478253,
  0.421987, 0.484354, 0.498546, 0.458699, 0.454159, 0.472748,
  0.487516, 0.515601, 0.469062, 0.444556, 0.512202, 0.467378,
  0.476591, 0.490334, 0.493614, 0.482219, 0.47992, 0.487136,
  0.469177, 0.458932, 0.470518, 0.44086, 0.46012, 0.483835,
  0.513214, 0.443071, 0.464164, 0.457503, 0.496528, 0.499249,
  0.483955, 0.480918, 0.46382, 0.482754, 0.450514, 0.478388,
  0.43553, 0.450719, 0.489762, 0.475093, 0.473463, 0.492036,
  0.473311, 0.437096, 0.465455, 0.469864, 0.459643, 0.45774,
  0.481779, 0.479867, 0.462064, 0.488603, 0.483635, 0.478178,
  0.506626, 0.500229, 0.541474, 0.489289, 0.486794, 0.466852,
  0.451566, 0.510345, 0.477271, 0.480121, 0.509856, 0.482701,
  0.535584, 0.487765, 0.436081, 0.470439}
```

Let the interest rate be

```
In[134]:= r = .05;
```

The *volatility constraint* is

```
In[135]:= η = .2;
```

The *budget constraint* is imposed through

```
In[136]:= μ = {1m};
```

and

```
In[137]:= ξ = {1};
```

The first thing to check is whether volatility constraint is compatible with the budget constraint, and if not, to relax it

```
In[138]:= η = If[η^2 - ξ.Inverse[μ.Inverse[σ.Transpose[σ]].
               Transpose[μ]].ξ > 0, η,
            1.1 √(ξ.Inverse[μ.Inverse[σ.Transpose[σ]].
                Transpose[μ]].ξ)]

Out[138]= 0.2
```

Next we compute the solution under the volatility and budget constraint only:

```
In[139]:= λ2 = (√((a - r).Inverse[σ.Transpose[σ]].(a - r) -
               (a - r).Inverse[σ.Transpose[σ]].
                Transpose[μ].Inverse[
                  μ.Inverse[σ.Transpose[σ]].Transpose[μ]].
                μ.Inverse[σ.Transpose[σ]].(a - r)))/
           (2 √(η^2 - ξ.Inverse[μ.Inverse[σ.Transpose[σ]].
                  Transpose[μ]].ξ))

Out[139]= 134.739
```

```
In[140]:= λ1 = Inverse[
             μ.Inverse[σ.Transpose[σ]].Transpose[μ]].μ.
            Inverse[σ.Transpose[σ]].(a - r) - 2 λ2 Inverse[
              μ.Inverse[σ.Transpose[σ]].Transpose[μ]].ξ

Out[140]= {0.0521941}
```

```
In[141]:= ComputedPortfolio =
           (Chop[#1, 1/10^6] &)[1/(2 λ2) (Inverse[σ.Transpose[σ]].
              ((a - r) - Transpose[μ].λ1))]

Out[141]= {-3.41249, -2.40213, -9.35374, -3.47871, -2.19655,
           -2.19291, 5.99264, -5.53546, -3.24952, 7.7113, -0.733307,
           2.1488, 0.621959, -7.01615, 0.598971, -3.04528, 10.2078,
           1.81103, 5.04638, -2.01686, 7.10732, -1.82224, -0.219214,
```

−8.32308, −8.06543, −0.820464, −3.84879, −2.16991,
−5.38069, 10.7277, −4.84036, −2.45128, 1.29179, −4.3567,
4.78009, −5.83315, 2.28682, −1.44159, 7.40329, 11.1366,
7.18222, −2.69287, −1.45342, −6.87963, 4.9006, 0.673596,
11.9365, −4.49803, −6.12593, −8.96772, −6.20439,
−0.640752, −0.840365, 1.56726, 0.31138, −5.01402, 3.75888,
−2.00567, −3.57149, 5.0066, −2.72815, 6.48688, −6.04004,
−13.8387, 1.1166, −3.8623, −2.47587, 1.35633, 2.31061,
−6.7839, 1.9376, −2.7512, 7.75346, −12.0293, 8.26845,
0.468731, 1.49942, −2.51688, −3.15798, −1.41649, 0.904216,
2.85816, 2.1586, 2.4395, −4.03277, 4.18002, 0.427073,
0.875927, 8.68921, −0.521623, 2.10891, −1.84859, 1.02911,
−0.633707, 11.081, 8.55483, 7.82122, 2.53065, −0.965836,
8.63763}

If the computed portfolio happens to satisfy the no-short-selling constraint as well, the problem is solved. If not, the affine *equality* constraint needs to be complemented (many times if necessary) in such a way so that the corresponding final computed portfolio satisfies the no-short-selling (inequality) constraint. So, actually the problem with inequality constraints is attempted to be solved by looking for an appropriate equality constraint and the corresponding solution. Notice however that in the following search there is no guarantee that the actual maximal solution is obtained. Compute

```
In[142]:= P = Catch[Do[y = BlockMatrix[
            ({e#1} &) /@ Transpose[Position[Negative /@
                ComputedPortfolio, True]][[1]];
          μ = BlockMatrix[{{μ}, {y}}]; ξ = Join[ξ,
            Table[0, {Length[μ] - Length[ξ]}]]; η =
           If[η^2 - ξ.Inverse[μ.Inverse[σ.Transpose[σ]].
                Transpose[μ]].ξ > 0, η,
             1.1 √ξ.Inverse[μ.Inverse[σ.Transpose[σ]].
               Transpose[μ]].ξ]; λ2 =
           (√((a - r).Inverse[σ.Transpose[σ]].(a - r) -
                (a - r).Inverse[σ.Transpose[σ]].
                Transpose[μ].Inverse[μ.Inverse[
                    σ.Transpose[σ]].Transpose[μ]].μ.
                Inverse[σ.Transpose[σ]].(a - r)))/
             (2 √(η^2 - ξ.Inverse[μ.Inverse[σ.
                      Transpose[σ]].Transpose[μ]].ξ));
```

```
λ1 = Inverse[μ.Inverse[σ.Transpose[σ]].
      Transpose[μ]].μ.
    Inverse[σ.Transpose[σ]].(a - r) -
   2 λ2 Inverse[μ.Inverse[σ.Transpose[σ]].
       Transpose[μ]].ξ;
ComputedPortfolio = (Chop[#1, 1/10^6] &)[
   1/(2 λ2) (Inverse[σ.Transpose[σ]].
      ((a - r) - Transpose[μ].λ1))];
If[And @@ (#1 ≥ 0 &) /@ ComputedPortfolio,
  Throw[ComputedPortfolio]], {1000}]];
```

yielding an optimal portfolio rule candidate

```
In[143]:= Π[X_, t_] = P X
```

Out[143]= {0, 0, 0, 0, 0, 0, 0, 0, 0, 0, 0, 0, 0, 0, 0, 0, 0, 0, 0.00432178 X, 0, 0, 0, 0, 0, 0, 0, 0, 0, 0, 0, 0, 0, 0, 0, 0, 0, 0, 0, 0, 0, 0.169314 X, 0, 0, 0, 0, 0, 0, 0, 0, 0, 0, 0, 0, 0, 0, 0, 0, 0, 0, 0.17778 X, 0, 0.0401088 X, 0, 0, 0, 0, 0, 0, 0, 0.00233172 X, 0, 0.134907 X, 0, 0.297205 X, 0, 0, 0, 0, 0, 0, 0.0371096 X, 0, 0, 0, 0, 0, 0, 0, 0, 0, 0, 0.136923 X, 0, 0, 0, 0, 0}

The reader may wish to evaluate μ and ξ at this point. Not being able to claim that this is *the* optimal portfolio rule makes it more important to be able to judge the quality of the computed rule. Since the wealth-volatility, although constrained, might have been changed (relaxed) during the optimization process, we take a look at it:

```
In[144]:= η
```

Out[144]= 0.2

while the obtained wealth-appreciation rate is equal to

```
In[145]:= r + 1/(2 λ2) ((a - r).Inverse[σ.Transpose[σ]].(a - r) -
            λ1.μ.Inverse[σ.Transpose[σ]].(a - r))
```

Out[145]= 0.576635

So even in the event that the computed portfolio rule is not the best possible, the above numbers can testify how good the rule is, in absolute if not relative terms. This, arguably, is more important anyway. It is interesting to compare the wealth-appreciation rate under the computed portfolio rule with the individual stocks appreciation rates:

In[146]:= `Sort[a]`

Out[146]= {−0.155424, −0.143033, −0.134402, −0.131625, −0.114565, −0.0978032, −0.0973385, −0.0898256, −0.0590306, −0.0395444, −0.0193958, −0.0134293, 0.000852499, 0.00334419, 0.00876824, 0.013559, 0.0203352, 0.02232, 0.0351249, 0.0502862, 0.0720054, 0.0800152, 0.0828667, 0.0998497, 0.112884, 0.113652, 0.118227, 0.120382, 0.126637, 0.136982, 0.14038, 0.14189, 0.164017, 0.175922, 0.178288, 0.20829, 0.214788, 0.216903, 0.218142, 0.219866, 0.227673, 0.240267, 0.240783, 0.247648, 0.253687, 0.261209, 0.262671, 0.269288, 0.283361, 0.283559, 0.285411, 0.293906, 0.300998, 0.309009, 0.310966, 0.315145, 0.340547, 0.342135, 0.345143, 0.350659, 0.359303, 0.362226, 0.362281, 0.368843, 0.371783, 0.389877, 0.397507, 0.402193, 0.404643, 0.404895, 0.414143, 0.437968, 0.442084, 0.442228, 0.443275, 0.448872, 0.464478, 0.469208, 0.472999, 0.480965, 0.498805, 0.501003, 0.515998, 0.521333, 0.523748, 0.529194, 0.532751, 0.540117, 0.540165, 0.540248, 0.542305, 0.552258, 0.56028, 0.56714, 0.56742, 0.569611, 0.573167, 0.57726, 0.58294, 0.596444}

as well as the portfolio's volatility η with the individual stock's volatilities:

In[147]:= `Sort[Volatilities]`

Out[147]= {0.421987, 0.43553, 0.436081, 0.437096, 0.438571, 0.44086, 0.443071, 0.444556, 0.450514, 0.450719, 0.451566, 0.453066, 0.454159, 0.454631, 0.457503, 0.45774, 0.458699, 0.458932, 0.459643, 0.46012, 0.462064, 0.46382, 0.464164, 0.464864, 0.465455, 0.466564, 0.466852, 0.467378, 0.468933, 0.469062, 0.469177, 0.469864, 0.470439, 0.470518, 0.471277, 0.472377, 0.472748, 0.473311, 0.473463, 0.474954, 0.475093, 0.475562, 0.476448, 0.476499, 0.476499, 0.476591, 0.477271, 0.478178, 0.478253, 0.478388, 0.478431, 0.478678, 0.479275, 0.479469, 0.479867, 0.47992, 0.480121, 0.480918, 0.481703, 0.481779, 0.482219, 0.48236, 0.482701, 0.482754, 0.483635, 0.483835, 0.483955, 0.484354, 0.486794, 0.487136, 0.487516, 0.487561, 0.48761, 0.487765, 0.488603, 0.489289, 0.489762, 0.490334, 0.492036, 0.493614, 0.494706, 0.495111, 0.496528, 0.498546,

0.498874, 0.499249, 0.500229, 0.50309, 0.506626, 0.508842,
0.509856, 0.510345, 0.510847, 0.512202, 0.513214, 0.515601,
0.51686, 0.526233, 0.535584, 0.541474}

Although there will be always at least one stock with a higher appreciation rate, the reduction in resulting portfolio balance volatility is well worth the modest reduction in the appreciation rate. It takes only moments to do the above calculation for hundreds of stocks considered for trading.

Many other problems can be solved similarly.

8 Advanced Trading Strategies

Reduced Monge–Ampère PDEs of Advanced Optimal Portfolio Hedging and Hypoelliptic Obstacle Problems of Optimal Momentum Trading

"... the portfolio-selection process needs to be purged of its reliance on 'static' optimization techniques, which are incapable, by their very nature, of evaluating intertemporal tradeoffs. A fully satisfactory method of portfolio selection must come to grips with the large nonlinear systems of multivariate partial differential equations of dynamic optimality...."

(**Robert C. Merton,** I.E. Block Community Lecture at the meeting of Society for Industrial and Applied Mathematics in Toronto in July 1998).

8.1 Remarks

As we have seen so far, mathematics empowered with *Mathematica*® can be used as a framework for answering many practical questions in the market analysis, investing, and trading of stocks and options. Chapter 5 and 6 present some sophisticated ways as to how to analyze the market from the point of view of estimating the perceived stock volatilities. In Chapter 7 it was shown how mathematics and *Mathematica*® can be used for synthesizing the available information about market dynamics, market uncertainty, and an investor's attitude towards uncertainty with respect to explicit trading and investment diversification decisions, provided that the market dynamics is simple enough.

The methodology for determining the implied volatility for European and American options is quite satisfactory. On the other hand, the obvious possible criticism of the stochastic control methods in portfolio management presented so far is that the underlying market dynamics model (the Log-Normal price dynamics), which by virtue of its simplicity has allowed beautiful and efficient symbolic solutions, is quite possibly too simple, and therefore cannot to a significant degree capture and exploit the important features of real life trading. What can be done to improve the portfolio optimization methodology? Can we extend the methodology described in Chapter 7 to the case where the assumed market dynamics is much more descriptive?

As it turns out, and as it was known for some time, to do this, one would first have to address successfully a fundamental mathematical problem: How to find quality numerical solutions of HJB and Monge–Ampère partial differential equations (in higher dimension)? Since dimension is of paramount significance in numerical solutions of PDEs, we shall show how to *reduce* (by one) the dimension of a PDE characterizing the value function in the optimal portfolio problem under price dependent price dynamics. We also announce that the problem of finding quality solutions for such Monge–Ampère type PDEs, and therefore also of HJB PDEs that can be reduced to the Monge–Ampère PDEs, has been settled. Those two methodologies combined, the *symbolic* dimension reduction and *numerical* solution of reduced equations set the stage, in my opinion, for a new phase in portfolio management technology and practice. We shall illustrate these advances in two computational examples: optimal portfolios of stocks and optimal hedging of options, both under appreciation rate reversing market dynamics.

Finally, as seen in Chapter 7, optimal portfolio hedging requires perpetual trading, which can be afforded only by big investing organizations. For small investors, each trade is paid for, and too frequent portfolio hedging does not seem to be a reasonable alternative. Different kinds of strategies therefore need to be developed. For either type of investors, big or small, one issue that seems worth addressing is: Can we employ advanced mathematics in designing optimal trading strategies in "momentum trading"? To address this problem, the first issue is to introduce a model for "momentum price dynamics", and the question then is: Do we need to move beyond first order scalar SDEs in Markov processes stock-price-dynamics modeling? Specifically, we were led to the second order equation, or which is the same, to a *degenerate* system of two first order SDEs, as a model for stock-price dynamics.

Finding optimal strategies in such a framework, as it turned out, has lead us again to the fundamental mathematical problem: How to find quality numerical solutions of hypoelliptic problems, and in particular, of hypoelliptic obstacle problems (where, at least in practical terms, the hypoellipticity of generators of Markov processes can be thought of as a *precise* framework for addressing the issue of degeneracy)? We announce in this closing chapter that this problem has been settled as well.

8.2 Reduced Monge–Ampère PDEs of Advanced Portfolio Hedging

8.2.1 Advanced Optimal Portfolio Hedging Problems

8.2.1.1 The Fundamental Trichotomy

The common feature in portfolio management problems solved so far has been the asset-price-dynamics independence of the "state variable", i.e., of the asset-prices themselves. Similarly, as in option pricing/hedging, in portfolio hedging, time-dependency is much easier to handle than the state, i.e. price-dependency. Therefore we shall refer to portfolio management problems that involve asset-price dynamics which are price-dependent (more precisely, appreciation rates and volatilities are allowed to be price-dependent) as *advanced* optimal portfolio management problems.

It has been understood for some time that solving such problems will involve numerical solutions of very complicated PDEs. What was less well understood is that instead of solving (in this chapter solving will often mean solving numerically) HJB PDEs, the productive path is to solve the associated Monge–Ampère PDEs. The complexity of both problems was discouraging for many, and probably this is the reason that advanced portfolio management problems were not solved very successfully before, and that they are still not applied in practice. What we shall present here is a simple but fundamental discovery that, in the advanced optimal portfolio hedging problems, instead of solving the Monge–Ampère equations that are directly derived from the HJB equations, one can go a step further and derive the *reduced Monge–Ampère equations of optimal portfolio hedging*, eliminating what has turned out to be a trivial variable. Such equations have one variable less to solve for, which of course is of great importance for computations. Moreover, using this approach, we can solve problems for (safety exponents) $\gamma >> 1$ (if there are no contradicting constraints), which is particularly important for reasonable investing, and which did not appear possible in the direct numerical value function approach. On the other hand, the direct numerical value function approach can solve problems for small and even $\gamma \leq 0$, but this seems to be more of a mathematical curiosity than of practical importance, so we decided not to go there in this presentation. So, instead of studying only HJB and Monge–Ampère PDEs, we establish here the trichotomy: HJB, Monge–Ampère, and *reduced Monge–Ampère PDEs*, as a framework for solving advanced optimal portfolio hedging problems. If the data is asset-price-dependent, there is no hope for further reduction/simplification—these equations contain information that has to be recovered by means of numerical solutions. This indeed can be done. Such a fortunate turn of events also confirms that the stochastic control approach to optimal portfolio management is a fundamentally correct framework for this very important practical problem.

8.2.1.2 Market Dynamics

It is a daily ritual for many, not only the professionals, to pay attention to the fluctuations of some of the major market indices, such as DOW, NAS, S&P, DAX, FTSE, CAC, NIK, etc., and for some, using this information to anticipate future market developments. The portfolio management techniques presented in Chapter 7 are not capable of utilizing such information to the full extent. We shall show here how the methods presented in Chapter 7 can be improved to the extent that such information can be used in a mathematically meaningful and correct fashion.

Consider an index, or a significant security such as QQQ, and suppose that its value/price $q(t)$ is governed by the SDE

$$d\!\!\!d\, q(t) = q(t)\, a_q(t, q(t))\, d\!\!\!d\, t + q(t)\, \sigma_q(t, q(t)).d\!\!\!d\, B(t) \tag{8.2.1}$$

where $B(t)$ is the n-dimensional Brownian motion, $\sigma_q(t, q)$ is an n-vector-valued, and $a_q(t, q)$ is a scalar-valued function. The obvious difference with the Log-Normal model, utilized heavily before, is that the appreciation rate and volatility are not constants. By the way, everything can be generalized to the case when q is vector-valued, i.e., in the case when instead of only one index being tracked, we track two (two is no problem) or more, but the numerical computational complexity quickly becomes prohibitive (at least for a laptop computer in the year 2002). This is why we have chosen to discuss the case when q is only scalar-valued.

Additionally, the stock market prices $S(t) = \{S_1(t), \ldots, S_m(t)\}$ are assumed to be governed by

$$dS(t) = S(t)\, a_s(t, q(t))\, dt + S(t)\, \sigma_s(t, q(t)).dB(t) \tag{8.2.2}$$

where $\sigma_s(t, q)$ is $m \times n$-matrix-valued and $a_s(t, q)$ is an m-vector-valued function. So, compared with the Chapter 7, the stock-price dynamics is allowed to depend on q, i.e., on the market index (as well as on time t). This is a significant improvement since so far we could solve only time-dependent problems. On the other hand, as discussed above, the number of indexes we can consider is very limited since otherwise the problem would become higher dimensional, and although conceptually this would not be an issue, computationally it is.

We notice that it is also possible to trade index itself. For example if we choose to track Nasdaq 100, we can instead track *and trade* QQQ. The above framework allows for that if we take, for example $S(t) = \{q(t), S_1(t), \ldots, S_m(t)\}$, and adjust accordingly $\sigma_s(t, q)$ and $a_s(t, q)$. In Section 8.2.4.2 we shall study in detail advanced hedging of options, which is a very similar problem: one tracks and trades the underlying stock.

As before, an investor has a cash account whose balance, positive or negative, $C(t)$ at the time t, evolves, in the absence of trading, according to the equation

$$dC(t) = r\, C(t)\, dt \tag{8.2.3}$$

where r is the interest rate (r could be a function $r = r(t, q)$—nothing changes below; for example, if an index runs up a lot over short period of time, one can expect interest rates to go up, and vice-versa).

8.2.1.3 Market Dynamics Example: Appreciation-Rate Reversing Model

```
In[1]:= Clear["Global`*"]
```

To fix ideas, we chose the data for the above described market model. The index volatility vector $\sigma_q(t, q)$ is chosen to be a constant vector:

```
In[2]:= σq[t_, θ_] =
          {0.208946, -0.106235, 0.284088, -0.120832};
```

and let the stocks-volatility matrix be

```
In[3]:= σs[t_, q_] =
         {{0.100948, 0.276666, 0.191967, -0.0273367},
          {-0.0591624, -0.156995, 0.184767, 0.227785},
          {0.0271511, 0.184511, 0.282641, 0.148763}}
```

Out[3]=
$$\begin{pmatrix} 0.100948 & 0.276666 & 0.191967 & -0.0273367 \\ -0.0591624 & -0.156995 & 0.184767 & 0.227785 \\ 0.0271511 & 0.184511 & 0.282641 & 0.148763 \end{pmatrix}$$

We introduce the q-dependency only for appreciation rates in order to isolate the phenomena when studying it. So the index appreciation rate is assumed to be

```
In[4]:= aq[t_, θ_] = .05 (50 - θ);
```

while the stock appreciation rates are

```
In[5]:= as[t_, θ_] = {1.2, -1, .3} aq[t, θ]
```

Out[5]= $\{0.06\,(50-\theta),\ -0.05\,(50-\theta),\ 0.015\,(50-\theta)\}$

All together, they look like:

```
In[6]:= Plot[
          Evaluate[Prepend[as[t, θ], aq[t, θ]]], {θ, 30, 70},
          PlotStyle → AbsoluteThickness /@ {2, 1, 1, 1},
          AxesLabel → {"q", "aq(t,q),as(t,q)"}];
```

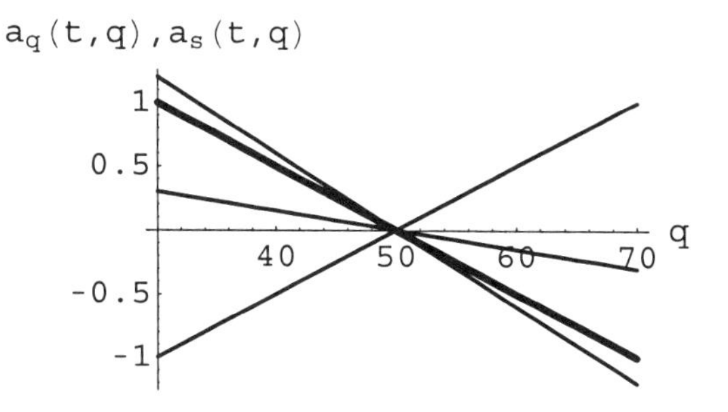

So, the index value $q = q_{\text{equi}} = 50$ is assumed to be a sort of a market equilibrium for the index during the time period when trading is considered. When the index is above its equilibrium, i.e., when $q > q_{\text{equi}}$, the index, as well as the first and third stock have a tendency to decline, while the opposite is true for the second stock; also, when $q < q_{\text{equi}}$, the reverse holds. Of course, the equilibrium q_{equi} does not have to be a constant; it could be a function of time.

For understanding this dynamics, we perform a Monte–Carlo simulation of the Markov process $\{q(t), S(t)\}$. We shall need

```
In[7]:= << "Statistics`NormalDistribution`"
```

and we merge the index and stocks data together:

```
In[8]:= A[t_, θ_] = Prepend[as[t, θ], aq[t, θ]]
```

Out[8]= $\{0.05\,(50-\theta),\ 0.06\,(50-\theta),\ -0.05\,(50-\theta),\ 0.015\,(50-\theta)\}$

in the case of appreciation rates, and

```
In[9]:= Σ[t_, q_] = Prepend[σs[t, q], σq[t, q]]
```

Out[9]=
$$\begin{pmatrix} 0.208946 & -0.106235 & 0.284088 & -0.120832 \\ 0.100948 & 0.276666 & 0.191967 & -0.0273367 \\ -0.0591624 & -0.156995 & 0.184767 & 0.227785 \\ 0.0271511 & 0.184511 & 0.282641 & 0.148763 \end{pmatrix}$$

for volatilities. The individual volatilities are equal to

```
In[10]:= Vol = (√Plus @@ #1² &) /@ Σ[t, q]
```

Out[10]= {0.387622, 0.352609, 0.337894, 0.369862}

Also relevant for judging whether this example is realistic, the cross-correlation matrix is equal to

```
In[11]:= Cor = Σ[t, q].Transpose[Σ[t, q]] / Outer[Times, Vol, Vol]
```

Out[11]=
$$\begin{pmatrix} 1. & 0.362454 & 0.223576 & 0.337536 \\ 0.362454 & 1. & -0.16925 & 0.79729 \\ 0.223576 & -0.16925 & 1. & 0.444372 \\ 0.337536 & 0.79729 & 0.444372 & 1. \end{pmatrix}$$

We notice here, that although we need $\Sigma(t, q)$, or equivalently $\sigma_q(t, q)$ and $\sigma_s(t, q)$, for the Monte–Carlo simulations below, $\Sigma(t, q)$ is not going to be needed nor, possibly, will it be available in real practice. What will always be available through statistics, as discussed in Chapter 4, are good estimates of $\Sigma(t, q).\Sigma(t, q)^T$. This is not a problem, because everything that follows will not depend on $\sigma_q(t, q)$ and $\sigma_s(t, q)$, but only on their products $\sigma_q(t, q).\sigma_q(t, q)$, $\sigma_q(t, q).\sigma_s(t, q)^T$, and $\sigma_s(t, q).\sigma_s(t, q)^T$, and they are all available in the matrix $\Sigma(t, q).\Sigma(t, q)^T$. Indeed,

$$\Sigma(t, q).\Sigma(t, q)^T = \begin{pmatrix} \sigma_q(t, q) \\ \sigma_s(t, q) \end{pmatrix} . \begin{pmatrix} \sigma_q(t, q) \\ \sigma_s(t, q) \end{pmatrix}^T = \begin{pmatrix} \sigma_q(t, q) \\ \sigma_s(t, q) \end{pmatrix} . (\,\sigma_q(t, q) \quad \sigma_s(t, q)^T\,)$$

$$= \begin{pmatrix} \sigma_q(t, q).\sigma_q(t, q) & \sigma_q(t, q).\sigma_s(t, q)^T \\ \sigma_s(t, q).\sigma_q(t, q) & \sigma_s(t, q).\sigma_s(t, q)^T \end{pmatrix}.$$

Continuing, let the initial values for the index and stocks be

```
In[12]:= K = 1000; t0 = 0; T = 1; dt = (T - t0)/K;
         n = Length[A[t, q]]; m = n - 1;
         stockprice = Table[Random[Real, {15, 110}], {n}]
```

Out[12]= {54.1401, 20.8636, 18.5337, 37.7796}

and compute the market evolution trajectory

```
In[13]:= dB = Table[Random[NormalDistribution[0, 1]],
             {K}, {n}] √dt ;
         G[{t_, S_}, dB_] := {dt + t,
            A[t, S[[1]]] dt S + Σ[t, S[[1]]].dB S + S};
         MarketEvolution = FoldList[G,
            {t0, stockprice}, dB];
```

The evolution over a time frame of one year turned out to be

```
In[16]:= ithStock[i_] :=
           ({#1[[1]], #1[[2, i]]} &) /@ MarketEvolution;
         PS = Prepend[Table[RGBColor[Random[], 0, Random[]],
            {m}], {AbsoluteThickness[1.5],
            RGBColor[0, 0, 0]}]; equi =
          Plot[50, {t, t0, T}, DisplayFunction → Identity,
           PlotRange → {0, Automatic},
           PlotStyle → RGBColor[1, 1/2, 0]];
         Show[equi, (ListPlot[ithStock[#1], PlotRange →
               {0, Max[(#1[[2]] &) /@ MarketEvolution]},
              PlotJoined → True, PlotStyle → PS[[#1]],
              DisplayFunction → Identity] &) /@ Range[n],
           DisplayFunction → $DisplayFunction];
```

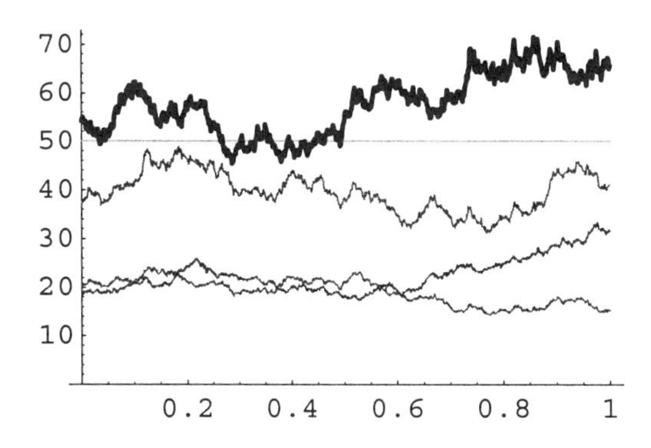

(the thick trajectory corresponds to the index price evolution).

So, the index has an equilibrium price, and that also influences other stock prices. Speaking of the equilibrium price, we shall see in the second part of this chapter, i.e., in Section 8.3, another stock-price model with an equilibrium price. The difference is going to be that such a model, in addition to equilibrium, has another parameter, the *period*, which can be thought of as an expected time to make an oscillation around the equilibrium. The present model does not have this feature. By experimenting with both models one can gain some intuition as to when one model is better suited to be applied than the other.

Having understood the market model, we proceed with the issue of trading in such a market. We want to use the data from this example later when we compute the optimal strategies. To this end let

```
In[19]:= {aq[t, θ], as[t, q], σq[t, θ], σs[t, q]} >>
         "AdvancedTrading"
```

8.2.1.4 Stochastic Control Problems

As before, $X(t)$ is the investor's cumulative wealth available at the time t: cash plus the cash value of all of the investments.

The problem is to design a trading strategy, i.e., any vector-valued function $\Pi(t, X, q) = \{\Pi_1(t, X, q), \Pi_2(t, X, q), ..., \Pi_m(t, X, q)\}$, where $\Pi_j(t, X, q)$ is the cash value, positive or negative, of the investment in the jth stock considered. So, as the stock prices, the trading strategy Π is allowed to depend on q. Then

$$X(t) = C(t) + \Pi(t, X(t), q(t)).\mathbb{1}_m$$

where $\mathbb{1}_m = \{1, ..., 1\}$, and as before,

$$\begin{aligned} dX(t) &= (\Pi(t, X(t), q(t)).(a_s(t, q(t)) - r) + r\,X(t))\,dt \\ &+\Pi(t, X(t), q(t)).\sigma_s(t, q(t)).dB(t). \end{aligned} \tag{8.2.4}$$

Any strategy $\Pi(t, X, q)$ determines a Markov process $\{X(t), q(t)\} = \{X^{\Pi}(t), q(t)\}$, the unique solution of the SDE system (8.2.1) and (8.2.4). Notice that the system is not coupled, since (8.2.1) does not depend on X. So, SDEs (8.2.1) and (8.2.4) describe the controlled system. The time horizon T is the constant time at which time we agree to measure the success or failure of investing/trading ("retirement time").

So, for a given safety exponent γ, the problem is to find a Markovian strategy $\Pi^{\star}(t, X, q)$ such that

$$\sup_{\Pi} E_{t,X,q}\,\psi_\gamma\,(X^{\Pi}(T)) = \operatorname*{Max}_{\Pi} E_{t,X,q}\,\psi_\gamma\,(X^{\Pi}(T)) = E_{t,X,q}\,\psi_\gamma\left(X^{\Pi^{\star}}(T)\right)$$

if there are no constraints on the portfolio, or

$$\begin{aligned} &\sup_{\Pi:\mu(t,q).\Pi(t,X,q)=\xi(t,X)\,X} E_{t,X,q}\,\psi_\gamma\,(X^{\Pi}(T)) \\ &= \operatorname*{Max}_{\Pi:\mu(t,q).\Pi(t,X,q)=\xi(t,X)\,X} E_{t,X,q}\,\psi_\gamma\,(X^{\Pi}(T)) = E_{t,X,q}\,\psi_\gamma\left(X^{\Pi^{\star}}(T)\right) \end{aligned}$$

in the case of a general affine constraint where ψ_γ is the utility of wealth function. We can consider numerical problems *in bounded domains* for any utility function, and in particular for utility functions ψ_γ for $\gamma \le 0$. But since we want to concentrate the attention of the reader to the below-introduced *reduced Monge–Ampère PDE* approach, we shall restrict our consideration to the case $\gamma > 0$, and for the most part even to $\gamma \geq 1$, noticing that from the practical point of view, as discussed in Chapter 7, this is by far the most important case.

As usual in stochastic control, the crucial object of study is the *value function*

$$\varphi_{1,\gamma}(t, X, q) = \sup_{\Pi} E_{t,X,q}\, \psi_\gamma\,(X^{\Pi}(T)) \tag{8.2.5}$$

in the case of no-constraints, and

$$\varphi_{2,\gamma}(t, X, q) = \sup_{\Pi:\mu(t,q).\Pi(t,X,q)-\xi(t,X)\, X} E_{t,X,q}\, \psi_\gamma\,(X^{\Pi}(T)) \tag{8.2.6}$$

in the case of a general affine constraint on the trading strategy. In order to find value functions, we need to characterize them first.

8.2.2 Derivation of the Monge–Ampère PDEs

8.2.2.1 No Constraints

We shall first derive the HJB and Monge–Ampère PDEs characterizing the value function $\varphi_{1,\gamma}$ in (8.2.5). The following formulas hold:

$$\begin{aligned}(dX(t))^2 = \Pi(t, X(t), q(t)).\sigma_s(t, q(t)).\sigma_s(t, q(t))^T\\ .\Pi(t, X(t), q(t))\, dt.\end{aligned} \tag{8.2.7}$$

$$(dq(t))^2 = q(t)^2\, \sigma_q(t, q(t)).\sigma_q(t, q(t))\, dt. \tag{8.2.8}$$

$$dX(t)\, dq(t) = q(t)\, \Pi(t, X(t), q(t)).\sigma_s(t, q(t)).\sigma_q(t, q(t))\, dt. \tag{8.2.9}$$

Let $\varphi = \varphi(t, X, q)$ be a differentiable function. Using the Itô chain rule

$$\begin{aligned}
d\varphi\,(t, X\,(t), q\,(t)) = \Big(&\varphi_t(t, X(t), q(t)) + \varphi_X(t, X(t), q(t))\,(\Pi(t, X(t), q(t))\\
&.(a_s(t, q(t)) - r) + r\, X(t)) + \varphi_q(t, X(t), q(t))\, q(t)\\
&a_q(t, q(t)) + \frac{1}{2}\,\varphi_{X,X}(t, X(t), q(t))\,\Pi(t, X(t), q(t)).\sigma_s(t, q(t))\\
&.\sigma_s(t, q(t))^T.\Pi(t, X(t), q(t)) + \varphi_{X,q}(t, X(t), q(t))\\
&\; q(t)\,\Pi(t, X(t), q(t)).\sigma_s(t, q(t)).\sigma_q(t, q(t))\\
&+ \frac{1}{2}\,\varphi_{q,q}(t, X(t), q(t))\, q(t)^2\,\sigma_q(t, q(t)).\sigma_q(t, q(t))\Big)\, dt\\
&+\varphi_X(t, X(t), q(t))\,\Pi(t, X(t), q(t)).\sigma_s(t, q(t)).dB(t)\\
&+\varphi_q\,(t, X\,(t), q\,(t))\, q\,(t)\,\sigma_q\,(t, q\,(t)).dB\,(t).
\end{aligned} \tag{8.2.10}$$

Therefore if φ solves the linear PDE (for fixed Π)

$$\begin{aligned}&\varphi_t(t, X, q) + \varphi_X(t, X, q)\,(\Pi(t, X, q).(a_s(t, q) - r) + r\,X)\\&+\varphi_q(t, X, q)\,q\,a_q(t, q) + \frac{1}{2}\,\varphi_{X,X}(t, X, q)\,\Pi(t, X, q).\sigma_s(t, q)\\&.\sigma_s\,(t, q)^T.\Pi\,(t, X, q) + \varphi_{X,q}(t, X, q)\,q\,\Pi(t, X, q).\sigma_s(t, q)\\&.\sigma_q(t, q) + \frac{1}{2}\,\varphi_{q,q}(t, X, q)\,q^2\,\sigma_q(t, q).\sigma_q(t, q) = 0\end{aligned} \tag{8.2.11}$$

in $Q = (t_0, T) \times (0, \infty)$, together with the terminal condition

$$\varphi(T, X, q) = \psi_\gamma(X), \tag{8.2.12}$$

then

$$\varphi(t, X, q) = E_{t,X,q}\,\psi_\gamma\,(X^\Pi(T)). \tag{8.2.13}$$

On the other hand, if φ is the solution of the HJB PDE

$$\begin{aligned}&\operatorname*{Max}_\Pi\big[\varphi_t(t, X, q) + \varphi_X(t, X, q)\,(\Pi(t, X, q).(a_s(t, q) - r) + r\,X)\\&\quad + \varphi_q(t, X, q)\,q\,a_q(t, q) + \frac{1}{2}\,\varphi_{X,X}(t, X, q)\,\Pi(t, X, q).\sigma_s(t, q)\\&\quad .\sigma_s(t, q)^T.\Pi(t, X, q) + \varphi_{X,q}(t, X, q)\,q\,\Pi(t, X, q).\sigma_s(t, q)\\&\qquad .\sigma_q(t, q) + \frac{1}{2}\,\varphi_{q,q}(t, X, q)\,q^2\,\sigma_q(t, q).\sigma_q(t, q)\big] = 0\end{aligned} \tag{8.2.14}$$

in Q, together with the terminal condition (8.2.12), then φ is the value function, i.e., it has a probabilistic representation (8.2.5). Solving the above stochastic control problem is equivalent to solving this HJB PDE. There is no hope of finding an explicit/symbolic solution to this equation—the only hope is try to find a numerical solution. Furthermore, a naive numerical approach would be a direct one—trying to construct a numerical solution of (8.2.14) directly. We shall introduce here by far a superior approach by deriving the associated Monge–Ampère PDE, then in the following sections, by reducing the dimension of that equation and deriving the *reduced* Monge–Ampère PDE, and finally solving numerically such an equation. Along the way, as the equations become simpler and simpler, the optimal strategies $\Pi^\star$ become more and more explicit.

As before, we maximize the expression in the HJB PDE. We first solve the problem without constraints. So, we maximize

$$\begin{aligned}f\,(\Pi) &= \varphi_X(t, X, q)\,\Pi.(a_s(t, q) - r) + \frac{1}{2}\,\varphi_{X,X}(t, X, q)\,\Pi.\sigma_s(t, q).\sigma_s(t, q)^T.\Pi\\&\quad+\varphi_{X,q}(t, X, q)\,q\,\Pi.\sigma_s(t, q).\sigma_q(t, q).\end{aligned}$$

Again, we see in order for the quadratic function f to have maximum, it is necessary and sufficient that

$$\varphi_{X,X}(t,\, X,\, q) < 0. \tag{8.2.15}$$

The necessary condition for the maximizer is $\nabla_\Pi f(\Pi^\star) = 0$, i.e. $\Pi^\star$ solves the equation

$$\begin{aligned}\nabla f(\Pi) &= \varphi_X(t,\, X,\, q)\,(a_s(t,\, q) - r)\\ &\quad + \varphi_{X,X}(t,\, X,\, q)\,\sigma_s(t,\, q).\sigma_s(t,\, q)^T.\Pi + \varphi_{X,q}(t,\, X,\, q)\,q\,\sigma_s(t,\, q).\sigma_q(t,\, q) = 0,\end{aligned}$$

which implies

$$\begin{aligned}\Pi^\star\,(t,\, X,\, q) &= -\frac{\varphi_X(t,\, X,\, q)}{\varphi_{X,X}(t,\, X,\, q)}\,(\sigma_s(t,\, q).\sigma_s(t,\, q)^T)^{-1}.(a_s(t,\, q) - r)\\ &\quad - \frac{\varphi_{X,q}(t,\, X,\, q)}{\varphi_{X,X}(t,\, X,\, q)}\,q\,(\sigma_s(t,\, q).\sigma_s(t,\, q)^T)^{-1}.\sigma_s(t,\, q).\sigma_q(t,\, q),\end{aligned} \tag{8.2.16}$$

or, which is the same and also useful,

$$\begin{aligned}&= -\frac{\varphi_X(t,\, X,\, q)}{\varphi_{X,X}(t,\, X,\, q)}\,(a_s(t,\, q) - r).(\sigma_s(t,\, q).\sigma_s(t,\, q)^T)^{-1}\\ &\quad - \frac{\varphi_{X,q}(t,\, X,\, q)}{\varphi_{X,X}(t,\, X,\, q)}\,q\,\sigma_q(t,\, q).\sigma_s(t,\, q)^T.(\sigma_s(t,\, q).\sigma_s(t,\, q)^T)^{-1}\end{aligned} \tag{8.2.17}$$

with φ still unknown. We plug $\Pi^\star$ back into the HJB PDE, which yields, after multiplication of the equation by $\varphi_{X,X}(t,\, X,\, q)$,

$$\begin{aligned}0 &= \varphi_{X,X}(t,\, X,\, q)\,\varphi_t(t,\, X,\, q) + \varphi_{X,X}(t,\, X,\, q)\,\varphi_X(t,\, X,\, q)\,r\,X\\ &\quad + \varphi_{X,X}(t,\, X,\, q)\,\varphi_q(t,\, X,\, q)\,q\,a_q(t,\, q)\\ &\quad + \frac{1}{2}\,\varphi_{X,X}(t,\, X,\, q)\,\varphi_{q,q}(t,\, X,\, q)\,q^2\,\sigma_q(t,\, q).\sigma_q(t,\, q)\\ &\quad + \varphi_{X,X}\,(t,\, X,\, q)\,\varphi_X(t,\, X,\, q)\,\Pi^\star.(a_s(t,\, q) - r)\\ &\quad + \frac{1}{2}\,\varphi_{X,X}(t,\, X,\, q)^2\,\Pi^\star.\sigma_s(t,\, q).\sigma_s(t,\, q)^T.\Pi^\star\\ &\quad + \varphi_{X,X}(t,\, X,\, q)\,\varphi_{X,q}(t,\, X,\, q)\,q\,\Pi^\star.\sigma_s(t,\, q).\sigma_q(t,\, q).\end{aligned} \tag{8.2.18}$$

This equations needs to be simplified, so we prepare formulas

$$\begin{aligned}&\varphi_{X,X}\,(t,\, X,\, q)\,\varphi_X(t,\, X,\, q)\,\Pi^\star.(a_s(t,\, q) - r)\\ &= -\varphi_X(t,\, X,\, q)^2\,(a_s(t,\, q) - r).(\sigma_s(t,\, q).\sigma_s(t,\, q)^T)^{-1}.(a_s(t,\, q) - r)\\ &\quad - \varphi_X(t,\, X,\, q)\,\varphi_{X,q}(t,\, X,\, q)\,q\,\sigma_q(t,\, q).\sigma_s(t,\, q)^T\\ &\quad .(\sigma_s(t,\, q).\sigma_s(t,\, q)^T)^{-1}.(a_s\,(t,\, q) - r)\end{aligned} \tag{8.2.19}$$

and

$$
\begin{aligned}
&\frac{1}{2}\,\varphi_{X,X}(t,\,X,\,q)^2\,\Pi^\star.\sigma_s(t,\,q).\sigma_s(t,\,q)^T.\Pi^\star\\
&=\frac{1}{2}\left(\varphi_X(t,\,X,\,q)^2\,(a_s(t,\,q)-r).(\sigma_s(t,\,q).\sigma_s(t,\,q)^T)^{-1}.(a_s(t,\,q)-r)\right)\\
&+q\,\varphi_X(t,\,X,\,q)\,\varphi_{X,q}(t,\,X,\,q)\,(a_s(t,\,q)-r).(\sigma_s(t,\,q).\sigma_s(t,\,q)^T)^{-1}\\
&.\sigma_s\,(t,\,q).\sigma_q\,(t,\,q)+\frac{1}{2}\,q^2\,\varphi_{X,q}(t,\,X,\,q)^2\,\sigma_q(t,\,q).\sigma_s(t,\,q)^T\\
&\qquad.(\sigma_s(t,\,q).\sigma_s(t,\,q)^T)^{-1}.\sigma_s(t,\,q).\sigma_q(t,\,q)
\end{aligned}
$$

and

$$
\begin{aligned}
&\varphi_{X,X}\,(t,\,X,\,q)\,\varphi_{X,q}(t,\,X,\,q)\,q\,\Pi^\star.\sigma_s(t,\,q).\sigma_q(t,\,q)\\
&=-q\,\varphi_{X,q}(t,\,X,\,q)\,\varphi_X(t,\,X,\,q)\,(a_s(t,\,q)-r).(\sigma_s(t,\,q).\sigma_s(t,\,q)^T)^{-1}\\
&\qquad.\sigma_s(t,\,q).\sigma_q(t,\,q)-\varphi_{X,q}(t,\,X,\,q)^2\,q^{\,2}\,\sigma_q(t,\,q).\sigma_s(t,\,q)^T\\
&\qquad.(\sigma_s(t,\,q).\sigma_s(t,\,q)^T)^{-1}.\sigma_s(t,\,q).\sigma_q(t,\,q).
\end{aligned}
\tag{8.2.21}
$$

Utilizing (8.2.19), (8.2.20) and (8.2.21) in (8.2.18), after some work, we arrive at

$$
\begin{aligned}
&\varphi_t(t,\,X,\,q)\,\varphi_{X,X}(t,\,X,\,q)+\varphi_{X,X}(t,\,X,\,q)\,\varphi_X(t,\,X,\,q)\,r\,X\\
&\quad+\varphi_{X,X}(t,\,X,\,q)\,\varphi_q(t,\,X,\,q)\,q\,a_q(t,\,q)+\frac{1}{2}\,\varphi_{X,X}(t,\,X,\,q)\\
&\quad\varphi_{q,q}(t,\,X,\,q)\,q^2\,\sigma_q(t,\,q).\sigma_q(t,\,q)-\frac{1}{2}\,\varphi_X(t,\,X,\,q)^2\\
&\qquad(a_s(t,\,q)-r).(\sigma_s(t,\,q).\sigma_s(t,\,q)^T)^{-1}.(a_s(t,\,q)-r)\\
&\quad-\varphi_X(t,\,X,\,q)\,\varphi_{X,q}(t,\,X,\,q)\,q\,\sigma_q(t,\,q).\sigma_s(t,\,q)^T.(\sigma_s(t,\,q)\\
&\qquad.\sigma_s(t,\,q)^T)^{-1}.(a_s(t,\,q)-r)-\frac{1}{2}\,q^2\,\varphi_{X,q}(t,\,X,\,q)^2\\
&\quad\sigma_q(t,\,q).\sigma_s(t,\,q)^T.(\sigma_s(t,\,q).\sigma_s(t,\,q)^T)^{-1}.\sigma_s(t,\,q).\sigma_q(t,\,q)=0
\end{aligned}
\tag{8.2.22}
$$

in Q. Equation (8.2.22), together with the backward parabolic boundary condition (8.2.12), and the additional geometric condition (8.2.15), can be referred to as the Monge–Ampère PDE associated with the stochastic control problem (8.2.5). Due to (8.2.16), the stochastic control problem is solved once equation (8.2.22) is solved. Equation (8.2.22) may be one of the simplest Monge–Ampère equations in the context of optimal portfolio hedging that is beyond the methods of Chapter 7. Nevertheless, it contains all the important features that are present in more complicated situations. One such more complicated problem is the problem having the same underlying dynamics, but with affine constraints on the portfolio, discussed next.

8.2.2.2 Affine Constraints

We shall now derive the Monge–Ampère PDEs characterizing the value function $\varphi_{2,\gamma}$ in (8.2.6), i.e., in the case of the general affine constraint

$$\mu(t, q).\Pi(t, X, q) = \Pi(t, X, q).\mu(t, q)^T = \xi(t, q)\, X \tag{8.2.23}$$

where $\mu(t, q)$ is $k \times m$ matrix-valued function, and $\xi(t, q)$ is k-vector-valued. Instead of (8.2.14), the HJB equation now reads as

$$\begin{aligned}
\operatorname*{Max}_{\Pi:\mu(t,q).\Pi(t,X,q)=\xi(t,q)\,X} &\big[\varphi_t(t, X, q) + \varphi_X(t, X, q)\,(\Pi(t, X, q) \\
&.(a_s(t, q) - r) + r\,X) + \varphi_q(t, X, q)\, q\, a_q(t, q) \\
&+ \frac{1}{2}\,\varphi_{X,X}(t, X, q)\,\Pi(t, X, q).\sigma_s(t, q).\sigma_s(t, q)^T.\Pi(t, X, q) \\
&+ \varphi_{X,q}(t, X, q)\, q\, \Pi(t, X, q).\sigma_s(t, q).\sigma_q(t, q) \\
&+ \frac{1}{2}\,\varphi_{q,q}(t, X, q)\, q^2\, \sigma_q(t, q).\sigma_q(t, q)\big] = 0.
\end{aligned} \tag{8.2.24}$$

So, for every (t, X, q), we need to maximize

$$\begin{aligned}
h(\Pi) = {}& \varphi_X(t, X, q)\,\Pi.(a_s(t, q) - r) + \frac{1}{2}\,\varphi_{X,X}(t, X, q)\,\Pi.\sigma_s(t, q) \\
& .\sigma_s(t, q)^T.\Pi + \varphi_{X,q}(t, X, q)\, q\, \Pi.\sigma_s(t, q).\sigma_q(t, q)
\end{aligned} \tag{8.2.25}$$

(again for that it is necessary for (8.2.15) to hold) under the constraint

$$g(\Pi) = \Pi.\mu(t, q)^T - \xi(t, q)\, X = 0. \tag{8.2.26}$$

Therefore, as before, there exists a k-vector λ, such that $\nabla_\Pi h(\Pi) = \nabla_\Pi g(\Pi).\lambda$, i.e.,

$$\begin{aligned}
\nabla_\Pi h(\Pi) = {}& \varphi_X(t, X, q)\,(a_s(t, q) - r) + \varphi_{X,X}(t, X, q)\,\sigma_s(t, q).\sigma_s(t, q)^T.\Pi \\
& + \varphi_{X,q}(t, X, q)\, q\, \sigma_s(t, q).\sigma_q(t, q) = \nabla_\Pi g(\Pi).\lambda = \mu(t, q)^T.\lambda(t, X, q),
\end{aligned}$$

and consequently

$$\begin{aligned}
\Pi^\star(t, X, q) = {}& \frac{1}{\varphi_{X,X}(t, X, q)}\,(\sigma_s(t, q).\sigma_s(t, q)^T)^{-1}.\mu(t, q)^T.\lambda(t, X, q) \\
& - \frac{\varphi_X(t, X, q)}{\varphi_{X,X}(t, X, q)}\,(\sigma_s(t, q).\sigma_s(t, q)^T)^{-1}.(a_s(t, q) - r) - \\
& \frac{\varphi_{X,q}(t, X, q)}{\varphi_{X,X}(t, X, q)}\, q\,(\sigma_s(t, q).\sigma_s(t, q)^T)^{-1}.\sigma_s(t, q).\sigma_q(t, q)
\end{aligned} \tag{8.2.27}$$

with λ and φ yet to be determined. Using the affine constraint

$$\mu(t, q).\left(\frac{1}{\varphi_{X,X}(t, X, q)} (\sigma_s(t, q).\sigma_s(t, q)^T)^{-1}.\mu(t, q)^T.\lambda(t, X, q) \right.$$
$$- \frac{\varphi_X(t, X, q)}{\varphi_{X,X}(t, X, q)} (\sigma_s(t, q).\sigma_s(t, q)^T)^{-1}.(a_s(t, q) - r)$$
$$\left. - \frac{\varphi_{X,q}(t, X, q)}{\varphi_{X,X}(t, X, q)} q\, (\sigma_s(t, q).\sigma_s(t, q)^T)^{-1}.\sigma_s(t, q).\sigma_q(t, q) \right) = \xi(t, q)\, X$$

we get

$$\mu(t, q).(\sigma_s(t, q).\sigma_s(t, q)^T)^{-1}.\mu(t, q)^T.\lambda(t, X, q) = \varphi_{X,X}\,(t, X, q)\, \xi\,(t, q)\, X$$
$$+\varphi_X\,(t, X, q)\, \mu(t, q).(\sigma_s(t, q).\sigma_s(t, q)^T)^{-1}.(a_s(t, q) - r)$$
$$+\varphi_{X,q}\,(t, X, q)\, q\, \mu(t, q).(\sigma_s(t, q).\sigma_s(t, q)^T)^{-1}.\sigma_s(t, q).\sigma_q(t, q).$$

Assuming that the matrix $\mu(t, q).(\sigma_s(t, q).\sigma_s(t, q)^T)^{-1}.\mu(t, q)^T$ is invertible, we conclude that

$$\begin{aligned}\lambda\,(t, X, q) &= \varphi_{X,X}(t, X, q) \left(\mu(t, q).(\sigma_s(t, q).\sigma_s(t, q)^T)^{-1}.\mu(t, q)^T\right)^{-1} \\ &.\xi\,(t, q)\, X + \varphi_X(t, X, q) \left(\mu(t, q).(\sigma_s(t, q).\sigma_s(t, q)^T)^{-1}.\mu(t, q)^T\right)^{-1} \\ &.\mu(t, q).(\sigma_s(t, q).\sigma_s(t, q)^T)^{-1}.(a_s(t, q) - r) + \varphi_{X,q}(t, X, q)\, q \left(\mu(t, q)\right. \\ &\quad .(\sigma_s(t, q).\sigma_s(t, q)^T)^{-1}.\mu(t, q)^T\left.\right)^{-1}.\mu(t, q).(\sigma_s(t, q) \\ &\quad .\sigma_s(t, q)^T)^{-1}.\sigma_s(t, q).\sigma_q(t, q).\end{aligned} \tag{8.2.28}$$

Going back into the expression for $\Pi^\star$, we get

$$\begin{aligned}\Pi^\star(t, X, q) &= (\sigma_s(t, q).\sigma_s(t, q)^T)^{-1}.\Big(\mu(t, q)^T.(\mu(t, q).(\sigma_s(t, q) \\ &\qquad .\sigma_s(t, q)^T)^{-1}.\mu(t, q)^T\big)^{-1}.\xi(t, q)\, X \\ &\quad + \frac{\varphi_X(t, X, q)}{\varphi_{X,X}(t, X, q)} \big(\mu(t, q)^T.(\mu(t, q).(\sigma_s(t, q).\sigma_s(t, q)^T)^{-1} \\ &\qquad .\mu(t, q)^T\big)^{-1}.\mu(t, q).(\sigma_s(t, q).\sigma_s(t, q)^T)^{-1} - \mathbb{I}_m\big) \\ &\quad .(a_s(t, q) - r) + \frac{\varphi_{X,q}(t, X, q)}{\varphi_{X,X}(t, X, q)}\, q \left(\mu(t, q)^T\right. \\ &\qquad .\big(\mu(t, q).(\sigma_s(t, q).\sigma_s(t, q)^T)^{-1}.\mu(t, q)^T\big)^{-1}.\mu(t, q) \\ &\qquad .(\sigma_s(t, q).\sigma_s(t, q)^T)^{-1} - \mathbb{I}_m\left.\right).\sigma_s(t, q).\sigma_q(t, q)\Big)\end{aligned} \tag{8.2.29}$$

where $\mathbb{I}_m$ is the m-dimensional identity matrix, which is the optimal trading strategy, with φ, the value function, yet to be determined. Going back to (8.2.24), after multiplying the equation by $\varphi_{X,X}(t, X, q)$, we get

$$\varphi_t(t,X,q)\,\varphi_{X,X}(t,X,q)+\varphi_X(t,X,q)\,\varphi_{X,X}(t,X,q)\,(\Pi^\star(t,X,q)$$
$$.(a_s(t,q)-r)+r\,X)+\varphi_{X,X}(t,X,q)\,\varphi_q(t,X,q)\,q\,a_q(t,q)$$
$$+\frac{1}{2}\,\varphi_{X,X}(t,X,q)^2\,\Pi^\star(t,X,q).\sigma_s(t,q).\sigma_s(t,q)^T.\Pi^\star(t,X,q)$$
$$+\varphi_{X,q}(t,X,q)\,\varphi_{X,X}(t,X,q)\,q\,\Pi^\star(t,X,q).\sigma_s(t,q).\sigma_q(t,q)$$
$$+\frac{1}{2}\,\varphi_{X,X}(t,X,q)\,\varphi_{q,q}(t,X,q)\,q^2\,\sigma_q(t,q).\sigma_q(t,q)=0.$$

We need to simplify this equation. So we prepare formulas

$$\begin{aligned}&\varphi_X\,\varphi_{X,X}\,\Pi^\star.(a_s-r)=\varphi_X\;\varphi_{X,X}\;X\;\xi.\big(\mu.(\sigma_s.\sigma_s{}^T)^{-1}.\mu^T\big)^{-1}.\mu.(\sigma_s.\sigma_s{}^T)^{-1}\\&.(a_s-r)+\varphi_X\;\varphi_X\;(a_s-r).\big((\sigma_s.\sigma_s{}^T)^{-1}.\mu^T.\big(\mu.(\sigma_s.\sigma_s{}^T)^{-1}.\mu^T\big)^{-1}.\mu\\&\qquad-\mathbb{I}_m\big).(\sigma_s.\sigma_s{}^T)^{-1}.(a_s-r)+\varphi_X\;\varphi_{X,q}\,q\;\sigma_q.\sigma_s{}^T\\&\qquad.\big((\sigma_s.\sigma_s{}^T)^{-1}.\mu^T.\big(\mu.(\sigma_s.\sigma_s{}^T)^{-1}.\mu^T\big)^{-1}.\mu-\mathbb{I}_m\big).(\sigma_s.\sigma_s{}^T)^{-1}.(a_s-r)\end{aligned}\tag{8.2.31}$$

as well as

$$\begin{aligned}&\frac{1}{2}\,\varphi_{X,X}{}^2\,\Pi^\star.\sigma_s.\sigma_s{}^T.\Pi^\star=\frac{1}{2}\,\varphi_{X,X}{}^2\,X^2\,\xi.\big(\mu.(\sigma_s.\sigma_s{}^T)^{-1}.\mu^T\big)^{-1}.\mu\\&.(\sigma_s.\sigma_s{}^T)^{-1}.\mu^T.\big(\mu.(\sigma_s.\sigma_s{}^T)^{-1}.\mu^T\big)^{-1}.\xi+\frac{1}{2}\,\varphi_X{}^2\,(a_s-r).\big((\sigma_s.\sigma_s{}^T)^{-1}\\&\qquad-(\sigma_s.\sigma_s{}^T)^{-1}.\mu^T.\big(\mu.(\sigma_s.\sigma_s{}^T)^{-1}.\mu^T\big)^{-1}.\mu.(\sigma_s.\sigma_s{}^T)^{-1}\big).(a_s-r)\\&+\frac{1}{2}\,\varphi_{X,q}{}^2\,q^2\,\sigma_q.\sigma_s{}^T.(\sigma_s.\sigma_s{}^T)^{-1}.\big(\mathbb{I}_m-\mu^T.\big(\mu.(\sigma_s.\sigma_s{}^T)^{-1}.\mu^T\big)^{-1}.\mu\\&\qquad.(\sigma_s.\sigma_s{}^T)^{-1}\big).\sigma_s.\sigma_q+\varphi_X\,\varphi_{X,q}\,q\,(a_s-r).(\sigma_s.\sigma_s{}^T)^{-1}\\&.\big(\mathbb{I}_m-\mu^T.\big(\mu.(\sigma_s.\sigma_s{}^T)^{-1}.\mu^T\big)^{-1}.\mu.(\sigma_s.\sigma_s{}^T)^{-1}\big).\sigma_s.\sigma_q\end{aligned}\tag{8.2.32}$$

and

$$\begin{aligned}&\varphi_{X,q}\,\varphi_{X,X}\,q\,\Pi^\star.\sigma_s.\sigma_q=\varphi_{X,q}\,\varphi_{X,X}\,q\,\xi.\big(\mu.(\sigma_s.\sigma_s{}^T)^{-1}.\mu^T\big)^{-1}.\mu\,.(\sigma_s.\sigma_s{}^T)^{-1}\\&\qquad.\sigma_s.\sigma_q\,X+\varphi_X\,\varphi_{X,q}\,q\,(a_s-r)\\&\qquad.\big((\sigma_s.\sigma_s{}^T)^{-1}.\mu^T.\big(\mu.(\sigma_s.\sigma_s{}^T)^{-1}.\mu^T\big)^{-1}.\mu-\mathbb{I}_m\big)\\&\qquad.(\sigma_s.\sigma_s{}^T)^{-1}.\sigma_s.\sigma_q+\varphi_{X,q}\,q\,\varphi_{X,q}\,q\,\sigma_q.\sigma_s{}^T\\&\qquad.\big((\sigma_s.\sigma_s{}^T)^{-1}.\mu^T.\big(\mu.(\sigma_s.\sigma_s{}^T)^{-1}.\mu^T\big)^{-1}.\mu-\mathbb{I}_m\big)\\&\qquad.(\sigma_s.\sigma_s{}^T)^{-1}.\sigma_s.\sigma_q.\end{aligned}\tag{8.2.33}$$

Plugging (8.2.31), (8.2.32), and (8.2.33) back into (8.2.30), we conclude, after more work,

$$
\begin{aligned}
&\varphi_t \varphi_{X,X} + \varphi_X \varphi_{X,X} X \left(r + \xi.\left(\mu.(\sigma_s.\sigma_s^T)^{-1}.\mu^T\right)^{-1}.\mu.(\sigma_s.\sigma_s^T)^{-1}.(a_s - r)\right) \\
&+\frac{1}{2}\varphi_X^2 (a_s - r).\left((\sigma_s.\sigma_s^T)^{-1}.\mu^T.\left(\mu.(\sigma_s.\sigma_s^T)^{-1}.\mu^T\right)^{-1}.\mu - \mathbb{I}_m\right).(\sigma_s.\sigma_s^T)^{-1} \\
&.(a_s - r) + \varphi_X \varphi_{X,q}\, q\, \sigma_q.\sigma_s^T.\left((\sigma_s.\sigma_s^T)^{-1}.\mu^T.\left(\mu.(\sigma_s.\sigma_s^T)^{-1}.\mu^T\right)^{-1}.\mu - \mathbb{I}_m\right) \\
&.(\sigma_s.\sigma_s^T)^{-1}.(a_s - r) + \varphi_{X,X} \varphi_q\, q\, a_q \\
&+\frac{1}{2}\varphi_{X,X}^2 X^2 \xi.\left(\mu.(\sigma_s.\sigma_s^T)^{-1}.\mu^T\right)^{-1}.\xi + \varphi_{X,q}\varphi_{X,X} X q \\
&\qquad \xi.\left(\mu.(\sigma_s.\sigma_s^T)^{-1}.\mu^T\right)^{-1}.\mu\,.(\sigma_s.\sigma_s^T)^{-1}.\sigma_s.\sigma_q + \frac{1}{2}\varphi_{X,q}^2 q^2 \\
&\qquad \sigma_q.\sigma_s^T.\left((\sigma_s.\sigma_s^T)^{-1}.\mu^T.\left(\mu.(\sigma_s.\sigma_s^T)^{-1}.\mu^T\right)^{-1}.\mu - \mathbb{I}_m\right) \\
&\qquad .(\sigma_s.\sigma_s^T)^{-1}.\sigma_s.\sigma_q + \frac{1}{2}\varphi_{X,X}\varphi_{q,q}\, q^2 \sigma_q.\sigma_q = 0
\end{aligned}
$$

which is yet another Monge–Ampère PDE. We notice a possibility of notation being confusing in (8.2.34) and in (8.2.22): a_q is the appreciation rate for the index price q, and similarly for σ_q, while φ_q is the partial derivative of the value function φ with respect to q. We shall write these equations very precisely below.

8.2.3 Reduced Monge–Ampère PDEs of Advanced Optimal Portfolio Hedging

8.2.3.1 No Constraints

In[20]:=
```
Remove["Global`*"]
```

We shall work on the above derived Monge–Ampère PDE (8.2.22) first. Let us rewrite (8.2.22) very precisely, in *Mathematica*® meaningful fashion:

In[21]:=
```
Eq3D =
  1/2 ∂{q,2} φ[t, X, q] ∂{X,2} φ[t, X, q] σq[t, q].σq[t, q] q^2 -
   1/2 (∂X,q φ[t, X, q])^2 σq[t, q].Transpose[σs[t, q]].
     Inverse[σs[t, q].Transpose[σs[t, q]]].
     σs[t, q].σq[t, q] q^2 - ∂X φ[t, X, q] ∂X,q φ[t, X, q]
    σq[t, q].Transpose[σs[t, q]].Inverse[
      σs[t, q].Transpose[σs[t, q]]].(as[t, q] - r) q +
   ∂q φ[t, X, q] ∂{X,2} φ[t, X, q] aq[t, q] q +
   ∂t φ[t, X, q] ∂{X,2} φ[t, X, q] +
   r X ∂X φ[t, X, q] ∂{X,2} φ[t, X, q] -
   1/2 (∂X φ[t, X, q])^2 (as[t, q] - r).Inverse[
      σs[t, q].Transpose[σs[t, q]]].(as[t, q] - r) == 0
```

Out[21]= $-\frac{1}{2}\sigma_q(t,q).\sigma_s(t,q)^T.(\sigma_s(t,q).\sigma_s(t,q)^T)^{-1}.\sigma_s(t,q).\sigma_q(t,q)$
$\varphi^{(0,1,1)}(t,X,q)^2 q^2 + \frac{1}{2}\sigma_q(t,q).\sigma_q(t,q)\,\varphi^{(0,0,2)}(t,X,q)$
$\varphi^{(0,2,0)}(t,X,q)\,q^2 - \sigma_q(t,q).\sigma_s(t,q)^T.(\sigma_s(t,q).\sigma_s(t,q)^T)^{-1}$
$.(a_s(t,q)-r)\,\varphi^{(0,1,0)}(t,X,q)\,\varphi^{(0,1,1)}(t,X,q)\,q + a_q(t,q)$
$\varphi^{(0,0,1)}(t,X,q)\,\varphi^{(0,2,0)}(t,X,q)\,q - \frac{1}{2}(a_s(t,q)-r).(\sigma_s(t,q)$
$.\sigma_s(t,q)^T)^{-1}.(a_s(t,q)-r)\,\varphi^{(0,1,0)}(t,X,q)^2$
$+r\,X\,\varphi^{(0,1,0)}(t,X,q)\,\varphi^{(0,2,0)}(t,X,q)$
$+\varphi^{(0,2,0)}(t,X,q)\,\varphi^{(1,0,0)}(t,X,q) == 0$

Experimenting with numerical solutions of the above equation, the following simple but, as it has turned out, a fundamental discovery was made: assuming that the terminal condition is the utility function $\psi_\gamma(X) = \frac{X^{1-\gamma}}{1-\gamma}$ for $1 \neq \gamma > 0$, and $\psi_1(X) = \log(X)$, we can look for a solution of the Eq3D in the form

In[22]:=
```
η[t_, X_, q_] := X^(1-γ) f[t, q] / (1 - γ)
```

if $1 \neq \gamma > 0$, in which case also the terminal condition translates into $f(T,q) = 1$. Also, in the case $\gamma = 1$, we can look for the solution in the form

In[23]:=
```
ζ[t_, X_, q_] := f[t, q] + Log[X]
```

in which case the terminal condition translates into $f(T,q) = 0$. Indeed, plugging them into the Monge–Ampère PDE (8.2.22), we get

In[24]:=
```
Eq2Dγ = Simplify[Eq3D /. φ → η]
```

Out[24]= $\frac{1}{2(\gamma-1)}\Big(X^{-2\gamma}\Big(-(\gamma-1)\big(2r\gamma + (a_s(t,q)-r).(\sigma_s(t,q).\sigma_s(t,q)^T)^{-1}$
$.(a_s(t,q)-r)\big)f(t,q)^2 + \big(\gamma\,(\sigma_q(t,q).\sigma_q(t,q)$
$f^{(0,2)}(t,q)\,q^2 + 2\,a_q(t,q)\,f^{(0,1)}(t,q)\,q$
$+2\,f^{(1,0)}(t,q)) - 2\,q\,(\gamma-1)\,\sigma_q(t,q).\sigma_s(t,q)^T$
$.(\sigma_s(t,q).\sigma_s(t,q)^T)^{-1}.(a_s(t,q)-r)\,f^{(0,1)}(t,q)\big)$
$f(t,q) - q^2(\gamma-1)\,\sigma_q(t,q).\sigma_s(t,q)^T.(\sigma_s(t,q).\sigma_s(t,q)^T)^{-1}$
$.\sigma_s(t,q).\sigma_q(t,q)\,f^{(0,1)}(t,q)^2\Big)\Big) == 0$

i.e.,

In[25]:=
```
ReducedMongeAmpere1 =
  Thread[(2 (γ - 1) / X^(-2 γ) #1 &)[Eq2Dγ], Equal]
```

Out[25]= $-(\gamma-1)\left(2r\gamma+(a_s(t,q)-r).(\sigma_s(t,q).\sigma_s(t,q)^T)^{-1}.(a_s(t,q)-r)\right) f(t,q)^2+\left(\gamma(\sigma_q(t,q).\sigma_q(t,q)f^{(0,2)}(t,q)q^2+2a_q(t,q) f^{(0,1)}(t,q)q+2f^{(1,0)}(t,q))-2q(\gamma-1)\sigma_q(t,q) .\sigma_s(t,q)^T.(\sigma_s(t,q).\sigma_s(t,q)^T)^{-1}.(a_s(t,q)-r)f^{(0,1)}(t,q)\right) f(t,q)-q^2(\gamma-1)\sigma_q(t,q).\sigma_s(t,q)^T.(\sigma_s(t,q).\sigma_s(t,q)^T)^{-1}.\sigma_s(t,q) .\sigma_q(t,q)f^{(0,1)}(t,q)^2==0$

in the case $1\neq\gamma>0$; in the case $\gamma=1$

In[26]:= `Eq2D1 = Simplify[Eq3D /. φ → ς]`

Out[26]= $\frac{1}{2X^2}\left(\sigma_q(t,q).\sigma_q(t,q)f^{(0,2)}(t,q)q^2+2a_q(t,q)f^{(0,1)}(t,q)q +2r+(a_s(t,q)-r).(\sigma_s(t,q).\sigma_s(t,q)^T)^{-1}.(a_s(t,q)-r) +2f^{(1,0)}(t,q)\right)==0$

In the case $1\neq\gamma>0$ the resulting equation is reminiscent of Monge–Ampère PDE. Nevertheless two regions for γ, $\gamma<1$ or $\gamma>1$, are quite different. What follows was carried out successfully only under the condition that $\gamma>1$ ($\gamma=1$ is trivial case as we shall see shortly). Recall that the case $\gamma>1$ is the most important case in practice. In this case the resulting equation is Monge–Ampère PDE but a simpler one than (8.2.22), since one of the variables X has been eliminated. Therefore we shall call the derived equation above the *reduced Monge–Ampère PDE of optimal portfolio hedging*. It has the form

$$f^{(1,0)}(t,q)f(t,q)+a_1(t,q)f(t,q)^2+a_2(t,q)f^{(0,2)}(t,q)f(t,q) +a_3(t,q)f^{(0,1)}(t,q)f(t,q)+a_4(t,q)f^{(0,1)}(t,q)^2=0 \quad (8.2.35)$$

and thereby we also introduce the notation $a_i(t,q)$, for $i=1,\ldots,4$, to be used below. In the case $\gamma=1$, the resulting equation is particularly simple—it is a *linear* backward parabolic equation

In[27]:= `Thread[(2 X^2 #1 &)[Eq2D1], Equal]`

Out[27]= $\sigma_q(t,q).\sigma_q(t,q)f^{(0,2)}(t,q)q^2+2a_q(t,q)f^{(0,1)}(t,q)q+2r+ (a_s(t,q)-r).(\sigma_s(t,q).\sigma_s(t,q)^T)^{-1}.(a_s(t,q)-r)+2f^{(1,0)}(t,q)==0$

together with the terminal condition $f(T,q)=0$. This equation is obviously solvable, and therefore in the case $\gamma=1$, the assumed form of the value function is justified. Moreover, since regardless of what exactly f is,

In[28]:= `∂X,q ς[t, X, q]`

Out[28]= 0

and moreover since

In[29]:= $\frac{\partial_X \zeta[t, X, q]}{\partial_{\{X,2\}} \zeta[t, X, q]}$

Out[29]= $-X$

using (8.2.16), the case $\gamma = 1$ is suddenly solved completely, without solving the above linear backward parabolic PDE, i.e., without ever figuring out what exactly f is:

$$\Pi_1^\star(t, X, q) = X\,(\sigma_s(t, q).\sigma_s(t, q)^T)^{-1}.(a_s(t, q) - r). \tag{8.2.36}$$

That is in the case $\gamma = 1$, nothing changes, the Merton optimal strategy applies. This is reminiscent of the case $\gamma = 1$ in the appreciation-rate-uncertainty case in Chapter 7, and it is another indication that, for the most reasonable investing, one has to use safety exponents $\gamma >> 1$.

So, consider the case $\gamma > 1$, and assume we can solve ReducedMongeAmpere1, together with the terminal condition $f(T, q) = 1$. We have

In[30]:= $\frac{\partial_X \eta[t, X, q]}{\partial_{\{X,2\}} \eta[t, X, q]}$

Out[30]= $-\frac{X}{\gamma}$

and

In[31]:= $\frac{\partial_{X,q} \eta[t, X, q]}{\partial_{\{X,2\}} \eta[t, X, q]}$

Out[31]= $-\frac{X\, f^{(0,1)}(t, q)}{\gamma\, f(t, q)}$

and therefore, in the case $\gamma > 1$, the optimal trading strategy is equal to

$$\Pi_\gamma^\star(t, X, q) = X\, P_\gamma^\star(t, q) = \frac{X}{\gamma}\,(\sigma_s(t, q).\sigma_s(t, q)^T)^{-1}.(a_s(t, q) - r) \\ + \frac{X\, f^{(0,1)}(t, q)}{\gamma\, f(t, q)}\, q\,(\sigma_s(t, q).\sigma_s(t, q)^T)^{-1}.\sigma_s(t, q).\sigma_q(t, q) \tag{8.2.37}$$

assuming $f = f_\gamma$ can be determined using the above *reduced Monge–Ampère PDE* of optimal portfolio hedging. We shall refer to f as the *reduced value function.*

As we shall see below, under the assumption that $\gamma > 1$, the reduced Monge–Ampère PDE can be solved numerically, i.e., $f = f_\gamma$ can be computed, and therefore what was achieved is quite a substantial advancement. Indeed, the dimension of the Monge–Ampère equation was reduced by one (the X variable was eliminated), and moreover the optimal strategy is now in terms of the first derivative of f instead of the second derivative of η—numerically this should have significance. In particular, as it turns out below, we can solve the reduced equation numerically for large values of γ (in the absence of contradicting constraints), which is very important, since those are the reasonable values for the safety exponent. The dimension reduction definitively has significance. So, as it appears, in optimal portfolio hedging it is the reduced value function f that contains no trivial information and not the value function η. The value function η contains the information that f does, but that information is mixed up with the trivial information—the X variable is trivial. The fact that the reduced Monge–Ampère equation of advanced portfolio hedging cannot be reduced any further can be inferred from the form of the linear equation derived in the simplest case when $\gamma = 1$—except in the trivial cases such equations have to be solved numerically.

Notice in (8.2.35) that if $a_1(t, q)$ is independent of q, i.e., $a_1(t, q) = a_1(t)$, and for simplicity let it be a constant, say $a_1(t, q) = a_1$, then the reduced Monge–Ampère PDE (8.2.22) is equivalent to

$$a_1 f(t, q)^2 + f^{(1,0)}(t, q) f(t, q) = 0$$

or, together with the terminal condition,

$$\begin{aligned} &a_1 f(t, q) + f^{(1,0)}(t, q) = 0 \\ &f(T, q) = 1 \end{aligned}$$

which is just an ODE having an explicit solution

$$f(t, q) = e^{a_1 (T-t)} = e^{-\frac{((\gamma-1)(2 r \gamma + (a_S - r).(\sigma_S.\sigma_S{}^T)^{-1}.(a_S - r)))(T-t)}{2\gamma}}.$$

This is trivial of course, bringing us back to the situation from Chapter 7, but it is important to have an explicit solution for verifying and calibrating the numerical algorithm. This was done successfully for the numerical algorithm used in solving examples below.

8.2.3.2 General Affine Constraints

We shall work on the above derived Monge–Ampère equation in the case of affine constraints. We rewrite (8.2.34) precisely, in a *Mathematica*® meaningful way as

```
In[32]:= Eq3DAffineConstraints =
    1/2 ∂{q,2} φ[t, X, q] ∂{X,2} φ[t, X, q] σq[t, q].σq[t, q] q^2 +
     1/2 (∂X,q φ[t, X, q])^2 σq[t, q].Transpose[σs[t, q]].
       (Inverse[σs[t, q].Transpose[σs[t, q]]].
          Transpose[μ[t, q]].Inverse[μ[t, q].
            Inverse[σs[t, q].Transpose[σs[t, q]]].
            Transpose[μ[t, q]]].μ[t, q] - 𝕀m).
       Inverse[σs[t, q].Transpose[σs[t, q]]].
       σs[t, q].σq[t, q] q^2 + ∂X φ[t, X, q]
     ∂X,q φ[t, X, q] σq[t, q].Transpose[σs[t, q]].
```

```
      (Inverse[σs[t, q].Transpose[σs[t, q]]].
         Transpose[μ[t, q]].Inverse[μ[t, q].
           Inverse[σs[t, q].Transpose[σs[t, q]]].
           Transpose[μ[t, q]]].μ[t, q] - 𝕀m).
     Inverse[σs[t, q].Transpose[σs[t, q]]].
     (as[t, q] - r) q +
  X ∂{X,2} φ[t, X, q] ∂X,q φ[t, X, q] ξ[t, q].Inverse[
      μ[t, q].Inverse[σs[t, q].Transpose[σs[t, q]]].
       Transpose[μ[t, q]]].μ[t, q].
     Inverse[σs[t, q].Transpose[σs[t, q]]].
     σs[t, q].σq[t, q] q +
  ∂q φ[t, X, q] ∂{X,2} φ[t, X, q] aq[t, q] q +
  ∂t φ[t, X, q] ∂{X,2} φ[t, X, q] +
  1/2 X^2 (∂{X,2} φ[t, X, q])^2 ξ[t, q].Inverse[
      μ[t, q].Inverse[σs[t, q].Transpose[σs[t, q]]].
       Transpose[μ[t, q]]].ξ[t, q] +
  1/2 (∂X φ[t, X, q])^2 (as[t, q] - r).
      (Inverse[σs[t, q].Transpose[σs[t, q]]].
         Transpose[μ[t, q]].Inverse[μ[t, q].
           Inverse[σs[t, q].Transpose[σs[t, q]]].
           Transpose[μ[t, q]]].μ[t, q] - 𝕀m).
     Inverse[σs[t, q].Transpose[σs[t, q]]].
     (as[t, q] - r) +
  X ∂X φ[t, X, q] ∂{X,2} φ[t, X, q]
    (r + ξ[t, q].Inverse[μ[t, q].Inverse[σs[t, q].
            Transpose[σs[t, q]]].Transpose[μ[t, q]]].
       μ[t, q].Inverse[σs[t, q].Transpose[σs[t, q]]].
       (as[t, q] - r)) == 0
```

Out[32]=

$$\frac{1}{2}\,\sigma_q(t,q).\sigma_s(t,q)^T.\Big((\sigma_s(t,q).\sigma_s(t,q)^T)^{-1}.\mu(t,q)^T.\big(\mu(t,q)$$
$$.(\sigma_s(t,q).\sigma_s(t,q)^T)^{-1}.\mu(t,q)^T\big)^{-1}.\mu(t,q)-\mathbb{I}_m\Big)$$
$$.(\sigma_s(t,q).\sigma_s(t,q)^T)^{-1}.\sigma_s(t,q).\sigma_q(t,q)\,\varphi^{(0,1,1)}(t,X,q)^2\,q^2$$
$$+\frac{1}{2}\,\sigma_q(t,q).\sigma_q(t,q)\,\varphi^{(0,0,2)}(t,X,q)\,\varphi^{(0,2,0)}(t,X,q)\,q^2$$
$$+\sigma_q(t,q).\sigma_s(t,q)^T.\Big((\sigma_s(t,q).\sigma_s(t,q)^T)^{-1}.\mu(t,q)^T.\big(\mu(t,q)$$
$$.(\sigma_s(t,q).\sigma_s(t,q)^T)^{-1}.\mu(t,q)^T\big)^{-1}.\mu(t,q)-\mathbb{I}_m\Big)$$
$$.(\sigma_s(t,q).\sigma_s(t,q)^T)^{-1}.(a_s(t,q)-r)\,\varphi^{(0,1,0)}(t,X,q)\,\varphi^{(0,1,1)}(t,X,q)$$

$$q + a_q(t, q)\,\varphi^{(0,0,1)}(t, X, q)\,\varphi^{(0,2,0)}(t, X, q)\,q + X\,\xi(t, q)$$
$$.\left(\mu(t, q).(\sigma_s(t, q).\sigma_s(t, q)^T)^{-1}.\mu(t, q)^T\right)^{-1}.\mu(t, q).(\sigma_s(t, q)$$
$$.\sigma_s(t, q)^T)^{-1}.\sigma_s(t, q).\sigma_q(t, q)\,\varphi^{(0,1,1)}(t, X, q)$$
$$\varphi^{(0,2,0)}(t, X, q)\,q + \frac{1}{2}\,(a_s(t, q) - r).\left((\sigma_s(t, q).\sigma_s(t, q)^T)^{-1}\right.$$
$$.\mu(t, q)^T.\left(\mu(t, q).(\sigma_s(t, q).\sigma_s(t, q)^T)^{-1}.\mu(t, q)^T\right)^{-1}$$
$$\left..\mu(t, q) - \mathbb{1}_m\right).(\sigma_s(t, q).\sigma_s(t, q)^T)^{-1}.(a_s(t, q) - r)$$
$$\varphi^{(0,1,0)}(t, X, q)^2 + \frac{1}{2}\,X^2\,\xi(t, q).\left(\mu(t, q).(\sigma_s(t, q).\sigma_s(t, q)^T)^{-1}\right.$$
$$\left..\mu(t, q)^T\right)^{-1}.\xi(t, q)\,\varphi^{(0,2,0)}(t, X, q)^2 + X\left(r + \xi(t, q)\right.$$
$$.\left(\mu(t, q).(\sigma_s(t, q).\sigma_s(t, q)^T)^{-1}.\mu(t, q)^T\right)^{-1}.\mu(t, q)$$
$$\left..(\sigma_s(t, q).\sigma_s(t, q)^T)^{-1}.(a_s(t, q) - r)\right)\varphi^{(0,1,0)}(t, X, q)$$
$$\varphi^{(0,2,0)}(t, X, q) + \varphi^{(0,2,0)}(t, X, q)\,\varphi^{(1,0,0)}(t, X, q) == 0$$

Assuming that the terminal condition is the utility function $\psi_\gamma(X) = \frac{X^{1-\gamma}}{1-\gamma}$, for $1 \neq \gamma > 0$, and $\psi_1(X) = \log(X)$, we can look again for a solution of the Eq3DAffine-Constraints in the same form as in the case of no-constraints, i.e., we use the above functions η and ζ. So,

```
In[33]:= Eq2DγAffineConstraint =
          Simplify[Eq3DAffineConstraints /. φ → η]
```

Out[33]=
$$\frac{1}{2(\gamma - 1)}\left(X^{-2\gamma}\left((\gamma - 1)\left(\xi(t, q).\left(\mu(t, q).(\sigma_s(t, q).\sigma_s(t, q)^T)^{-1}.\mu(t, q)^T\right)^{-1}\right.\right.\right.$$
$$.\xi(t, q)\,\gamma^2 - 2\left(r + \xi(t, q).\left(\mu(t, q).(\sigma_s(t, q).\sigma_s(t, q)^T)^{-1}\right.\right.$$
$$\left..\mu(t, q)^T\right)^{-1}.\mu(t, q).(\sigma_s(t, q).\sigma_s(t, q)^T)^{-1}$$
$$\left..(a_s(t, q) - r)\right)\gamma + (a_s(t, q) - r).\left((\sigma_s(t, q)\right.$$
$$.\sigma_s(t, q)^T)^{-1}.\mu(t, q)^T.\left(\mu(t, q).(\sigma_s(t, q)\right.$$
$$\left..\sigma_s(t, q)^T)^{-1}.\mu(t, q)^T\right)^{-1}.\mu(t, q) - \mathbb{1}_m\Big)$$
$$\left..(\sigma_s(t, q).\sigma_s(t, q)^T)^{-1}.(a_s(t, q) - r)\right)f(t, q)^2$$
$$+\left(2\,q\,(\gamma - 1)\,\sigma_q(t, q).\sigma_s(t, q)^T.\left((\sigma_s(t, q).\sigma_s(t, q)^T)^{-1}\right.\right.$$
$$.\mu(t, q)^T.\left(\mu(t, q).(\sigma_s(t, q).\sigma_s(t, q)^T)^{-1}.\mu(t, q)^T\right)^{-1}$$
$$\left..\mu(t, q) - \mathbb{1}_m\right).(\sigma_s(t, q).\sigma_s(t, q)^T)^{-1}.(a_s(t, q)$$
$$- r)\,f^{(0,1)}(t, q) + \gamma\left(\sigma_q(t, q).\sigma_q(t, q)\right.$$
$$f^{(0,2)}(t, q)\,q^2 - 2\,(\gamma - 1)\,\xi(t, q).\left(\mu(t, q)\right.$$
$$.(\sigma_s(t, q).\sigma_s(t, q)^T)^{-1}.\mu(t, q)^T\Big)^{-1}.\mu(t, q)$$

$$.(\sigma_s(t,q).\sigma_s(t,q)^T)^{-1}.\sigma_s(t,q).\sigma_q(t,q)$$
$$f^{(0,1)}(t,q)\,q + 2\,a_q(t,q)\,f^{(0,1)}(t,q)\,q$$
$$+ 2\,f^{(1,0)}(t,q)))\,f(t,q) + q^2\,(\gamma-1)\,\sigma_q(t,q)$$
$$.\sigma_s(t,q)^T.((\sigma_s(t,q).\sigma_s(t,q)^T)^{-1}.\mu(t,q)^T.(\mu(t,q)$$
$$.(\sigma_s(t,q).\sigma_s(t,q)^T)^{-1}.\mu(t,q)^T)^{-1}.\mu(t,q) - \mathbb{I}_m)$$
$$.(\sigma_s(t,q).\sigma_s(t,q)^T)^{-1}.\sigma_s(t,q).\sigma_q(t,q)\,f^{(0,1)}(t,q)^2)) == 0$$

i.e.,

```
In[34]:= ReducedMongeAmpere2 =
  Thread[( 2 (γ - 1) / X^(-2 γ)  #1 &) [Eq2DγAffineConstraint],
   Equal] >> "ReducedMongeAmpere2"
```

in the case $1 \neq \gamma > 0$ (we shall be able to solve this equation in the case $\gamma > 1$). It is safer not to try to simplify this equation by hand—indeed symbolic capabilities are *necessary* for successful application of the proposed reduced value function approach to optimal portfolio hedging; nevertheless we state that again the equation has the form (8.2.35), with different coefficients of course. In the case $\gamma = 1$, we have

```
In[35]:= Eq2D1AffineConstraint =
  Simplify[Eq3DAffineConstraints /. φ → ζ]
```

Out[35]=
$$\frac{1}{2X^2}\big(-\sigma_q(t,q).\sigma_q(t,q)\,f^{(0,2)}(t,q)\,q^2 - 2\,a_q(t,q)\,f^{(0,1)}(t,q)\,q$$
$$-2r + \xi(t,q).(\mu(t,q).(\sigma_s(t,q).\sigma_s(t,q)^T)^{-1}.\mu(t,q)^T)^{-1}$$
$$.\xi(t,q) + (a_s(t,q)-r).((\sigma_s(t,q).\sigma_s(t,q)^T)^{-1}.\mu(t,q)^T$$
$$.(\mu(t,q).(\sigma_s(t,q).\sigma_s(t,q)^T)^{-1}.\mu(t,q)^T)^{-1}.\mu(t,q) - \mathbb{I}_m)$$
$$.(\sigma_s(t,q).\sigma_s(t,q)^T)^{-1}.(a_s(t,q)-r) - 2\,\xi(t,q).(\mu(t,q)$$
$$.(\sigma_s(t,q).\sigma_s(t,q)^T)^{-1}.\mu(t,q)^T)^{-1}.\mu(t,q)$$
$$.(\sigma_s(t,q).\sigma_s(t,q)^T)^{-1}.(a_s(t,q)-r) - 2\,f^{(1,0)}(t,q)\big) == 0$$

Again we were able to eliminate the X variable. In the case $\gamma = 1$, since the above equation for $f = f_1$ is linear backward parabolic, and therefore solvable, indeed the value function *is* in the form $\zeta(t,X,q) = f_1(t,q) + \log(X)$, and consequently as before $\frac{\partial^2\zeta(t,X,q)}{\partial X\,\partial q} = 0$, and $\frac{\partial\zeta(t,X,q)}{\partial X} \Big/ \frac{\partial^2\zeta(t,X,q)}{\partial X^2} = -X$, yielding the optimal hedging strategy, without any numerical computation,

$$\begin{aligned}\Pi_1^\star(t, X, q) = X\, Q_1^\star(t, q) = X\, (\sigma_s(t, q).\sigma_s(t, q)^T)^{-1}.\big(\mu(t, q)^T \\ .\big(\mu(t, q).(\sigma_s(t, q).\sigma_s(t, q)^T)^{-1}.\mu(t, q)^T\big)^{-1}.\xi(t, q) \\ - \big(\mu(t, q)^T.\big(\mu(t, q).(\sigma_s(t, q).\sigma_s(t, q)^T)^{-1}.\mu(t, q)^T\big)^{-1} \\ .\mu(t, q).(\sigma_s(t, q).\sigma_s(t, q)^T)^{-1} - \mathbb{I}_m\big).(a_s(t, q) - r)\big)\end{aligned} \tag{8.2.38}$$

which is exactly equal to the strategy derived in Chapter 7.

On the other hand, for appropriately taken data, the above derived reduced Monge–Ampère PDE, can be solved numerically, which infers the existence of such $f = f_\gamma$, and consequently the existence of the value function in the form $\xi(t, X, q) = \frac{X^{1-\gamma} f_\gamma(t,q)}{1-\gamma}$. This yields, as in the above case with no constraints, that $\frac{\partial \xi(t,X,q)}{\partial X} \Big/ \frac{\partial^2 \xi(t,X,q)}{\partial X^2} = \frac{-X}{\gamma}$, and furthermore that $\frac{\partial^2 \xi(t,X,q)}{\partial X\, \partial q} \Big/ \frac{\partial^2 \xi(t,X,q)}{\partial X^2} = -\frac{X\, f_\gamma^{(0,1)}(t,q)}{\gamma\, f_\gamma(t,q)}$ and therefore

$$\begin{aligned}\Pi_\gamma^\star(t, X, q) = X\, Q_\gamma^\star(t, q) = X\, (\sigma_s(t, q).\sigma_s(t, q)^T)^{-1}.\Big(\mu(t, q)^T.\big(\mu(t, q) \\ .(\sigma_s(t, q).\sigma_s(t, q)^T)^{-1}.\mu(t, q)^T\big)^{-1}.\xi(t, q) - \frac{1}{\gamma} \\ \big(\mu(t, q)^T.\big(\mu(t, q).(\sigma_s(t, q).\sigma_s(t, q)^T)^{-1}.\mu(t, q)^T\big)^{-1} \\ .\mu(t, q).(\sigma_s(t, q).\sigma_s(t, q)^T)^{-1} - \mathbb{I}_m\big).(a_s(t, q) - r) \\ - \frac{f_\gamma^{(0,1)}(t, q)}{\gamma\, f_\gamma(t, q)}\, q\, \big(\mu(t, q)^T.\big(\mu(t, q).(\sigma_s(t, q).\sigma_s(t, q)^T)^{-1} \\ .\mu(t, q)^T\big)^{-1}.\mu(t, q).(\sigma_s(t, q).\sigma_s(t, q)^T)^{-1} \\ - \mathbb{I}_m\big).\sigma_s(t, q).\sigma_q(t, q)\Big)\end{aligned} \tag{8.2.39}$$

where $f = f_\gamma$ is the solution of the ReducedMongeAmpere2, together with the terminal condition $f(T, q) = 1$.

Notice that, as before, if $a_1(t, q)$ is constant, $a_1(t, q) = a_1$, then the reduced Monge–Ampère equation (8.2.34) has an explicit solution

$$\begin{aligned}e^{(T-t)a_1} = \operatorname{Exp}\Big[(T - t)\, \frac{(\gamma - 1)}{2\gamma}\, \big(\xi.\big(\mu.(\sigma_s.\sigma_s{}^T)^{-1}.\mu^T\big)^{-1}.\xi\, \gamma^2 \\ - 2\,\big(r + \xi.\big(\mu.(\sigma_s.\sigma_s{}^T)^{-1}.\mu^T\big)^{-1}.\mu.(\sigma_s.\sigma_s{}^T)^{-1}.(a_s - r)\big)\,\gamma + (a_s - r) \\ .\big((\sigma_s.\sigma_s{}^T)^{-1}.\mu^T.\big(\mu.(\sigma_s.\sigma_s{}^T)^{-1}.\mu^T\big)^{-1}.\mu - \mathbb{I}_m\big).(\sigma_s.\sigma_s{}^T)^{-1}.(a_s - r)\big)\Big].\end{aligned}$$

8.2.4 Computational Examples of Advanced Portfolio Hedging

8.2.4.1 Advanced Portfolios of Stocks with and without Affine Constraints

We shall use the definitions of Section 8.2.3. Also, we recall the coefficients for the market evolution SDE system from Section 8.2.1.3:

```
In[36]:= {aq[t_, θ_], as[t_, q_], σq[t_, θ_], σs[t_, q_]} =
           << "AdvancedTrading";
```

Also let

```
In[37]:= r = .05; 𝕀m_ := IdentityMatrix[m]; γ = 30; m = 3;
```

and let the (non-constant) affine constraint be given by

```
In[38]:= ξ[t_, q_] = {5 - q/10}; μ[t_, q_] = {Table[1, {m}]}
Out[38]= (1  1  1)
```

We have chosen the particular non-constant constraint so that the solutions to the no-constraint problem and constraint problem are close enough. One can see that by comparing the two reduced Monge–Ampère equations (actually the right-hand sides only):

```
In[39]:= Eq1 = Simplify[ReducedMongeAmpere1[[1]]]
```

Out[39]= $(-2.15347\,q^2 + 209.216\,q - 5175.56)\,f(t,q)^2 + (4.50753\, f^{(0,2)}(t,q)\,q^2 + (1.08393\,q - 40.2499)\,f^{(0,1)}(t,q)\,q + 60\,f^{(1,0)}(t,q))\,f(t,q) - 4.34566\,q^2\,f^{(0,1)}(t,q)^2$

and

```
In[40]:= Eq2 = Simplify[ReducedMongeAmpere2[[1]]]
```

Out[40]= $(-2.00914\,q^2 + 192.214\,q - 4674.85)\,f(t,q)^2 + (4.50753\, f^{(0,2)}(t,q)\,q^2 + (2.65105\,q - 132.553)\,f^{(0,1)}(t,q)\,q + 60\,f^{(1,0)}(t,q))\,f(t,q) - 0.0917517\,q^2\,f^{(0,1)}(t,q)^2$

Indeed, one can see that all the coefficients have the same sign, which is not going to be the case in general.

We now proceed to solve these equations. Since we solve these two equations, corresponding to the problem with no constraints and to the problem with affine constraints respectively, in parallel, we shall need two different notations. So function f solving Eq1 = 0 will be denoted h, while function f solving Eq2 = 0 is going to be denoted g.

We first compute the coefficients $a_1(t, q)$, ..., $a_4(t, q)$ from (8.2.35) corresponding to the equation Eq1 $= 0$:

```
In[41]:= Coffs =
    Function[j, (Coefficient[j, #1] &) /@ {f^(0,1)[t, q]^2,
        f[t, q] f^(0,2)[t, q], f[t, q] f^(0,1)[t, q],
        f[t, q]^2, f[t, q] f^(1,0)[t, q]}][Eq1];
  {a1[t_, q_], a2[t_, q_], a3[t_, q_], a4[t_, q_]} =
   ({#1[[4]]/#1[[5]], #1[[2]]/#1[[5]], #1[[3]]/#1[[5]], #1[[1]]/#1[[5]]} &)[Coffs]
```

Out[41]= $\left\{\frac{1}{60}(-2.15347\,q^2+209.216\,q-5175.56),\ 0.0751255\,q^2,\ \frac{1}{60}\,q\,(1.08393\,q-40.2499),\ -0.0724277\,q^2\right\}$

So the first equation to solve is

$$\begin{aligned}&f^{(1,0)}(t,q)\,f(t,q)\\&+\frac{1}{60}(-2.15347\,q^2+209.216\,q-5175.56)\,f(t,q)^2\\&+0.0751255\,q^2\,f^{(0,2)}(t,q)\,f(t,q)\\&+\frac{1}{60}\,q\,(1.08393\,q-40.2499)\,f^{(0,1)}(t,q)\,f(t,q)\\&-0.0724277\,q^2\,f^{(0,1)}(t,q)^2=0\end{aligned}\tag{8.2.40}$$

for $t_0 = 0 < t < 1 = T,\ 0 < q < \infty$, together with the terminal condition

$$f(T, q) = 1. \tag{8.2.41}$$

No boundary condition is needed nor possible to impose at $q = 0$ since the process $q(t)$ can never (theoretically) become zero. The numerical solution $h(t, q)$ is going to be computed in a truncated domain $q_0 = 28 < q < 70 = q_1$, and with a truncated, "non-interfering" boundary condition at $q = q_0$ and $q = q_1$.

What kind of a truncated, "non-interfering" boundary condition is used? Since it always holds that $0 < f(t, q) \le 1$, for $t \le T$ and $0 < q < \infty$, and since therefore f ($= h$) is flat in q direction when f is either close to 0 or 1, the following nonlinear boundary condition seems to work very well:

$$\frac{\partial f(t, q_i)}{\partial \nu} = -\alpha\, f(t, q_i)\,(1 - f(t, q_i)) \tag{8.2.42}$$

for $i = 0, 1$ and $0 < t < T$, where $\frac{\partial}{\partial \nu}$ is the exterior normal derivative, i.e., $\frac{\partial}{\partial \nu} = -\frac{\partial}{\partial q}$ at $q = q_0$ and $\frac{\partial}{\partial \nu} = \frac{\partial}{\partial q}$ for $q = q_1$; $\alpha \ge 0$ is chosen experimentally to be $\alpha = 0.05$. Possibly the best argument for the boundary condition (8.2.42) is to compare the solution of the above problem:

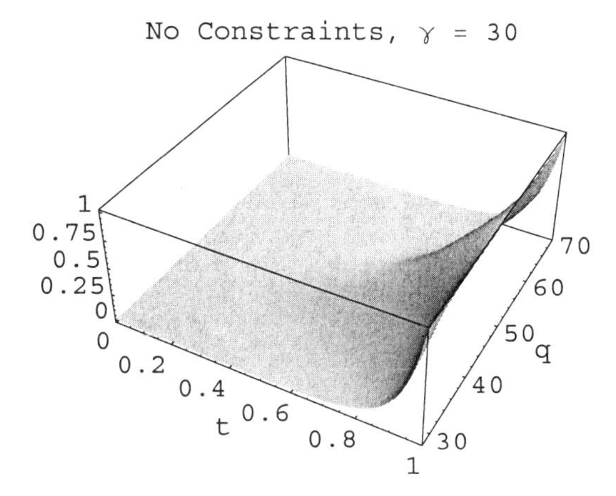

with the one obtained using a simple Dirichlet boundary condition

$$f(t, q_i) = 1 \tag{8.2.43}$$

for $i = 0, 1$ and $0 < t < T$. If (8.2.43) is imposed instead of (8.2.42), then the solution looks like this:

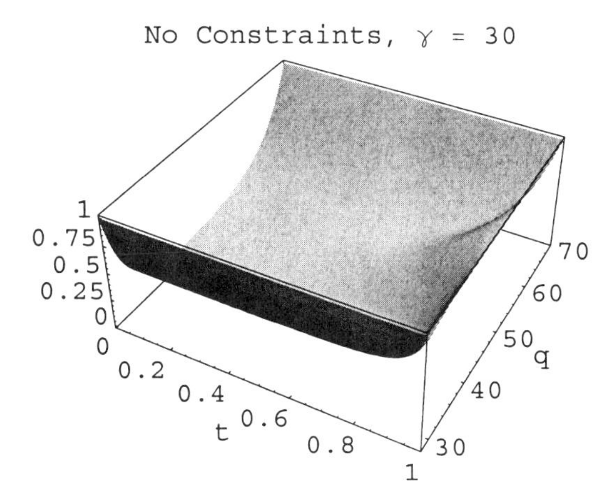

Obviously, the influence of a boundary condition such as (8.2.43) is limited, or at least a solution is not *driven* by the boundary condition, and since its influence is also artificial, it needs to be minimized. The above non-linear boundary condition works very well to this end. We notice the lack of theoretical results for Monge–Ampère PDEs above with non-linear boundary condition (8.2.42)—nevertheless, the numerical evidence is conclusive that such problems are well posed.

Similarly, we now prepare the coefficients $a_1(t, q)$, ..., $a_4(t, q)$ from (8.2.35) corresponding to the affine constraint problem:

```
In[42]:= Coffs =
   Function[j, (Coefficient[j, #1] &) /@ {f^(0,1)[t, q]^2,
       f[t, q] f^(0,2)[t, q], f[t, q] f^(0,1)[t, q],
       f[t, q]^2, f[t, q] f^(1,0)[t, q]}][Eq2];
  {a1[t_, q_], a2[t_, q_], a3[t_, q_], a4[t_, q_]} =
   ({#1[[4]]/#1[[5]], #1[[2]]/#1[[5]], #1[[3]]/#1[[5]], #1[[1]]/#1[[5]]} &)[Coffs]
```

Out[42]= $\{\frac{1}{60}(-2.00914\, q^2 + 192.214\, q - 4674.85), 0.0751255\, q^2, \frac{1}{60}\, q\,(2.65105\, q - 132.553), -0.0015292\, q^2\}$

and then compute the solution in the equation in the case with affine constraints:

$$\begin{aligned} & f^{(1,0)}(t, q)\, f(t, q) \\ & + \frac{1}{60}(-2.00914\, q^2 + 192.214\, q - 4674.85)\, f(t, q)^2 \\ & + 0.0751255\, q^2\, f^{(0,2)}(t, q)\, f(t, q) \\ & + \frac{1}{60}\, q\,(2.65105\, q - 132.553)\, f^{(0,1)}(t, q)\, f(t, q) \\ & - 0.0015292\, q^2\, f^{(0,1)}(t, q)^2 = 0 \end{aligned} \tag{8.2.44}$$

for $t_0 < t < T$, together with the above terminal condition (8.2.41). The numerical solution $g(t, q)$, in a truncated domain $q_0 < q < q_1$, and with same kind of a truncated boundary condition (8.2.42), was computed and it looks like this:

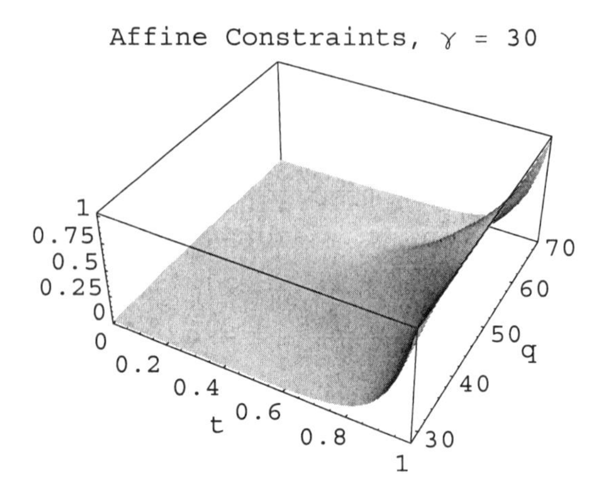

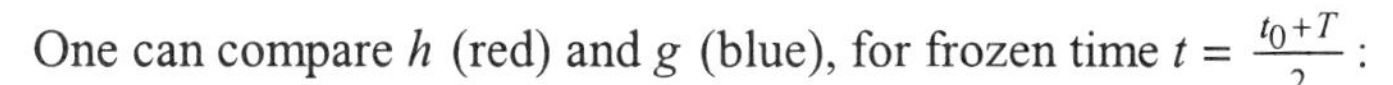

One can compare h (red) and g (blue), for frozen time $t = \frac{t_0+T}{2}$:

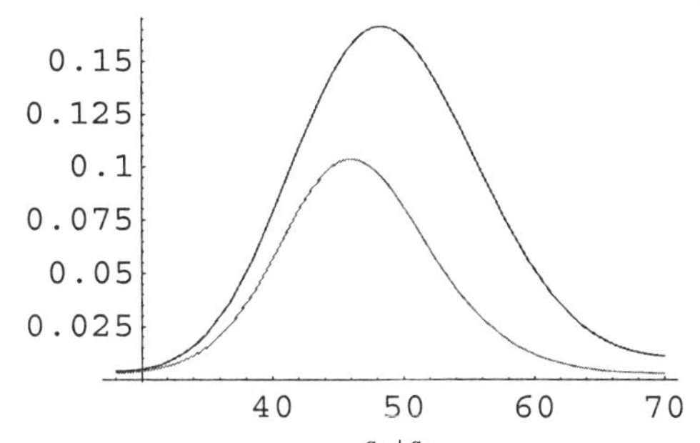

as well as for frozen index value $q = \frac{q_0+q_1}{2}$:

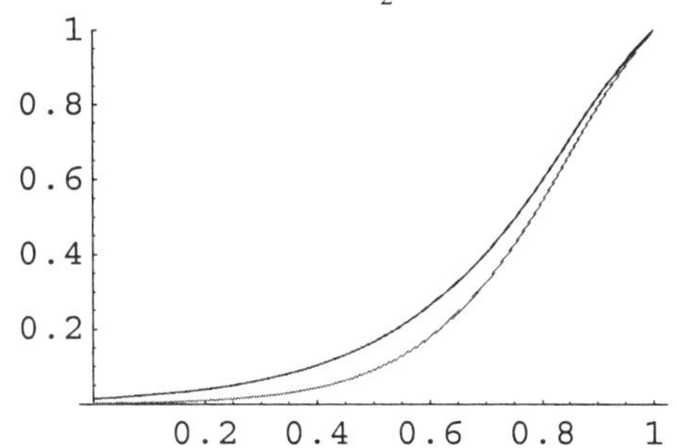

Notice that the smaller the reduced value function f, for $\gamma > 1$, i.e., for $1-\gamma < 0$, the bigger the value function $\frac{x^{1-\gamma}}{1-\gamma} f(t, q)$. The above result makes perfect sense. Since the constraint reduces the payoff, it increases the reduced value function: $h < g$, i.e., $\frac{x^{1-\gamma}}{1-\gamma} h(t, q) > \frac{x^{1-\gamma}}{1-\gamma} g(t, q)$.

One could argue that the reduced value functions h and g are more interesting than the value functions $\frac{x^{1-\gamma}}{1-\gamma} h(t, q)$ and $\frac{x^{1-\gamma}}{1-\gamma} g(t, q)$, which can be now seen (for some fixed time, e.g., $t = \frac{1}{2}$) precisely:

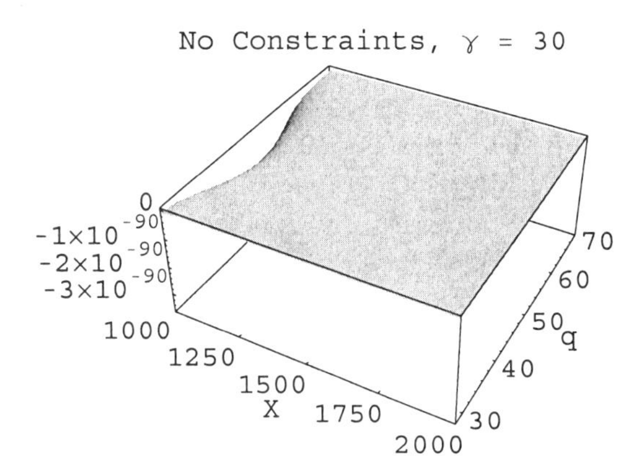

and

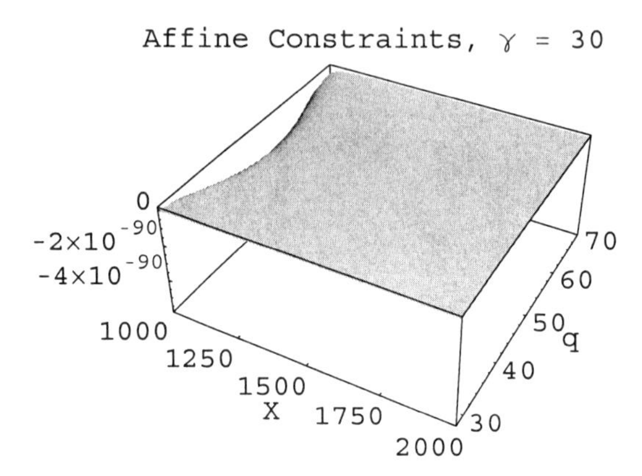

Notice how extremely small (negative) values are taken by the value functions, and consequently how flat they are. Consequently, especially since the optimal portfolio rule would have to be computed using the first and second derivatives of the value functions, serious numerical challenges would have to be faced if the direct numerical solutions for value functions are attempted instead of solving for reduced value functions, as proposed here.

Our next goal is to compute the optimal portfolio hedging strategy without any constraints:

$$\Pi_\gamma^\star(t, X, q) = X\, P_\gamma(t, q) = X\,(P_{\gamma,1}(t, q) + P_{\gamma,2}(t, q)) \tag{8.2.45}$$

for $\gamma > 1$, where

```
In[43]:= Pγ,1[t_, q_] =
           1/γ (Inverse[σs[t, q].Transpose[σs[t, q]]].
               (as[t, q] - r)) // Simplify

Out[43]= {5.74685 - 0.122517 q, 2.88125 - 0.0630012 q, 0.115046 q -
           5.36729}
```

where also

$$P_{\gamma,2}(t, q) = \frac{h^{(0,1)}(t, q)\, q\, (\sigma_s(t, q).\sigma_s(t, q)^T)^{-1}.\sigma_s(t, q).\sigma_q(t, q)}{\gamma\, h(t, q)}, \tag{8.2.46}$$

and where h is the solution of (8.2.40)–(8.2.42).

We also compute the optimal portfolio hedging strategy in the case of affine constraints:

$$Q_\gamma^\star(t, X, q) = X\, Q_\gamma(t, q) = X\,(Q_{\gamma,1}(t, q) + Q_{\gamma,2}(t, q)) \tag{8.2.47}$$

for $\gamma > 1$, where

```
In[44]:= Qγ,1[t_, q_] =
  Simplify[Inverse[σs[t, q].Transpose[σs[t, q]]].
    (Transpose[μ[t, q]].Inverse[
      μ[t, q].Inverse[σs[t, q].Transpose[σs[t, q]]].
       Transpose[μ[t, q]]].ξ[t, q] -
     1/γ ((Transpose[μ[t, q]].Inverse[μ[t, q].
          Inverse[σs[t, q].Transpose[σs[t, q]]].
          Transpose[μ[t, q]]].μ[t, q].
         Inverse[σs[t, q].Transpose[σs[t, q]]] -
       Im).(as[t, q] - r)))]

Out[44]= {8.2551 - 0.165102 q, 4.66031 - 0.0932063 q, 0.158308 q -
   7.91542}
```

where also

$$\begin{aligned} Q_{\gamma,2}(t, q) = {} & (\sigma_s(t, q).\sigma_s(t, q)^T)^{-1}.\Big(-\frac{1}{\gamma\, g(t, q)}\,\big(g^{(0,1)}(t, q)\, q \\ & \big(\mu(t, q)^T.\big(\mu(t, q).(\sigma_s(t, q).\sigma_s(t, q)^T)^{-1}.\mu(t, q)^T\big)^{-1}.\mu(t, q) \\ & .(\sigma_s(t, q).\sigma_s(t, q)^T)^{-1} - \mathbb{I}_m\big).\sigma_s(t, q).\sigma_q(t, q)\big)\Big), \end{aligned} \tag{8.2.48}$$

and where g is the solution of (8.2.44), (8.2.41), (8.2.42).

In particular, we are interested in getting some evidence of the relative significance of $P_{\gamma,1}(t, q)$ and $P_{\gamma,2}(t, q)$, i.e., between Merton's strategy $P_{\gamma,1}(t, q)$, and $P_{\gamma,2}(t, q)$ provided by this methodology, and similarly for $Q_{\gamma,1}(t, q)$ and $Q_{\gamma,2}(t, q)$—if $P_{\gamma,2}(t, q)$ and $Q_{\gamma,2}(t, q)$ are very small, then all this work would be, at least from the pragmatic point of view, for nothing.

We come to the highlight of this book. Does "advanced portfolio hedging" bring something substantially new? Is the "correction" term $P_{\gamma,2}^\star(t, q)$ significant in size, compared to $P_{\gamma,1}^\star(t, q)$? In the following plot, the first row is $P_{\gamma,1}^\star(t, q)$, the second row is $P_{\gamma,2}^\star(t, q)$, and the third row is the cumulative complete strategy $P_\gamma^\star(t, q) = P_{\gamma,1}^\star(t, q) + P_{\gamma,2}^\star(t, q)$:

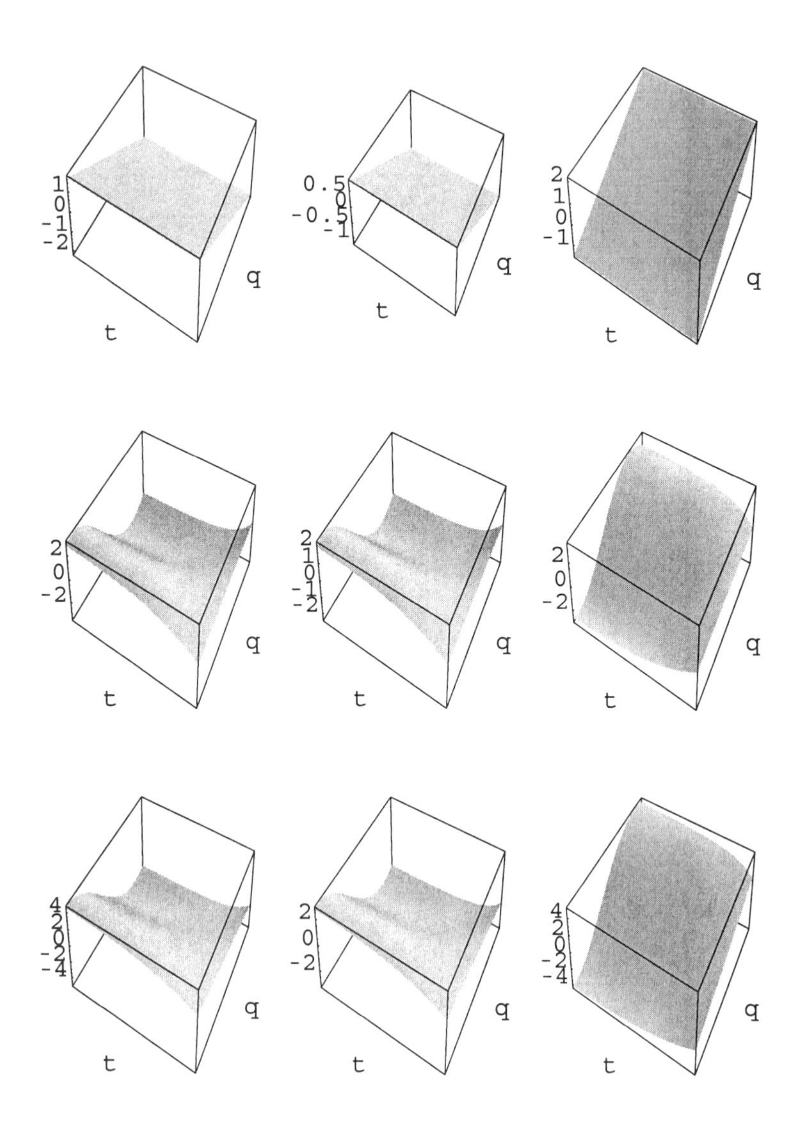

For the current data, the relative size of the middle row, i.e., of the non-trivial numerical part of the optimal portfolio rule is quite substantial. So, the answer to the above question is definitely affirmative. It is also interesting to notice that there exists an intrinsic time-dependency—even though the data is not time-dependent, as opposed to $P^{\star}_{\gamma,1}(t, q)$, $P^{\star}_{\gamma,2}(t, q)$ and consequently $P^{\star}_{\gamma}(t, q)$ exhibit the "intertemporal tradeoffs".

In the case of the affine constraint, the optimal portfolio rules look like:

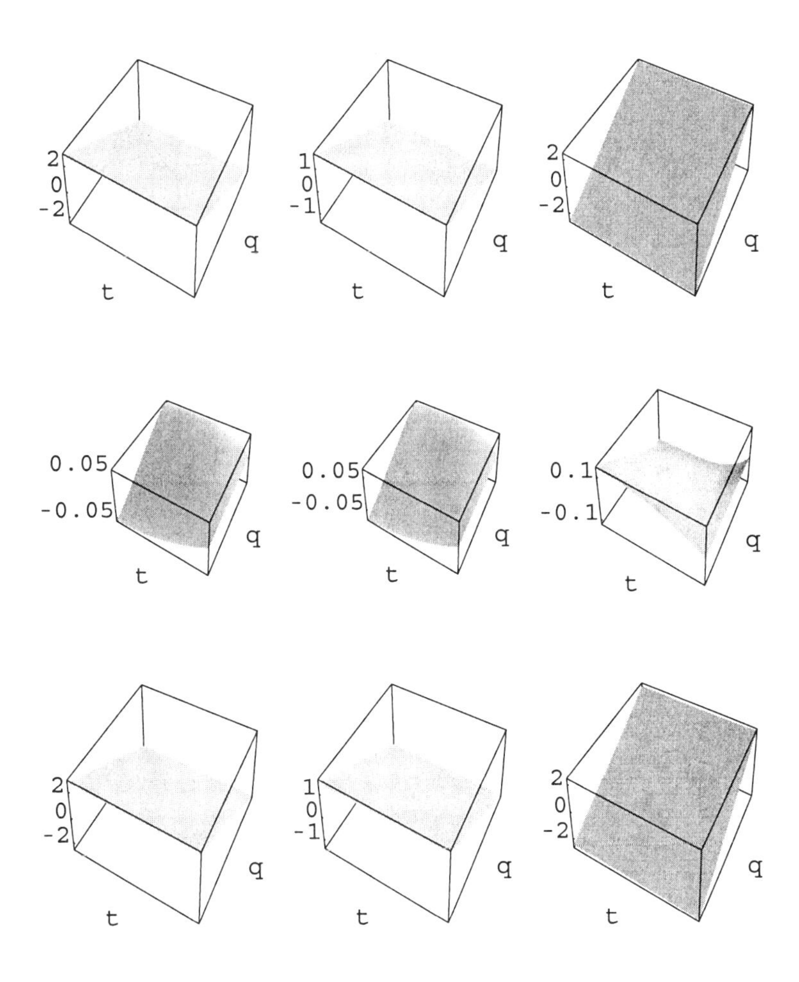

Finally, notice that the computational complexity is not increasing if m, the number of stocks considered, increases. Thousands of stocks can be considered, as in Chapter 7. The computational complexity increases only when the number of the recorded indexes increases.

8.2.4.2 Advanced Hedging of Options

In[45]:=
```
Remove["Global`*"]
```

We now consider a very different problem, but in the same general framework—the problem of advanced options hedging. So the role of an index is given to an underlying stock, which is at the same time not just tracked but also traded. Many different problems can be considered. We choose probably the simplest one, but it will be obvious how many more different problems can be considered similarly. So, consider an underlying stock whose price q evolves according to

$$dq(t) = q(t)\, a_q(t, q(t))\, dt + q(t)\, \sigma_q(t, q(t)).dB(t) \tag{8.2.49}$$

where $B(t)$ is a 2-dimensional Brownian motion, $\sigma_q(t, q)$ is a 2-vector-valued function

```
In[46]:= σq[t_, q_] = {σq,1[t, q], σq,2[t, q]};
```

and $a_q(t, q)$ is a scalar-valued function. There are only 2 assets considered for trading: the underlying stock and, either call or put options at the prescribed strike-price k, and fixed expiration time T. We assume that the option market, on average, observes a particular option pricing methodology, e.g., the Black–Scholes formula, according to which the option price at the time t and at the underlying price $q(t)$ is equal to $V(t, q(t))$. So the assets considered for trading have a 2-vector-valued price:

$$S(t) = \{q(t), V(t, q(t))\}. \tag{8.2.50}$$

The dynamics for $S(t)$ can be derived by means of the Itô chain rule, according to which

$$\begin{aligned}
&dV(t, q(t)) = V^{(1,0)}(t, q(t))\, dt + V^{(0,1)}(t, q(t))\, dq(t) \\
&+\frac{1}{2} V^{(0,2)}(t, q(t))\, (dq(t))^2 = V^{(1,0)}(t, q(t))\, dt \\
&+V^{(0,1)}(t, q(t))\, (q(t)\, a_q(t, q(t))\, dt + q(t)\, \sigma_q(t, q(t)).dB(t)) \\
&+\frac{1}{2} V^{(0,2)}(t, q(t))\, q(t)^2\, \sigma_q(t, q(t)).\sigma_q(t, q(t))\, dt \\
&= \Big(V^{(1,0)}(t, q(t)) + V^{(0,1)}(t, q(t))\, q(t)\, a_q(t, q(t)) \\
&\qquad + \frac{1}{2} V^{(0,2)}(t, q(t))\, q(t)^2\, \sigma_q(t, q(t)).\sigma_q(t, q(t))\Big)\, dt \\
&\quad + V^{(0,1)}(t, q(t))\, q(t)\, \sigma_q(t, q(t)).dB(t)
\end{aligned} \tag{8.2.51}$$

and therefore

$$\begin{aligned}
&\frac{dV(t, q(t))}{V(t, q(t))} \\
&= \frac{1}{V(t, q(t))} \Big(V^{(1,0)}(t, q(t)) + V^{(0,1)}(t, q(t))\, q(t)\, a_q(t, q(t))
\end{aligned}$$

$$+ \frac{1}{2} V^{(0,2)}(t, q(t))\, q(t)^2\, \sigma_q(t, q(t)).\sigma_q(t, q(t))\Big) dt$$
$$+ \frac{V^{(0,1)}(t, q(t))\, q(t)\, \sigma_q(t, q(t))}{V(t, q(t))}.dB(t)$$
$$= A(t, q(t))\, dt + \beta(t, q(t)).dB(t)$$

i.e., the option-appreciation rate is equal to

```
In[47]:= A[t_, q_] = 1/V[t, q]
           (V^(1,0)[t, q] + V^(0,1)[t, q] q a_q[t, q] +
             1/2 V^(0,2)[t, q] q^2 σ_q[t, q].σ_q[t, q])
```

$$\text{Out[47]=}\quad \frac{1}{V(t, q)}\left(\frac{1}{2}\,(\sigma_{q,1}(t, q)^2 + \sigma_{q,2}(t, q)^2)\, V^{(0,2)}(t, q)\, q^2 + a_q(t, q)\right.$$
$$\left. V^{(0,1)}(t, q)\, q + V^{(1,0)}(t, q)\right)$$

while the option vector-volatility is equal to

$$\beta(t, q) = \frac{V^{(0,1)}(t, q)\, q}{V(t, q)}\, \sigma_q(t, q) = \frac{V^{(0,1)}(t, q)\, q}{V(t, q)}\, \{\sigma_{q,1}(t, q), \sigma_{q,2}(t, q)\}. \tag{8.2.53}$$

This is the case if there would be a perfect correlation between the realized option and underlying prices. On the other hand, what is easily observed on the option market is that the spread between bid and ask prices, in relative terms, is much larger than in the case of stocks. This motivates the introduction of another layer of randomness (for most of the results below we shall, although we don't have to, eventually take $\epsilon \to 0$). So we augment, at least temporarily, the above dynamics to

$$\frac{dV(t, q(t))}{V(t, q(t))} = A(t, q(t))\, dt + (\beta(t, q(t)) + \epsilon).dB(t) \tag{8.2.54}$$

where ϵ is a small vector independent of $\beta(t, q)$, i.e., independent of $\sigma_q(t, q)$. For example, if $\sigma_q(t, q) = \{s_q(t, q), 0\}$ for some scalar function $s_q(t, q)$, then ϵ can be taken to be $\epsilon = \{0, \epsilon_0\}$ for some small ϵ_0. Written together, the dynamics for the underlying stock and the considered option is equal to

$$dS(t) = S(t)\, a_s(t, q(t))\, dt + S(t)\, \sigma_s(t, q(t)).dB(t) \tag{8.2.55}$$

where $\sigma_s(t, q)$ is the 2×2-matrix-valued function

```
In[48]:= σs[t_, q_] = {{σq,1[t, q], σq,2[t, q]},
           {ϵ1 + (V^(0,1)[t, q] q) σq,1[t, q] / V[t, q],
            ϵ2 + (V^(0,1)[t, q] q) σq,2[t, q] / V[t, q]}}
```

Out[48]= $\begin{pmatrix} \sigma_{q,1}(t, q) & \sigma_{q,2}(t, q) \\ \epsilon_1 + \frac{q\,\sigma_{q,1}(t,q)\,V^{(0,1)}(t,q)}{V(t,q)} & \epsilon_2 + \frac{q\,\sigma_{q,2}(t,q)\,V^{(0,1)}(t,q)}{V(t,q)} \end{pmatrix}$

and where $a_s(t, q)$ is the 2-vector-valued function

```
In[49]:= as[t, q] = {aq[t, q], A[t, q]}
```

Out[49]= $\left\{a_q(t, q), \frac{1}{V(t, q)} \left(\frac{1}{2} (\sigma_{q,1}(t, q)^2 + \sigma_{q,2}(t, q)^2)\, V^{(0,2)}(t, q)\, q^2 + a_q(t, q)\, V^{(0,1)}(t, q)\, q + V^{(1,0)}(t, q)\right)\right\}$

Of course, we consider also a cash account whose balance, positive or negative, $C(t)$ at the time t, evolves, in the absence of trading, according to the equation

$$d C(t) = r\, C(t)\, d t. \tag{8.2.56}$$

We introduce considered strategies

$$\Pi(t, X, q) = \{\Pi_1(t, X, q), \Pi_2(t, X, q)\} \tag{8.2.57}$$

where the components are the cash value in the underlying stock and the cash value in the considered option, respectively. To mimic the kind of hedging that was described when the Black–Scholes formula was derived in Chapter 3, and yet to put the problem in the present context, we introduce an affine constraint

$$\begin{aligned} \{\Pi_2\,(t, X, q)\} &= (\,0 \quad 1\,).\Pi(t, X, q) = \mu.\Pi(t, X, q) \\ &= \xi\, X = \{\kappa\}\, X = \{\kappa\, X\} \end{aligned} \tag{8.2.58}$$

where, obviously,

```
In[50]:= ξ[t_, q_] = {κ}; μ[t_, q_] = {{0, 1}}
```

Out[50]= $(\,0 \quad 1\,)$

This means that the fixed portion (κ) of the available wealth is invested in the considered option (instead of holding a single option as in Chapter 3)—the question is how to hedge such an investment? As opposed to a single strategy which was derived previously, we would like to develop a whole spectrum of possible hedges to be applied depending on the investor's attitude towards risk-taking.

At one end of that spectrum lies the wealth-volatility-minimization strategy. Indeed, according to (7.5.5), the wealth-volatility-minimization strategy is equal to

$$\Pi_\infty(t, X, q) = X\, P_\infty(t, q) \tag{8.2.59}$$

where $P_\infty(t, q)$ can be computed quickly as

```
In[51]:= P∞[t_, q_] =
          Apart[Inverse[σs[t, q].Transpose[σs[t, q]]].
            Transpose[μ[t, q]].Inverse[
             μ[t, q].Inverse[σs[t, q].Transpose[σs[t, q]]].
              Transpose[μ[t, q]]].ξ[t, q]]
```

Out[51]= $\left\{-\frac{\kappa\,(\epsilon_1\,\sigma_{q,1}(t,q)+\epsilon_2\,\sigma_{q,2}(t,q))}{\sigma_{q,1}(t,q)^2+\sigma_{q,2}(t,q)^2}-\frac{q\,\kappa\,V^{(0,1)}(t,q)}{V(t,q)},\ \kappa\right\}$

In particular, if either ϵ and $\sigma_q(t, q)$ are orthogonal, or if $\epsilon \to 0$, we have

$$\Pi_\infty(t, X, q) = X\, P_\infty(t, q) = X\,\kappa\left\{-q\,\frac{V^{(0,1)}(t,q)}{V(t,q)},\ 1\right\} \tag{8.2.60}$$

which is the Black–Scholes hedging strategy. Indeed, as derived in Chapter 3, one option worth $V(t, q)$ is hedged by $-V^{(0,1)}(t, q)$ stocks, which is worth the amount of $-q\,V^{(0,1)}(t, q)$. Similarly, $X\,\kappa$ worth of options is hedged proportionally.

So the Black–Scholes hedging is the most conservative hedging ($\gamma = \infty$). What happens in the case $\gamma < \infty$? To get more meaningful and shorter formulas consider just a little bit less general case when

```
In[52]:= {σq,1[t_, q_], σq,2[t_, q_]} = {sq[t, q], 0};
         {ϵ1, ϵ2} = {0, ϵ}; 𝕀m = {{1, 0}, {0, 1}};
```

Then the case $\gamma = 1$ is also solved immediately as $\Pi_1(t, X, q) = X\, P_1(t, q)$, and where (compare with (8.2.60))

```
In[53]:= P1[t_, q_] =
          Simplify[Inverse[σs[t, q].Transpose[σs[t, q]]].
            (Transpose[μ[t, q]].Inverse[
               μ[t, q].Inverse[σs[t, q].Transpose[σs[t, q]]].
                Transpose[μ[t, q]]].ξ[t, q] -
             (Transpose[μ[t, q]].Inverse[μ[t, q].
                  Inverse[σs[t, q].Transpose[σs[t, q]]].
                  Transpose[μ[t, q]]].μ[t, q].
                Inverse[σs[t, q].Transpose[σs[t, q]]] -
               𝕀m).(as[t, q] - r))]
```

Out[53]= $\left\{\frac{a_q(t, q) - r}{s_q(t, q)^2} - \frac{q\,\kappa\, V^{(0,1)}(t, q)}{V(t, q)}, \kappa\right\}$

while in the case $1 < \gamma < \infty$, we have $\Pi_\gamma(t, X, q) = X\, P_\gamma(t, q)$, for

```
In[54]:= Pγ_[t_, q_] =
    Simplify[Inverse[σs[t, q].Transpose[σs[t, q]]].
      (Transpose[μ[t, q]].Inverse[
          μ[t, q].Inverse[σs[t, q].Transpose[σs[t, q]]].
           Transpose[μ[t, q]]].ξ[t, q] -
        1/γ ((Transpose[μ[t, q].Inverse[μ[t, q].
                 Inverse[σs[t, q].Transpose[σs[t, q]]].
                 Transpose[μ[t, q]]].μ[t, q].Inverse[
                σs[t, q].Transpose[σs[t, q]]] - Im).
            (as[t, q] - r)) - 1/(γ fγ[t, q]) (fγ^(0,1)[t, q]
           q (Transpose[μ[t, q]].Inverse[μ[t, q].
                  Inverse[σs[t, q].Transpose[σs[t, q]]].
                  Transpose[μ[t, q]]].μ[t, q].
                Inverse[σs[t, q].Transpose[σs[t, q]]] -
              Im).σs[t, q].σq[t, q]))]
```

Out[54]= $\left\{\frac{a_q(t, q) - r}{\gamma\, s_q(t, q)^2} - \frac{q\,\kappa\, V^{(0,1)}(t, q)}{V(t, q)} + \frac{q\, f_\gamma^{(0,1)}(t, q)}{\gamma\, f_\gamma(t, q)}, \kappa\right\}$

where the reduced value function $f_\gamma(t, q)$ is a solution of (if for example $\epsilon \to 0$)

```
In[55]:= BlackScholesReducedMongeAmpereEquation =
    Thread[(V[t, q] sq[t, q]^2 # /. ϵ → 0 &)[
      Simplify[<< "ReducedMongeAmpere2"]], Equal]
```

Out[55]=
$$-q^2 (\gamma - 1)\, V(t, q)\, f^{(0,1)}(t, q)^2\, s_q(t, q)^4$$
$$+ f(t, q)\, V(t, q)\, (2\, q\, (r\, (\gamma - 1) + a_q(t, q))\, f^{(0,1)}(t, q)$$
$$+ \gamma\, (q^2\, f^{(0,2)}(t, q)\, s_q(t, q)^2 + 2\, f^{(1,0)}(t, q)))\, s_q(t, q)^2$$
$$-(\gamma - 1)\, f(t, q)^2\, (\gamma\, \kappa\, (q^2\, V^{(0,2)}(t, q)\, s_q(t, q)^2$$
$$+ 2\, q\, r\, V^{(0,1)}(t, q) + 2\, V^{(1,0)}(t, q))\, s_q(t, q)^2$$
$$+ V(t, q)\, (r^2 - 2\, \gamma\, (\kappa - 1)\, s_q(t, q)^2\, r - 2\, a_q(t, q)\, r + a_q(t, q)^2))$$
$$== 0$$

Let's fix ideas furthermore by assuming fixed volatility $s_q(t, q) = \sigma$. Also we need to fix ideas in regard to the option type and pricing. So for example, if we consider a put, and use the Black–Scholes formula, we have

```
In[56]:= << "CFMLab`BlackScholes`"
         U[t_, q_] := VP[t, q, T, k, r, σ]
         BSMA = Simplify[
           BlackSholesReducedMongeAmpereEquation /.
            {s_q[t, q] → σ, V → U}]
```

Out[58]=

$$\frac{1}{2}\left(e^{r(t-T)}\, k \operatorname{erfc}\left(\frac{2\log\left(\frac{q}{k}\right)-(t-T)(2r-\sigma^2)}{2\sqrt{2}\sqrt{(T-t)\sigma^2}}\right) - q\operatorname{erfc}\left(\frac{2\log\left(\frac{q}{k}\right)-(t-T)(\sigma^2+2r)}{2\sqrt{2}\sqrt{(T-t)\sigma^2}}\right)\right)$$
$$\left(-q^2(\gamma-1)f^{(0,1)}(t,q)^2\sigma^4 + f(t,q)\left(2q\left(r(\gamma-1)+a_q(t,q)\right)f^{(0,1)}(t,q) + \gamma\left(q^2 f^{(0,2)}(t,q)\sigma^2 + 2f^{(1,0)}(t,q)\right)\right)\sigma^2 - (\gamma-1)f(t,q)^2\left(a_q(t,q)^2 - 2r\,a_q(t,q) + r(2\gamma\sigma^2+r)\right)\right)$$
$$== 0$$

Further simplifying we arrive at

```
In[59]:= SEqn = Thread[# /
           (1/2 (e^(r (t-T)) k Erfc[(2 Log[q/k] - (t - T) (2 r - σ^2)) / (2 √2 √((T - t) σ^2))] -
              q Erfc[(2 Log[q/k] - (t - T) (σ^2 + 2 r)) / (2 √2 √((T - t) σ^2))])) &[
         BSMA], Equal]
```

Out[59]=

$$-q^2(\gamma-1)f^{(0,1)}(t,q)^2\sigma^4 + f(t,q)\left(2q\left(r(\gamma-1)+a_q(t,q)\right) f^{(0,1)}(t,q) + \gamma\left(q^2 f^{(0,2)}(t,q)\sigma^2 + 2f^{(1,0)}(t,q)\right)\right)\sigma^2 -(\gamma-1)f(t,q)^2\left(a_q(t,q)^2 - 2r\,a_q(t,q) + r(2\gamma\sigma^2+r)\right) == 0$$

which is the equation we have to solve numerically. Notice that $V(t, q)$, the Black–Scholes option-pricing formula and its derivatives are eliminated! This equation holds for any stock-price appreciation rate $a_q(t, q)$, and as it can be checked the same equation also holds in the case of calls. We also notice that if $a_q(t, q) = \text{const.} = a$, then SEqn reduces to an ODE:

```
In[60]:= SODE = SEqn /. {aq[t, q] → a,
           Derivative[α_, β_ /; β > 0][f][t, q] → 0,
           Derivative[α_, 0][f][t, q] →
            Derivative[α][f][t], f[t, q] → f[t]}
```

Out[60]= $2\gamma\sigma^2 f(t) f'(t) - (\gamma-1)(a^2 - 2ra + r(2\gamma\sigma^2 + r)) f(t)^2 == 0$

with an explicit solution

```
In[61]:= Off[DSolve::"bvnul"];
         g[t_, q_] =
          f[t] /. DSolve[{SODE, f[T] == 1}, f[t], t][[1]] //
           Simplify
```

Out[62]= $e^{\frac{(t-T)(\gamma-1)\left(a^2-2ra+r\left(2\gamma\sigma^2+r\right)\right)}{2\gamma\sigma^2}}$

This implies that if the underlying stock obeys the Log-Normal dynamics, the optimal hedging of options is very simple: $\Pi_\gamma(t, X, q) = X\, P_\gamma(t, q)$ where $P_\gamma(t, q)$ is equal to

```
In[63]:= Pγ[t, q] /. {aq[t, q] → a, sq[t, q] → σ, fγ → g}
```

Out[63]= $\left\{\frac{a-r}{\gamma\sigma^2} - \frac{q\,\kappa\, V^{(0,1)}(t, q)}{V(t, q)}, \kappa\right\}$

In general, when the underlying stock appreciation rate $a_q(t, q)$ does depend on q, the problem is more interesting since we have to find a numerical solution of SEqn. We notice that SEqn can be written in the form (8.2.35), identifying coefficients $a_1(t, q)$, ..., $a_4(t, q)$:

```
In[64]:= Coffs =
          Function[j, (Coefficient[j, #1] &) /@ {f^(0,1)[t, q]^2,
              f[t, q] f^(0,2)[t, q], f[t, q] f^(0,1)[t, q],
              f[t, q]^2, f[t, q] f^(1,0)[t, q]}][SEqn[[1]]];
         {a1[t_, q_], a2[t_, q_], a3[t_, q_], a4[t_, q_]} =
          FullSimplify[
           ({#1[[4]]/#1[[5]], #1[[2]]/#1[[5]], #1[[3]]/#1[[5]], #1[[1]]/#1[[5]]} &)[Coffs]]
```

Out[64]= $\left\{-\frac{(\gamma-1)(a_q(t, q)^2 - 2r\,a_q(t, q) + r(2\gamma\sigma^2 + r))}{2\gamma\sigma^2}, \frac{q^2\sigma^2}{2}, \frac{q\,(r(\gamma-1) + a_q(t, q))}{\gamma}, -\frac{q^2(\gamma-1)\sigma^2}{2\gamma}\right\}$

Finally, we choose the appreciation rate for the underlying stock; notice that it is not the same as the index-appreciation-rate in the previous example—still by experimenting with the market trajectories one can see that it is not unrealistic:

```
In[65]:= aq[t_, l_] := .5 (50 - l)
```

Further, specifying the data, the safety exponent, interest rate, volatility of the underlying stock, the expiration time, strike price, the portion of the wealth in the market, the current time, and the option expiration time (for no particular reason we choose the underlying stock equilibrium price to be the same as the option strike price k):

```
In[66]:= γ = 3; r = .04; σ = .5; k = 50; κ = .5; t0 = 0; T = 3 / 12;
```

So the equation to solve is

$$\begin{aligned}
&f^{(1,0)}(t, q)\, f(t, q) \\
&+(-0.333333\, q^2 + 33.28\, q - 830.749)\, f(t, q)^2 \\
&+0.125\, q^2 f^{(0,2)}(t, q)\, f(t, q) \\
&+(8.36 - 0.166667\, q)\, q\, f^{(0,1)}(t, q)\, f(t, q) \\
&-0.0833333\, q^2 f^{(0,1)}(t, q)^2 = 0
\end{aligned} \tag{8.2.61}$$

for $t_0 = 0 < t < \frac{3}{12} = T$, together with the terminal condition $f(T, q) = 1$. The numerical solution $f(t, q)$, for the above parameters in a truncated domain $q_0 = 28 < q < 70 = q_1$, and with the same truncated boundary condition (8.2.42), was computed and it looks like this (we also choose more aggressive investing: $\gamma = 3$):

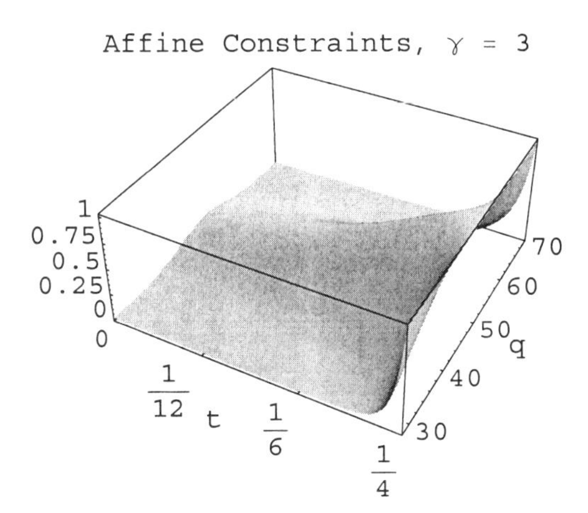

The optimal hedging rule $\frac{a_q(t,q)-r}{\gamma\sigma^2} - \frac{q\,\kappa\, V^{(0,1)}(t,q)}{V(t,q)} + \frac{q\, f_\gamma^{(0,1)}(t,q)}{\gamma\, f_\gamma(t,q)}$ can be analyzed in a similar fashion as before, where $M_\gamma(t, q) = \frac{a_q(t,q)-r}{\gamma\sigma^2} - \frac{q\,\kappa\, V^{(0,1)}(t,q)}{V(t,q)}$ can be considered as an analogue of Merton's strategy, while $N_\gamma(t, q) = \frac{q\, f_\gamma^{(0,1)}(t,q)}{\gamma\, f_\gamma(t,q)}$ is the new term computed numerically. So, $M_\gamma(t, q)$ is easy to compute symbolically (since it does not depend on the numerical solution $f_\gamma(t, q)$):

```
In[67]:= Plot3D[(a_q[t, l] - r)/(γ σ^2) - (l κ U^(0,1)[t, l])/(U[t, l]), {l, 35, 65},
   {t, t0, T - .02}, Mesh → False, PlotPoints → 60,
   PlotLabel → StringJoin["γ = ", ToString[γ]],
   AxesLabel → {"q", "t", ""}];
```

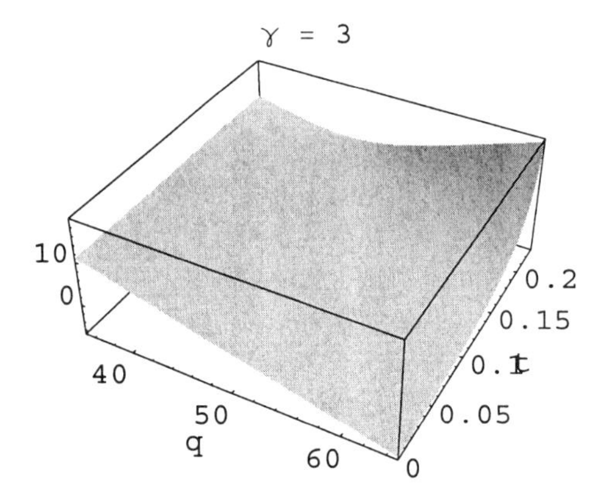

Although $N_1(t, q) = 0$, in general $N_\gamma(t, q) \neq 0$ for $\gamma > 1$. The question is whether $M_\gamma(t, q)$ yields a correct strategy under the above market conditions, or which is the same, whether the numerical term $N_\gamma(t, q)$ is negligible? In the following plot we can see that the contribution of the numerical term $N_3(t, q)$ is, on the contrary, under above market conditions, quite substantial:

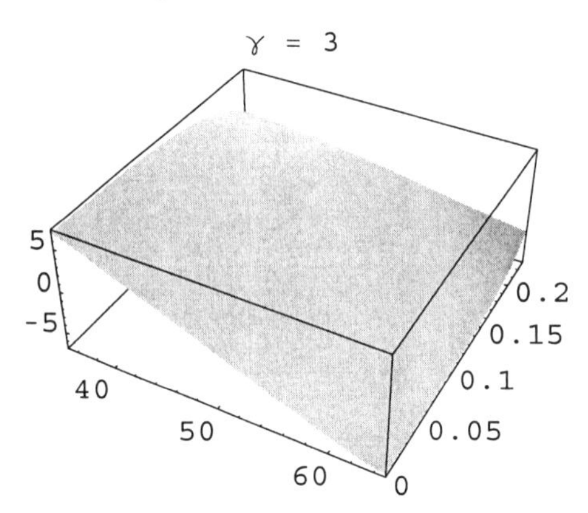

The cumulative optimal hedging strategy $M_\gamma(t, q) + N_\gamma(t, q)$ is then equal to

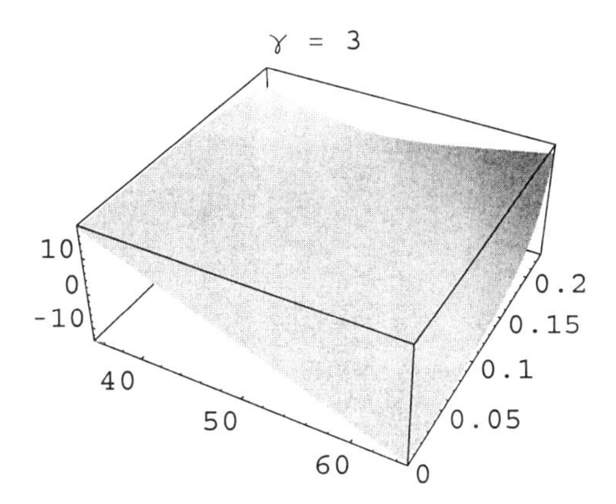

Obviously, this success sets the stage for the reexamination, testing, precise calibration, etc., of all the hedging strategies, such as combinations of calls and puts with the same or different strikes, under all kinds of different dynamic scenarios for the underlying, and under the options market circumstances such as options being priced correctly, or they are underpriced or overpriced (implied volatility equal to the underlying volatility, or whether it is below or above). Moreover if q is two-dimensional, in which case the reduced Monge–Ampère PDE is 3-dimensional (including time variable), then one can consider all those problems for two underlying stocks and their options, and see what is the interaction.

8.3 Hypoelliptic Obstacle Problems in Optimal Momentum Trading

8.3.1 Problems

Occasionally, the stock or index prices appear to oscillate around an equilibrium price. For example, between Dec 1st, 2000, and Feb 21st, 2001 (Jan 1st, 2001 being represented as $t = 1$), the price evolution of QQQ looked like this:

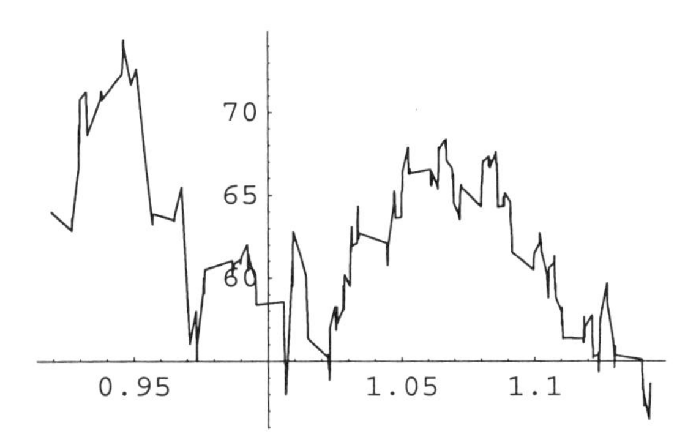

In Section 8.2 we saw a model that can handle the existence of an equilibrium price. On the other hand, if the feeling is that in the short run, in addition to the existence of an equilibrium, prices exhibit a stronger kind of oscillation that entails an expected *period* of oscillation, then that model is not suitable. We shall therefore propose a (class of) model(s) for momentum price dynamics based on second order ODEs and SDEs.

The previous discussion of continuous portfolio hedging is quite satisfactory. The same framework can be applied in many more situations. Nevertheless perpetual trading is not that common for small investors. Here we shall address a completely different kind of trading. We shall ask the questions such as when to sell the held security, stock or option? When to buy, if we plan to sell? When to sell if we plan to buy and then to sell again? ... So there will be a very limited number of transactions, as opposed to the previous kind of trading, when indeed the number of transactions was (in the limit) infinite.

In answering these questions, due to the postulated price-dynamics, we shall address removing yet another omnipresent and quite often unnatural assumption in financial mathematics, as well as in stochastic control in general: the non-degeneracy assumption. Solving the above problem will lead us to degenerate PDEs and obstacle problems, and more precisely, to the issue of *hypoellipticity*. So, the methodology introduced for, and used here to solve numerically the present problem, has much wider applicability.

8.3.2 SDE Model: Price/Trend Process

8.3.2.1 ODE Motivation

In[68]:= `Clear["Global`*"]`

Oscillations are very well understood in the theory of ODEs. The simplest ODE modeling oscillations can be written as (emphasizing two parameters we are interested in)

$$y''(t) + \left(\frac{2\pi}{p}\right)^2 y(t) = \left(\frac{2\pi}{p}\right)^2 e \tag{8.3.1}$$

with initial conditions $y(t_0) = y_0$ (initial position) and $y'(t_0) = y_1$ (initial velocity), and with the parameters e the equilibrium, and p the *period*. The solution is simple to compute. For example,

In[69]:=
```
z_{e_,p_}[t_] = y[t] /.
    Simplify[DSolve[{y[t] (2π/p)^2 + y''[t] == (2π/p)^2 e,
        y[0] == e, y'[0] == 200}, y[t], t][[1]]]
```

Out[69]= $e + \frac{100\, p \sin\left(\frac{2\pi t}{p}\right)}{\pi}$

For some different periods and equilibriums we have the following:

```
In[70]:= Plot[{50, z_{50,1/2}[t], 20, z_{20,1}[t]}, {t, 0, 1}];
```

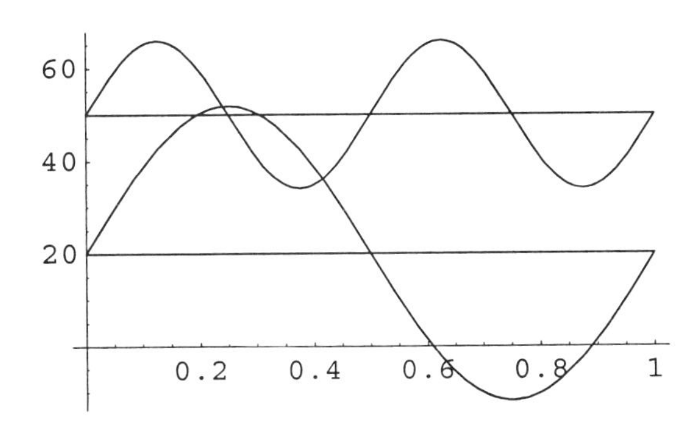

By the way, notice that, depending on the data, there is nothing in the equation to keep the solution above zero. So, if this equation, or its stochastic modification to be introduced below, is to be used for modeling of stock-price dynamics, the model could possibly be only carefully applied in the short run, and in bounded price regions.

More generally, we can consider an ODE ($b \geq 0$ is usually referred to as damping)

$$\frac{d^2 Y(t)}{dt^2} + b(t)\frac{dY(t)}{dt} + \left(\frac{2\pi}{p(t)}\right)^2 Y(t) = \left(\frac{2\pi}{p(t)}\right)^2 e(t) \tag{8.3.2}$$

where all the parameters are functions of time t. This equation is obviously equivalent to the 1st order system

$$\begin{aligned} \frac{dY(t)}{dt} &= Z(t) \\ \frac{dZ(t)}{dt} &= -b(t)Z(t) + \left(\frac{2\pi}{p(t)}\right)^2 (e(t) - Y(t)). \end{aligned} \tag{8.3.3}$$

Notice that the *state space* is now two dimensional, and that $Y(t)$ is a *position*, and $Z(t)$ is the *velocity*. We can solve this system directly, i.e., without any reference to (8.3.2). For example, in the simplest case when $b(t) = 0$, $e(t) = e$ and $p(t) = p$

```
In[71]:= w_{e_,p_}[t_] = Simplify[{Y[t], Z[t]} /.
           DSolve[{Y'[t] == Z[t], Z'[t] == (2π/p)^2 (e - Y[t]),
             Y[0] == e, Z[0] == 200}, {Y[t], Z[t]}, t][[1]]]
```

Out[71]= $\left\{\frac{\pi e + 50 i e^{-\frac{2 i \pi t}{p}} p - 50 i e^{\frac{2 i \pi t}{p}} p}{\pi}, 100 e^{-\frac{2 i \pi t}{p}} \left(1 + e^{\frac{4 i \pi t}{p}}\right)\right\}$

The following plot is sometimes called a *phase plane* plot, and its understanding is going to be important for what follows:

In[72]:=
```
Off[ParametricPlot::"ppcom"];
ParametricPlot[{w_{50, 1/2}[t], w_{20,1}[t]}, {t, 0, 1},
 AxesOrigin → {50, 0}, AxesLabel → {"Y", "Z"}];
```

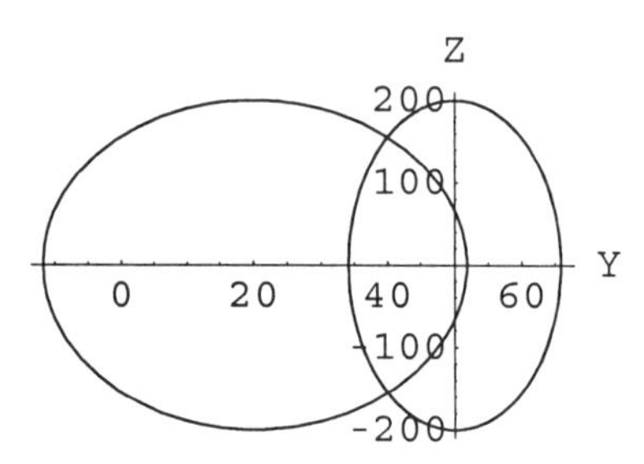

Compare the above two plots. They represent the same solutions. The first plot shows position and time and the second plot shows position and velocity, ignoring time. Which type of a plot is more informative depends on what we are interested in.

Finally, for subsequent purposes, it is convenient to write the above system in its *differential form*. For example, in the simplest case: without damping, and with constant equilibrium and constant period we have

$$\begin{aligned} dY(t) &= Z(t)\, dt \\ dZ(t) &= \left(\frac{2\pi}{p}\right)^2 (e - Y(t))\, dt. \end{aligned} \tag{8.3.4}$$

Our model for the *momentum stock-price dynamics* will be this system augmented by a source of randomness to account for stock volatility.

8.3.2.2 Price/Trend Process

In[73]:=
```
<< "Statistics`ContinuousDistributions`"
```

We augment the system (8.3.4) with a source of randomness:

$$\begin{aligned} dY(t) &= Z(t)\, dt + \sigma_0\, dB(t) \\ dZ(t) &= \left(\frac{2\pi}{p}\right)^2 (e - Y(t))\, dt. \end{aligned} \tag{8.3.5}$$

The (coupled) stochastic system (8.3.5) is going to be our basic model for price dynamics in momentum trading. We shall refer to $Y(t)$, obviously, as *price*, and to $Z(t)$ as the *trend*. So (8.3.5) is going to be referred to as the Price/Trend SDE.

Notice that σ_0 is not the volatility—$\sigma_0 / Y(t)$ is. Also trend $Z(t)$ is not the appreciation rate—$Z(t) / Y(t)$ is. Notice that we have introduced randomness only in the price component. The reason is that we want some level of stability in the trend, and the randomness to be manifested in the price only. This will imply many interesting consequences since the lack of the randomness in the second component implied the *degeneracy* of the random process $\{Y(t), Z(t)\}$, and more precisely, the *hypoellipticity* of many PDE problems to follow.

Also notice, as pointed out already above in the case of the ODE, negative values for price $Y(t)$ are possible. This is possible to rectify by modifying the model to conform more to the price model used so far (involving volatility and appreciation rate instead of the price-diffusion and trend), but the equations become more complicated, so here we choose to consider this simple, imperfect, model.

To gain some intuition about the Price/Trend SDE, we perform the Monte–Carlo simulations. Let

```
In[74]:= gσ_,e_,p_[{t_, {Y_, Z_}}, dB_] :=
           {t + dt, {Y + Z dt + σ dB, Z + (2π/p)^2 (e - Y) dt}}
```

and define

```
In[75]:= Experiment[e_, p_, σ_, k_, Y0_, Z0_, T_] := (dt = T/k;
           dB = Table[Random[NormalDistribution[0, √dt]],
             {k}]; SolutionList =
            FoldList[gσ,e,p[#1, #2] &, {0, {Y0, Z0}}, dB];
           xc = Transpose[SolutionList]〚2〛;
           pp = Show[(Graphics[{RGBColor[0,
                    Position[Reverse[xc], #1]〚1, 1〛 / Length[xc], 0],
                 Point[#1]}] &) /@xc, Frame → True,
             FrameLabel → {"Price Y", "Trend Z"},
             Axes → True, AxesOrigin → {e, 0}];)
```

We can now perform many experiments, such as

```
In[76]:= e = 50; p = 1/3; Y0 = 40; Z0 = 10; T = 2 p;
         Experiment[e, p, 25, 3000, Y0, Z0, T]
```

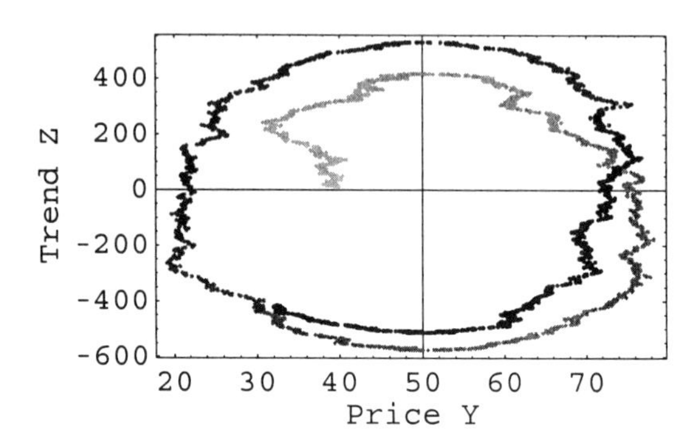

The phase-plane plot does not refer to time. So, in order to see time dependence, we need two plots instead of the one above. Compute also the solution of the corresponding (2nd order) ODE (i.e., solution of the SDE system corresponding to $\sigma_0 = 0$)

```
In[77]:= z[t_] = Chop[Simplify[
           y[t] /. DSolve[{y[t] (2 π / p)^2 + y''[t] == (2 π / p)^2 e,
             y[0] == Y0, y'[0] == Z0}, y[t], t][[1]]]]
```

Out[77]= $-10\cos(6\pi t) + \frac{5\sin(6\pi t)}{3\pi} + 50$

We plot the price trajectory, together with the corresponding ODE solution and the equilibrium:

```
In[78]:= lp = ListPlot[({#1[[1]], #1[[2, 1]]} &) /@ SolutionList,
           DisplayFunction → Identity];

In[79]:= p2 = Plot[{e, z[t]},
           {t, 0, T}, PlotRange → All, PlotStyle →
            {{RGBColor[0, 1, 0], AbsoluteThickness[2]},
             {AbsoluteThickness[3], RGBColor[1, 1, 0]},
             {RGBColor[0, 0, 0], AbsoluteThickness[1]}},
           AxesLabel → {"t", "Y[t]"}, PlotLabel → "Price",
           DisplayFunction → Identity];

In[80]:= Show[p2, lp, DisplayFunction → $DisplayFunction];
```

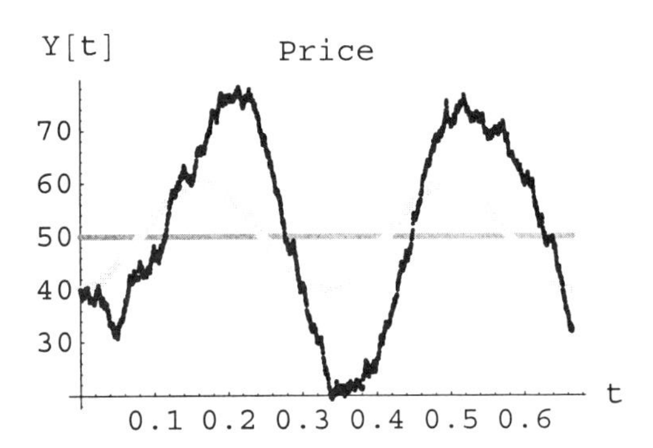

as well as the trend trajectory, together with the (second component) of the corresponding ODE solution, and its equilibrium (zero):

```
In[81]:= lp = ListPlot[({#1[[1]], #1[[2, 2]]} &) /@ SolutionList,
         DisplayFunction → Identity];

In[82]:= p2 = Plot[Evaluate[{0, ∂t z[t]}],
         {t, 0, T}, PlotRange → All, PlotStyle →
          {{RGBColor[0, 1, 0], AbsoluteThickness[2]},
           {AbsoluteThickness[3], RGBColor[1, 1, 0]},
           {RGBColor[0, 0, 0], AbsoluteThickness[1]}},
         AxesLabel → {"t", "Z[t]"}, PlotLabel → "Trend",
         DisplayFunction → Identity];

In[83]:= Show[p2, lp, DisplayFunction → $DisplayFunction];
```

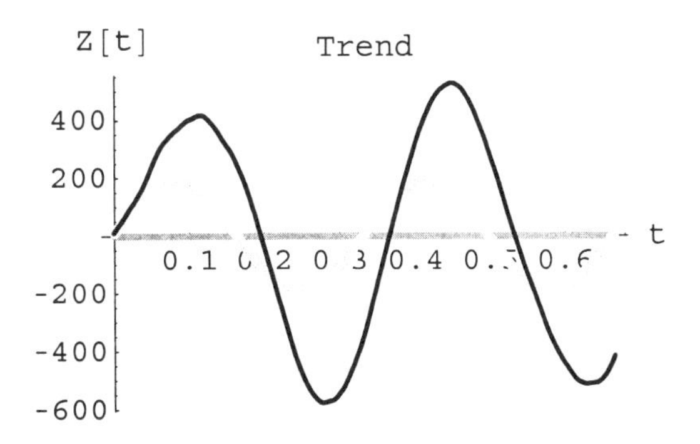

We emphasize again that the trend $Z(t)$ is not the same as the appreciation rate, say $A(t)$. The relationship would be $Z(t) = Y(t)\,A(t)$, or $A(t) = Z(t)\,/\,Y(t)$. Also, it appears that for short term trading the trend is a more relevant quantity to consider than the appreciation rate, since it refers to absolute change in prices as opposed to relative ones. It may be interesting to see the order of magnitude of the appreciation rates $A(t)$, implied by the present model:

```
In[84]:= ListPlot[({#1[[1]], #1[[2, 2]]/#1[[2, 1]]} &) /@ SolutionList,
    AxesLabel → {"t", "A(t)"}];
```

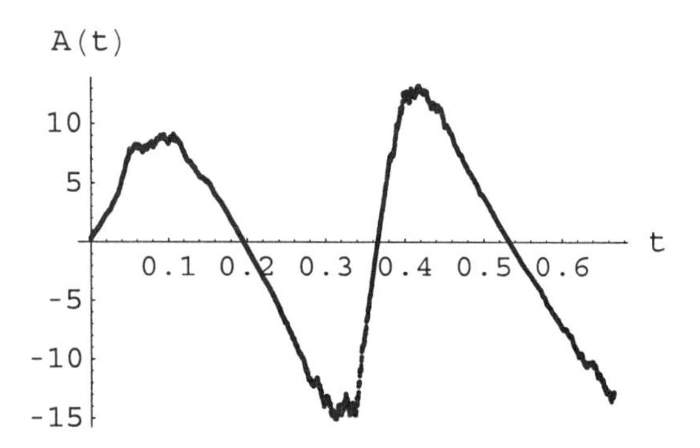

Notice that while the trend $Z(t)$ is smooth, the appreciation, i.e., the growth rate $A(t)$ is not.

In practice, if this model is to be used, one has to estimate statistically the parameters e, p and σ_0. Additionally, one should notice a serious difficulty. The trend $Z(t)$, which is not a parameter, but rather a component of the state of the system, is not observable. Only the stock price $Y(t)$ is observed on the stock market. One has to use the price trajectory, as well as the estimates on e, p and σ_0, to estimate $Z(t)$, i.e., to compute

$$m(t) = E[Z(t) \mid Y(s),\ 0 \le s \le t].$$

As was discussed in Section 4.7, $m(t)$ can be computed from the recursive equation (4.7.13), which now becomes

$$d\,m(t) = \left(\frac{2\pi}{p}\right)^2 (e - Y(t))\,d\,t + \frac{g_0^2}{{\sigma_0}^2 + t\,g_0^2}\,(d\,Y(t) - m(t)\,d\,t)$$
$$m(0) = m_0 \tag{8.3.6}$$

where $Z(0) \sim N(m_0, g_0)$. Ideally, the (mean square) error

$$E[(Z(t) - m(t))^2 \mid Y(s),\ 0 \le s \le t] = \frac{{\sigma_0}^2\, g_0^2}{{\sigma_0}^2 + t\,g_0^2} \tag{8.3.7}$$

would have to be taken into account when designing a trading strategy, but we shall not do that here. We shall assume that $Z(t)$ is known precisely by the investor, or equivalently, that $g_0 = 0$.

8.3.3 Hypoellipticity of the Infinitesimal Generator of the Price/Trend Process

8.3.3.1 $\mathcal{L}$: Infinitesimal Generator of the Price/Trend Process

Consider the Price/Trend SDE system (8.3.5). Notice that $(dY(t))^2 = \sigma_0^2\, dt$, $dY(t)\,dZ(t)$, and $(dZ(t))^2 = 0$. Therefore, if we suppose $\varphi = \varphi(t, Y, Z)$ is a differentiable function, then by the Itô chain rule

$$
\begin{aligned}
& d\varphi(t, Y(t), Z(t)) = \varphi_t(t, Y(t), Z(t))\,dt + \varphi_Y(t, Y(t), Z(t))\,dY(t) \\
& +\varphi_Z(t, Y(t), Z(t))\,dZ(t) + \frac{1}{2}\,\varphi_{Y,Y}(t, Y(t), Z(t))\,(dY(t))^2 \\
& +\varphi_{Y,Z}(t, Y(t), Z(t))\,dY(t)\,dZ(t) + \frac{1}{2}\,\varphi_{Z,Z}(t, Y(t), Z(t))\,(dZ(t))^2 \\
& = \varphi_t(t, Y(t), Z(t))\,dt + \varphi_Y(t, Y(t), Z(t))\,(Z(t)\,dt + \sigma_0\,dB(t)) \\
& +\varphi_Z\,(t, Y\,(t), Z\,(t))\left(\frac{2\,\pi}{p}\right)^2 (e - Y(t))\,dt \\
& +\frac{1}{2}\,\varphi_{Y,Y}\,(t, Y\,(t), Z(t))\sigma_0{}^2 dt = \Bigg(\varphi_t(t, Y(t), Z(t)) \\
& \qquad + Z(t)\,\varphi_Y(t, Y(t), Z(t)) + \left(\frac{2\,\pi}{p}\right)^2 (e - Y(t))\;\varphi_Z(t, Y(t), Z(t)) \\
& \qquad + \frac{1}{2}\,\sigma_0{}^2\,\varphi_{Y,Y}(t, Y(t), Z(t))\Bigg)dt \\
& +\varphi_Y\,(t, Y\,(t), Z\,(t))\,\sigma_0\,dB\,(t) = (\varphi_t(t, Y(t), Z(t)) \\
& \qquad +(\mathcal{L}\,\varphi)\,(t, Y(t), Z(t)))\,dt + \varphi_Y(t, Y(t), Z(t))\sigma_0 dB(t)
\end{aligned}
\tag{8.3.8}
$$

where the differential operator $\mathcal{L}$ is given by

$$
\mathcal{L} = \frac{1}{2}\,\sigma_0{}^2\,\frac{\partial^2}{\partial\,Y^2} + Z\,\frac{\partial}{\partial\,Y} - (Y - e)\left(\frac{2\,\pi}{p}\right)^2 \frac{\partial}{\partial Z}. \tag{8.3.9}
$$

Integrating (8.3.8), we get

$$
\begin{aligned}
& \varphi(T, Y(T), Z(T)) - \varphi(s, Y(s), Z(s)) \\
& = \int_s^T (\varphi_t(t, Y(t), Z(t)) + (\mathcal{L}\,\varphi)\,(t, Y(t), Z(t)))\,dt \\
& + \int_s^T \varphi_Y(t, Y(t), Z(t))\,\sigma_0\,dB(t),
\end{aligned}
$$

and consequently

$$E_{s,Y,Z}\,\varphi(T,\,Y(T),\,Z(T)) = \varphi(s,\,Y,\,Z)$$
$$+E_{s,Y,Z}\int_s^T (\varphi_t(t,\,Y(t),\,Z(t)) + (\mathcal{L}\,\varphi)\,(t,\,Y(t),\,Z(t)))\,dt. \tag{8.3.10}$$

So, if $\varphi = \varphi(t,\,Y,\,Z)$ is such that

$$\varphi_t(t,\,Y,\,Z) + \mathcal{L}\,\varphi(t,\,Y,\,Z) = 0, \tag{8.3.11}$$

and the terminal condition

$$\varphi(T,\,Y,\,Z) = \psi(Y,\,Z) \tag{8.3.12}$$

holds, then

$$\varphi(s,\,Y,\,Z) = E_{s,Y,Z}\,\psi(Y(T),\,Z(T)). \tag{8.3.13}$$

Alternatively, if we consider a problem in a bounded $Y,\,Z$-domain Ω, and if T is the first exit time from Ω

$$T = T_\Omega = \mathrm{Min}(t,\,\{Y(t),\,Z(t)\} \notin \Omega) \tag{8.3.14}$$

then (8.3.10) still holds. So, if φ is a function of only Y and Z, and analogously to (8.3.11) we have

$$\mathcal{L}\,\varphi(Y,\,Z) = 0 \tag{8.3.15}$$

inside Ω, and the boundary condition

$$\varphi(Y,\,Z) = \psi(Y,\,Z) \tag{8.3.16}$$

for any $\{Y,\,Z\} \in \partial_{\mathcal{L}}\Omega$, where $\partial_{\mathcal{L}}\Omega$ is an appropriate subset of the boundary $\partial\Omega$ (the *hypoelliptic boundary* corresponding to the operator $\mathcal{L}$) of Ω, then

$$\varphi(Y,\,Z) = E_{Y,Z}\,\psi(Y(T),\,Z(T)). \tag{8.3.17}$$

The hypoelliptic boundary $\partial_{\mathcal{L}}\Omega$ can be thought of as the set of all points on the boundary of Ω that could possibly be reached by the process generated by $\mathcal{L}$, starting from the inside of Ω—due to degeneracy of the process, part of the boundary $\partial\Omega$ can never be reached. Sometimes, the above facts are referred to by saying that the differential operator $\mathcal{L}$ is the *infinitesimal generator* of the Markov process $\{Y(t),\,Z(t)\}$.

8.3.3.2 Hypoellipticity and Probability: Simple Examples

What is the meaning of hypoellipticity? The concept has a very prominent position in the theoretical study of PDEs. By the classical definition, the operator $\mathcal{H}$ is hypoelliptic if $\mathcal{H}\,u \in C^\infty \Rightarrow u \in C^\infty$. We shall try here to provide some possibly new insight using only elementary means, by connecting intuitively hypoellipticity and probability.

It is a fundamental theorem of L. Hörmander (see [30]), that if $\mathcal{H}$ is a linear differential operator of second order in $\mathbb{R}^n$, and if $\mathcal{H}$ has a representation

$$\mathcal{H} = \sum_{i=1}^{k} X_i^2 + X_0 \tag{8.3.18}$$

where $X_0, X_1, \ldots, X_k$ are first order linear operators with C^∞ coefficients, and if moreover

$$\begin{aligned} &\mathrm{Span}[\mathrm{Lie}[\{X_0, X_1, \ldots, X_k\}]](x) \\ &= \mathrm{Span}[\{X_0, X_1, \ldots, X_k, [X_0, X_1], \ldots, [X_0, X_k], \ldots, [X_0, [X_0, X_1]], \ldots\}](x) \\ &\quad = T_x \mathbb{R}^n \end{aligned} \tag{8.3.19}$$

for any x, where $[.,.]$ is the commutator, i.e., $[X_0, X_1] = X_0 X_1 - X_1 X_0$, then $\mathcal{H}$ is hypoelliptic. Loosely speaking, condition (8.3.19) implies that the PDE $\mathcal{H} u = f$ is truly n-dimensional, as opposed to being a family of lower dimensional PDEs. Alternatively, in a probabilistic context, the condition is that the Markov process generated by $\mathcal{H}$ diffuses, although not necessarily in all directions, throughout a genuine n-dimensional region. We illustrate that observation by a very simple example.

Consider the process Ξ, given by $\Xi = \{X(t), Y(t)\}_t$ where

$$\begin{aligned} dX(t) &= dB(t) \\ dY(t) &= dt + dB(t) \end{aligned}$$

and where $B(t)$ is the (same) standard Brownian motion. The process Ξ has an infinitesimal generator

$$\mathcal{H} = \frac{1}{2}\frac{\partial^2}{\partial x^2} + \frac{\partial^2}{\partial x \partial y} + \frac{1}{2}\frac{\partial^2}{\partial y^2} + \frac{\partial}{\partial y} = \frac{1}{2}\left(\frac{\partial}{\partial x} + \frac{\partial}{\partial y}\right)^2 + \frac{\partial}{\partial y}.$$

We shall show that $\mathcal{H}$ is hypoelliptic using the Hörmander theorem and we shall provide some examples of explicit solutions; we shall do the same for an operator $\mathcal{A}$ which not hypoelliptic, and then finally we shall compare those two operators using probability.

So, let $X_1 = \frac{1}{\sqrt{2}}\left(\frac{\partial}{\partial x} + \frac{\partial}{\partial y}\right)$, and let $X_0 = \frac{\partial}{\partial y}$. Then $\mathcal{H} = X_1^2 + X_0$, and

$$\begin{aligned} &\mathrm{Span}[\mathrm{Lie}[X_0, X_1]][x, y] \\ &= \mathrm{Span}[X_0, X_1][x, y] = \mathrm{Span}\left[\frac{\partial}{\partial x}, \frac{\partial}{\partial y}\right][x, y] = T_{(x,y)} \mathbb{R}^2. \end{aligned}$$

According to Hörmander's theorem $\mathcal{H}$ is hypoelliptic. Actually, by separating variables x and $y - x$, i.e., by looking for solutions in a form $u(x, y) = X(x)\, Z(z) = X(x)\, Z(y - x)$, we can identify a class of explicit solutions of the PDE $\mathcal{H} u = 0$:

```
In[85]:= uλ_,c1_,c2_[x_, y_] =
           If[2 λ + 1 < 0, (c1 e^-x Sin[√-(2 λ + 1) x] +
               c2 e^-x Cos[√-(2 λ + 1) x]) e^(y-x) λ,
            (c1 e^-x e^(√(2 λ+1) x) + c2 e^-x e^(-√(2 λ+1) x)) e^(y-x) λ];
```

For example,

```
In[86]:= u1,1,1[x, y]
```

Out[86]= $e^{y-x}\left(e^{-\sqrt{3}\,x-x}+e^{\sqrt{3}\,x-x}\right)$

which indeed is, obviously, a C^{∞} function. Let's check the equation; if

```
In[87]:= ℋ[v_] := 1/2 ∂{x,2} v[x, y] +
           ∂x,y v[x, y] + 1/2 ∂{y,2} v[x, y] + ∂x v[x, y]
```

then

```
In[88]:= Simplify[ℋ[u1,1,1]]
```

Out[88]= 0

Now, we construct a non-hypoelliptic operator. Consider process Ψ given by $\Psi = \{X(t), Y(t)\}_t$ and

$$dX(t) = dB(t)$$
$$dY(t) = dB(t)$$

where $B(t)$ is the standard Brownian motion. Process Ψ has an infinitesimal generator

$$\mathcal{A} = \frac{1}{2}\frac{\partial^2}{\partial x^2} + \frac{\partial^2}{\partial x\,\partial y} + \frac{1}{2}\frac{\partial^2}{\partial y^2} = \frac{1}{2}\left(\frac{\partial}{\partial x} + \frac{\partial}{\partial y}\right)^2$$

which is not hypoelliptic ($\mathcal{H} = \mathcal{A} + \frac{\partial}{\partial y}$). Indeed, let $X_1 = \frac{1}{\sqrt{2}}\left(\frac{\partial}{\partial x} + \frac{\partial}{\partial y}\right)$. Then $\mathcal{A} = X_1^2$, and

$$\text{Span}[\text{Lie}[X_1]][x, y] = \text{Span}[X_1][x, y] = \text{Span}\left[\frac{\partial}{\partial x} + \frac{\partial}{\partial y}\right][x, y] \neq T_{(x,y)}\,\mathbb{R}^2$$

and Hörmander's condition is not satisfied. Again by separating variables x and $y - x$, we can identify a class of explicit solutions

```
In[89]:= u2c1_,c2_,Z_[x_, y_] = (c1 x + c2) Z[y - x]
```

Out[89]= $(c2 + c1\,x)\,Z(y - x)$

for *any* twice continuously differentiable function Z. For example, if $Z(z) = z^3$, we get

```
In[90]:= w[x_, y_] = u2_{1,1,#1^3&}[x, y]
```

Out[90]= $(x+1)(y-x)^3$

and checking the equation for the operator

```
In[91]:= 𝒜[v_] := 1/2 ∂_{x,2} v[x, y] + ∂_{x,y} v[x, y] + 1/2 ∂_{y,2} v[x, y]
```

we get

```
In[92]:= Simplify[𝒜[u2_{1,1,#1^3&}]]
```

Out[92]= 0

Since w is C^∞, we still do not have a verification of non-hypoellipticity of $\mathcal{A}$. To this end, modify Z by cutting-off the negative values, i.e., set $Z(z) = z^3\,\theta(z)$, where θ is the unit step function. Then Z is still C^2, but not more than that, and therefore

```
In[93]:= w[x_, y_] = u2_{1,1,UnitStep[#1] #1^3&}[x, y]
```

Out[93]= $(x+1)(y-x)^3\,\theta(y-x)$

is the solution of $\mathcal{A}w = 0 \in C^\infty$, and $w \notin C^3$ and consequently $w \notin C^\infty$, and therefore, by definition, $\mathcal{A}$ is not hypoelliptic.

The connection with the probability is that the 2-dimensional Markov process Ξ, generated by $\mathcal{H}$, spreads over a genuine 2-dimensional region (although due to degeneracy not in all directions), while the 2-dimensional Markov process Ψ, generated by $\mathcal{A}$, spreads only over a 1-dimensional region. Indeed, let the randomization be

```
In[94]:= Clear["Global`*"]
         << "Statistics`NormalDistribution`"
         T = 3; K = 5000; dt = T/K;
         x1 = -1; x2 = 1; y1 = -1; y2 = 1;
         bdry = Graphics[{Thickness[.02], Line[{{x1, y1},
               {x2, y1}, {x2, y2}, {x1, y2}, {x1, y1}}]}];
         IC = {0, 0}; dB = Table[Random[
            NormalDistribution[0, √dt]], {K}];
```

We compute the trajectory of the $\mathcal{H}$-process:

```
In[97]:= 𝓗solution =
    Transpose[Drop[Transpose[NestWhileList[
        {#1⟦1⟧ + #1⟦3, 1⟧, dt + #1⟦2⟧ + #1⟦3, 1⟧,
          Drop[#1⟦3⟧, 1]} &, Append[IC, dB],
        x1 ≤ #1⟦1⟧ ≤ x2 && y1 ≤ #1⟦2⟧ ≤ y2 &,
        1, Length[dB]]], -1]];
```

and then the trajectory of the $\mathscr{A}$-process, corresponding to the same randomization,

```
In[98]:= Off[General::"spell1"]
```

```
In[99]:= 𝒜solution =
    Transpose[Drop[Transpose[NestWhileList[
        {#1⟦1⟧ + #1⟦3, 1⟧, #1⟦2⟧ + #1⟦3, 1⟧,
          Drop[#1⟦3⟧, 1]} &, Append[IC, dB],
        x1 ≤ #1⟦1⟧ ≤ x2 && y1 ≤ #1⟦2⟧ ≤ y2 &,
        1, Length[dB]]], -1]];
```

The trajectories look like:

```
In[100]:= Function[sol,
    Show[bdry, Graphics[{AbsolutePointSize[8],
       RGBColor[0, 0, 1], Point[𝓗solution⟦1⟧]}],
     (Graphics[Point[#1]] &) /@ sol, PlotRange → All,
     AspectRatio → 1]] /@ {𝓗solution, 𝒜solution};
```

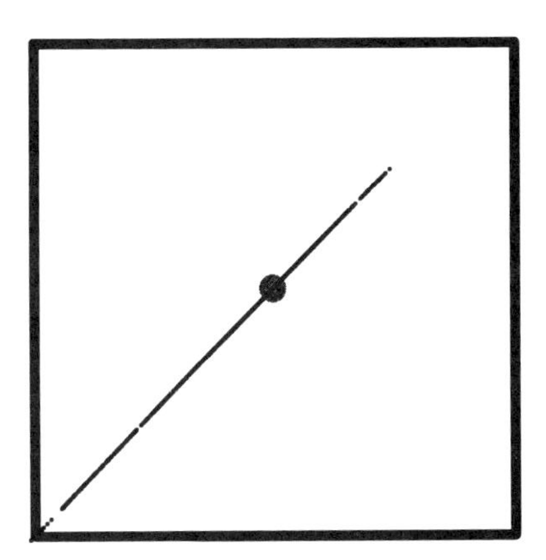

respectively. So the process with an n-dimensional hypoelliptic generator spreads over a genuine n-dimensional region (although due to the degeneracy not in all directions), while the process with an n-dimensional infinitesimal generator that fails to be hypoelliptic spreads (locally) only over an $(n-1)$-or-less-dimensional region. Above $n = 2$.

8.3.3.3 Hypoellipticity of the Infinitesimal Generator $\mathcal{L}$

Above, we have computed the generator $\mathcal{L}$ of the momentum Price/Trend process. The hypoellipticity of $\mathcal{L}$ is easy enough. Indeed, $\mathcal{L} = X_1^2 + X_0$ for

$$X_1 = X_{\mathcal{L},1} = \frac{\sigma_0}{\sqrt{2}} \frac{\partial}{\partial Y} \tag{8.3.20}$$

and

$$X_0 = X_{\mathcal{L},0} = Z \frac{\partial}{\partial Y} - (Y-e)\left(\frac{2\pi}{p}\right)^2 \frac{\partial}{\partial Z} \tag{8.3.21}$$

and the Lie bracket, the commutator,

$$\begin{aligned}
&[X_{\mathcal{L},0}, X_{\mathcal{L},1}] = [X_0, X_1] = X_0 X_1 - X_1 X_0 \\
&= Z \frac{\partial}{\partial Y} \frac{\sigma_0}{\sqrt{2}} \frac{\partial}{\partial Y} - (Y-e)\left(\frac{2\pi}{p}\right)^2 \frac{\partial}{\partial Z} \frac{\sigma_0}{\sqrt{2}} \frac{\partial}{\partial Y} \\
&- \frac{\sigma_0}{\sqrt{2}} \frac{\partial}{\partial Y} Z \frac{\partial}{\partial Y} + \frac{\sigma_0}{\sqrt{2}} \frac{\partial}{\partial Y} (Y-e)\left(\frac{2\pi}{p}\right)^2 \frac{\partial}{\partial Z} \\
&= \frac{\sigma_0}{\sqrt{2}} \left(\frac{2\pi}{p}\right)^2 \frac{\partial}{\partial Z}.
\end{aligned} \tag{8.3.22}$$

Therefore, even though $Z \frac{\partial}{\partial Y} - (Y-e)\left(\frac{2\pi}{p}\right)^2 \frac{\partial}{\partial Z} = Z \frac{\partial}{\partial Y}$ on $Y = e$,

$$\text{Span}[\text{Lie}[X_0, X_1]][Y, Z] = \text{Span}[\frac{\partial}{\partial Y}, \frac{\partial}{\partial Z}] = T_{(Y,Z)}\,\mathbb{R}^2 \tag{8.3.23}$$

and Hörmander's condition holds, yielding the hypoellipticity in $\mathbb{R}^2$. It can be checked that the hypoellipticity in $\mathbb{R}^3$ is true for $\frac{\partial}{\partial t} \pm \mathcal{L}$. Notice that the hypoelliptic-ity is confirmed already by the Monte–Carlo experiments done before: the Price/-Trend process was indeed spreading over a genuine 2-dimensional Price/Trend region.

8.3.3.4 Dirichlet Problem for $\mathcal{L}$

We are interested in a solution u to the problem

$$\mathcal{L}u = \frac{1}{2}\sigma_0^2 \frac{\partial^2 u}{\partial Y^2} + Z\frac{\partial u}{\partial Y} - (Y - e)\left(\frac{2\pi}{p}\right)^2 \frac{\partial u}{\partial Z} = 0 \tag{8.3.24}$$

in a domain Ω, with the boundary condition

$$u(Y, Z) = u_b(Y, Z) \tag{8.3.25}$$

on $\partial_{\mathcal{L}}\Omega$ (the $\mathcal{L}$-hypoelliptic boundary of Ω). This problem is called the Dirichlet problem for $\mathcal{L}$. Specifying operator $\mathcal{L}$, we choose the equilibrium price and the period for the price oscilations to be $e = 50$, $p = \frac{1}{6}$; also $\sigma_0 = 30$. Recall that σ_0 is not the volatility—the volatility is equal to $\sigma_0 / Y(t) \approx \sigma_0 / e = 3/5$. Also, we shall consider very simple Ω's: $\Omega = (y_1, y_2) \times (-b, b)$, and for example let $y_1 = \frac{e}{2}$, $y_2 = 3\,\frac{e}{2}$, $b = 750$. In such a case the hypoelliptic boundary $\partial_{\mathcal{L}}\Omega$, i.e., the set of all points on the boundary $\partial\Omega$ that could possibly be reached by the Markov process $\{Y(t), Z(t)\}$ starting from the inside Ω, can be easily identified. In the following picture it is displayed in bold blue:

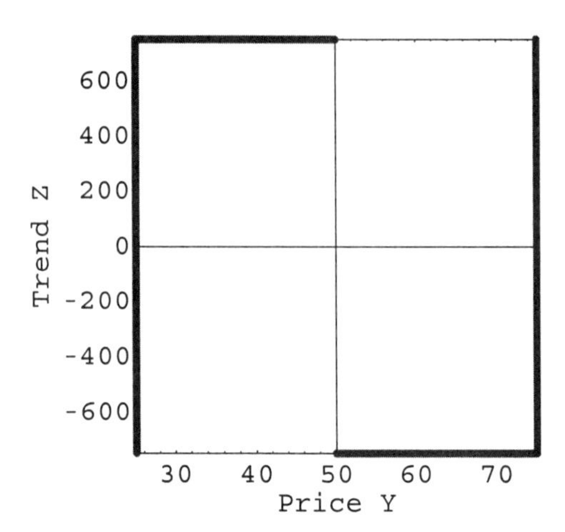

The solution of the above Dirichet problem has a probabilistic representation

$$u(Y, Z) = E_{Y,Z}\, u_b(Y(T_\Omega), Z(T_\Omega)) \tag{8.3.26}$$

(see (8.3.14)). If furthermore the boundary value is chosen to be $u_b(Y, Z) = Y$, then the numerical solution is computed and it looks like this:

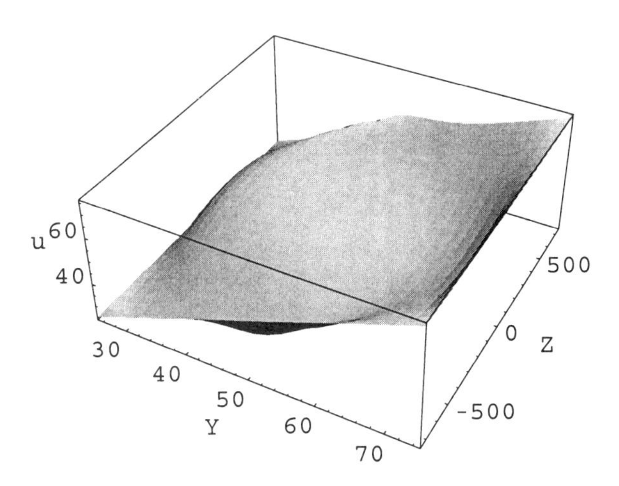

Recall, by hypoellipticity, the solution u is a $C^\infty(\Omega)$-function.

$\mathcal{L}$ has an interesting structure. If Z is considered as "time" for a moment, then the above equation can be considered as backward-parabolic in $\left(\frac{e}{2}, e\right)\times(-b, b)$, and forward-parabolic in $\left(e, 3\,\frac{e}{2}\right)\times(-b, b)$, with the additional condition that the solutions of those two equations match smoothly (even C^∞) at $e\times(-b, b)$. It is then not surprising that, due to the degeneracy of the equation, i.e., due to this special kind of parabolicity, that the boundary value is not taken either at $\left(\frac{e}{2}, e\right)\times(-b)$, or at $\left(e, 3\,\frac{e}{2}\right)\times b$. So, $\partial_{\mathcal{L}}\Omega = \partial\Omega \setminus \left(\left(\frac{e}{2}, e\right)\times(-b) \cup \left(e, 3\,\frac{e}{2}\right)\times b\right)$. This is equivalent to the fact that the process $\{Y(t), Z(t)\}_t$, due to degeneracy, i.e., due to the absence of the second source of randomness, has no chance of hitting $\partial\Omega\setminus\partial_{\mathcal{L}}\Omega$.

8.3.4 Optimal Momentum Trading of Stocks

8.3.4.1 Obstacle Problem for $\mathcal{L}$: When is it optimal to sell a stock?

Consider now a related obstacle problem arising in finance. As for the most parts of this book, from the theoretical point of view our arguments are formal, but they are intuitive and backed up by very successful numerical computations. For new theoretical results related somewhat to this problem one can consult [14].

Assume the above stock-price dynamics, and suppose an investor owns such a stock, and he/she is committed to sell once the {price, trend} exits the considered region Ω, or before. This is an optimal stopping problem with the value function u characterized as a solution of the hypoelliptic obstacle problem written as a fully non-linear PDE,

$$\mathrm{Max}[(\mathcal{L}\,u)\,(Y,\,Z),\;Y - u(Y,\,Z)] = 0 \tag{8.3.27}$$

i.e.,

$$\begin{aligned}\mathrm{Max}\Big[\frac{1}{2}\,{\sigma_0}^2\,\frac{\partial^2 u(Y,Z)}{\partial Y^2} + Z\,\frac{\partial u(Y,Z)}{\partial Y} - (Y-e)\left(\frac{2\pi}{p}\right)^2 \\ \frac{\partial u(Y,Z)}{\partial Z},\; Y - u(Y,Z)\Big] = 0\end{aligned} \tag{8.3.28}$$

in $\Omega = \left(\frac{e}{2},\, 3\,\frac{e}{2}\right)\times(-b,\,b)$, together with the boundary condition

$$u(Y,\,Z) = Y \tag{8.3.29}$$

on $\partial_{\mathcal{L}}\Omega\;\left(=\partial\Omega\setminus\left(\left(\frac{e}{2},\,e\right)\times(-b)\cup\left(e,\,3\,\frac{e}{2}\right)\times b\right)\right)$. Problem (8.3.28) can be written as a complementarity problem

$$\frac{1}{2}\,{\sigma_0}^2\,\frac{\partial^2 u(Y,Z)}{\partial Y^2} + Z\,\frac{\partial u(Y,Z)}{\partial Y} - (Y-e)\left(\frac{2\pi}{p}\right)^2\frac{\partial u(Y,Z)}{\partial Z} \le 0 \tag{8.3.30}$$

$$u(Y,\,Z) \ge Y \tag{8.3.31}$$

$$\begin{aligned}\Big(\frac{1}{2}\,{\sigma_0}^2\,\frac{\partial^2 u(Y,Z)}{\partial Y^2} + Z\,\frac{\partial u(Y,Z)}{\partial Y} \\ - (Y-e)\left(\frac{2\pi}{p}\right)^2\frac{\partial u(Y,Z)}{\partial Z}\Big)(Y - u(Y,Z)) = 0.\end{aligned} \tag{8.3.32}$$

As stated above, the solution u is the value function of the optimal stopping problem, i.e., it has a probabilistic representation

$$u(Y,\,Z) = \sup_{\tau\le T_\Omega} E_{Y,Z}\,Y(\tau) = \mathrm{Max}_{\tau\le T_\Omega}\,E_{Y,Z}\,Y(\tau) = E_{Y,Z}\,Y(\tau_{\mathrm{opt}}) \tag{8.3.33}$$

where τ are stopping times, and T_Ω is the first exit time from Ω. Notice that we are implicitly using the "utility of wealth function" $\psi_0(Y,\,Z) = Y$. This is OK, since we consider a problem in a bounded region; also only one trade is allowed (below will be only few trades allowed).

So, u is equal to the value function of the optimal stopping problem, the optimal stopping time being equal to

$$\tau_{\mathrm{opt}} = \mathrm{Min}(t,\; Y(t) = u(Y(t),\,Z(t))) = \mathrm{Min}(t,\; \{Y(t),\,Z(t)\} \in \Lambda_{\mathrm{sell}}) \tag{8.3.34}$$

where $\Lambda_{\mathrm{sell}} \subset \Omega$ is the coincidence set

$$\Lambda_{\text{sell}} = \{\{Y, Z\} \in \Omega,\ Y = u(Y, Z)\}. \tag{8.3.35}$$

Solving the optimal stopping problem (8.3.33) is then equivalent to finding its value function u, i.e., to solving the hypoelliptic obstacle problem (8.3.28)–(8.3.29). Of course finding an explicit/symbolic solution of (8.3.28)–(8.3.29) is not possible—only numerical solutions are. Before we compute the value function u we make an observation about the coincidence set Λ_{sell}.

In the coincidence set Λ_{sell} we have $Y = u(Y, Z)$ and therefore $\frac{\partial u(Y,Z)}{\partial Y} = 1$, $\frac{\partial^2 u(Y,Z)}{\partial Y^2} = 0$, and $\frac{\partial u(Y,Z)}{\partial Z} = 0$ in the interior of Λ_{sell}, and consequently due to (8.3.30) if $\{Y, Z\} \in \Lambda_{\text{sell}}$ then $0 \geq \mathcal{L}\, u(Y, Z) = Z$. This implies that

$$\Lambda_{\text{sell}} \subset \{\{Y, Z\},\ Z \leq 0\}. \tag{8.3.36}$$

Even though (8.3.36) makes trading sense, it does not appear to be much. On the contrary, the property of the coincidence set (8.3.36) might be very useful for the verification of numerical algorithms.

We compute now the value function under the same data as in the case of the equation above. The solution looks like

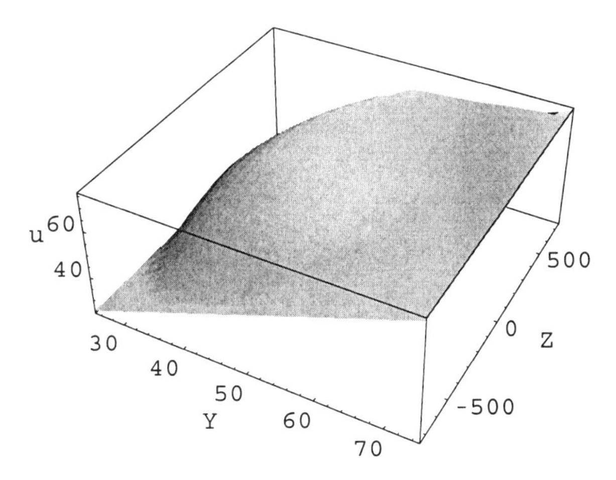

Also, in the YZ-plane the coincidence region Λ_{sell}, free boundary and the non-coincidence region can be seen more precisely (by identifying the coincidence region):

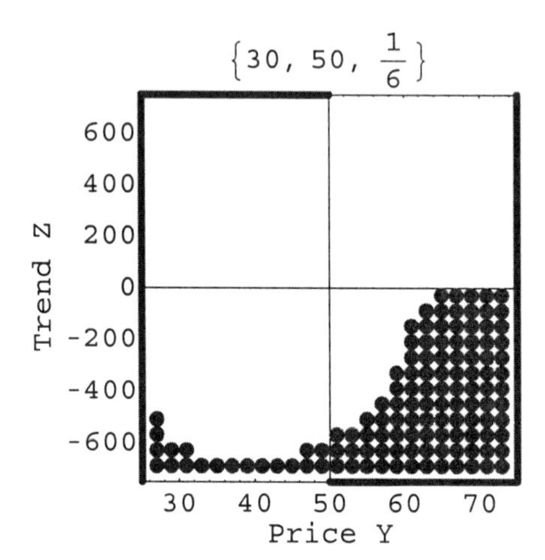

So, this is the region in the price/trend domain where, under the conditions described above, and in particular, under the risky utility function $\psi_0(Y, Z) = Y$, one needs to sell the stock (and retire—no trading afterwards). Notice that (8.3.36) holds, and that the *inclusion is tight*, and therefore indeed useful for the verification of the numerical solution.

8.3.4.2 "Implicit" Obstacle Problems for $\mathcal{L}$

When is it optimal to buy a stock?

The following problem can be classified in stochastic control theory as an optimal switching problem (see [40] for the theoretical results in the case when the governing operators are *uniformly elliptic*—the mathematical novelty, although we don't prove theorems but only state some claims supported by computational examples, of the current problem is that the governing operator $\mathcal{L}$ is merely hypoelliptic.)

Denote the previously computed value function as u_1. The previous problem addressed the issue of selling the stock. That stock, on the other hand, had to be bought first. The question is when it was optimal to buy the stock that is to be sold according to the above optimal strategy? When is it optimal to spend Y, the current price of the stock, in order to be in a position to expect the previously computed payoff, i.e., the value function u_1? So the present obstacle is $u_1(Y, Z) - Y$ (no transaction costs). The alternative is to do nothing, and wait until the stock price/trend leaves Ω, when the speculation ends, yielding the boundary condition $u_2 = 0$ on $\partial_{\mathcal{L}}\Omega$.

So the problem is to find function u_2, such that

$$\text{Max}[(\mathcal{L}\, u_2)\,(Y, Z),\, u_1(Y, Z) - Y - u_2(Y, Z)] = 0$$

i.e.,

$$\mathrm{Max}\Big[\frac{1}{2}\,\sigma_0{}^2\,\frac{\partial^2 u_2(Y,Z)}{\partial Y^2}+Z\,\frac{\partial u_2(Y,Z)}{\partial Y}$$
$$-(Y-e)\left(\frac{2\,\pi}{p}\right)^2\,\frac{\partial u_2(Y,Z)}{\partial Z},\,u_1(Y,Z)-Y-u_2(Y,Z)\Big]=0$$

in $\Omega=\left(\frac{e}{2},\,3\,\frac{e}{2}\right)\times(-b,\,b)$, together with the boundary condition

$$u_2(Y,Z)=0$$

on $\partial_{\mathcal{L}}\Omega$.

The numerical solution u_2 looks like this:

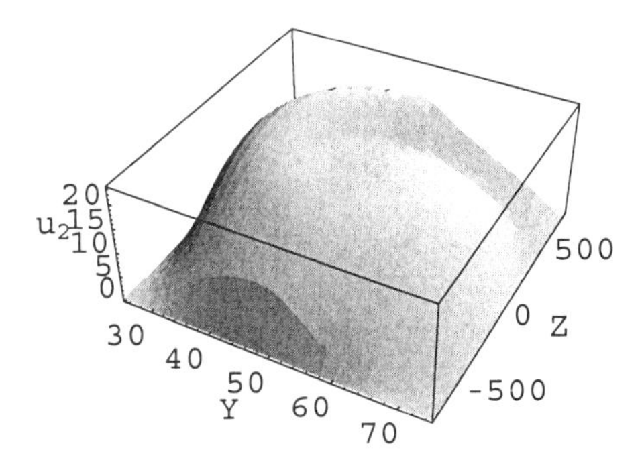

Notice above the numerical evidence that indeed the boundary value is *not* taken at $\partial\Omega\backslash\partial_{\mathcal{L}}\Omega$. The value function u_2 can be also considered as the expected profit under the above market situation if the optimal trading, i.e., optimal buying and selling strategy is adopted. The coincidence set Λ_2, identifying the optimal buying price/trend region looks like this:

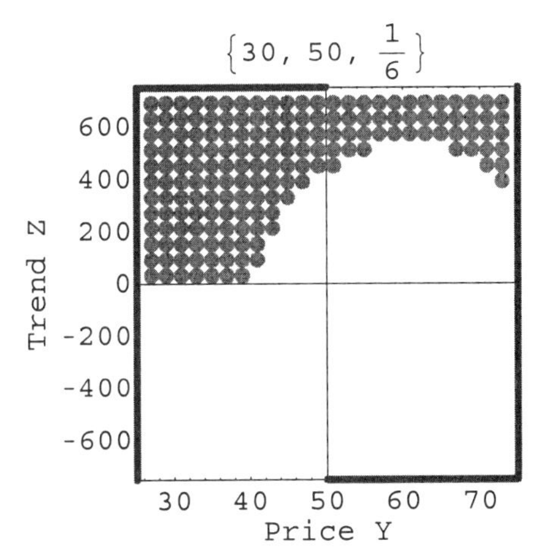

The buying Λ_2 and selling ("retiring") Λ_1 regions can be seen also together

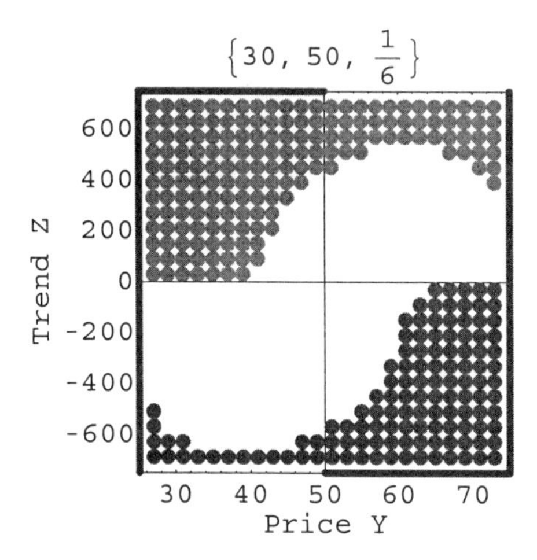

This solves the problem, if only one buy/sell cycle is considered. On the other hand, if two buy/sell cycles are considered, without going into any details, we can compute additionally u_3, u_4, and corresponding coincidence sets Λ_3, Λ_4. In particular, the payoff function u_4, i.e., the expected profit if two buy/sell cycles are considered looks like this:

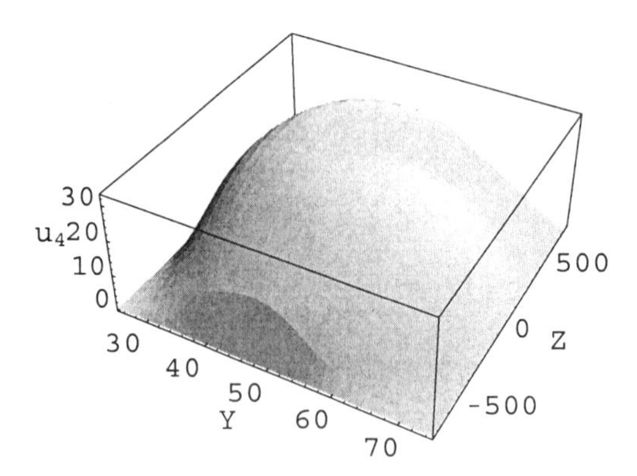

Compare u_2 and u_4: one would expect that the expected profit is larger if two trading cycles are considered instead of only one. Indeed, as one can see above, $u_4 > u_2$. The optimal trading regions Λ_1, Λ_2, Λ_3, Λ_4, superimposed on each other, look like

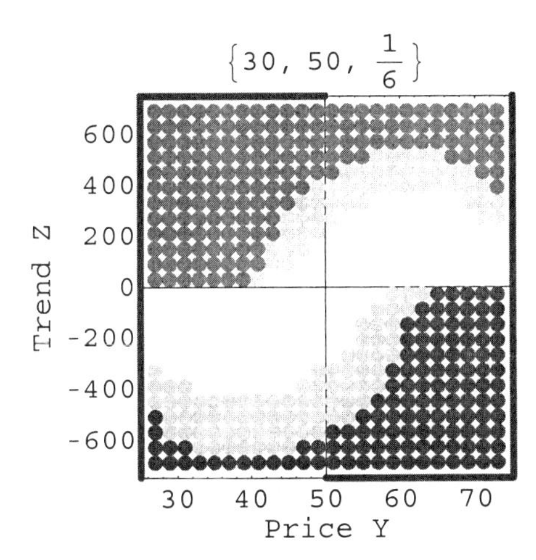

Here the light red is the first buying region Λ_4, dark red is the second, i.e., the last buying region Λ_2, and similarly for blue selling regions Λ_3, Λ_1. One can see that

$$\Lambda_4 \supset \Lambda_2$$

and

$$\Lambda_3 \supset \Lambda_1.$$

These inclusions are plausible: one will buy or sell less discriminatively if another trading cycle is considered. Also notice that in the above we did not take the cost of the transactions into account, i.e., we assumed it is equal to zero. It is easy to modify the computation in order to have the transaction cost taken into account, as we shall do in thc next problem. Also many more buy/sell cycles can be considered easily.

8.3.5 Optimal Momentum Trading of Stock Options

We close this book with a short description of the solution of the problem of optimal momentum trading of stock options. We use Price/Trend dynamics for the underlying stock as a view of the market of an options speculator. The market, on the other hand, has its own more conventional view of itself (say Log-Normal stock-price dynamics), which translates into the option pricing mechanism (say Black–Scholes formula) that is separate from Price/Trend dynamics.

8.3.5.1 Obstacle Problem for $\frac{\partial}{\partial t} + \mathcal{L}$: When to sell a call/put option?

Assume the above described basic Price/Trend market dynamics for the considered stock price $Y(t)$:

$$\begin{aligned} dY(t) &= Z(t)\,dt + \sigma_0\,dB(t) \\ dZ(t) &= \left(\frac{2\pi}{p}\right)^2 (e - Y(t))\,dt. \end{aligned} \tag{8.3.37}$$

Instead of trading in such a stock, the investor considers *speculating* in a particular option (European call; this is not essential and it is just to fix ideas). We assume the Black–Scholes option pricing formulas $V_{T,k,r,\sigma}(t, Y)$ where the variables are Y, the stock price and t the current time, in addition to the parameters k the strike price, T the expiration time, σ the stock volatility, and r the interest rate. We also allow for a transaction cost $c \geq 0$ to be taken into account (in the computational examples below $c = 0.3$).

More specifically, we assume that the market makers have the view of the market described by the "Log-Normal" stock-price model,

$$dY(t) = Y(t)\,a(t, Y(t))\,dt + Y(t)\,\sigma\,dB_1(t) \tag{8.3.38}$$

and therefore Black–Scholes formulas apply, while the user of this methodology has a private view of the market described by the Price/Trend process. Notice the difference between σ and σ_0 above—σ is the price-*volatility* assumed by the market-makers and σ_0 is the price-*diffusion* assumed by the trader. The issue is, assuming that the private view of the market is indeed correct, how to maximize the profit expectation.

More precisely, the problem is to design a strategy consisting of two stopping times τ_{buy} and τ_{sell} when it is optimal to buy and subsequently to sell the considered option (unless the Price/Trend process gets out of the considered price/trend region Ω, in which case the position is cashed in, and the speculation stops). One first considers the problem of selling. The sale is executed only if, after the transaction costs c are accounted for, there is a positive balance. Therefore the sale payoff at time t is equal to

$$\psi_{\text{sell}}(t, Y) = \psi(t, Y) = \text{Max}[V_{T,k,r,\sigma}(t, Y) - c, 0].$$

Consider the backward evolution operator $\mathcal{S} = \frac{\partial u}{\partial t} + \mathcal{L}$, i.e.,

$$\mathcal{S} = \frac{\partial}{\partial t} + \frac{1}{2}\sigma_0{}^2 \frac{\partial^2}{\partial Y^2} + Z\frac{\partial}{\partial Y} - (Y - e)\left(\frac{2\pi}{p}\right)^2 \frac{\partial}{\partial Z} \tag{8.3.39}$$

$\mathcal{S}$ is hypoelliptic. Now, consider an evolution obstacle problem: Find $u_{\text{buy}} = u$, such that

$$\text{Max}[(\mathcal{S}\,u)(t, Y, Z), \psi(t, Y) - u(t, Y, Z)] = 0$$

i.e.,

$$\text{Max}\Big[\frac{\partial u(t,Y,Z)}{\partial t}+\frac{1}{2}\sigma_0{}^2\frac{\partial^2 u(t,Y,Z)}{\partial Y^2}+Z\frac{\partial u(t,Y,Z)}{\partial Y}$$
$$-(Y-e)\left(\frac{2\pi}{p}\right)^2\frac{\partial u(t,Y,Z)}{\partial Z},\ \psi(t,Y)-u(t,Y,Z)\Big]=0 \tag{8.3.40}$$

in $Q_T=\Omega\times(0,T)=\left(\frac{e}{2},3\,\frac{e}{2}\right)\times(-b,b)\times(0,T)$, together with the boundary condition

$$u(t,Y,Z)=\psi(t,Y)$$

on $\partial_S Q_T$ (the hypoelliptic boundary of Q_T). More explicitly, (8.3.30) can be rewritten as a complementarity problem

$$\frac{\partial u(t,Y,Z)}{\partial t}+\frac{1}{2}\sigma_0{}^2\frac{\partial^2 u(t,Y,Z)}{\partial Y^2}+Z\frac{\partial u(t,Y,Z)}{\partial Y}$$
$$-(Y-e)\left(\frac{2\pi}{p}\right)^2\frac{\partial u(t,Y,Z)}{\partial Z}\le 0$$

$$u(t,Y,Z)\ge\psi(t,Y)$$

$$\left(\frac{\partial u(t,Y,Z)}{\partial t}+\frac{1}{2}\sigma_0{}^2\frac{\partial^2 u(t,Y,Z)}{\partial Y^2}+Z\frac{\partial u(t,Y,Z)}{\partial Y}\right.$$
$$\left.-(Y-e)\left(\frac{2\pi}{p}\right)^2\frac{\partial u(t,Y,Z)}{\partial Z}\right)(\psi(t,Y)-u(t,Y,Z))=0.$$

The solution of the above evolution hypoelliptic obstacle problem has the probabilistic representation

$$u(t,Y,Z)=\sup\nolimits_{t\le\tau\le T}E_{t,Y,Z}\,\psi(\tau,Y(\tau))$$
$$=\text{Max}_{t\le\tau\le T}E_{t,Y,Z}\,\psi(\tau,Y(\tau))=E_{t,Y,Z}\,\psi(\tau_{\text{opt}},Y(\tau_{\text{opt}}))$$

i.e., it is equal to the value function of the optimal stopping problem, the optimal stopping time being equal to

$$\tau_{\text{opt}}=\text{Min}(t,\psi(t,Y(t))=u(Y(t),Z(t),t))=\text{Min}(t,\{Y(t),Z(t),t\}\in\Lambda_{\text{sell}})$$

where Λ_{sell} is the coincidence set

$$\Lambda_{\text{sell}}=\{\{t,Y,Z\}\in Q_T,\ \psi(t,Y)=u(t,Y,Z)\}.$$

The problem is solved (approximately) by solving a sequence of stationary obstacle problems (backward in time): One starts with the terminal condition

$$u(Y,Z,T)=\psi(T,Y)$$

and then inductively solves *stationary* hypoelliptic obstacle problems:

$$\mathrm{Max}\Big[\frac{1}{2}\,\sigma_0{}^2\,\frac{\partial^2 u(t,Y,Z)}{\partial Y^2}+Z\,\frac{\partial u(t,Y,Z)}{\partial Y}$$
$$-(Y-e)\left(\frac{2\pi}{p}\right)^2\frac{\partial u(t,Y,Z)}{\partial Z}-\frac{1}{dt}\,u(t,Y,Z)$$
$$+\frac{u(t+dt,Y,Z)}{dt},\ \psi(t,Y)-u(t,Y,Z)\Big]=0$$

in $\Omega=\left(\frac{e}{2},3\,\frac{e}{2}\right)\times(-b,b)$, together with the boundary condition

$$u(t,Y,Z)=\psi(t,Y)$$

on $\partial_{\mathcal{L}}\Omega$. Computing the numerical solution for $0<t<T=\frac{30}{365}$ (there are 30 days until the expiration of the considered call option), and for the interest rate $r=0.03$, stock-price volatility assumed by the market-makers $\sigma=0.5$, call-option strike $k=45$, and with private market view discribed with $\sigma_0=30$, $e=50$, $p=\frac{1}{6}$, we get the coincidence, i.e. the selling set, of the computed solution for $t=0$ to look like

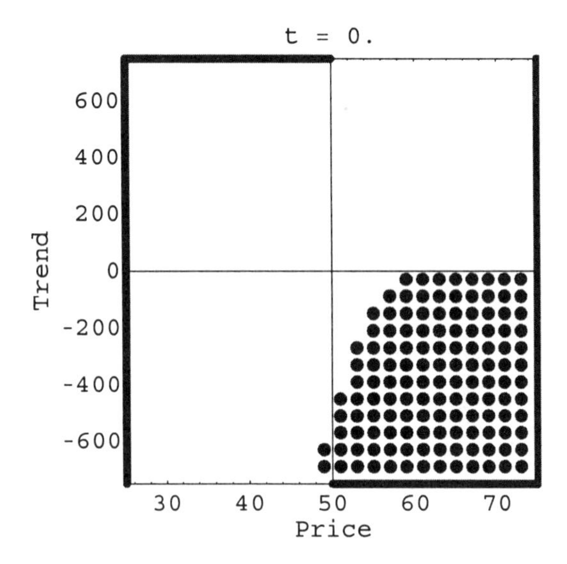

In the electronic form of this book one can double click to see the evolution for $0<t<T$. The value function (again for $t=0$) looks like

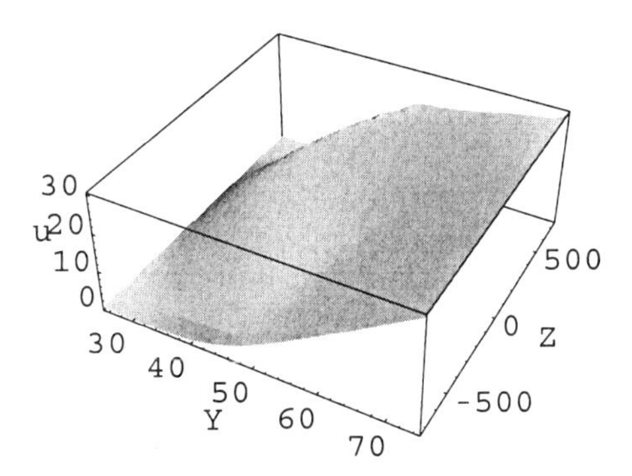

This shows the solution of the problem of finding the optimal time to sell an option under the momentum dynamics for the underlying stock price.

8.3.5.2 Implicit Obstacle Problem for $\frac{\partial}{\partial t} + \mathcal{L}$: When to buy a call/put option?

Denote the value function computed above u_{sell}. Now let the obstacle be

$$\psi_{\text{buy}}(t,\, Y) = u_{\text{sell}}(t,\, Y,\, Z) - V_{T,k,r,\sigma}(t,\, Y) - c$$

and consider the evolution obstacle problem

$$\text{Max}\Big[\frac{\partial u(t,\, Y,\, Z)}{\partial t} + (\mathcal{L}\, u)\,(t,\, Y,\, Z),\, \psi_{\text{buy}}(t,\, Y) - u(t,\, Y,\, Z)\Big] = 0$$

i.e.,

$$\begin{aligned}\text{Max}\Big[&\frac{\partial u(t,\, Y,\, Z)}{\partial t} + \frac{1}{2}\,{\sigma_0}^2\, \frac{\partial^2 u(t,\, Y,\, Z)}{\partial Y^2} + Z\, \frac{\partial u(t,\, Y,\, Z)}{\partial Y} \\ &- (Y - e)\left(\frac{2\,\pi}{p}\right)^2 \frac{\partial u(t,\, Y,\, Z)}{\partial Z},\, \psi_{\text{buy}}(t,\, Y) - u(t,\, Y,\, Z)\Big] = 0\end{aligned} \tag{8.3.41}$$

together with the boundary condition $u = 0$ at $\partial_S Q_T$ (do nothing).

The solution of the above evolution hypoelliptic obstacle problem has the probabilistic representation

$$\begin{aligned}u(t,\, Y,\, Z) &= u_{\text{buy}}(t,\, Y,\, Z) \\ &= \text{Max}_{t \le \tau \le T}\, E_{t,Y,Z}\, \psi_{\text{buy}}(\tau,\, Y(\tau)) = E_{t,Y,Z}\, \psi_{\text{buy}}(\tau_{\text{opt}},\, Y(\tau_{\text{opt}}))\end{aligned}$$

i.e., it is equal to the value function of the optimal stopping problem of deciding when to buy the option, the optimal stopping time being equal to

$$\tau_{\text{buy}} = \text{Min}(t,\, \psi_{\text{buy}}(t,\, Y(t)) = u_{\text{buy}}(Y(t),\, Z(t),\, t)) = \text{Min}(t,\, (Y(t),\, Z(t),\, t) \in \Lambda_{\text{buy}})$$

where Λ_{buy} is the coincidence set

$$\Lambda_{\text{buy}} = \{(t, Y, Z) \in Q_T, \psi_{\text{buy}}(t, Y) = u_{\text{buy}}(t, Y, Z)\}.$$

We solve this problem immediately.

The value function, i.e., the expected profit in the present case, looks like this (again for $t = 0$):

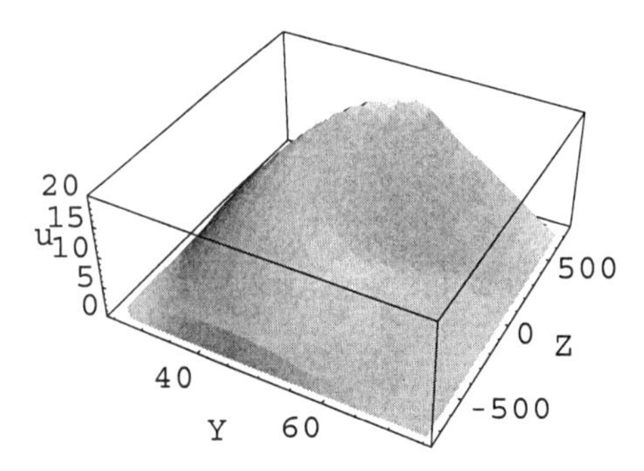

Finally, the optimal buying (red), and selling (blue) regions, superimposed with the option strike price, for $t = 0$, look like

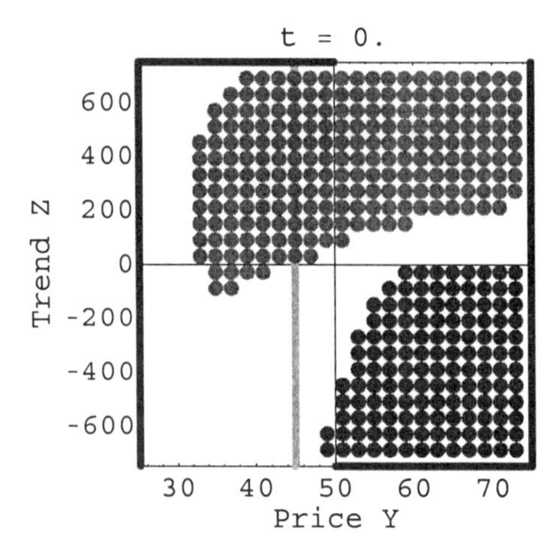

We can see the plausible result: option is bought in the positive trend region (or close to it), but not unless the difference between the option strike price and the current stock price is not too big (otherwise, option is expected to expire before the difference is made up). One should keep in perspective that the wealth utility function implicitly used in this problem corresponds to very aggressive investing.

Bibliography

[1] Barbu V., *Optimal Control of Variational Inequalities*, Pitman, Boston, 1984.

[2] Bardi M. and Dolcetta I.C., *Optimal Control and Viscosity Solutions of Hamilton–Jacobi–Bellman Equations*, Birkhäuser, Boston, 1997.

[3] Basawa I.V. and Prakasa-Rao B.L.S., *Statistical Inference for Stochastic Processes*, Academic Press, New York, 1980.

[4] Baxter M. and Rennie A., *Financial Calculus*, Cambridge, Cambridge UK, 1996.

[5] Bensoussan A. and Lions J.L., *Applications des Inéquations Variationnelles en Contrôle Stochastique*, Dunod, Paris, 1978.

[6] Bensoussan A., *Stochastic Control by Functional Analysis Methods*, North-Holland, Amsterdam, 1982.

[7] Black F. and Scholes M., The pricing of options and corporate liabilities, *J. Political Econ.* 81 (1973), 637–659.

[8] Bouchouev, I. and Isakov V., Uniqueness, stability and numerical methods for the inverse problem that arises in financial markets, *Inverse Problems*, 1999, R95–R116.

[9] Caffarelli L.A. and Cabre X., *Fully Nonlinear Elliptic Equations*, AMS, Providence, 1995.

[10] Caffarelli L.A., Regularity of free boundaries in higher dimension, *Acta Math.* 139 (1977), 155–184.

[11] Carlin B.P. and Louis T.A., *Bayes and Empirical Bayes Methods for Data Analysis*, Chapman & Hall/CRC, Boca raton, 2000.

[12] Cea J., *Optimization—Theory and Algorithms*, Springer-Verlag, Berlin, 1978.

[13] Chow Y.S., Robbins H., and Teicher H., *Great Expectations: The Theory of Optimal Stopping*, Houghton Mifflin, Boston, 1972.

[14] Danielli D., Garofalo N., and Salsa S., Variational inequalities with lack of ellipticity Part I: Optimal interior regularity and non-degeneracy of the free boundary, *Indiana Univ. Math. J.* (to appear).

[15] DiBenedetto E., *Partial Differential Equations*, Birkhäuser, Boston, 1995.

[16] Dupire B., Pricing with a smile, *RISK* 7 (1994), 18–20.

[17] Evans L.C., *Partial Differential Equations*, AMS, Providence, 1998.

[18] Elliot R. J., *Stochastic Calculus and Applications*, Springer-Verlag, New York, 1982.

[19] Fleming W.H. and Rishel R.W., *Deterministic and Stochastic Optimal Control*, Springer-Verlag, New York, 1975.

[20] Fleming W.H. and Soner H.M., *Controlled Markov Processes and Viscosity Solutions*, Springer-Verlag, New York, 1993.

[21] Friedman A., *Partial Differential Equations of Parabolic Type*, Prentice Hall, Englewood Cliffs, 1964.

[22] Friedman A., *Stochastic Differential Equations, Vol 1 & 2*, Academic Press, New York, 1975.

[23] Friedman A., *Variational Principles and Free Boundary Problems*, Wiley, New York, 1982.

[24] Gilbarg D. and Trudinger N.S., *Elliptic Partial Differential Equations of Second Order*, Springer-Verlag, New York, 1983.

[25] Glowinski R., *Lectures on Numerical Methods for Non-Linear Variational Problems*, Springer-Verlag, New York, 1981.

[26] Glowinski R., Lions J.L., and Tremolières R., *Analyse Numerique des Inéquations Variationnelles*, Dunod, Paris, 1976.

[27] Glynn J. and Gray T., *The Beginner's Guide to Mathematica Version 4*, Cambridge, Cambridge UK.

[28] Goodman V. and Stampffi J., *The Mathematics of Finance: Modeling and Hedging*, Brooks Cole, Pacific Grove, 2001.

[29] Gutiérrez C.E., *The Monge–Ampère Equations*, Birkhäuser, Boston, 2001.

[30] Hörmander L., Hypoelliptic second order differential equations, *Acta Math.* 119 (1967), 147–171.

[31] Hull J., *Options, Futures, and Other Derivatives*, Prentice Hall, Englewood Cliffs, 1997.

[32] Karatzas I. and Shreve S.E., *Brownian Motion and Stochastic Processes*, Springer-Verlag, New York, 1996.

[33] Karatzas I. and Shreve S.E., *Methods of Mathematical Finance*, Springer-Verlag, New York, 1998.

[34] Karatzas I., *Lectures on the Mathematics of Finance*, AMS, Providence, 1997.

[35] Kaufmann S., *A Crash Course in Mathematica*, Birkhäuser, Boston, 1999.

[36] Kinderlehrer D. and Stampacchia G., *An Introduction to Variational Inequalities and their Applications*, Academic Press, New York, 1980.

[37] Kopp P.E. and Elliott R.J., *Mathematics of Financial Markets*, Springer-Verlag, New York, 1999.

[38] Krylov N.V., *Controlled Diffusion Processes*, Springer-Verlag, New York, 1980.

[39] Krylov N.V., *Nonlinear Elliptic and Parabolic Equations of the Second Order*, Reidel, Dordrecht, 1987.

[40] Lenhart S.M. and Belbas S.A., A system of nonlinear partial differential equations arising in the optimal control of stochastic systems with switching costs, *SIAM J. Appl. Math.*, **43** (1983), 465-475.

[41] Lions J.L., *Optimal Control of Systems Governed by Partial Differential Equations*, Springer-Verlag, New York, 1971.

[42] Liptser R.S. and Shiryayev A.N., *Statistics of Random Processes, Vol 1 & 2*, Springer-Verlag, New York, 1977.

[43] Li X. and Yong J., *Optimal Control Theory for Infinite Dimensional Systems*, Birkhäuser, Boston, 1995.

[44] Maeder R., *Programming in Mathematica*, Addison-Wesley, 1996.

[45] Ma J. and Yong J., *Forward-Backward Stochastic Differential Equations and Their Applications*, Springer-Verlag, New York, 1999.

[46] Merton R.C., *Continuous-Time Finance*, Blackwell, Cambridge MA, 1990.

[47] Merton R.C., Theory of rational option pricing, *Bell J. Econ. Manage. Sci.* 4 (1973) 141-183.

[48] Musiela M. and Rutkowski M., *Martingale Methods in Financial Modelling*, Springer-Verlag, New York, 1998.

[49] Oksendal B., *Stochastic Differential Equations*, Springer-Verlag, New York, 1998.

[50] Pliska S., *Introduction to Mathematical Finance: Discrete Time Models*, Blackwell, Cambridge MA, 1997.

[51] Revuz D. and Yor M., *Continuous Martingales and Brownian Motion*, Springer-Verlag, New York, 1998.

[52] Shaw W.T., *Modelling Financial Derivatives with Mathematica*, Cambridge, Cambridge UK, 1998.

[53] Shaw W.T. and Tigg J., *Applied Mathematica: Getting Started, Getting it Done*, Addison-Wesley, 1994.

[54] Shiryayev A.N., *Optimal Stopping Rules*, Springer-Verlag, New York, 1978.

[55] Stojanovic S., Implied volatility for American options via optimal control and fast numerical solutions of obstacle problems, *Differential Equations and Control Theory*, Aizicovici and Pavel Eds., Marcel Dekker (2001), 277–294.

[56] Stojanovic S., Perturbation formula for regular free boundaries in elliptic and parabolic obstacle problems, *SIAM J. Control and Optimiz.*, 35 (1997), 2086–2100.

[57] Stojanovic S., Remarks on $W^{2,p}$-solutions of bilateral obstacle problems, *IMA Preprint* #1318, University of Minnesota, 1995.

[58] Wilmott P., Howison S., and Dewynne J., *The Mathematics of Financial Derivatives*, Cambridge, Cambridge UK, 1995.

[59] Wolfram S., *The Mathematica Book*, Cambridge, Cambridge UK, 1999.

[60] Zhou X.Y. and Yong J., *Stochastic Controls: Hamiltonian Systems and HJB Equations*, Springer-Verlag, New York, 1999.

Index